BUILDING SCIENCE
FOR A COLD CLIMATE

Neil B. Hutcheon, B.E. (Mech.), Ph.D.
Former Director
Division of Building Research
National Research Council of Canada

Gustav O.P. Handegord, B.E. (Mech.), M.Sc.
Principal Research Officer
Division of Building Research
National Research Council of Canada

SI Metric

First printing 1983
Second printing 1989
Third printing 1995

Canadian Cataloguing in Publication Data

Hutcheon, N.B.
Building science for a cold climate

Includes index.
ISBN 0-9694366-0-2

1. Building - Cold weather conditions. I. Handegord, Gustav O. II. Title.

TH153.H87 690 C83-098599-9

BUILDING SCIENCE
FOR A COLD CLIMATE

Contents

Foreword

Canada has come a long way since Voltaire called it "quelques arpents de neige." Although bitterly resented by the residents of the time, this remark foreshadowed the difficulty of housing a nation in a cold climate. Canadians had to devise economical solutions for year-round indoor comfort under harsh outdoor conditions.

The control of air, heat and moisture flow in a severe climate calls for great ingenuity from designers and builders. Beginning in the late 1940's, the National Research Council's Division of Building Research tackled a vast array of construction problems. The Division's work contributed to energy-efficient designs, evolved over years of learning in the laboratory and in the field, from many failures and a few solid successes.

After some thirty years, it was realized that builders were frequently unaware of the theoretical knowledge gained from past lessons. As techniques and materials changed, builders faced new problems, because what made the old solutions work was never properly understood. Neil B. Hutcheon, Director of the Division from 1969 to 1974, and Gustav O. Handegord, wrote a text that summarized the new discipline of building science. *Building Science for a Cold Climate* incorporated many of the Division's own findings, along with contributions from colleagues in virtually all countries that share the problems of a cold climate.

The Division, now the Institute for Research in Construction, partners with industry associations, government agencies and other publishers to make technical information, such as contained in this book, widely accessible. They share our conviction that good building science can greatly increase productivity and improve building performance.

This re-printing of *Building Science for a Cold Climate* is a re-issue of the 1983 classic, in response to continued demand from consultants and educators across Canada. No attempt has been made to update the text at this time. Of course, at the Institute, as a result of our research, we are constantly bringing new information to the fore. With organizations around the world, we introduce new techniques and building systems, and publish our results on a continuous basis. For the latest advances in building science, we suggest that you contact us.

G. Seaden
Director General
Institute for Research in Construction
National Research Council of Canada

Preface

This book is in many ways a pioneer effort. It was written to define and describe a body of knowledge that consists of some old science and much applied science and engineering information that has been developed, held, and used in separate branches and specialties serving the building industry over many years. To this we have added the results of studies by the Division of Building Research and those carried out by others since World War II, which cannot be claimed as the property of a particular discipline, and which form a growing body of knowledge increasingly being described as "building science".

In our view, the material that we have selected is essential to an understanding of many aspects of buildings, and forms a background to decision making on many difficult contemporary technical issues. Indeed, we believe it must be recognized as a fundamental part of technical and professional competence in the building field. It is hoped that this book will serve as a text for those in training and as a review and reference volume for those already in practice.

The information and examples presented relate mainly to Canadian buildings, Canadian conditions, and the effects of exposure to Canadian winters. Most of the material is relevant to other cold countries as well as to those with less severe winters. Practitioners elsewhere must judge how much of what we have written on the basis of Canadian experience can be adapted successfully for their purposes. For this reason, and to identify accurately the subject areas in building science we wish to address, we have defined "a cold climate" as one with an outdoor winter design temperature lower than $-7°C$.

We have adhered to SI throughout, recalculating tables, converting equations, and redrawing graphs when necessary. Metrication posed many difficulties for us with preferred nomenclature. When there were differences in the symbols used by the specialties from which information was being drawn, we chose the symbols and usage that we believed to be the clearest. Examples throughout the text are intended to assist the reader in changing to SI, as well as to clarify the material presented.

Extensive references to existing sources of information are included. We hope readers will be encouraged to acquire some of these books and reports for their own libraries and in this way to extend further their knowledge of building.

We are grateful for the help of all those who have contributed to the development of this book, especially those who have given permission to use copyright material or whose works have been cited. The contribution of Margaret Gerard in editing and arranging for the preparation of manuscripts and co-ordinating the editorial and publication process was invaluable. The dedicated efforts of Douglas Scott and Doreen Charron in preparing the illustrations are greatly appreciated. Sincere thanks are offered to George Tamura and Alan Dalgliesh and other members of the Division of Building Research, National Research Council, for their helpful technical review, and to Mr. John Breckenridge of the

United States Army Research Institute of Environmental Medicine and Professor Ole Fanger of the Technical University of Denmark for their help and comments. The encouragement and support of Carl Crawford, Director of the Division of Building Research, is most gratefully acknowledged.

This book is dedicated to the improvement of the design, construction, and operation of buildings.

N.B. Hutcheon
G.O.P. Handegord

1
People and Their Building Requirements

1.1 Introduction

Humans, unlike other species, consciously manipulate their environment. They have by their intelligence been able to extend greatly their ability to live and work under an increasingly wide range of climatic conditions. This capability did not develop rapidly, but grew slowly over the centuries to be passed on as a social heritage from generation to generation.

In the distant past, the struggle against nature was primarily for survival. Food was the most demanding requirement. Air and water are also essential to life, but water could be found relatively easily and air was everywhere in abundance except at great altitude. Protection from natural enemies was probably more important than shelter from the elements, but at times both would have been sacrificed in the interests of finding enough food.

As long as man lived by hunting, he was forced to go where game was to be found. This precluded the use of permanent shelters other than natural caves, but did permit the use of fire and development of clothing and such forms of shelter as could be dismantled for transport or improvised from materials available at the new site. The traditions of centuries can be seen in the practices of native tribes today in many parts of the world. Invariably, readily available materials are used, and structures are often well adapted to local needs and conditions.

The Indians of North America who relied mainly on hunting lived in tents or teepees made of skins. In forested areas, bark was used as a covering and, like the skins, could be rolled up for transport. Where trees were scarce, the teepee poles also had to be transported. Furs were used for clothing, although it is recorded that the Indians were able to go about in extremely cold weather with their bodies only partially covered. The Inuit in the Arctic still make extensive use of clothing made from furs, and, being hunters, they rely on temporary or movable housing: igloos made from snow in the winter [**1.1**] and shelters made from skins in other seasons. For centuries, the Mongol tribes of Asia depended on their herds of domesticated animals but, like the hunters, were nomadic because they had continually to seek new grazing for their animals in the arid regions they inhabited. Horses and other animals were used for transportation of their dwellings, called yurts, which consisted of a round latticed framework covered with felt. For hundreds of years, gypsy tribes lived in wagons drawn by horses; in contemporary times, people move about the country in mobile homes, vans, trailers, or campers, the products of a society that has become nomadic again in its search for recreation.

The shift from hunting to farming began about 10 000 years ago in Europe and Western Asia and gradually led to the establishment of settlements [**1.2**] and thus to the increasing use of fixed and more durable forms of shelter. As wealth was accumulated, protection could be provided beyond the requirements of bare survival. Dwellings could be enlarged to accommodate increased domestic activities. At later stages, separate buildings could be constructed to house animals, to store food and fodder, and to accommodate crafts and manufacturing. The range of uses for buildings was thus gradually extended consistent with cultural, technological, and economic development. This development is still continuing today.

About one-sixth of the total national effort in a country with a cold climate such as Canada is directed today to construction, about 55% of which will be used for buildings [**1.3**]. Half of the expenditure on buildings will be for residential construction, and the remainder for industrial, commercial, institutional, and other construction. These buildings may vary from unheated bus-stop shelters providing some protection from the weather to special factories and laboratories providing precisely controlled indoor environmental conditions. In all cases, the basic requirement is the same: to assist in the provision of a suitable physical environment for the people, animals, plants, and processes involved in the occupancy that the building is intended to accommodate.

1.2 Buildings and Environment

Unlike clothing (which is also used for protection from the weather), buildings provide space for our activities within them. They create an indoor environment different from that of the outdoors. Simple buildings such as bus-stop shelters may provide only partial protection from sun, wind, and precipitation, depending on the completeness of the enclosure provided. These structures must still be considered as buildings even if they have only a roof to intercept rain, snow, and sun. Most buildings intended for human occupancy in countries with cold climates provide complete enclosures made up of walls, roofs, floors, windows, and doors. Although these components are barriers to any objects of appreciable size, they can seldom be constructed with sufficient tightness to prevent completely the entrance of insects and dust. Similarly, air, water, water vapour, and other gases may pass through joints and cracks, or directly through porous materials, at rates dependent on the nature and extent of the leakage paths and the flow potentials that exist.

The flow of energy through enclosures is also dependent on the constructions used. Solar energy, consisting of both visible and nonvisible radiation, may pass through some materials (notably glass) providing light and a heat gain to the interior. Heat, or thermal energy, is transmitted whenever there is a temperature difference. Such flows can be restricted, but never entirely prevented, by using constructions having high resistances to heat flow.

Every building loses and gains heat, moisture, and air as a result of differences between indoor and outdoor conditions. There may also be substantial contributions of heat, water vapour, carbon dioxide, and other air contaminants from the occupants and the activities within the building as well as substantial heat gains from lights and other equipment. When buildings are operated in a passive role—without means of adjusting the flows of mass and energy—the indoor conditions may vary markedly with time as they adjust to levels at which the losses and gains will be balanced.

Temperatures within a closed, unheated, shuttered building will be moderated by the building and will be higher in winter and lower in summer than those outdoors, while a building with many unshaded windows or with occupancies generating a great deal of heat may be markedly hotter than the outdoors on summer days.

The shift toward an active role for buildings involves the deliberate adjustment of the building and its equipment so that certain features of the indoor environment can be kept constant or held within required limits. It has long been common practice to incorporate means of manual adjustment of natural lighting levels and ventilation and to provide supplementary lighting and heating. Today, these and other adjustments are usually carried out automatically with the aid of electrical and mechanical equipment using electricity and other sources of energy for heating, lighting, air conditioning, and the movement and distribution of air and water. The calculated hourly rates at which heat, air, and water vapour may have to be added or removed to effect control constitute the heating, cooling, and ventilating loads for the building. Different adjustments may be required in different parts of the building, and these may vary with time and in relation to one another. It is common for the core of a large building to require cooling at all times, while the perimeter rooms require heating in winter and cooling in summer. The mechanical and electrical equipment in many modern buildings accounts for more than one-third of their initial cost, as well as a substantial annual operating cost. This equipment and these expenditures are required in order to provide the building with an active environmental control capability.

1.3 Building Practice and Building Science

The design of buildings has been, and still is to a large extent, based on building practice. Changes have been slow and, in the main, have come about through an evolutionary process of trial and error. Building practice has been fundamentally an inheritance from the past, modified by factors such as climate, economy, social habits, local aesthetic values, and local resources of materials and skills. The evolutionary process works slowly under the influence of new factors; it is equally slow in rejecting the obsolete.

The growth of scientific knowledge has led to great advances in the analysis and rational design of the purely structural functions of a building. There has also been a great deal of development in individual materials and components. As yet, there have been relatively small advances in dealing adequately with all the combinations of elements and with the complex interrelationships of phenomena involved in the performance of an entire building. The reasons are not hard to find. It is sufficient to note that, even now, contemporary building design draws on the knowledge and experience of almost every branch of engineering science.

We have long since passed the point where we are content to rely on the "trial-by-use" method of assessing changes in design, materials, and construction.

Many new and interesting materials, systems, and methods of design and construction are offered each year. Those responsible for assessing and screening such new developments realize only too well the relative inadequacy of our present knowledge of the suitability of any given material or method. In addition, our standards of performance are continually being raised. As we reduce our major difficulties in turn, minor ones assume greater relative proportions, and we clamour for their reduction or elimination also, in the name of progress. The existing state of knowledge appears less and less adequate as the demands upon it increase.

In the past, the architect or master builder provided the direction and supervision of all aspects of building construction. The introduction and steady development of science and engineering led gradually to the division of professional responsibility. The first great change came about with the development of structural engineering, a field concerned with the soundness and structural sufficiency, or safety, of buildings. Building codes were developed as municipalities sought to provide guidance for safe practices in structural, fire and health considerations. Design of building services, including heating, ventilating, air conditioning, and the transport of fluids, has grown into another large and important professional specialization. Electrical engineering became involved in the application of electric power. Mechanical and electrical services together now account for as much as 50% of the cost of some buildings.

Lighting, acoustics, foundations, and fire safety have more recently developed as recognized professional specialties. Materials science, concerned with the performance and durability of materials, may be the next discipline to develop. One of the consequences of this fractionation of responsibilities has been a strong tendency for basic knowledge about buildings to be developed and held in the first instance by each specialized group. This has delayed recognition of the large and growing body of knowledge represented by the sum of the individual contributions which ought to be the common property of all who are seriously involved with the design, construction, and operation of buildings.

"Building science" is a term now widely used, for want of a better one, to describe the growing body of knowledge about the relevant physical science and its application. This knowledge has already become so extensive that it is impossible to avoid subdivision, if only for convenience. Subdivision has occurred partly in response to the sensory impressions and the concerns of human occupants—lighting, acoustics, the thermal environment, health, fire, and structural safety. These factors involve most of the stimuli that can have an influence through the senses. Odours and sensations of motion arising from deflections and vibrations are now considered as topics needing special attention.

The reactions of materials and other inanimate things are legitimate interests of the building scientist. The study of the reactions of people and other living organisms to the factors in the physical environment is largely the responsibility of the life scientist, but both must become involved in matters of comfort, health, behaviour, performance, and the general well-being of people and animals. Practitioners in the design and operation of buildings as well as building scientists must appreciate the contribution of the life sciences, and those seeking to study the behaviour of living subjects in buildings must also appreciate the importance and relevance of the physical environment in buildings.

1.4 The Design of Buildings

The design of buildings is a formidable and demanding task. The need for specialized knowledge is discussed at length in a 1974 report from which the following paragraphs have been selected [**1.4**]:

> This broader view of the construction industry, for purposes of dealing with its communication, information, knowledge and research needs, leads logically to a recognition of the three principal functional roles which are characteristic of it. These are, in concept, the owner, the designer, and the contractor. There is always an owner, who may be a private individual, a group or an agency, a corporation, or a government agency. The construction that is to be provided must serve his enterprise, whether it be residential, institutional, commercial, or engineering construction. The designer, whether he be one man or a design team, must identify the owner's needs and then devise a total scheme to satisfy the owner's needs and must describe them in plans and specifications which provide a basis for a contract between the owner and the contractor.
>
> Although the situation becomes very complex if viewed in any detail, the broad lines of information flow, as well as their general nature, can be envisaged quite usefully by reference again to the owner, designer, and contractor and their relationship to the three major sources of influence upon their decisions, as shown in Figure 1.1. The owner's requirements grow out of the needs of his enterprise or objective. He is therefore concerned in the first instance with what to build. Public and user reactions of various kinds will influence his decision in general terms as well as in matters of detail. Many of these will exert a marked influence also on the designer's decisions; some may even affect the contractor's decisions.

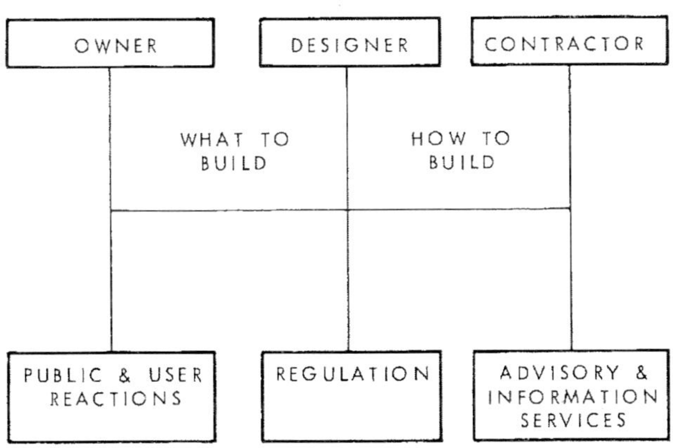

FIGURE 1.1 Information paths between principal agents and influences. (From Hutcheon, N.B., *Research for construction*, NRCC 14005, 1974, Fig. I-1, p. 3.)[**1.4**]

> Regulation, taken to include governmental monetary aids and concessions of all kinds as well as regulations imposed by law, may very substantially influence what the owner will decide to build. Municipal building bylaws and provincial safety codes are among the more common and more important of these governmental influences. The designer and the contractor must also conform to those that are mandatory and may be influenced in what they do by other aids and concessions.

The contractor is mainly concerned with how to build. He is influenced by information and advice coming to him from many sources. These may range from special-interest groups, both public and private, including manufacturers and their industry associations who would like to see their products used, to individuals with special interests, to libraries, and other agencies, such as independent research laboratories, which are regularly in the business of disseminating information or promoting knowledge. Product information and prices may be of interest to the designer as well as to the contractor, as will a very substantial amount of other technical information and advice on how to build.

The designer or, more properly, the design professionals, occupy an important central position at the professional level of participation. The design process is now being recognized as a common discipline of all professions and is defined as an iterative decision-making process. Decisions must be made at various stages to be tested in the light of the implications that may only be discovered as the design studies proceed. To choose, to decide, requires prediction as to the consequences of the choice that is to be made. The ability to predict grows out of knowledge. There is always a need for as much knowledge as possible on the part of the designer.

1.5 Sources of Information for Building

The various sources to which practitioners can turn for information are shown schematically in Figure 1.2, along with the lines of communication, which together form what has been called a knowledge system. A paper discussing the construction industry needs in general terms [**1.5**] and others identifying the situation in Canada may be consulted for details and further discussion [**1.4, 1.6–1.8**].

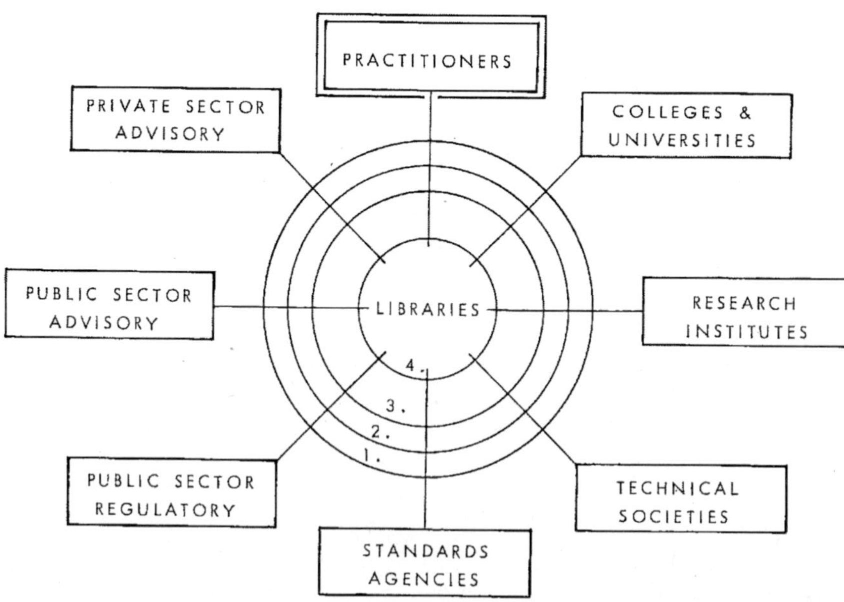

FIGURE 1.2 Knowledge system for building practitioners. Lines of communication: 1. personal inquiry; 2. lectures, symposia, formal discussion; 3. current periodicals, books, and papers; 4. libraries.

1.6 Building Research Institutes

Prior to 1940 there were two buildings research institutes in the world. The best known was the Building Research Station in the United Kingdom, which was established in 1921; the other was in the USSR. By 1945 the Building Research Division of the National Bureau of Standards in the United States was also doing notable work on the evaluation of materials and structures. As other governments became involved in the promotion of housing and building in the postwar rehabilitation period, they established building research institutes or divisions. Canada's Division of Building Research (DBR) was set up in 1947 by the National Research Council of Canada (NRCC).

There are now 70 or more such institutes in more than 40 countries that engage in national programs, exchange information, and arrange joint activities through membership in two international organizations, Conseil International du Bâtiment pour la Recherche, l'Étude et la Documentation (CIB) and Réunion des Laboratoires d'Essais et de Recherches sur les Matériaux et les Constructions (RILEM) [**1.6**]. As a result of these developments and activities in the private and public sectors there has been a substantial increase in the information about buildings in most developing and industrialized countries. For various reasons the actual improvement in buildings has not always matched the growth in knowledge. Changes driven by economic or other forces often take place before the necessary knowledge is available, thus creating new problems.

Building institutes have generally served as information agencies as well as research agencies, and have often performed an advisory role. Canada's DBR, for example, is technical advisor to the Canada Mortgage and Housing Corporation, is represented on more than 400 technical advisory and standards committees throughout Canada and the United States, and provides secretariat services for the preparation, distribution, and servicing of the model National Building Code of Canada and its companion, the National Fire Code of Canada. Both codes are offered for voluntary adoption by municipal, provincial, and federal governments or their agencies.

1.7 Basic Requirements of Buildings

The function of walls, floors, and roofs is to enclose space in such a way that some or all of the physical environmental conditions, either inside or outside the enclosed space, can be regulated within acceptable limits. Physical environment must be regarded in the broadest sense to include not only weather factors such as temperature, air movement, humidity, rain, snow, and light, but also dust, odours, noise, and perhaps nuclear radiation. Control of the physical environment may include consideration of all forms of energy and all forms of matter that can occur in space.

The term "environment" implies a subject, or subjects, which may be living plants and animals, including humans, or inanimate materials, such as goods in process or in storage. The space inside or outside a building may be occupied by any or all of these. A building is normally used for the control of the physical environment within the enclosed space, but it may also be used to contain or control substances, organisms, or radiation which might otherwise become a nuisance or a hazard outside the building.

Consideration of the broad function of the space enclosure serves two purposes: it emphasizes the range of consideration that may be involved in establishing the specific functional requirements in any given case, and it serves as a background for a logical approach to practical discussions of "envelope" design.

An exterior envelope may be regarded as a large membrane separating indoor and outdoor environments. In fulfilling this function the envelope must possess adequate strength and rigidity, requirements which vary with the extent to which the envelope must contribute to the overall structural strength and rigidity of the building. If there are no significant differences between indoor and outdoor environments, a separator may be unnecessary. It is important to note that the separator may be subjected to variations in almost all the factors that make up the physical environment.

The differences in these factors represent the use which is being made of the separator. The differences to be maintained will determine the properties the separator must have. The variation in the magnitude of the conditions throughout the separator from time to time will determine the conditions under which the materials making up the separator must function.

Environmental factors can originate within and on the separator itself as a result of matter and energy transformations. The materials may, for example, provide a source of odours or of undesirable gases that are not found outdoors. Radiant energy may be absorbed by an enclosure and transmitted as heat. Water may enter the outside of an enclosure and be transformed by changes in temperature and, with the uptake of heat, into vapour.

Similarly, water vapour originating either indoors or outdoors may enter a separator and condense to liquid or ice within it. Wind blowing on the outside of a structure may produce vibrations that cause noise within the structure. Mechanical disturbances at one point may cause vibrations and noise at some distant point. Such transformations create many problems, since the possibility of their occurrence is frequently difficult to predict.*

The basic overall requirement of a building is to contribute to the owner's enterprise and accommodate the occupancy or occupancies as required.

A building must satisfy several general requirements. It must be

- safe in respect of structure, fire, and health
- economical in initial cost and operating cost
- aesthetically pleasing, inoffensive to the senses, and an aid in sensory tasks.

To achieve safety it must provide

1. structural strength and rigidity
2. resistance to initiation and spread of fire
3. control of air and water quality and means for waste disposal

* The first seven paragraphs of section 1.7 were written in 1953 [**1.9**] as an introduction to a statement of requirements for walls, following several years of discussions in the postwar period with prospective homeowners seeking improved alternatives to wood-frame wall construction. Similar versions have been presented in many discussions of aspects of enclosure design by the authors and their colleagues at DBR/NRCC over the years [**1.10–1.12**]. Here we adapt them to the development of basic requirements for buildings.

To achieve economy it must

4. be well matched to its purpose
5. have durable materials and components
6. have reasonable maintenance and operating costs

To be inoffensive and an aid in sensory tasks it must provide control of

7. odours
8. light
9. sound and vibrations

To function as a moderator of the environment and to satisfy all other requirements, it must provide for control of

10. heat flow
11. air flow
12. movement of water as vapour and as liquid
13. solar and other radiation.

Control of heat flow is an important element in the conservation of energy in summer and winter. The control of the energy involved in the transmitted and absorbed solar radiation is also an important consideration, as is control of air flow.

The ways in which the four requirements for flow control are satisfied will in combination determine the thermal and moisture environment or, more simply, the thermal environment for the occupancies. They will also determine the thermal environment for the materials and components; if they promote premature failure and degradation they will become involved in durability, cost, and safety considerations. In the event of fire, air movement provides oxygen and is a carrier of hot toxic gases of combustion [**1.13**]. It is seldom possible to deal with radiation, heat, air, and water vapour individually, and taken together they form the major part of an important area of study, the thermal environment.

1.8 Buildings for Cold Weather

The thermal environment varies from point to point throughout the enclosed space and all parts of a building as well as immediately outdoors. It is important to be able to understand and thus to predict these variations and how, when necessary, they can be suitably modified or controlled. The thermal environment is always strongly influenced by the weather; it becomes vitally important in considerations of the performance of buildings in cold weather.

References

1.1 *Down but not out.* Canadian Forces Publication 217. Ottawa: Information Canada, 1970.

1.2 Sandström, G.E. *Man the builder.* Stockholm: International Book Production, 1970.

1.3 *Canada Year Book, 1980–81.* Chap. 9. Ottawa: Statistics Canada.

1.4 Hutcheon, N.B. *Research for construction.* DBR/NRCC, March 1974. (NRCC 14005)

1.5 Granum, H. *Communication in building*. DBR/NRCC, Oct. 1971. (NRCC 12187)

1.6 *The first 25 years, 1947–1972*. DBR/NRCC, July 1973. (NRCC 13240)

1.7 Carson, E.R., and Williams, G.P. *Publications on Canadian building technology*. Canadian Building Digest 196. DBR/NRCC, May 1978.

1.8 Crawford, C.B. *Building technology and its use*. Canadian Building Digest 200. DBR/NRCC, Nov. 1978.

1.9 Hutcheon, N.B. Fundamental considerations in the design of exterior walls for buildings. *Engineering Journal*, 1953, *36* (6) pp. 687–98, 706. (NRCC 3057)

1.10 Hutcheon, N.B. *Requirements for exterior walls*. Canadian Building Digest 48. DBR/NRCC, Dec. 1963.

1.11 Latta, J.K. *Walls, windows and roofs for the Canadian climate*. DBR/NRCC, Oct. 1973. (NRCC 13487)

1.12 Baker, M.C. *Roofs: Design, application and maintenance*. Montreal: Multi-Science Publications Ltd., 1980.

1.13 *ASHRAE Handbook 1980 Systems*. Chap. 41.

2

Weather and Climate

2.1 Introduction

The discussion of people and buildings presented in Chapter 1 emphasized the importance of the weather as a major determinant of the physical environment. Temperature, precipitation, wind, and sunshine have a direct and immediate influence on everything exposed to them. Surrounding objects may concentrate, moderate, transform, or divert solar radiation, wind, or precipitation. A deliberate choice of surroundings may permit us to take advantage of the elements that make the environment more favourable, and clothing and manmade shelter may be used to influence further the direct effects of the weather.

Further evidence of the importance of the weather is seen in the efforts of every country of the world to record and forecast weather for a variety of purposes. The principal weather factors are measured at appropriate intervals using standardized instruments and techniques at as many locations throughout a country as need and economics dictate. Data collected over a period of time provide a record from which statistical measures such as mean, maximum, minimum, and variation can be obtained for the individual weather factors. The long-time course of the weather is known as the climate of a locality or region.

Climatic characteristics obtained from weather records are useful in a number of ways. When considered in conjunction with the nature of the soil in a region, they may indicate general suitability for agriculture. Precipitation records provide an indication of the availability of water. Temperature records provide a measure of space heating requirements; rainfall records provide a basis for estimating run-off rates for sizing drains and culverts; wind and snow records provide a basis for estimating wind and snow loads on building structures. The probabilities of rain, snow, frost, freeze-up, and break-up derived from weather records can be of great value in planning construction operations, and similar information can be used to advantage in many commercial ventures.

The prediction of what the weather will be at any particular time, or over a period of time, can be highly beneficial in support of decisions about many matters. Although statistical measures obtained from long-time weather records form a good basis for prediction of weather characteristics over correspondingly long periods in the future, they are not helpful in predicting the weather for a particular hour, day, or week. Consequently, much attention has been directed to the study of meteorology as related to weather forecasting. The greatest inducement for this study has been the needs of aviation, an industry which has always been critically dependent on the weather. With a greatly improved understanding of the atmosphere it has been possible to identify factors and conditions that are indicative of weather conditions at some future time. Although far from perfect, present-day forecasting provides sufficient improvement over statistically based approaches to justify substantial continuing expenditures. The principal technique used is the collation of meteorological observations from a large number of stations on a weather map so that one can judge the movement and interaction of air masses and the attendant weather patterns that are likely to result. Further information is obtained from pilots in flight and from photographs of cloud cover taken from earth satellites.

The science of meteorology has much in common with building science. Some of the topics of common interest will now be developed; others will be discussed later, in the sequence best suited to the purposes of this presentation. Some specialized aspects of common interest will not be elaborated; they are covered in many of the excellent books and papers on the subject [**2.1**–**2.4**].

2.2 The Earth in the Solar System

The sun is a great nuclear reactor at the centre of the solar system. It is the primary source of the energy of the atmosphere, and all natural meteorological phenomena can be traced to it. This energy is transmitted as solar radiation through space, and can be captured by receiving surfaces that can "see" the sun. It is important to recall the way in which the earth moves in relation to the sun; this movement has a major influence on the amount of solar radiation directed toward any given part of the earth, and gives rise to daily and seasonal variations in the weather.

The most familiar aspect of the earth's motion is its rotation about an axis through the poles once every 24 hours, producing alternate periods of day and night. The resulting cycles of heating and cooling produce daily, or diurnal, changes in the weather.

The earth also moves in an elliptical path around the sun, making one revolution in 365.25 days. This ellipse has a small eccentricity which results in variations in orbital radius and velocity in accord with Kepler's laws of planetary motion. The variations must be taken into account when making precise calculations of the sun's position relative to the earth. The sun is at one focus of the ellipse, so the earth is closest when at one end of the major axis in December and farthest when at the other in June. The mean distance is 1.496×10^8 km. The intensity of solar radiation, which varies inversely as the square of the distance from the source, varies by 3.5% about a mean value of 1353 W/m^2 at the earth. This has only a small seasonal influence.

2.3 Seasonal Changes

Seasonal changes experienced everywhere on earth result from the combination of the revolution of the earth and the inclination of the earth's axis. If the earth's axis were at right angles to the orbital plane, there would be equal periods of day and night everywhere. Its inclination in a fixed direction at an angle of 23.5 degrees from the perpendicular to the plane of the orbit results in a marked yearly variation in the sun's direction as seen from the earth and in the length of the daytime period during which any particular location on the earth is in a position to receive solar radiation.

Figure 2.1 provides the basis for a more detailed explanation of the combined effects of the earth's rotation and revolution. As the direction toward which the earth's axis is inclined lies nearly in the major axis of the ellipse, the solstices, which are the points at which the inclination is toward the sun, are near the ends of the major axis. The winter solstice for the northern hemisphere occurs about December 22 each year. At that time, the sun is directly overhead at noon in latitude 23.5° S. The summer solstice occurs about June 22 when the sun is directly overhead at noon in latitude 23.5° N. These latitudes are known as the Tropic of Capricorn and the Tropic of Cancer.

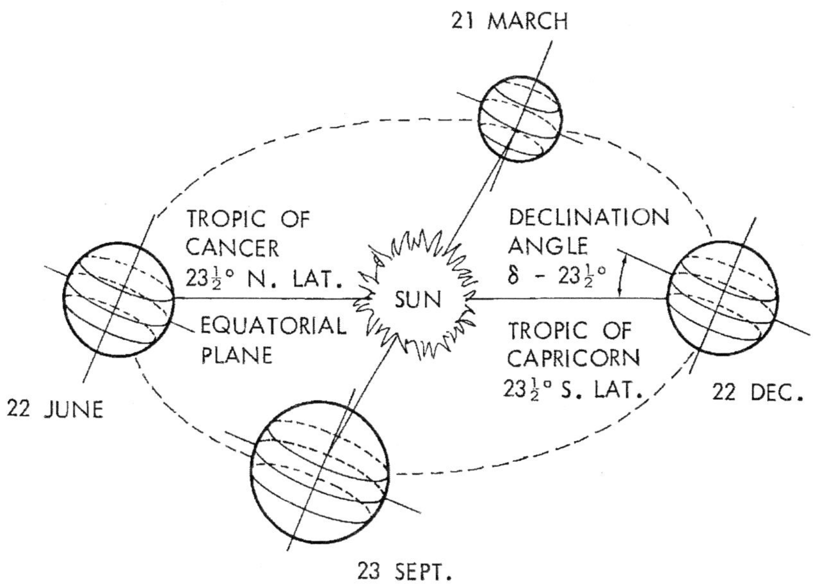

FIGURE 2.1 Earth's motion around the sun.
(*ASHRAE Handbook 1981 Fundamentals*, Fig. 1, p. 27.2)[**2.8**]

At orbital positions midway between the solstices, the relative sun position is in the equatorial plane of the earth, resulting in equal periods of day and night. These are known as the vernal equinox, occurring about March 21, and the autumnal equinox, occurring about September 23. (The dates vary slightly because of the one-day correction of the calendar each leap year.)

Days and nights at the equator are always 12 hours long. At locations above and below the equator the days and nights (which are of equal length at the equinoxes) show their greatest differences at the solstices. At the winter solstice (northern hemisphere) as shown in Figure 2.2, the days are of minimum length, decreasing with increasing northern latitude until at 66.5° N (the Arctic Circle) the sun does not appear above the horizon. At still more northerly latitudes an increasing number of days near the winter solstice have no direct sunlight; at the North Pole there are six months of darkness. The situation in the southern hemisphere is the same as is experienced in northern latitudes but with six months' difference in time.

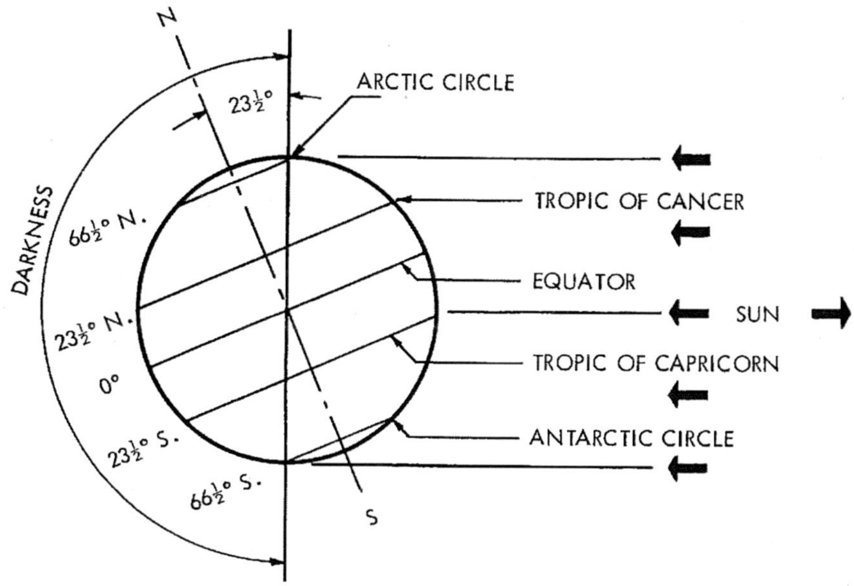

FIGURE 2.2 Earth relative to sun at winter solstice.

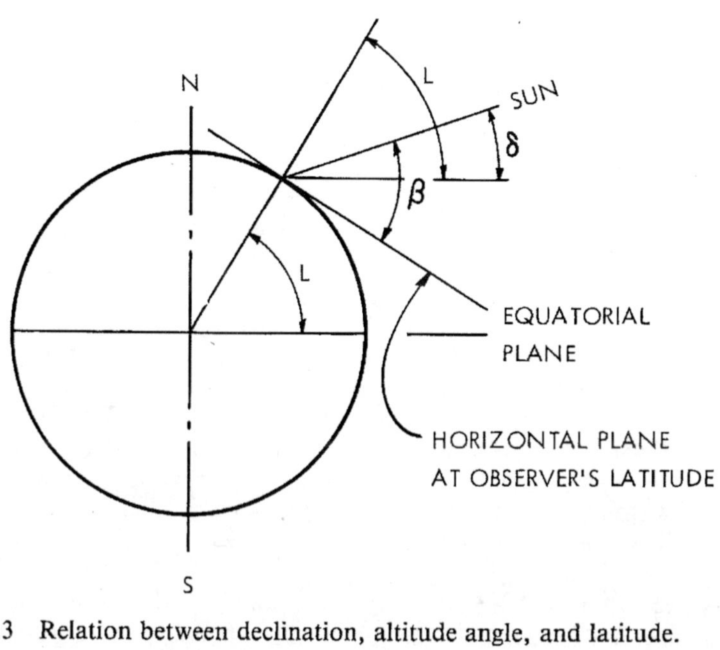

FIGURE 2.3 Relation between declination, altitude angle, and latitude.

The relationship of the altitude angle, β, of the sun at solar noon measured relative to the tangent plane at the point of observation at latitude, L, and the declination of the sun is shown in Figure 2.3. The solar declination, δ, is the angle between the earth-sun line and the equatorial plane. It varies throughout the year from +23.5 degrees at the summer solstice to 0° at the equinoxes to −23.5 degrees at the winter solstice.

The altitude angle is given by

$$\beta = 90 - L + \delta. \tag{2.1}$$

The altitude angles for Ottawa, Canada, for example, at latitude 45.5° N are 68° and 21° at noon at the solstices.

Since latitude defines a unique procession of values of sun altitude throughout the year, the measurement of the sun's altitude at noon can be used to define latitude (as is, in fact, widely done in surveying and navigation).

The principal factor directly influencing the seasons is the annual cycle in the length of day: the longer the day, the greater the amount of sunshine (and therefore energy) that can be expected. The energy received on a horizontal surface is dependent on the angle of the sun's rays, which varies throughout the day and the year. Outside the earth's atmosphere the intensity of solar radiation varies by about 7% owing to the changing orbital radius. The earth's atmosphere produces large variations in the radiation reaching the earth's surface. Clouds may block much of the direct solar radiation by reflection and absorption. Some of the absorbed energy that is reradiated and some of the reflected radiation may still reach the earth's surface. Even clear skies absorb a small but significant portion of the incoming radiation.

The length of the atmospheric path is increased when the sun's rays are oblique to the earth's surface. In addition, a bundle of rays representing a given amount of energy is spread over a greater surface area resulting in reduced energy per unit of surface area in proportion to the cosine of the angle of incidence. Variations in solar energy received throughout the year are therefore minimal at the equator and increase with increasing latitudes, giving rise to corresponding seasonal changes.

It is customary to define the seasons in terms of the earth's position in its orbit. Spring is the period from vernal equinox to summer solstice; summer begins with the summer solstice, autumn with the autumnal equinox, and winter with the winter solstice. It might be expected that the height of summer rather than its beginning would occur at summer solstice when solar radiation is greatest. The earth, however, cannot change temperature quickly in response to changes in solar radiation. Energy must be absorbed and stored before temperature can increase. There is thus a lag between energy supplied and temperature change because of the relatively great heat-storage capacity of the earth's surface. The peak of the mean daily air temperature, which reflects the thermal interaction between the earth's surface and the sun with the atmosphere as an active intermediary, occurs over most of Canada on an average of about one month after the longest day has been reached.

The oceans act as great thermal storage reservoirs for the earth as a whole. They prevent abrupt temperature changes in adjoining regions, and also transport and distribute thermal energy from the sun over the earth on a slow seasonal basis. The atmosphere of the earth is by comparison much smaller in energy-storage capacity but is highly mobile and highly variable in its response to the periodic influences of the earth's rotation and revolution on solar input and to the moderating influences of the thermal capacity of the land and the oceans. The atmosphere is thus largely responsible for the highly changeable nature of the weather.

2.4 The Atmosphere

The influence of the atmosphere can be appreciated by considering first the hypothetical case of an earth with no atmosphere: there would be regular diurnal and annual cycles of temperature at the surface; surface temperatures would rise and fall much more than they do now; and the surface temperatures at the equator would be much higher than at the poles. The atmosphere acts as a storage medium for thermal energy and as an agent in heat exchange at the earth's surface. It can also, by its circulation, transport heat from one region to another. Its greatest influence, however, arises from its ability to intercept, absorb, and redirect incoming solar radiation.

The atmosphere consists of a mixture of gases held by gravitational attraction to the earth. It is compressed under its own weight, and varies in density with altitude and with temperature and water content. The composition by volume of dry air at ground level is 78% nitrogen, 21% oxygen, 0.8% argon, and 0.03% carbon dioxide, with small amounts of neon, helium, methane, krypton, hydrogen, carbon monoxide, xenon, ozone, and radon totalling less than 0.003%. The atmosphere contains water vapour in amounts up to 4% by weight. This amount is highly variable, depending greatly on temperature and previous history of contact with water sources. Water is a highly influential and important component, since it can exist as vapour, liquid, or solid within the range of atmospheric temperatures encountered: clouds are formed when excess vapour condenses producing suspended droplets and ice crystals or precipitation in the form of snow, hail, or rain. Changes in the state of the water are accompanied by changes in energy content which add to the complexity of the role of the atmosphere in energy exchanges.

Water is able to absorb solar radiation, which is often referred to as short-wave radiation. Clouds also reflect this radiation, so that the amount reaching the ground varies from 35% to 70% of that arriving from the sun. The earth radiates energy at longer wavelengths corresponding to its lower temperature, and the atmosphere with its clouds also absorbs and radiates energy at long wavelengths. Incoming and outgoing radiation above the atmosphere are about equal, which is reasonable since the earth would change in temperature if its energy losses and gains were not in balance. More energy is absorbed at the equator than at the poles, but this is offset in part by the effects of atmospheric circulation.

2.5 Atmospheric Circulation

Heating of the ground by radiation creates rising warm air currents. Cooler air must flow in at ground level as a replacement and must itself be replaced as part of a pattern of circulation. Black fields, which are good radiation absorbers, produce the thermals used by the glider pilot to gain altitude. The land warms and cools more quickly than the ocean, so landward breezes develop during the day and seaward breezes during the night. The equatorial zone receives more heat than the poles, creating a general circulation consisting of a flow of warm air from the equator to the poles and a return flow of cooler air toward the equator. The rotation of the earth introduces further forces, producing curving flows with east–west components. The net result is a complex pattern of pressures and winds which make up the general air circulation.

The general global circulation consists of three belts or cells between the equator and the poles, each characterized by its own pattern of poleward and east-west flows as shown in Figure 2.4. Rising warm air at the equator, creating a zone of low pressure, moves northward and southward, falling as it cools until it descends to form two zones of high pressure at 30° N and S. Part of it flows back to the equator along the surface with a strong westerly component. The resulting winds are known as the trade winds.

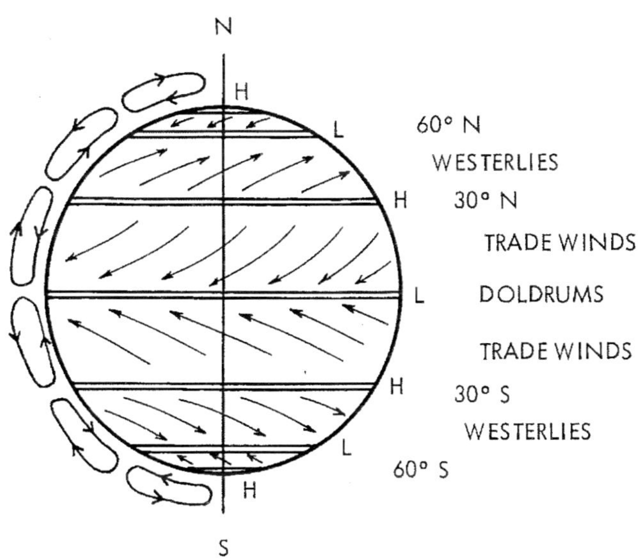

FIGURE 2.4 General circulation of the atmosphere.

The second belt between 30° and 60° N and S involves a poleward flow at the surface and a return flow from the poles at altitude, with a strong easterly component. These surface winds are known as the westerlies. The third belt, between 60° and the poles, produces surface winds known as the polar easterlies with a component of flow at the surface away from the poles. The zones separating these three belts are known as the doldrums, the horse latitudes, and the polar front. The first two terms are well known from the days of sailing ships as zones of weak and uncertain surface winds. The general pattern of the surface winds, particularly for latitudes beyond 30°, is modified by the global seasonal influences and (on a smaller scale) by topography and variable thermal effects.

2.6 Atmospheric Variation with Altitude

The density of the atmosphere decreases exponentially with altitude. Half of its weight lies below 6 km and three-quarters below 10 km. Vertical circulation produces vigorous mixing sufficient to maintain a uniform composition up to about 75 km. Beyond that, the composition changes until hydrogen, helium, and charged particles predominate. The ionized layers provide the reflectors for radio transmission around the earth.

The temperature of the upper atmosphere varies markedly with height according to a pattern which suggests that there are layers possessing different properties. When instruments became available for sounding the temperature at considerable height, it was found that the temperature decreased fairly steadily from ground level to about 10 km at a rate of about 6 K/km, and above this height remained fairly constant [**2.1**]. The first layer, with changing temperature, was named the troposphere and the second the stratosphere; the boundary between them was called the tropopause. It is now known that the altitude of the tropopause varies with latitude, being higher over the equator than at the poles. The tropopause varies with the seasons and the temperature gradients are not constant, and at times no marked change of gradient at the tropopause can be found. The concepts and the terms are useful, and it is desirable for many purposes to agree on some standard set of characteristics for the atmosphere. The International Civil Aviation Organization (ICAO) has adopted a standard [**2.2**], based on average conditions in the lower 20 km at about 40° N latitude, for use in the calibration of altimeters. Some of the features are:

(a) Mean sea level (MSL) pressure: 101.325 kPa
(b) MSL temperature: 15°C
(c) Rate of decrease of temperature with height (lapse rate) in the troposphere: 6.5 K/km
(d) Tropopause height: 11 km
(e) Tropopause temperature: –56.5°C
(f) Temperature constant with height to the top of the standard atmosphere at 20 km.

The stratosphere is now known to extend to about 25 km; above this height the temperature is no longer constant with height but increases to about 50 km, decreases to 75 km, and then increases markedly to 300 km and above. There are several standards for use in rocketry and space flight covering these greater heights, but most are in agreement with the values adopted by ICAO for the first 20 km.

It was believed that since storms and clouds occurred only in the troposphere, there would be no weather influences originating in the stratosphere. Studies have now shown that the upper layers can also have an effect on weather.

2.7 The Climate of Canada

Canada's climate is determined by its geographic location and topography. The westerlies provide the basic influence. The Western Cordilleras interfere with the general westerly flow of air from the Pacific Ocean and encourage intrusions of cold Arctic air and warm, moist air from the Gulf of Mexico. The interference between these three streams produces a succession of cyclones and anticyclones [**2.5**]. The cyclones, characterized by low pressure and counterclockwise rotation, are the result of the northern projection of tongues of warm air. The anticyclones with high pressure and clockwise rotation result from the intrusion of air from the North. The cold, high-pressure areas dominate the interior of the continent in winter, while in summer the migrant low-pressure areas travel across on more northerly paths. The forecasting of weather is based on plotting and predicting the movements of these pressure centres and fronts, which are the boundaries between contrasting air masses.

Additional features of particular regions interacting with the basic flow pattern serve to define six general climatic regions (see Figure 2.5): Arctic, Northern, Pacific, Cordillera, Prairie, and Southeastern [**2.5**] Significant differences occur on a more detailed basis and are reflected in the weather observations taken at a large number of stations across Canada.

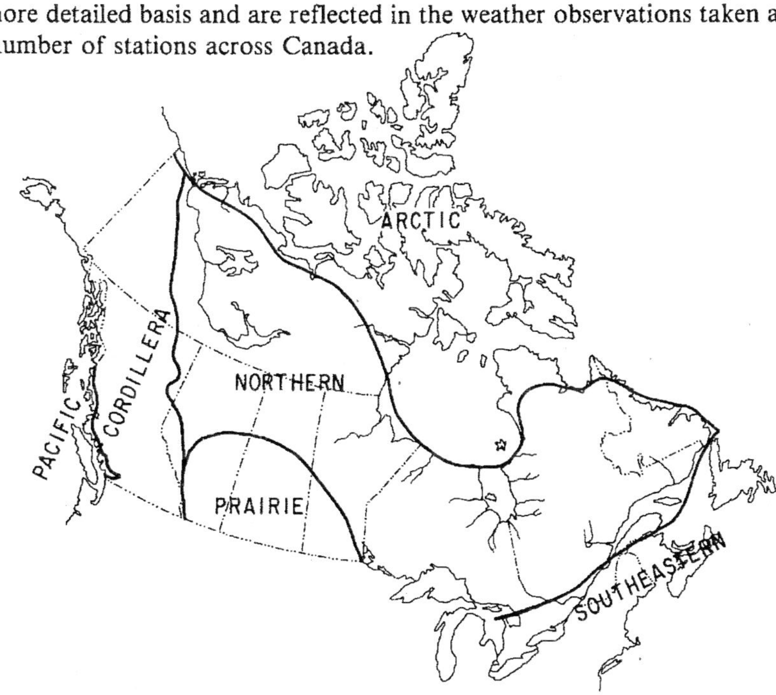

FIGURE 2.5 Climate regions of Canada. (*The climate of Canada*, Meteorological Branch, Department of Transport, 1962, Fig. 4)[**2.5**]

2.8 Meteorological Services

The Atmospheric Environment Service of Environment Canada in Ottawa (formerly the Meteorological Branch, Department of Transport) is similar in many respects to the weather services in other countries. It maintains several networks of weather-observing stations [**2.6**]. Many of these are linked and receive weather data from across the country from which maps are plotted and forecasts made region by region. Most of the first-order stations, located mainly at aerodromes, make complete observations every six hours for international exchange and recording.

Monthly records are provided by more than 2000 stations from which climatic summaries averaged over 10 to 30 or more years are prepared and then published. Climate data for 45 stations are given in *The climate of Canada* [**2.5**].

Measurements for basic records are made with relatively simple instruments. Rain and snow gauges are placed at ground level. Temperature and humidity measurements are made in standard boxes 1.2 m above ground. Winds are measured with anemographs recording miles of wind and direction. Hours of sunshine are recorded with simple burning-strip apparatus; solar radiation is measured at a few selected stations with radiometers. All instruments must be reliable, and they and the observation procedures, including times and locations, must always be standard if records are to be comparable.

Canadian activities in support of meteorology have been extensive. Those of the Atmospheric Environment Service are described in numerous publications which include both regular and special climatic records. All publications on the Canadian climate are listed in excellent bibliographies [**2.3**, **2.4**].

2.9 Climatic Data for Building

Climatic information is essential for building. It is needed for every location in which buildings are to be erected and is used in the planning, design, construction, and operation of buildings and in the regulation of building. The Supplement to the National Building Code 1980 [**2.7**] sets out the necessary climatic information in ways best suited for code use, and is useful for other purposes as well. Climatic information required for thermal environmental engineering is set out in the ASHRAE Handbook 1981 Fundamentals [**2.8**].

2.10 Climatic Data for Thermal Considerations

The climatic information needed in connection with the thermal performance of buildings is that describing the thermal environment, which includes temperature, humidity, wind, and sunshine. Building regulations have been concerned principally with the adequacy of heating equipment and thus with the lowest temperatures likely to be encountered. The lowest temperature on record is seldom of great significance, since what is wanted is a design value which disregards infrequent extremes. Designers are thus presented with a choice of design values designated 1% and 2.5%, which are the percentages of the hours in January at which the temperature will be below that given. They may then make a choice, keeping in mind the requirements of the design. Humidity and winds are not given for winter, but the moisture content of the air is implicit in the wet-bulb temperature given for July.

Degree days, which are a measure of yearly heating requirements, are given by the sum of the departures of the daily mean temperature from 18°C for each day on which the temperature falls below that value. Although degree days take no account of wind or sunshine, they correlate well with annual heating requirements. The base of 18°C is selected on the assumption that no heating is required until the outdoor mean temperature falls below that value.

2.11 Climatic Data for Structural Considerations

The weather-based forces acting on buildings are wind and accumulations of snow and rain. The hourly wind pressures given in the NBC Supplement for use in Canada are based on a probability of occurrence once in 10, 20, or 30 years. The ways in which these and other values are to be used in structural design for code purposes is set out in the NBC Supplement [**2.9**].

Snow loads representing an extreme accumulation of snow and rain are related not only to the weather but also to site conditions and building geometry and heat loss. The basic data are expressed as a ground snow load which is based on the maximum recorded depth of snow. To this is added the probable rainfall into the snow cover. The values thus found are modified as determined by site and building design [**2.9**].

Rainfall by itself is only occasionally the cause of structural loads when accumulations occur if the rainfall rate exceeds the capacity of drains. It is, however, a pertinent consideration in the sizing of drains, sewers and culverts.

2.12 Conclusion

There are a great many applications of climatic data in building, and more will be found. It is almost certain that in the future more refined information will be required on the gust structure of wind for use in detailed glazing and cladding design. There will undoubtedly be increased attention paid to the effects of weather on materials in specific locations on a building. Wetting and freezing can be particularly damaging; degree of saturation is another consideration, and humidity is of prime importance in corrosion. It may be that those effects resulting from particular combinations of factors will have to be the subject of special observations, but it is also possible to derive useful information on such situations from weather data currently recorded. In other cases, where the building itself or some part of it interacts with the weather, the problems may no longer be strictly meteorological ones although they may lead to increased use of weather data as ways are found to resolve them.

Many activities involving the use and operation of buildings are influenced by the weather and can be aided by the use of weather data. Information on climate may assist in selecting locations for particular business ventures, while local weather forecasts may be of value in anticipating demands on business or recreational facilities. Contractors who must transport workers and materials and engage in outdoor, on-site manufacturing and assembly are often greatly dependent on the weather. Weather information can be particularly valuable to them when planning construction in a part of the country which is not well known to them. Monthly averages of precipitation, temperature, and sunshine can assist in gauging the conditions to be encountered. Frost-free periods and dates of freeze-up and break-up may be useful indicators of preferred dates for many operations in northern areas.

The extensive use that can be made of climate and weather information is not really surprising. The weather is, after all, the principal determinant of the outdoor environment and has a pervasive influence on a wide range of human activity. It is highly relevant for buildings used to provide indoor environments differing from the outdoors.

References

2.1 Byers, H.R. *General meteorology*. 4th Ed. New York: McGraw-Hill, 1974.

2.2 *Weather ways*. 3d Ed. Meteorological Branch, Department of Transport. Ottawa: Information Canada, 1961.

2.3 Thomas, M.K. *A bibliography of Canadian climate, 1763-1957*. Meteorological Branch, Department of Transport and DBR/NRCC. Ottawa: Information Canada, 1961.

2.4 Thomas, M.K. *A bibliography of Canadian climate, 1958-1971*. Atmospheric Environmental Service, Department of the Environment. Ottawa: Information Canada, 1973.

2.5 *The climate of Canada*. Meteorological Branch, Department of Transport. Ottawa: Information Canada, 1962. (Reprinted from *Canada Year Book*, 1959 and 1960.)

2.6 *Canada Year Book*, 1980-81. Ottawa: Statistics Canada.

2.7 *Climatic information for building design in Canada*. Chap. 1, The Supplement to the National Building Code of Canada 1980. Ottawa: National Research Council of Canada, 1980, pp. 3-21. (NRCC 17724)

2.8 *ASHRAE Handbook 1981 Fundamentals*. Chap. 23, pp. 16-17.

2.9 *Commentaries on Part 4 of the National Building Code of Canada 1980*. Chap. 4, The Supplement to the National Building Code of Canada 1980. Ottawa: National Research Council of Canada, 1980, pp. 139-280. (NRCC 17724)

3

Molecules, Gases, Heat, and Radiation

3.1 Introduction

Much of the basis for understanding the physical environment is to be found in physical science. Brief reference will be made to some of the more important topics here, but this can appropriately be supplemented by private study of one or more good books on fundamentals [**3.1**-**3.3**]. Some elementary definitions and propositions serve as a starting point.

Elements always combine in the same definite proportions by weight to form chemical compounds. The atomic weight scale is based on the carbon atom being equal to 12. A gram atom of any element contains the same number of atoms as a gram atom of any other element as given by Avogadro's number, whose value is 6.02×10^{23} atoms per gram atom.

Molecular weight is the ratio of weight of any of the molecules of a compound and the weight of the most common carbon atom where the latter is assigned the value 12. A mole of any compound is that amount of it whose mass is equal to its molecular weight expressed in grams. A mole of any substance contains the same number of molecules as a mole of any other substance, as given also by Avogadro's number.

Atoms consist of a nucleus of protons and neutrons. The number of unit-positive charges on the atomic nucleus of an element is equal to the number of protons in the atomic nucleus, and is called the atomic number of the element. An element is a substance all of whose atoms have the same atomic number.

The number of electrons surrounding the nucleus of each uncharged atom is equal to the number of protons in the nucleus. If an atom loses an electron, it has a positive charge; if it gains one or more electrons, it has a negative charge. An ion is an atom, a molecule, a group of molecules, or any particle that has acquired a charge.

Electrons normally travel in preferred planetary orbits around the nucleus under the influence of strong forces of attraction. Atoms and the molecules they form in combination act as if they possess strong elastic shells of definite size and shape, and rebound elastically when they strike one another.

3.2 Kinetic Theory of Gases

The molecules of a gas travel at high velocity, striking one another with great frequency. It is possible to assign values to the diameter, the mass, the average speed at a given temperature, and the number of collisions with other molecules per second. Many of the physical properties of gases can be predicted on the basis of kinetic theory. It is assumed in kinetic theory that gas molecules are small compared with the distance between them, that they exert practically no force on one another except when they collide, and that they collide without loss of kinetic energy. Both theory and experiment lead to a second assumption: that the absolute temperature of a gas is directly proportional to the average kinetic energy of its molecules.

3.3 Temperature

Temperature is defined for gases as the average kinetic energy of translation of the gas molecules. A difference in temperature leads to the transfer of kinetic energy, which is a transfer of heat. Heat is related to the energy of molecular motion, which can include vibration and rotation as well as translation. If heat were associated only with the energy of translation, all gases would require the same amount of heat per mole for a temperature increase of one degree.

The pressure exerted by a gas is the result of the change in momentum associated with the elastic rebound of molecules. It can be shown to be

$$p = \frac{1}{3}\frac{Nm's^2}{V} \tag{3.1}$$

where

N = number of molecules
m' = mass of 1 molecule
V = volume
s = average rms speed of the molecules.

It may be recognized from Equation 3.1 that $m's^2$ is equal to twice the average kinetic energy of molecules and thus is proportional to absolute temperature. At a given temperature and pressure N/V is constant, confirming Avogadro's hypothesis that at any temperature and pressure the number of molecules per unit volume is the same for all gases. Avogadro's number gives the molecules per mole: 1 mol of a gas at 0°C and standard atmospheric pressure of 101 325 Pa has a volume of 22.414 L.

3.4 Gas Laws

The relationships between pressure, temperature, and volume for gases were established by early experiments. Boyle found that when temperature was kept constant the pressure-to-volume relationship was

$$pV = p'V' = \text{constant.} \tag{3.2}$$

Gay-Lussac expressed the results of varying temperature at constant pressure in the form

$$V = V_o(1 + \alpha_p t) \tag{3.3}$$

V is the volume at t°C and V_o the volume at 0°C. The value of α_p was later established as $1/273$ when the temperature, t, was expressed on the Celsius scale. This was followed by Charles' law, which stated that at constant pressure, volume varied as the absolute temperature.

These relationships lead to the combined form:

$$\frac{pV}{T} = C \tag{3.4}$$

where

C is a constant and T is the absolute temperature.

For any two conditions 1 and 2 for a given mass of gas:

$$\frac{p_1V_1}{T_1} = \frac{p_2V_2}{T_2}. \tag{3.5}$$

The constant C in Equation 3.4 applies to n moles of gas for which the volume is V. Introducing a new constant $R = C/n$ and writing $v_m = V/n$ leads to

$$pV = nRT \tag{3.6}$$

and

$$pv_m = RT. \tag{3.7}$$

These are different forms of the Equation of State. The constant, R, refers to one mole, that is, to an amount of substance having a mass in grams equal to the "molecular weight". For example, the atomic weight of carbon is 12 and of oxygen 16. Thus the molecular weight of CO_2 is 44 and 1 mol of CO_2 = 44 g, or 0.044 kg.

Example 3.1

Find the value of R from one consistent set of values as follows:

$$\begin{aligned} p &= 1 \text{ atmo} = 101\ 300 \text{ Pa } (101.3 \text{ kPa}) \\ n &= 1 \text{ mol} \\ T &= 273 \text{ K} \\ V_o &= 22.41 \text{ L} = 22.41 \times 10^{-3} \text{ m}^3 \\ R &= \frac{pV}{nT} = \frac{(101\ 300 \text{ Pa}) \times (22.41 \times 10^{-3} \text{ m}^3)}{1 \text{ mol} \times 273 \text{ K}} \\ &= 8.314 \text{ J/(mol} \bullet \text{K)} \end{aligned}$$

The units for R in Example 3.1 indicate that it is an energy or work term. In physical terms it is the external work done by 1 mol of gas increased 1 K at constant pressure. Engineers and some meteorologists have found it convenient to deal with grams or pounds of substance rather than moles. This creates confusion when the same symbol R is used without differentiation. For engineering purposes Equation 3.6 is commonly written in the form

$$pV = w\mathrm{R_g}T. \tag{3.8}$$

$\mathrm{R_g}$ as given in the equation will be used in this text to differentiate it from R. It now applies to a unit weight or mass of gas in the units assigned to w and not to 1 mol as for R. A subscript g will be used as a reminder that the value of $\mathrm{R_g}$ is no longer for 1 mol. Other subscripts will be used when particular gases such as air ($\mathrm{R_a}$) and water vapour ($\mathrm{R_w}$) are involved. The units used in equations 3.6 and 3.8 must always be consistent with those used in establishing R or $\mathrm{R_g}$. The examples to follow will provide clarification.

Example 3.2

Calculate in SI units the values of $\mathrm{R_g}$ for air and water vapour, kg basis. R (from Example 3.1) = 8.314 J/(mol • K), M = 0.028 96 kg/mol for air and M = 0.018 02 kg/mol for water vapour.

$$\mathrm{R_a} = \frac{\mathrm{R}}{\mathrm{M}} = \frac{8.314}{0.028\ 96} = 287.1\ \mathrm{J/(kg \cdot K)}\ \text{for air}$$

$$\mathrm{R_w} = \frac{\mathrm{R}}{\mathrm{M}} = \frac{8.314}{0.018\ 02} = 461.5\ \mathrm{J/(kg \cdot K)}\ \text{for water vapour}$$

Example 3.3

Using SI units find the mass of water vapour in 1 $\mathrm{m^3}$ at a pressure of 0.875 in. Hg and temperature of 23°C.

$$1\ \text{in. Hg} = 3.38\ \text{kPa}$$

$$\mathrm{R_w}\ \text{from Example 3.2} = 461.5\ \mathrm{J/(kg \cdot K)}$$

$$w = \frac{pV}{\mathrm{R_w}T} = \frac{(0.875 \times 3.38 \times 10^3\ \mathrm{Pa}) \times 1}{461.5 \times (273 + 23)} = 0.0216\ \text{kg}.$$

It will be convenient in later chapters to use the form of the Equation of State given by Equation 3.8. Note the values of $\mathrm{R_g}$ for air and water vapour calculated in Example 3.2, which will be frequently used.

3.5 Atmospheric Pressure

The atmosphere, which varies in pressure and density with altitude, provides a useful case for application of the gas laws. It will be recalled that the ICAO Standard Atmosphere is taken to be 15°C at mean sea level with a lapse rate of 6.5°C per kilometre up to the tropopause at 11 000 m, where the temperature is -56.5°C. The temperature in the stratosphere is assumed to be -56.5° up to 20 km.

The temperature at an elevation of h m for a lapse rate of β is

$$T = T_a - \beta h \tag{3.9}$$

so that

$$t = 15 - 0.0065\, h$$

and

$$T = 273 + 15 - 0.0065\, h. \qquad \text{(K)}$$

Dealing first with the case of an isothermal atmosphere, the basic relationship between pressure and altitude can be established with the aid of Figure 3.1 as

$$dp = -g\rho\, dh \text{ or } dh = -\frac{dp}{g\rho}$$

where ρ = mass density.

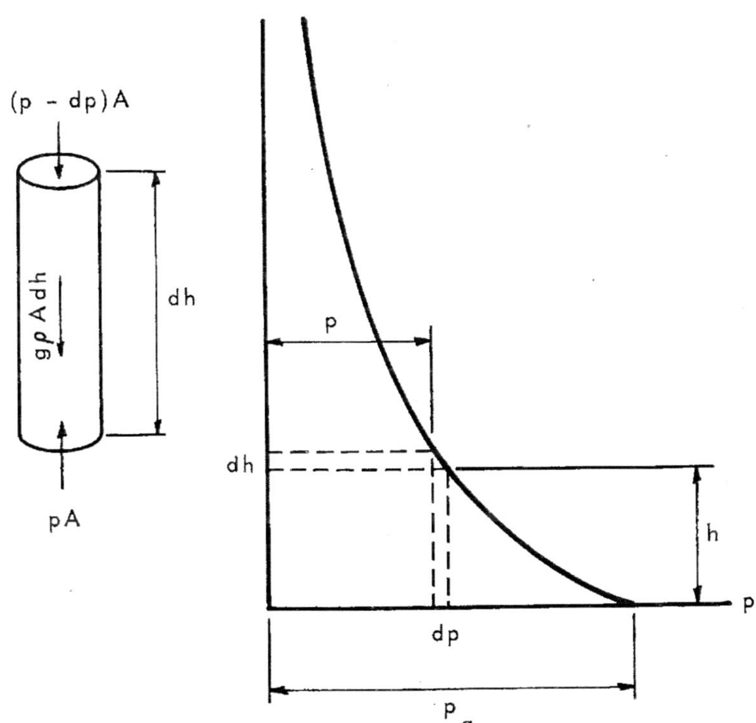

FIGURE 3.1 Atmospheric pressure versus altitude.

From Boyle's law it follows that

$$\frac{g\rho}{p} = \frac{g\rho_a}{p_a} \quad \text{and} \quad g\rho = g\rho_a \cdot \frac{p}{p_a}$$

so that

$$dh = -\frac{dp}{g\rho} = -\frac{p_a}{g\rho_a} \cdot \frac{dp}{p} .$$

Integrating from pressure p_a at zero altitude to pressure p at greater altitude, and rearranging, it can be shown **[3.4, pp. 17-21]** that the pressure at any altitude h in terms of pressure p_a at zero altitude is

$$p = p_a \, e^{-[\frac{h\,g\rho_a}{p_a}]} \tag{3.10}$$

Considering the case of a uniform temperature gradient, $-\beta$, the basic relationship is, as before,

$$dp = -g\rho\, dh = -\frac{g}{v}\, dh$$

where v is in m³/kg.

Now, from Equation 3.8, and noting that $v = \frac{V}{W}$ and $R_g = R_a$

$$pv = R_a T \text{ and } v = \frac{R_a T}{p} = \frac{R_a}{p}(T_a - \beta h) .$$

By selecting a suitable form for integration, integrating, and rearranging, it can be shown that

$$\frac{p}{p_a} = (1 - \frac{\beta h}{T_a})^{[\frac{g}{R_a \beta}]} . \tag{3.11}$$

Now, for the ICAO Standard Atmosphere,

$$p_a = 101\,300 \text{ Pa, } R_a = 287.1\text{, and } \beta = 0.0065 \text{ up to } 11\,000 \text{ m,}$$

and

$$T_a = 273.2 + 15 = 288.2 \text{ K, } g = 9.807 \text{ m/s}^2$$

$$pv = R_a T \text{ and } v = \frac{287.1 \times 288.2}{101\,300} = 0.816 \text{ m}^3/\text{kg}$$

so that

$$\rho_a = \frac{1}{0.816} = 1.225 \text{ kg/m}^3.$$

Substituting appropriately in Equation 3.11,

$$p = 101\,300\,(1 - 22.6 \times 10^{-6}\,h)^{5.256}. \qquad \text{(Pa)} \qquad (3.12)$$

Equation 3.12 applies up to $h = 11\,000$ m. Above that, the temperature is constant to 20 km.

The density at elevation h is given by

$$\frac{\rho}{\rho_a} = \frac{p}{p_a}\frac{T_a}{T} = \frac{T_a}{T}\left(1 - \frac{\beta h}{T_a}\right)^{\left[\frac{g}{R_a \beta}\right]}.$$

Substituting

$$T = T_a - \beta h$$

then

$$\rho = \rho_a\left(1 - \frac{\beta h}{T_a}\right)^{\left[\frac{g}{R_a \beta} - 1\right]} \qquad (3.13)$$

and for standard air

$$\rho = 1.225\,(1 - 22.6 \times 10^{-6}\,h)^{4.256}. \qquad (\text{kg/m}^3) \qquad (3.14)$$

Example 3.4

Find the temperature, pressure, and density for the standard atmosphere at 1000 m above sea level.

$$T = 288.2 - 6.5 = 281.7 \text{ K or } 8.5°\text{C}$$

and from Equation 3.12,

$$p = 101\,300\,(1 - 22.6 \times 10^{-6} \times 1000)^{5.256}$$

$$= 101\,300 \times 0.8868 = 89\,860 \text{ Pa}.$$

From Equation 3.14,

$$\rho = 1.225\,(1 - 22.6 \times 10^{-6} \times 1000)^{4.256} = 1.111 \text{ kg/m}^3.$$

Example 3.5

Find the pressure and density at the tropopause.

$$h = 11\ 000 \text{ m}, \ T = -56.5 + 273 = 216.7 \text{ K}.$$

From Equation 3.12,

$$p = 101\ 300\ (1 - 22.6 \times 10^{-6} \times 11\ 000)^{5.256}$$

$$= 101\ 300 \times 0.7514^{5.256} = 22\ 600 \text{ Pa}$$

$$\rho = 1225\ (1 - 22.6 \times 10^{-6} \times 11\ 000)^{4.256}$$

$$= 1225 \times 0.296 = 0.363 \text{ kg/m}^3.$$

3.6 Barometric Corrections

The altimeters used in aviation to determine altitude are essentially aneroid barometers. Since the primary need is to know the height above ground, it is necessary to consider the differences in barometric readings at various places so that a pilot can interpret altimeter readings accordingly. The correction should account for differences in elevation with due regard for differing air densities from station to station and the horizontal gradients in barometric pressure at common elevations above sea level. This can be a highly complex situation and the standard practice is to base the correction on the conditions of the standard atmosphere, always with reference to sea level.

Barometric corrections for meteorological purposes serve a somewhat different purpose. They are used to establish the horizontal pressure gradients which are indicative of the strength and direction of winds for forecast purposes. When there is a difference in elevation between observing stations it is necessary to make a correction for the added height of the air column contributing to the ground pressure at the station of lower elevation. The practice is to correct all station readings in accordance with a complicated set of rules so that they will then indicate on a comparative basis the effective pressure differences between stations. The barometric pressures quoted by weather stations for public use are usually already corrected to sea level. This information can be misleading, and anyone who requires a true atmospheric pressure reading must ascertain exactly what is being quoted. The potential discrepancy is related to elevation, being about 10 kPa per 1000 m of elevation (see Example 3.4). Differences for a location such as Montreal at 57 m above sea level will be small. Barometric pressure changes and corrections are also important in determining approximate altitudes in surveying.

3.7 Liquids and Solids

Liquids and solids, unlike gases, are relatively incompressible, indicating that the molecules are already close together and that forces of attraction between molecules have become important. These forces are so strong in solids that the molecules can no longer move about, although oscillatory motion is still possible.

Temperature in gases is related only to translational kinetic energy. A rise in temperature in liquids also means an increase in the average speed of translational motion. A rise in temperature of solids means an increase in vibrational motion of particles about their fixed position. Heat will flow under temperature differences by the transfer of kinetic energy of molecules resulting from collisions.

A change from liquid to gas represents an increase in the potential energy of the molecules. This results from the increased separation during evaporation, thereby lessening the forces of attraction, and may be likened to the raising of a stone against earth's gravity which increases its potential energy. The change of state from liquid to vapour takes place at a constant temperature and therefore without any change in kinetic energy, but involves the uptake of a definite amount of heat—the latent heat of evaporation—which provides the increased potential energy. When solids change to liquids the molecules must also gain potential energy, requiring that the latent heat of fusion, or heat of melting, be supplied.

When water in its liquid state is placed in an evacuated flask the molecules at the water surface with high kinetic energy can escape the forces of attraction of their fellow molecules in the liquid and gradually permeate the space above the water. As the concentration in the vapour state increases, the number of molecules recaptured on striking the liquid surface will increase until an equilibrium vapour concentration is reached that is characteristic of the temperature of the liquid. Heat is required to account for the evaporation, and is obtained by a reduction of the temperature of the liquid or transfer of heat from surroundings or both. The concentration of molecules bombarding one another and the walls of the container at a given mean speed characteristic of their temperature produces a definite and reproducible equilibrium vapour pressure.

An increase in temperature resulting from increased energy levels in both liquid and vapour leads to a higher concentration at equilibrium and thus to a higher vapour pressure. The equilibrium condition is known as saturation, and the vapour at equilibrium is referred to as saturated vapour. The vapour conditions for concentrations less than saturation are referred to as unsaturated. They are also said to be superheated, since the same conditions could have been reached by producing a saturated vapour at some lower temperature and then raising its temperature.

When water boils, the escape of molecules is accompanied by vigorous bubbling; this does not mean that evaporation cannot take place at lower temperatures. The vapour pressure at the boiling temperature is equal to the atmospheric pressure above, and vapour bubbles are able to grow inside the liquid.

Evaporation from the surface of water in its liquid state can take place at all temperatures above the freezing point and can continue from the surface of the solid at lower temperatures, the equilibrium conditions then being established between the vapour and solid states. The triple point of water, at which water, ice, and vapour coexist, provides a reference point on the temperature scale and has now been established at 273.16 K.

Water is not always frozen at temperatures below the normal freezing point. It is possible to produce supercooled water if no suitable conditions exist to produce nucleation of crystal formation. Ice crystals themselves or other particulate matter of comparable structure will induce nucleation, as is done in cloud-seeding to cause precipitation. Supercooling is commonly associated with water in a finely divided state, either as fine droplets in the air or absorbed in porous materials. In these cases the supercooled liquid can remain liquid at temperatures well below the normal freezing point. Some supercooling (to the extent of a few tenths of a degree) can exist even in bulk water, under certain conditions.

Supercooling, subcooling, or undercooling (as it is variously called) commonly occurs in the atmosphere. It is now being recognized as an important condition in the freezing of materials and in the heaving of soils. When water is supercooled, the vapour pressure over it corresponds to equilibrium values that are higher than those over ice at the same temperature, in amounts proportional to the number of degrees below freezing.

The importance of water in the environment, which has already been emphasized, warrants a detailed discussion. Its characteristics and behaviour are taken up in greater detail in Chapter 5, where the relevant data on its properties are presented.

3.8 Radiation

Radiation, like temperature and heat, originates with atomic structure and, after travelling through space, can be intercepted and then reappear as atomic and molecular energy. The sun is the most familiar example of a radiating source, but because it is unique it is not the best starting point for discussion.

A wide spectrum of electromagnetic waves ranges in frequency from 10^4 to greater than 3×10^{19} Hz. Since all electromagnetic radiation travels at the speed of light, 3×10^8 m/s, the wavelengths found by dividing speed by frequency range from 3×10^4 m for the lowest-frequency radio waves in use to less than 10^{-11} m. The commonly recognized divisions are shown in Table 3.1.

TABLE 3.1

Electromagnetic Radiation Spectrum

Type	Wavelength
Radio	30 km — 300 mm
Radar	300 mm — 300 μm
Infrared	300 μm — 800 nm
Visible	800 nm — 400 nm
Ultraviolet	400 nm — 1000 pm
X-rays	1000 pm — 10 pm
Gamma rays	< 10 pm

Electromagnetic radiation is produced by the acceleration of electric charges. The radio and television range, from 10^4 m down to about 0.3 m, uses electromagnetic radiation produced and received in oscillating electric circuits. Radar, from 30 cm down to 3 mm, uses electrons in resonant cavities instead of tuned electric circuits. The highest experimental frequency for radar of 10^{12} Hz corresponds to the beginning of the infrared range with wavelengths of 300 μm down to 0.8 μm.

Infrared radiation is associated with the accelerations and decelerations of electrons on charged particles or ions largely induced by the collisions of molecules. Violent agitations of electrons are necessary to produce radiation in the visible range from 800 nm down to 400 nm. This is a relatively narrow band, and it is interesting that the peak wavelength for solar intensity and about half of the energy of the sun fall in this range. Bodies must be very hot, about 1300 K, before they begin to radiate appreciably in the visible range. The sun is believed to radiate at an effective temperature of about 6000 K. Such strong heating produces changes in the electrons in orbit around atomic nuclei, some of which are raised to higher energy levels, releasing electromagnetic energy when they revert to original, or lower, levels of energy. The same effects can be produced by electric fields or by electromagnetic radiation, producing radiation in the visible and ultraviolet wavelengths. The higher-frequency X-rays are produced when fast electrons bombard atoms; gamma rays result from the disturbance of atomic nuclei.

Electromagnetic radiation can be reflected, transmitted, and absorbed in varying degrees. Gases and strongly heated solids emit and absorb at particular wavelengths corresponding to the discrete frequencies at which electrons may be caused to resonate and to jump to higher-energy orbits. Certain frequencies in the ultraviolet range are capable of disrupting chemical bonds, causing some materials to deteriorate.

Radiation absorbed on or in materials becomes heat when it produces increased molecular activity. This applies to all thermal radiation which "includes a portion of the long-wave fringe of the ultraviolet, the visible light region which extends from wave lengths of approximately 0.4 to 0.7 μm and the infrared region which extends from beyond the red end of the visible spectrum to about $\lambda = 1000$ μm. The infrared region is sometimes divided into the near infrared, extending from the visible region to about $\lambda = 25$ μm, and the far infrared, composed of the longer wave length portions of the infrared spectrum" [**3.5**].

The following chapter will deal in more detail with thermal radiation and the ways in which it becomes involved as a highly important environmental factor.

References

3.1 Krauskoff, C. and Beiser, A. *Fundamentals of physical science*. 6th Ed. New York: McGraw-Hill, 1975.

3.2 Marshall, J.S., Pounder, E.R., and Stewart, R.W. *Physics*. 2d Ed. Toronto: Macmillan of Canada, 1967.

3.3 Miller, F. Jr. *College physics*. 2d Ed. New York: Harcourt, Brace and World, 1972.

3.4 Dodge, R.A., and Thompson, M.J. *Fluid mechanics*. 1st Ed. New York: McGraw-Hill, 1937.

3.5 Siegel, R., and Howell, J.R. *Thermal radiation heat transfer*. New York: McGraw-Hill, 1972.

4

Thermal Radiation in the Environment

4.1 Introduction

Thermal radiation is an ever-present factor in the environment. It includes the infrared range and the narrow visible range, and extends into the ultraviolet. All bodies emit and receive thermal radiation which they reflect, absorb, and transmit in conformity with their properties. There is thus a continuing exchange of energy by radiation in the thermal range between each body and the bodies that surround it.

4.2 Blackbody Radiation

The molecular activity that gives rise to thermal radiation determines the amount and distribution of energy by wavelength. If the body is a perfect radiator, known as a blackbody, it will radiate in accord with Planck's equation derived from quantum theory considerations [**4.1, 4.2**]:

$$W_{b\lambda} = \frac{3.741\lambda^{-5} \times 10^{8}}{e^{14\,400/\lambda T} - 1} \; (\mathrm{W/(m^2 \cdot \mu m)}) \tag{4.1}$$

where $W_{b\lambda}$ is the blackbody monochromatic emissive power for the unit wavelength interval of 1 μm at wavelength λ in micrometres for the temperature T in kelvins.

The distribution of radiated energy by wavelength at various temperatures as given by Planck's equation is shown in Figure 4.1. It will be noted that the energy increases markedly with temperature and that the peak output occurs at shorter wavelengths as the temperature increases. Wien's equation giving the wavelength of maximum intensity can be found from Equation 4.1 to be

$$\lambda_m T = 2898 \tag{4.2}$$

where

λ_m = wavelength of maximum intensity, micrometres
T = kelvins.

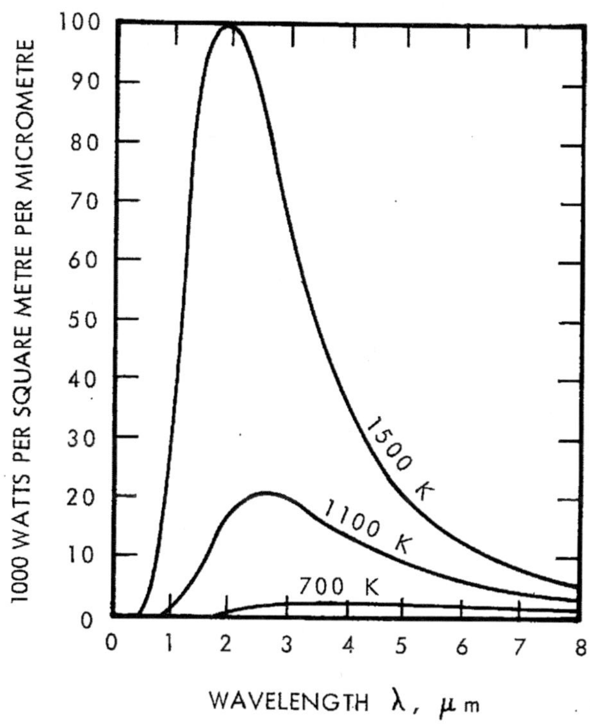

FIGURE 4.1 Monochromatic radiation from a blackbody at various temperatures.

The area under any of the curves of Figure 4.1 will have the dimensions of watts per square metre. The value, which can be found from Equation 4.1 by integration for increments of wavelength $d\lambda$ from $\lambda = 0$ to $\lambda = \infty$, is given by the Stefan-Boltzmann equation:

$$W_b = \sigma T^4 \qquad (4.3)$$

where

W_b is in units W/m^2 and is the total rate of energy emission per unit of surface area over all wavelengths from a blackbody at a temperature T, kelvins. The constant σ has the value 5.670×10^{-8} $W/(m^2 \cdot K^4)$.

A blackbody is conceived as one that emits at the maximum value for each wavelength as given by Equation 4.1. It is also one that will absorb all the radiation falling on it. The term "black" does not refer to colour but to the absence of reflected radiation. All real surfaces reflect some radiation. A practical blackbody condition is provided by the opening into a spherical cavity as shown in Figure 4.2. All radiation falling on the opening will be totally absorbed by multiple reflections in the cavity. The radiation emitted will also have the characteristics of that emitted from a blackbody.

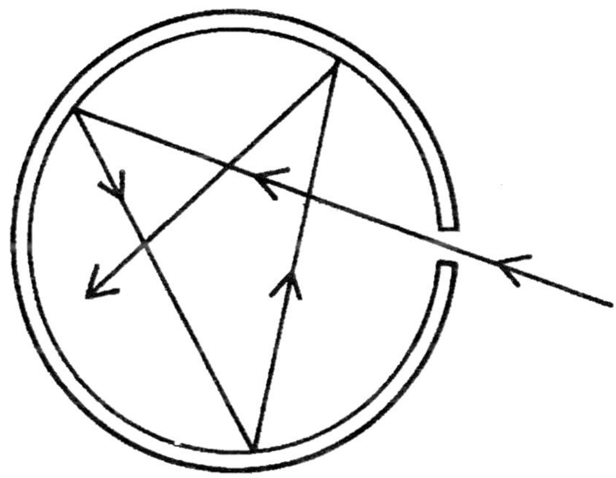

FIGURE 4.2 A practical blackbody.

Most real bodies will radiate at a rate less than that of a blackbody at a given temperature. The emissive power of a nonblack surface radiating to the hemispherical region above it is given by

$$W = \epsilon W_b = \epsilon \sigma T^4 \tag{4.4}$$

where

ϵ is the hemispherical emittance, with a value which will be less than unity.

The monochromatic emissive power of a nonblack surface will be given correspondingly by

$$W_\lambda = \epsilon_\lambda W_{b\lambda} \tag{4.5}$$

where

ϵ_λ = the monochromatic hemispherical emittance referring to a particular wavelength.

If ϵ_λ varies from wavelength to wavelength it becomes necessary to integrate over the range of wavelengths to find ϵ. The situation is easier to deal with if the monochromatic hemispherical emittance, though less than that of a blackbody, is the same for all wavelengths so that $\epsilon = \epsilon_\lambda$. A surface with this characteristic is said to be grey. This assumption is often made when it is known to be a reasonable approximation or when the relationship between ϵ_λ and λ is not known. There is a risk of error at high temperatures and for some materials.

When radiation falls on surfaces that are not black, it may be absorbed, reflected, or transmitted through the material, and

$$\alpha + \tau + \rho = 1 \tag{4.6}$$

where

$$\alpha = \text{fraction absorbed} = \text{absorptance}$$
$$\tau = \text{fraction transmitted} = \text{transmittance}$$
$$\rho = \text{fraction reflected} = \text{reflectance}$$

For opaque surfaces $\alpha + \rho = 1$; for black surfaces $\alpha = 1$.

The absorptance of a nonblackbody is related to its emissivity. Kirchhoff's law states that at a given temperature the emittance ϵ and the absorptance α are equal. If the surface is grey, ϵ and α will be the same for all wavelengths.

Values of ϵ and α for some selected materials are given in Table 4.1. Several show marked differences between ϵ and α because of the difference in wavelength distribution between solar energy and that emitted by bodies at more normal temperatures, thus indicating that they are not grey surfaces.

TABLE 4.1

Emittances and Absorptances for Some Surfaces
(From *ASHRAE Handbook 1981 Fundamentals*, Table 3, p. 2.8) [**4.1**]

Surface	Fraction of blackbody radiation at 10 to 38°C	Fraction of blackbody radiation at 540°C	Absorptivity for solar radiation
1. Small hole in an enclosure	0.97–0.99	0.97–0.99	0.97–0.99
2. Black, nonmetallic surfaces	0.90–0.98	0.90–0.98	0.85–0.98
3. Red brick and tile, stone and concrete, rusted iron and dark paints	0.85–0.95	0.75–0.90	0.65–0.80
4. Yellow and buff building materials	0.85–0.95	0.70–0.85	0.50–0.70
5. White or light cream surfaces	0.85–0.95	0.60–0.75	0.30–0.50
6. Glass	0.90–0.95	—	transparent (8% reflected)
7. Bright aluminum paint	0.40–0.60	—	0.30–0.50
8. Dull brass, copper, aluminum, polished iron	0.20–0.30	0.30–0.50	0.40–0.65
9. Polished brass, copper	0.02–0.05	0.05–0.15	0.30–0.50
10. Highly polished tin, aluminum, nickel, chrome	0.02–0.04	0.05–0.10	0.10–0.40

Note that highly polished metal surfaces have emittances of 0.02 to 0.04 at ordinary temperatures. Their absorptance for solar radiation increases to as much as 0.4, almost the same as white paint, which has an emittance at normal temperatures of 0.92.

All the relationships just given are for total hemispherical radiation without regard for any directional variation in intensity. Lambert's law states that the intensity of radiant energy at a hemispherical surface above the emitting surface is given by

$$I = I_n \cos\phi. \quad (4.7)$$

The angle ϕ is the angle between the normal to the radiating surface and the radial line to the point on the hemispherical surface. Radiation so distributed is said to be diffuse and is given exactly by black surfaces. Analyses of radiation are commonly based on the assumption of black or of grey-diffuse radiation.

4.3 Heat Exchange by Radiation

The rate at which radiant energy is emitted by a black plane surface, 1, with area A_1 is found from Equation 4.3, as

$$q_1 = W_1 A_1 = \sigma A_1 T_1^4.$$

The rate at which radiant energy is emitted by a parallel black surface of area A_2 will be

$$q_2 = W_2 A_2 = \sigma A_2 T_2^4.$$

If the planes are large in relation to the distance between them, and of equal size, the radiation that escapes at the edges will be small, and it will be possible to write

$$q_{1-2} = q_1 - q_2 = \sigma A_1 (T_1^4 - T_2^4) \quad (4.8)$$

where

q_{1-2} = the net radiant exchange rate between the two surfaces.

Example 4.1

Calculate the net radiant exchange between two closely spaced parallel planes with black surfaces at temperatures of 30°C and 20°C.

$$q_{1-2} = 5.670 \times 10^{-8} \times A_1 [(273 + 30)^4 - (273 + 20)^4].$$

The calculation can be simplified by noting that $(1/100)^4 = 10^{-8}$ so that

$$q_{1-2} = 5.670 \times A_1 [(\frac{303}{100})^4 - (\frac{293}{100})^4] = 5.670 A_1 (84.3 - 73.7) = 60.1 A_1$$

and

$$\frac{q_{1-2}}{A_1} = 60.1 \text{ W/m}^2.$$

Consider next parallel planes as before, but with emissivities ϵ_1 and ϵ_2. This case is illustrated in Figure 4.3. The rate of energy emission per square metre by plane 1 will be

$$W_1 = \epsilon_1 \sigma T_1^4.$$

Part of this energy

$$\epsilon_2 W_1 = \epsilon_1 \epsilon_2 \sigma T_1^4$$

will be absorbed by plane 2, and the balance $(1 - \epsilon_2)(\epsilon_1 \sigma T_1^4)$ will be reflected back to plane 1. This in turn will be partly absorbed (ϵ_1) and partly reflected $(1 - \epsilon_1)$.

The successive reflections can be followed in Figure 4.3. In a corresponding way the energy emitted by plane 2 per square metre will be $\epsilon_2 W_2$. Part of this will be absorbed (ϵ_1) at plane 1 and part reflected $(1 - \epsilon_1)$. The reflected part will be partly absorbed at plane 2 and partly reflected, and so on. The summing and simplification are completed in Figure 4.3 after recognizing that the net heat exchange is given by the emissions and reflections from plane 1 minus the absorptions. The net radiation exchange rate turns out to be:

$$q_{1-2} = W_{1-2} A_1 = \frac{\sigma A_1}{1/\epsilon_1 + 1/\epsilon_2 - 1} (T_1^4 - T_2^4). \quad (4.9)$$

This can also be written

$$q_{1-2} = W_{1-2} A_1 = F_E \sigma A_1 (T_1^4 - T_2^4) \quad (4.10)$$

where

F_E is an emissivity factor accounting for the departure from blackbody emissivities and has the value, in this case, of

$$F_E = \frac{1}{1/\epsilon_1 + 1/\epsilon_2 - 1}. \quad (4.11)$$

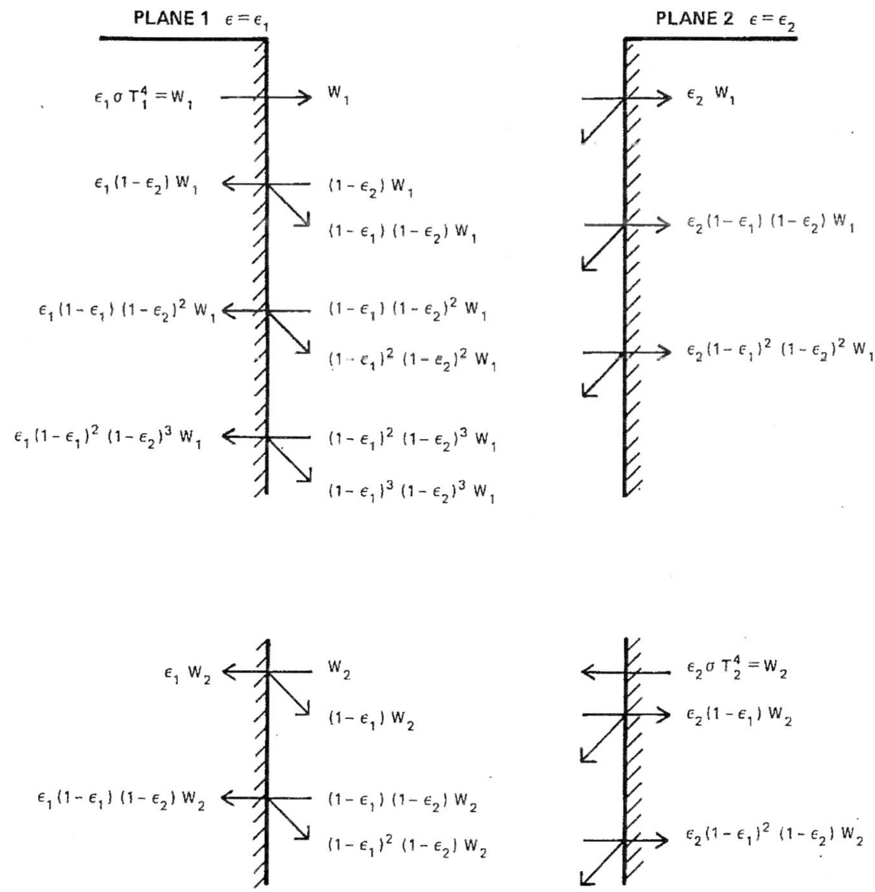

ENERGY LOST FROM PLANE 1

$$= W_1 - \epsilon_1 (1-\epsilon_2) W_1 - \epsilon_1 (1-\epsilon_1)(1-\epsilon_2)^2 W_1 - \epsilon_1 (1-\epsilon_1)^2 (1-\epsilon_2)^3 W_1 \text{ — — — —}$$

ENERGY GAINED FROM PLANE 2

$$= \epsilon_1 W_2 + \epsilon_1 (1-\epsilon_1)(1-\epsilon_2) W_2 + \epsilon_1 (1-\epsilon_1)^2 (1-\epsilon_2)^2 W_2 \text{ — — — —}$$

COMBINING AND PUTTING $Z = (1-\epsilon_1)(1-\epsilon_2)$ AND RECOGNIZING THE SERIES

$$1 + Z + Z^2 + Z^3 \text{ — — — } = \frac{1}{1-Z} \quad \text{WHERE } Z < 1$$

$$\text{THEN } W_{1\text{-}2} = W_1 \left[1 - \frac{\epsilon_1 (1-\epsilon_2)}{1-Z} \right] - W_2 \frac{\epsilon_1}{1-Z} = \frac{\sigma}{\dfrac{1}{\epsilon_1} + \dfrac{1}{\epsilon_2} - 1} (T_1^4 - T_2^4)$$

FIGURE 4.3 Radiation exchange between parallel planes.

Example 4.2

After noting that $F_E = 1$ when $\epsilon_1 = \epsilon_2 = 1$, calculate F_E for other combinations of ordinary surfaces $\epsilon = 0.90$ and reflective surfaces $\epsilon = 0.05$ for parallel planes.

For $\epsilon_1 = \epsilon_2 = 0.90$

$$F_E = \frac{1}{1/0.90 + 1/0.90 - 1} = 0.81.$$

For $\epsilon_1 = 0.90$, $\epsilon_2 = 0.05$

$$F_E = \frac{1}{1/0.90 + 1/0.05 - 1} = 0.05.$$

For $\epsilon_1 = 0.05$, $\epsilon_2 = 0.05$

$$F_E = \frac{1}{1/0.05 + 1/0.05 - 1} = 0.03.$$

Note that it does not matter which plane has the lower emissivity. Note also that the value of F_E for the first two cases is approximately the same as the value of $\epsilon_1 \times \epsilon_2$ which is used as the emissivity factor F_E in certain other situations.

Consider now the case of a small body, 1, in an enclosure, 2, with energy being exchanged between them. When the enclosure is large relative to the small body it acts as a cavity with little interference from the small body and so will have the characteristics of a blackbody with an effective emissivity $\epsilon_2 = 1$. The energy-exchange rate is then

$$q_{1-2} = \epsilon_1 \sigma A_1(T_1^4 - T_2^4). \tag{4.12}$$

The emissivity factor F_E for this case turns out to be simply ϵ_1. There is one condition to be satisfied: body 1 must be convex throughout and body 2 concave throughout.

In the case of parallel planes considered above, all energy leaving plane 1 reached plane 2, and all energy leaving plane 2 reached plane 1. In the case of the large enclosure, all radiation from body 1 reached body 2 and was ultimately totally absorbed. There is clearly a more general case to be considered in which the bodies are not enclosures and are arranged so that some of the emitted radiation does not reach the other body.

Consider the elementary black surfaces dA_1 and dA_2 at temperatures T_1 and T_2 disposed to one another as shown in Figure 4.4(a). The example is more readily visualized if dA_2 is considered to be rotated about the line joining dA_1 and dA_2 until the normals to dA_1 and dA_2 fall into the same plane as shown in Figure 4.4(b). This plane intersects the hemispherical surface of radius r centred on dA_1

which will receive radiation. The elementary area dA_2 lies with its centre on this surface but inclined at an angle ϕ_2.

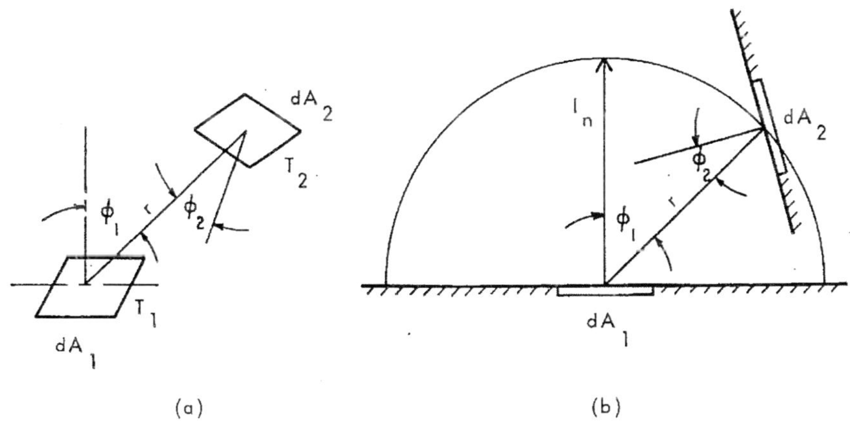

FIGURE 4.4 General case of radiation exchange.

The rate at which energy leaves dA_1 is given by $\epsilon_1 dA_1 \sigma T_1^4$. The rate at which it reaches dA_2 can be found by integration but can also be visualized [**4.1, 4.2**]. Lambert's law states that the intensity of radiation at an angle ϕ to the normal to an emitting surface is given by $I = I_n \cos\phi$. If I and I_n are intensities on the hemispherical surface at radius r, the value of I_n turns out to be the total rate of energy leaving dA_1, divided by πr^2, as if it were uniformly distributed over the base of the hemisphere centred on dA_1.

Consider now the rate at which energy is received by dA_2 which has an area normal to the radiation from dA_1 of $dA_2 \cos\phi_2$. The rate of energy leaving dA_1 is

$$dq = dA_1 \, \sigma \, T_1^4.$$

The intensity of energy normal to dA_1 at radius r_1 is

$$I_n = \frac{dA_1 \, \sigma \, T_1^4}{\pi r^2}.$$

The intensity of energy at angle ϕ_1 is

$$I = I_n \cos\phi_1 = \frac{dA_1 \, \sigma \, T_1^4}{\pi r^2} \cos\phi_1.$$

The rate at which radiation is received by dA_2, which has a projected area on the hemispherical surface of $dA_2 \cos\phi_2$, will be:

$$dq = \frac{dA_1 \, dA_2}{\pi r^2} \cos\phi_1 \cos\phi_2 \, \sigma \, T_1^4.$$

By exactly similar reasoning, considering the radiation to a hemispherical surface centred on dA_2, the radiation from dA_2 to dA_1 will be given by an expression identical with the above except that T_2 will replace T_1 and the net exchange rate will be

$$dq_{1-2} = \frac{dA_1\, dA_2}{\pi r^2} \cos\phi_1 \cos\phi_2\, \sigma\,(T_1^4 - T_2^4). \tag{4.13}$$

Equation 4.13 has been derived from consideration of the case of blackbodies so that $\epsilon_1 = \epsilon_2 = 1$. Thus no radiation was reflected from dA_2 and partly returned to dA_1 where it would again be partly absorbed and partly reflected. No account was taken of other surfaces contributing directly or as intermediaries in the radiation exchange. Equation 4.13 gives only the *direct* radiation exchange between black surfaces. It leads to a recognition of the influence of geometric relationships between surfaces exchanging radiation.

The area relationships can be dealt with by a new factor called the "angle factor", which may be seen from Equation 4.13 to be

$$F_A = dA_2 \frac{\cos\phi_1}{\pi r^2} \cos\phi_2$$

for the present case. Consideration of parallel planes earlier led to the concept of the factor F_E to account for departures from black surfaces. A practical form of the equation for the radiant heat-exchange rate can be written:

$$q_{1-2} = \sigma A_1 F_A F_E (T_1^4 - T_2^4). \tag{4.14}$$

Appropriate values of F_A and F_E must be determined for particular cases, and can be found in textbooks on heat transfer. Contemporary practice has moved toward the use of other methods for certain types of problems, but the concept of angle factors remains unchanged.

4.4 Angle Factors and Their Use

The angle factors, F_A, are also called configuration factors or geometric factors. They account for the proportion of energy leaving one surface which reaches directly the other surface involved in an exchange, and so will have values between 0 and 1. For a many-sided enclosure with sides at different temperatures, the angle factors may have to be calculated for each surface considered in turn with each of the other surfaces involved. A closed rectangular room, for example, will have 36 possible combinations. Fortunately, some simplifications are possible. Since a plane surface cannot see itself, factors such as F_{1-1} and F_{2-2} become zero. Advantage can be taken of reciprocity situations so that it is possible to write that

$$F_{1-2} A_1 = F_{2-1} A_2$$

and thus to find F_{2-1} knowing the factor F_{1-2} for exchange in the reverse direction. If this relationship were not true it would mean that there was a net exchange between A_1 and A_2 when at the same temperature, which is impossible. It is possible to exploit the fact that all the angle factors for a given surface exchanging with other surfaces that make up an enclosure must sum to unity. It is also possible to take advantage of identical relationships between pairs of surfaces, and in these various ways to reduce the number of angle factors that may have to be evaluated by other means.

Angle-factor relationships for many common situations have been worked out [**4.3**, **4.4**]. Tables and charts of values are given in most books on heat transfer, and computation methods have been described [**4.2**, **4.5**]. Fanger [**4.6**] deals at some length with the case of radiant exchange between the human body and the surrounding room surfaces.

Many cases of radiant exchange involve air spaces between parallel surfaces, and so present little difficulty. The angle factor can be taken as unity and F_E can be found from Equation 4.11 for use in Equation 4.14. If the planes are relatively far apart, the emissivity factor F_E is more appropriately taken as $\epsilon_1\epsilon_2$. The angle factor decreases progressively as the distance between planes increases, as shown by Figure 4.5 for rectangles.

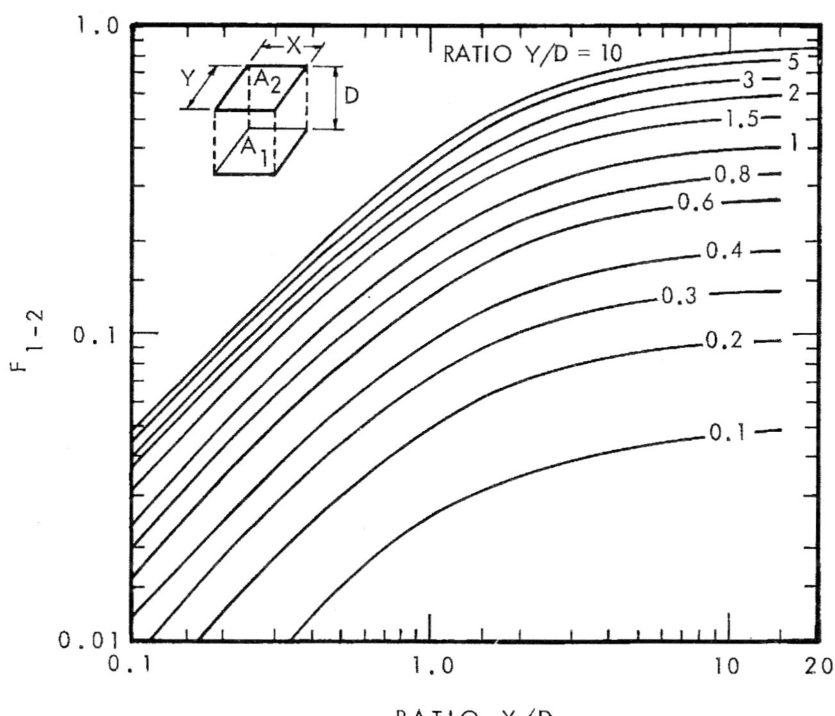

FIGURE 4.5 Radiation angle factor for radiation between parallel rectangles (From *ASHRAE Handbook 1981 Fundamentals*, Fig. 6B, p. 2.9)[**4.1**]

Example 4.3

Find the direct radiant-heat exchange between the ceiling and the heated floor of a room with radiant-panel heating. The dimensions are 6 m × 4 m × 2.5 m and the emissivities are $\epsilon_1 = \epsilon_2 = 0.90$. The temperatures are 35°C for the floor and 24°C for the ceiling.

$$T_1 = 273 + 35 = 308 \text{ K}; \; T_2 = 273 + 24 = 297 \text{ K}.$$

$$q_{1-2} = \sigma A_1 F_A F_E (T_1^4 - T_2^4)$$

where $\sigma = 5.670 \times 10^{-8}$.

Assume $F_E = \epsilon_1 \epsilon_2 = 0.90 \times 0.90 = 0.81$ and $A_1 = A_2 = 24 \text{ m}^2$.

To find F_A, use Figure 4.5 for rectangles 6 by 4 m which are 2.5 m apart. $\frac{X}{D} = 2.4$ and $\frac{Y}{D} = 1.6$. $F_A = 0.40$.

$$\begin{aligned} q_{1-2} &= 5.670 \times 24 \times 0.40 \times 0.81 \left[\left(\frac{308}{100}\right)^4 - \left(\frac{297}{100}\right)^4\right] \\ &= 44.1\,(90.0 - 77.8) \\ &= 538 \text{ watts, or for unit area } 22.4 \text{ W/m}^2. \end{aligned}$$

The solution in Example 4.3 does not take into account the walls, which may also receive radiation from the floor and reradiate to the ceiling, thus contributing to the exchange if they have surface temperatures intermediate between those of the floor and ceiling. In the special case of reradiating nonconducting walls, the energy received as a net exchange with the floor will provide for an equal net exchange between walls and ceiling. The walls will operate at some temperature between those of the floor and ceiling at which this equality will be maintained, provided there are no heat losses or gains by conduction through the walls themselves. The total energy transfer from floor to ceiling will then be equal to that exchanged directly plus that reaching the ceiling by way of the walls, providing a value about 50% higher than that found in Example 4.3.

Although the further complications of conductive enclosing walls cannot be examined in detail at this stage, it may be noted in passing that energy exchanges also take place by convection. These must be considered in combination with the radiation exchanges in finding the total energy exchanges between surfaces of rooms [**4.7**]. In addition, the air absorbs some radiation and thus introduces a further coupling between radiation and convection, which is often ignored in the case of rooms. Further complications in rooms may result from additional radiant-energy sources, such as lighting and building equipment, and from solar energy entering at windows. Fortunately, it is often possible to avoid such complications by the judicious use of approximations and judgements based on the more tractable situations.

Returning to Example 4.3: since the walls and ceiling form a complete enclosure with the floor, the angle factors for exchange from floor to walls and ceiling should sum to unity. Thus, if they are all at the same temperature, the energy lost from the floor by radiant exchange with walls and ceiling can be found from

$$q_{1-2} = \epsilon_1 \epsilon_2 A_1 \sigma (T_1^4 - T_2^4)$$
$$= 0.90 \times 0.90 \times 24 \times 5.670(12.2) = 1345 \text{ watts.}$$

Figure 4.6 can be used to find the angle factors for adjacent perpendicular planes. The angle factors for the floor to the four walls of Example 4.3 will be found to sum to 0.58 which, when added to the value previously found for floor to ceiling of 0.40, gives 0.98 or approximately 1, as expected.

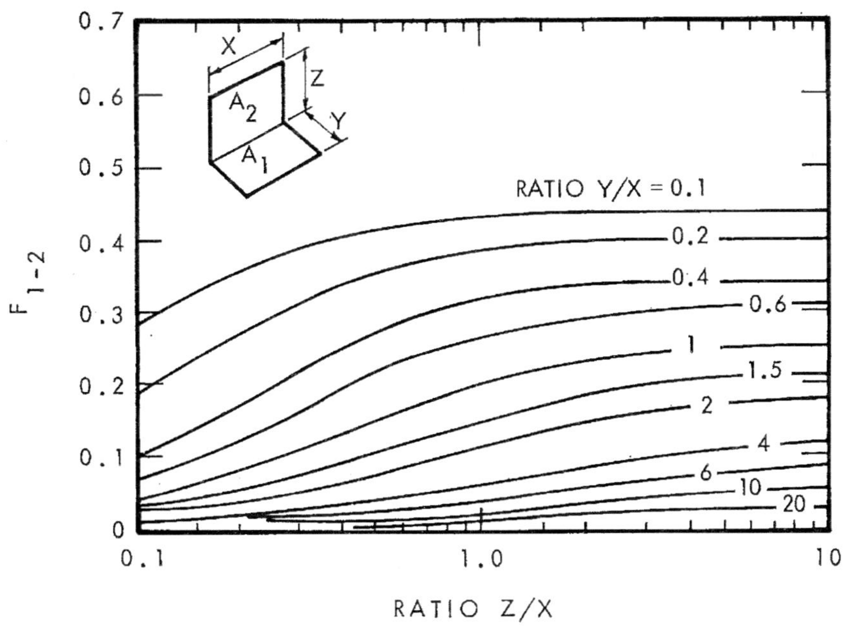

FIGURE 4.6 Radiation angle factors for radiation between perpendicular rectangles with a common edge.
(From *ASHRAE Handbook 1981 Fundamentals*, Fig. 6A, p. 2.9)[**4.1**]

4.5 Radiant Energy from the Sun

The temperature at the centre of the sun, maintained by nuclear reactions, is thought to be about 15×10^6 K. The first layer of the sun's atmosphere, known as the photosphere, is the immediate source of most solar radiation. Its effective blackbody temperature, judged from the radiation emitted, is about 5800 K.

The wavelength for maximum intensity given by Wien's equation for the effective temperature of 5800 K is

$$\lambda_m = 2898/5800 = 0.50 \ \mu\text{m}.$$

(Some sources give the value as 0.48 μm.) The corresponding wavelength for radiation from the earth at about 300 K, also known as terrestrial radiation, is 10 μm, or more than 20 times that of the solar radiation. The term "short-wave" is commonly applied to solar radiation to differentiate it from the "long-wave" radiation emitted by surfaces at ordinary temperatures. The need for this differentiation has already been recognized in the earlier discussion of absorptance of surfaces for solar radiation as shown in Table 4.1.

The distribution of wavelengths in solar radiation outside the earth's atmosphere is shown in Figure 4.7 [**4.8**]. About 9% of the total energy is contained between 0.29 and 0.40 μm. The visible region between 0.40 and 0.70 μm contains 40%; the near infrared region between 0.7 and 3.5 μm accounts for the remaining 51%. The spectrum is substantially modified in passing through the atmosphere and has the form at the earth's surface shown by the second curve in Figure 4.7. The reasons for this difference will be discussed later.

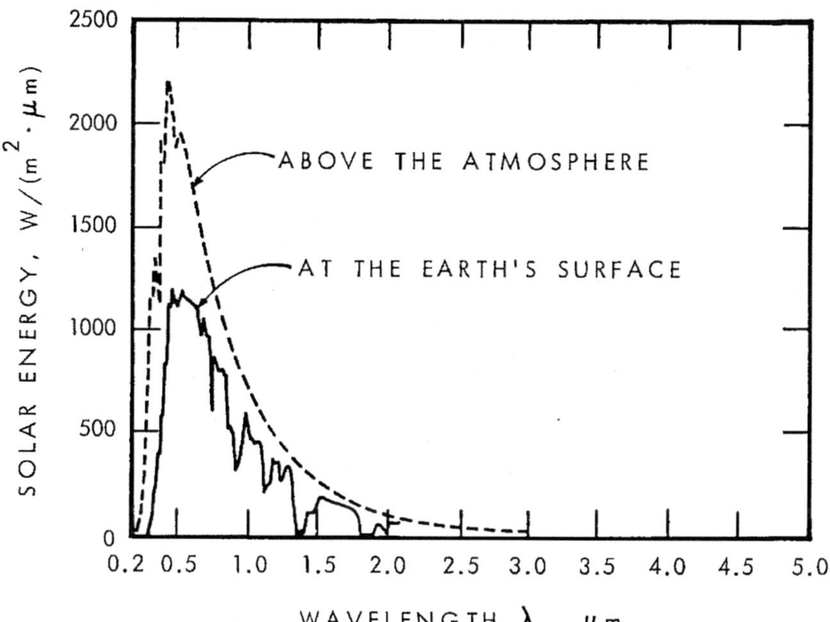

FIGURE 4.7 Solar radiation by micrometre bands.
(From Gates, D.M., *Science*, Vol. 151, Fig. 1, p. 524 [**4.8**]. Copyright 1966 by the American Association for the Advancement of Science.)

The distribution of wavelengths in the solar radiation is approximately that predicted from the Planck equation. While it is probably an oversimplification, it is instructive to consider the sun as a classical hotbody. The application of the Stefan-Boltzmann equation yields a value of rate of energy emitted of

$$q/A = \sigma T_1^4 = 5.670 \times 10^{-8} (5800)^4 = 64.2 \times 10^6 \text{ W/m}^2.$$

The sun's radius is about 696 000 km, and the earth-to-sun distance of 1.5×10^8 km is about 215 times as great. Noting that the inverse square law should apply, the intensity of the energy received at the earth should then be

$$I = 64.2 \times 10^6 \times \frac{1}{215^2} = 1387 \text{ W/m}^2.$$

This may be compared with the accepted value of 1353 W/m^2, or 426.7 Btu/(ft^2/hr) now widely used in air-conditioning calculations. The meteorologist uses the equivalent value of 1.94 cal/(cm^2/min), or 194 langleys per minute.

4.6 Radiation in the Atmosphere

The marked reduction in the intensity of solar energy in passing through the atmosphere as shown by Figure 4.7 occurs because of scattering and absorption. Scattering of some direct radiation is produced by molecules and minute particles, resulting in diffuse radiation travelling in all directions. This effect is quite marked at the blue end of the visible spectrum and gives the sky its blue colour. Sky radiation is largely the result of scattering. The redness of morning and evening light occurs when the atmospheric path is longest, producing the maximum reduction in blue light and causing red to predominate.

Gases absorb and reradiate, but unlike most non-opaque solids they do so in particular wavelengths, corresponding to their discrete energy states. This is characteristic of the gaseous state. The incoming solar radiation encounters ozone in the atmosphere at about 20 km which, although extremely rarified, is a strong absorber of ultraviolet radiation at 0.1 and 0.3 μm. This reduction is of great significance, since without it the level of ultraviolet radiation reaching the earth would be high enough to destroy organic material at a much higher rate than at present, thus rendering life impossible [**4.9**].

The two principal gases influencing atmospheric energy absorption are water and carbon dioxide. They are also involved in the radiant heat transfer in furnaces with atmospheres containing the products of combustion from hydrocarbon fuels. The water vapour in the atmosphere absorbs nearly six times as much solar radiation as all other gases combined [**4.10**]. Water has absorption bands at 1.38, 1.87, 2.7, and 6.3 μm. Carbon dioxide has absorption bands at 2.7, 4.2, and 14 μm. Only the bands below 2.5 μm will influence solar radiation directly, while those centred at about 10 μm will have a marked influence on radiation from the earth's surface and from clouds.

Clouds are the most important absorbers of radiation in the atmosphere at all wavelengths. They reflect a high percentage of direct solar radiation. They also absorb and reradiate terrestrial radiation. If they are not present at night, the earth's surface can radiate with a high net exchange upward which, in the absence of incoming solar radiation, results in a marked cooling effect by radiation to the night sky.

4.7 Radiation at the Earth's Surface

Although it is possible to assign a fixed value to the solar radiation received above the earth's atmosphere at right angles to the sun's rays, the intensity of radiation received at the earth's surface varies widely for several reasons. The continuously changing direction of the sun's rays relative to the observer's position, as already discussed in Chapter 2, varies with latitude, time of year, and time of day. It will be recalled that the intensity of radiation, I, on a surface at an angle to the sun's rays, is given by

$$I = I_{DN} \cos\theta \tag{4.15}$$

where I_{DN} is the direct normal intensity at ground level and θ is the angle of incidence. Thus, the radiation received is dependent on the orientation of the receiving surface and the altitude and azimuth of the sun.

The atmosphere moderates the direct solar beam by scattering and reflection, producing a diffuse short-wave component, and by absorption by gases and clouds. The direct and diffuse components constitute the solar radiation which for meteorological purposes is measured for a horizontal surface with instruments known as pyrheliometers.

The absorption and reradiation by clouds result in some conversion of short-wave to long-wave radiation directed both upward and downward to the earth's surface. The clouds become involved in a long-wave radiation exchange with objects at the earth's surface, which must be calculated for individual cases. The meteorologist, interested in the atmospheric energy balance, measures the net radiation on a horizontal surface instrumentally. These measurements, like those of solar radiation, are related to the particular radiation-exchange situation created by the design and operation of the instrument being used.

Values of solar radiation measured with an Eppley pyrheliometer and of net radiation measured with a CSIRO net radiometer at Ottawa on a clear day and a cloudy day are shown in Figure 4.8. The cloudy day was overcast in the morning with variable cloudiness in the afternoon. Negative values of net radiation indicate an excess of outgoing over incoming radiation, as occurs on clear nights.

Instruments for radiation measurements are usually elevated above an open site so that their surroundings will affect them as little as possible. Surfaces of most objects may receive some short-wave radiation by reflection from the ground or from adjacent roof or wall surfaces in addition to that received directly. The total solar radiation I_t received by a surface is

$$I_t = I_{DN} \cos\theta + I_d + I_r \tag{4.16}$$

where I_d is the diffuse and I_r the reflected short-wave intensity, and I_{DN} and θ are defined as in Equation 4.15. The evaluation of these components for purposes of air-conditioning calculations is given in detail elsewhere [**4.1**]; further discussion will be included in following chapters.

When the energy balance or the temperature at the surface of a practical body is of interest, it is necessary to consider both long- and short-wave contributions. The long-wave radiation exchange can be calculated by using Equation 4.14 as discussed earlier. This will be relatively simple if all surrounding surfaces are at the same temperature. The angle factor, F_A, can be taken as unity in an exchange calculation considering the surroundings as one surface area. When various surfaces are at different temperatures, and particularly when a view of the sky is included, it will be necessary to deal with these several areas individually. Low-level

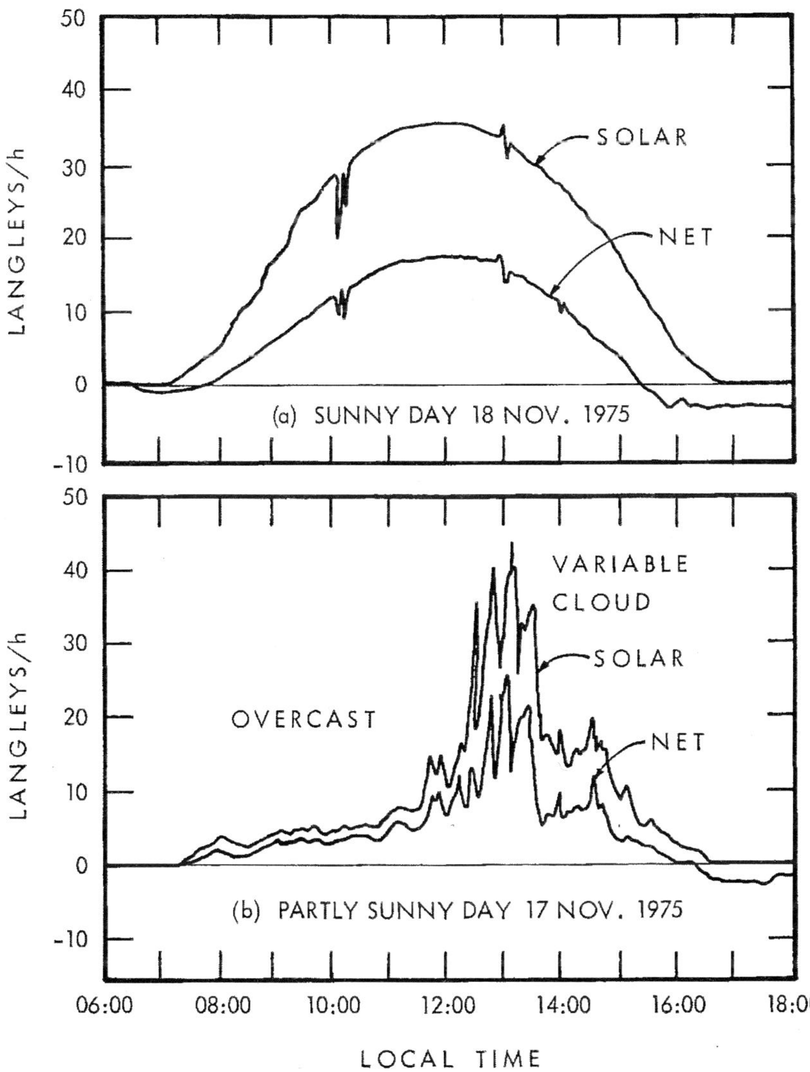

FIGURE 4.8 Solar and net radiation at Ottawa on a horizontal surface.

clouds act much like blackbodies at ground-level air temperature. Under clear-sky conditions the effective radiating temperature of the sky may be as much as 20 K below air temperature.

When surface temperatures differ from the temperature of the air in contact with them, an exchange of heat by convection occurs. There may also be a flow of heat into the body itself. In cases where the surface is moist or wetted there may be a transfer of heat from the surface by evaporation, as happens with perspiration on the human body. The elements in the energy balance that will determine the surface temperature for an opaque body are shown in Figure 4.9.

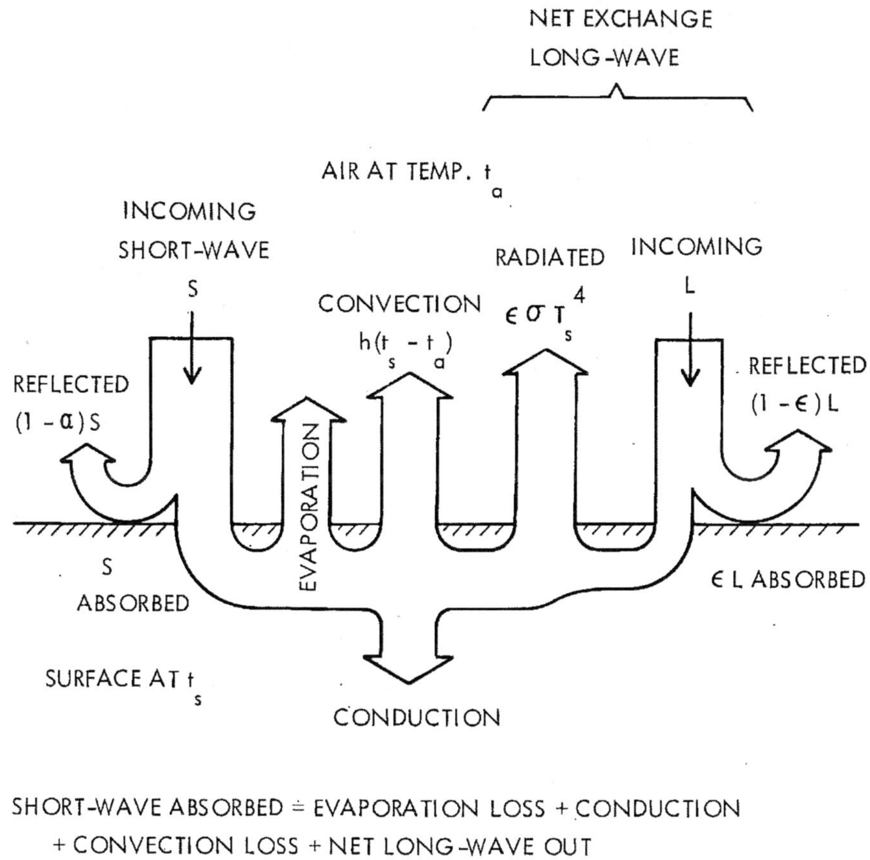

FIGURE 4.9 Heat balance at surface of opaque body receiving solar radiation and exchanging with surroundings.

Windows present a different situation, since glass is transparent to short-wave but not to long-wave radiation. The solar radiation will be partly reflected, partly absorbed, and partly transmitted in passing through the glass [**4.11**, **4.12**]. The long-wave radiation exchange with the outside surface can be combined with the convection exchange in the calculation of heat conduction through the window, as will be discussed in subsequent chapters.

4.8 The Visible Spectrum

The visible spectrum, it will be recalled, is bracketed between the infrared and the short-wavelength ultraviolet bands. The characteristic colour wavelengths for visible light are given in Table 4.2.

The colour temperatures of iron as it is heated have been widely used as a practical means of gauging its temperature in metal working. One source of information gives the values shown in Table 4.3.

TABLE 4.2

Wavelengths for Various Colours

Red	0.65 μm	Green	0.52 μm
Orange	0.60 μm	Blue	0.47 μm
Yellow	0.58 μm	Violet	0.41 μm

TABLE 4.3

Colour Temperatures in Metal Working

	°C	K	λ_m
Blood-red	566	839	3.45
Cherry, full red	746	1019	2.84
Orange	899	1172	2.47
Yellow	996	1269	2.28
White	1204	1477	1.96

The values of λ_m in Table 4.3, the wavelength of maximum emission, have been calculated from Equation 4.2. They are seen to vary from 2 μm to 3.5 and are thus considerably larger than the wavelengths for the spectral colours. The colour temperatures are not the temperatures for maximum emission at the colour wavelength, but those at which there is a substantial and dominating emission in the visible range corresponding to the stated colour. All substances that emit according to Planck's equation may be expected to have similar wavelength distributions at various temperatures and thus to have the same colour temperatures as iron.

The dilemma of incandescent lamps may now be appreciated. The highest practical temperature at which a tungsten-metal filament can be operated (about 3000 K) is still well below that for maximum emission in the visible range. The proportion of the energy output that is in the visible range is small compared with the heat that radiates from the electrically heated incandescent filament. This is further illustrated in Figure 4.10, although the log scales used make it difficult to compare areas directly. The curve for tungsten at 3000 K differs from the blackbody curve and also from the greybody curve because its emissivity is different for different wavelengths; it is called a selective radiator [**4.13**]. A carbon arc provides some improvement in visible radiation output because it can be operated up to 3800 K.

There is a further complication in considering ways to produce light, since radiation emitted in the range of 400 to 800 nm is not equally valuable as visible light, but must be weighted on the basis of wavelength as to the visual sensation produced in the human eye.

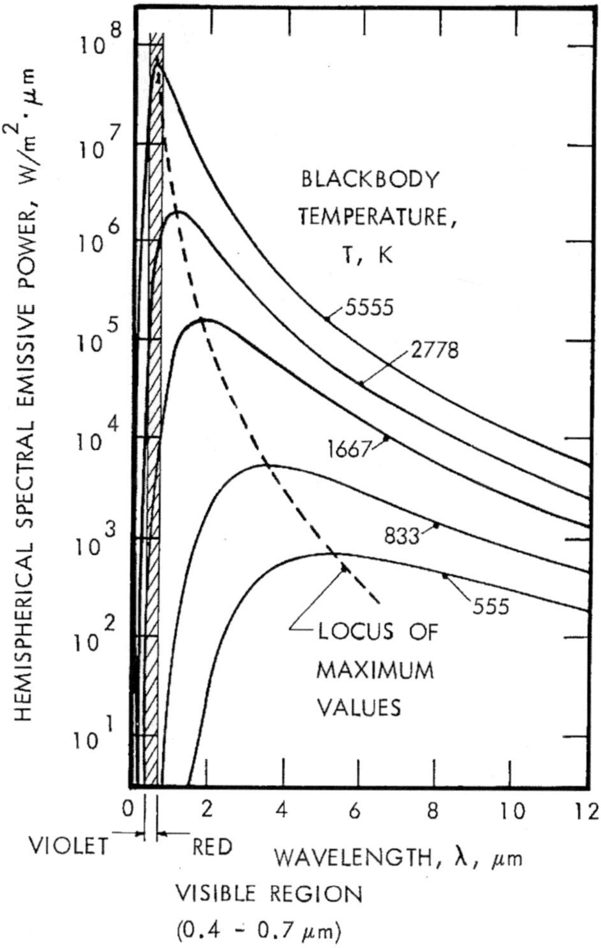

FIGURE 4.10 Hemispherical spectral emissive power of blackbody for several different temperatures.
(From Siegel and Howell, *Thermal radiation heat transfer,* 5th Ed. New York: McGraw-Hill, 1972, Fig. 2-6, p. 21)[**4.5**]

4.9 Light

Light is defined as radiant energy that is capable of exciting the retina in the human eye and producing a visual sensation. The visible portion of the spectrum extends from 380 to 770 nm [**4.13**]. Radiant energy can be defined and measured in physical terms, but light must be measured in terms of the response of the human eye. Because this response differs from person to person and with factors such as time, age, and state of health, an average representative relationship has been agreed upon for international use. The human eye has two kinds of receptors, called rods and cones. The principal receptors are the cones, which provide photopic, or day, vision. The relative response of cones to various wavelengths in the visible range is shown by Figure 4.11. The value is given as unity at the wavelength of greatest efficiency, which is 555 nm corresponding to a wave frequency of 540 × 10^{12} Hz. Scotopic, or night, vision based on the rod receptors in the eye peaks at about 510 nm and adapts to lighting levels that are as low as 1% of those for which photopic vision is best adapted. Relative numerical values are given in Table 4.4 [**4.13**].

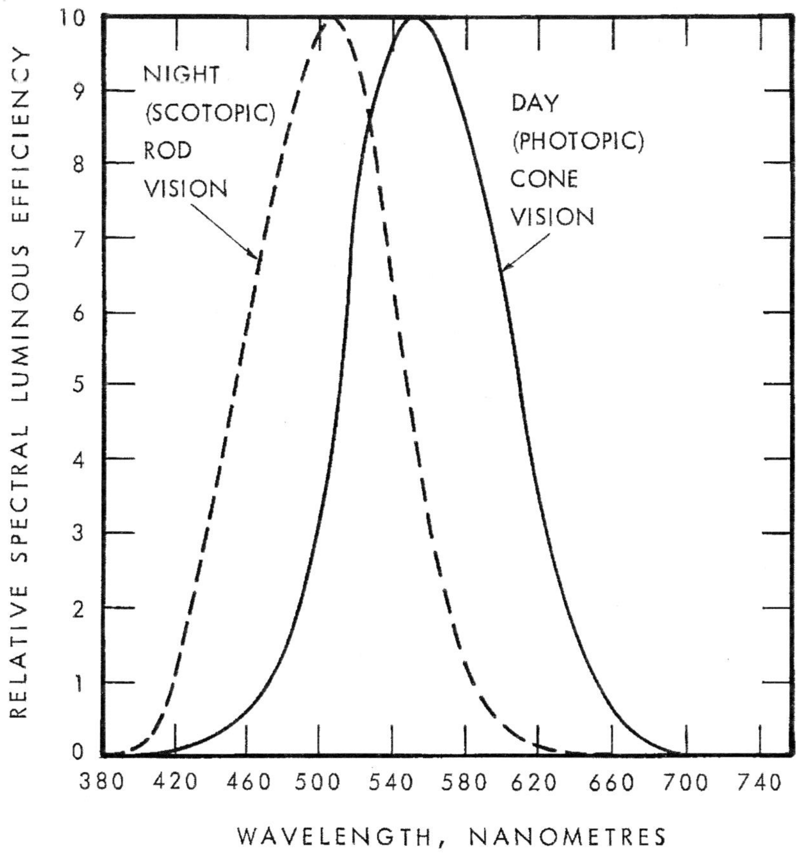

FIGURE 4.11 Relative spectral luminous efficiency curves for photopic and scotopic vision.
(From *IES Lighting Handbook,* 5th Ed., Fig. 3-8, p. 3-6)[**4.13**]

It was necessary to devise a separate system of units to describe the properties of light, because, despite a close relationship, those used in dealing with radiant energy no longer apply. An earlier system defined lighting in terms of a standard candle, but this has now been replaced, with international agreement, by the candela, the basic unit of luminous intensity.

TABLE 4.4

Spectral Luminous Efficiency
(From IES Lighting Handbook, 5th Ed., Fig. 3-7, p. 3-5) [**4.13**]

Wavelength,	Value	
nm	Photopic	Scotopic
380	0.000 04	0.000 589
420	0.004 0	0.096 6
460	0.060	0.567
500	0.323	0.982
540	0.954	0.650
580	0.870	0.121 2
620	0.381	0.007 37
660	0.061	0.000 313
700	0.004 1	0.000 018
740	0.000 25	0.000 001 4
780	—	0.139×10^{-6}

4.10 Standard System of Units for Light

The candela is defined as the luminous intensity in a given direction of a source that emits monochromatic radiation of frequency 540×10^{12} Hz and of which the radiant intensity in that direction is 1/683 watts per steradian.

The steradian is the unit of solid angle which subtends an area on the surface of a sphere equal to the square of the sphere radius. Since radiant energy, including light, varies inversely as the square of the distance from the source, the steradian provides a useful way of describing flux density.

Example 4.4

A radiant heater provides a beam with a radiant intensity of 1050 watts per steradian. Find the irradiance at 6 m from the source.

The spherical surface area subtending a solid angle of one steradian at 6 m is equal to $6^2 = 36$ m^2.

The irradiance, or radiant flux density at 6 m is $1050/36 = 29.2$ W/m^2.

The definition of the candela gives its radiant equivalent as 1/683 watts at 540×10^{12} Hz so that with this value and with the aid of Figure 4.11 or Table 4.4 it is possible to establish the weighting values by which the light equivalent of a radiant beam with a given wavelength distribution in the visible range can be calculated. Although this equivalence is usually established in practice by measurement, it does provide a useful focal point for an appreciation of the system of light units.

The watt is the measure of the radiant flux and the corresponding unit for luminous flux is the *lumen*.

Irradiance is measured in watts per square metre; illuminance is measured in lumens per square metre.

One lumen per square metre provides an illuminance of one *lux*.

The unit of luminous intensity, the *candela*, is one lumen per steradian.

The overall relationship between the light emitted by a lamp and its electric power input is expressed as lumens per watt and is called the *luminous efficacy*. It ranges from as little as 10 to 18 lumens per watt for incandescent to 30 to 80 for fluorescent and over 100 for some special types of lamps.

The relationship between candelas, lumens, lux and footcandles is described in Figure 4.12 [**4.13, 4.14**].

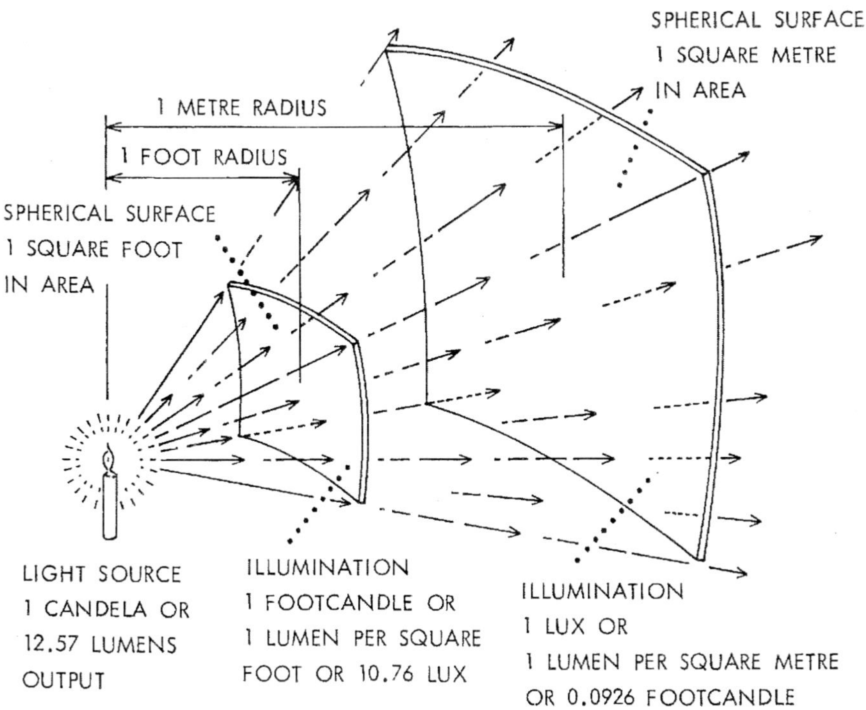

FIGURE 4.12 Relationship between candelas or lumens to footcandles or lux. (From Harris, D.A., et al., *Planning and designing the office environment*. New York: Van Nostrand Reinhold, 1981, Fig. 3-5, p. 91 [**4.14**] (General Electric TP-118)

4.11 Conclusion

Thermal radiation is of interest in many ways in the context of buildings. It is the means by which energy comes from the sun. It becomes involved in complex ways in producing the weather. It is generated by every substance at the expense of its molecular energy, or heat content, and when absorbed appears again as heat in the absorbing agent. When produced at one temperature, thermal radiation has a spectral distribution characteristic of that temperature. Upon absorption, it is radiated again at another temperature with its characteristic spectral distribution.

This ready convertibility means that it can seldom be considered by itself in practical problems involving energy transfer but must be taken in combination with other modes such as evaporation, conduction, and convection. The role of radiation in the thermal environment will be explored later in this text in connection with heat transfer through building sections, in the consideration of solar heat gains through glass, and in further discussion of the indoor thermal environment. Thermal radiation is involved even more widely in lighting and daylighting, as an agent in the deterioration of organic materials, and as an important determinant of the maximum temperatures reached by materials exposed outdoors.

References

4.1 *ASHRAE Handbook 1981 Fundamentals*. Chap. 2.

4.2 Gebhart, B. *Heat transfer*. 2d Ed. New York: McGraw-Hill, 1971.

4.3 Hottel, H.C. Radiant heat transmission. *Mechanical Engineering*, 1930, *52* (7), pp. 699–704.

4.4 Mackey, C.O., Wright, L.J.,Jr., Clark, R.E., and Gray, N.R., *Radiant heating and cooling, Part 1*. Bulletin 32, Cornell University Engineering Experiment Station, 1943.

4.5 Siegel, R., and Howell, J.R., *Thermal radiation heat transfer*. New York: McGraw-Hill, 1972.

4.6 Fanger, P.O. *Thermal comfort*. New York: McGraw-Hill, 1972.

4.7 Min, T.C., Schutrum, L.F., Parmelee, G.V. and Vouris, J.D. Natural convection and radiation in a panel-heated room. *ASHAE Transactions*, 1956, *62*, pp. 337–58.

4.8 Gates, D.M. Spectral distribution of solar radiation at the earth's surface. *Science*, 1966, *151* (3710), pp. 523-29.

4.9 Ashton, H.E. *Irradiation effects on organic materials*. Canadian Building Digest 121. DBR/NRCC, Jan. 1970.

4.10 Byers, H.R. *General meteorology*. New York: McGraw-Hill, 1974.

4.11 Garden, G.K. *Characteristics of window glass*. Canadian Building Digest 60. DBR/NRCC, Dec. 1964.

4.12 Stephenson, D.G. *Reflective glazing units*. Canadian Building Digest 101. DBR/NRCC, May 1968.

4.13 *IES Lighting Handbook*. 5th Ed. New York: Illuminating Engineering Society of North America, 1972.

4.14 Harris, D.A., Palmer, A.E., Lewis, M.S., Munson, D.L., Meckler, G., and Gerdes, R. *Planning and designing the office environment*. New York: Van Nostrand Reinhold, 1981.

5

Water in the Environment

5.1 Introduction

The role of water in the environment was introduced in Chapter 1. It is already evident from the further discussions of the weather in Chapter 2, the brief review of the nature of matter in Chapter 3, and the more detailed discussions in Chapter 4 of the influence of water in radiant-energy transfer that it exerts a broad and powerful influence and must be well understood as a basis for understanding the physical environment.

The widespread occurrence of water is one of its outstanding features, and it enters into a great many natural organic and inorganic processes. It also has certain unique properties that contribute to its influence. It occurs as a solid, a liquid, and a vapour within the normal range of environmental conditions and it has relatively high heats of evaporation and fusion. It also has a high specific heat. For these reasons water is a common and influential factor in energy transfer, storage, and transformation. Its atomic structure gives it the ability to dissolve many other substances. It also has its maximum density as a liquid at a temperature of 4°C, close to but above the freezing point. Thereafter, on further cooling, it decreases in density. The solid state is less dense than the liquid, so that ice floats, and lakes and rivers freeze from the top down. If this were not so, lakes would be coldest at depth, with great implications for all freshwater plants and animal life, as well as for many of man's activities.

Each scientific discipline emphasizes those aspects of water that are important for its own purposes. The applications and the treatment of the subject may differ, but the basic science remains the same. The physicist [**5.1**], the biologist [**5.2**], the soil scientist [**5.3**], the meteorologist [**5.4**], the thermal-environmental engineer [**5.5**], and others have contributed much to an understanding of water in their particular fields. The knowledge needs of building have something in common with all of these and much of value can be learned from them. The Division of Building Research of the National Research Council of Canada has been concerned with the many aspects of water in building. The topic has been discussed in several hundred papers, which can be identified from the DBR/NRCC Publications List. Reference to the extensive world literature can be found in many of the DBR papers and in library reference material.

5.2 Vapour Pressure and Change of Phase

The molecular activities characteristic of water in its liquid and gaseous states and the related topics of evaporation and saturation pressures have already been reviewed briefly in Chapter 3. Table 5.1 gives values of saturation pressures for water at various temperatures. The marked increase of vapour pressure with temperature should be noted. Water normally changes to the solid state at 0°C, and vapour pressures at various temperatures below freezing ought to be those for equilibrium over ice. As discussed in Chapter 3, it is possible for water to remain liquid on cooling below 0°C when it is present in the atmosphere or in

TABLE 5.1

Water-Vapour Pressures at Saturation at Various Temperatures over Plane Surfaces of Pure Water and Pure Ice

	Pressure, Pa			Pressure, Pa					
Temp., °C	Over ice	Over water	Temp., °C	Over ice	Over water	Temp., °C	Press., kPa	Temp., °C	Press., kPa
−50	3.935	6.409	−22	85.02	105.4	5	0.8719	33	5.031
−49	4.449	7.124	−21	93.70	115.0	6	0.9347	34	5.320
−48	5.026	7.975	−20	103.2	125.4	7	1.001	35	5.624
−47	5.671	8.918	−19	113.5	136.6	8	1.072	36	5.942
−46	6.393	9.961	−18	124.8	148.8	9	1.147	37	6.276
−45	7.198	11.11	−17	137.1	161.9	10	1.227	38	6.626
−44	8.097	12.39	−16	150.6	176.0	11	1.312	39	6.993
−43	9.098	13.79	−15	165.2	191.2	12	1.402	40	7.378
−42	10.21	15.34	−14	181.1	207.6	13	1.497	41	7.780
−41	11.45	17.04	−13	198.4	225.2	14	1.598	42	8.202
−40	12.83	18.91	−12	217.2	244.1	15	1.704	43	8.642
−39	14.36	20.97	−11	237.6	264.4	16	1.817	44	9.103
−38	16.06	23.23	−10	259.7	286.3	17	1.937	45	9.586
−37	17.94	25.71	−9	283.7	309.7	18	2.063	46	10.09
−36	20.02	28.42	−8	309.7	334.8	19	2.196	47	10.62
−35	22.33	31.39	−7	337.9	361.8	20	2.337	48	11.17
−34	24.88	34.63	−6	368.5	390.6	21	2.486	49	11.74
−33	27.69	38.18	−5	401.5	421.5	22	2.643	50	12.33
−32	30.79	42.05	−4	437.2	454.5	23	2.809	51	12.96
−31	34.21	46.28	−3	475.7	489.8	24	2.983	52	13.61
−30	37.98	50.88	−2	517.3	527.5	25	3.167	53	14.29
−29	42.13	55.89	−1	562.3	567.8	26	3.361	54	15.00
−28	46.69	61.39	0	610.8	610.8	27	3.565	55	15.74
−27	51.70	67.27	+0.01	Triple point of water		28	3.780	56	16.51
−26	57.20	73.71	1	—	656.6	29	4.006	57	17.31
−25	63.23	80.70	2	—	705.5	30	4.243	58	19.15
−24	69.85	88.27	3	—	757.5	31	4.493	59	19.02
−23	77.09	96.49	4	—	812.9	32	4.755	60	19.92

porous materials in a finely divided state. Thus, the vapour pressures over liquid below 0°C are of more than just academic interest, as will be evident later in this chapter. Values are given for vapour pressures over ice and over water below 0°C. The temperature of 0.01°C shown in the table is the triple point of water at which the solid, liquid, and vapour phases can coexist without change. At temperatures below freezing the vapour and solid phases can exist without liquid being present, and evaporation from the solid phase directly to vapour and condensation in the reverse direction can take place. This is known as sublimation.

The thermodynamic properties of water and its vapour at temperatures above 100°C, the boiling point at atmospheric pressure at sea level, are of great interest in heat-power engineering. Such values are given in Table 5.2 along with values for pressures below atmospheric, which will be of use later in air-conditioning calculations.

Saturation vapour conditions are often encountered, but partial saturation is more common in the ambient air. A common way of measuring and describing partial saturation is in terms of the ratio of the actual partial pressure of the water

TABLE 5.2

Properties of Water Vapour at Saturation

(From *ASHRAE Handbook 1981 Fundamentals*) (Table 3, p. 6.9–6.11)[5.5]

Temp., °C	Absolute pressure, kPa	Specific volume sat. vapour, m³/kg	Enthalpy, kJ/kg Liquid	Evap.	Sat. vapour
t	p	v_g	h_f	h_{fg}	h_g
0	0.6112	206.1	−0.04	2500.8	2500.8
10	1.228	106.3	42.0	2477.1	2519.1
20	2.339	57.77	83.9	2453.5	2537.4
30	4.246	32.89	125.7	2429.8	2555.5
40	7.384	19.52	167.5	2406.0	2573.5
50	12.35	12.03	209.3	2381.9	2591.3
60	19.94	7.67	251.2	2357.6	2608.8
70	31.20	5.04	293.1	2333.0	2626.0
80	47.41	3.404	335.0	2307.9	2642.9
90	70.18	2.359	377.0	2282.4	2659.4
100	101.4	1.672	419.2	2256.3	2675.4
110	143.4	1.209	461.4	2229.5	2690.9
120	198.7	0.891	503.8	2202.0	2705.8
130	270.3	0.668	546.4	2173.7	2720.0
140	361.6	0.509	589.2	2144.3	2733.5
150	476.2	0.3925	632.2	2113.9	2746.1
160	618.3	0.3069	675.5	2082.3	2757.8
170	792.2	0.2427	719.1	2049.3	2768.5
180	1002.9	0.1939	763.1	2014.9	2778.0
190	1255.4	0.1564	807.5	1978.7	2786.2
200	1555.1	0.1272	852.3	1940.8	2793.1

vapour to the saturation pressure at the same temperature. Scientists write this as p/p_o and refer to it as the relative vapour pressure. This ratio, expressed as a percentage, is commonly called relative humidity (rh).

The interaction of air and water molecules can be neglected, without serious error, for many purposes. Calculations can be made separately on the dry air and the water vapour in accordance with the gas laws and combined as required. The total or barometric pressure, p_t, is the sum of the actual partial pressures of the air and water vapour according to Dalton's law of partial pressures. Thus,

$$p_t = p_a + p_w. \tag{5.1}$$

It can be assumed for many purposes that water vapour behaves independently of the dry air present, but when there is bulk movement or when heat is being transferred the water vapour behaves as a component of the air. (More detailed consideration of air-vapour mixtures will be given in the next chapter.)

Example 5.1

Find the partial pressures of dry air and water vapour for an atmosphere at a condition of 50% rh at standard atmospheric pressure and temperature of 23°C.

Standard atmospheric pressure = 101.3 kPa.
From Table 5.1,
saturation pressure of water at 23°C = 2.809 kPa.

So

$$p_w = \frac{50}{100} \times 2.809 = 1.405 \text{ kPa}$$

$$p_a = p_t - p_w = 101.3 - 1.405 = 99.9 \text{ kPa.}$$

Example 5.2

Find the densities of air and water vapour in the air-vapour mixture of Example 5.1.

From Equation 3.8 (Chapter 3):

$$pV = w\mathrm{R_g}T$$

and from Example 3.2:

$$\mathrm{R_a} = 287.1 \text{ J/(kg} \cdot \text{K)} \quad \text{(for air)}$$

and

$$\mathrm{R_w} = 461.5 \text{ J/(kg} \cdot \text{K)} \quad \text{(for water vapour)}$$

when p is in Pa, and w in kg.

So

$$\text{air present per cubic metre} = \frac{w}{V} = \frac{P_a}{R_a T}$$

$$\frac{w}{V} = \frac{99\,900}{287.1\ (273 + 23)} = 1.176 \text{ kg/m}^3$$

and for water vapour

$$\frac{w}{V} = \frac{1405}{461.5 \times 296} = 0.010\,29 \text{ kg/m}^3.$$

Density of the air-vapour mixture is

$$1.176 + 0.0103 = 1.186 \text{ kg/m}^3.$$

5.3 Water in Materials

Most materials are exposed during manufacture, storage, and use to the water vapour in the ambient air unless protected, for example, by special packaging that excludes water vapour. The surfaces of most materials exert forces of attraction on molecules of water vapour, which are thus drawn to and adhere to the surface, forming invisible films varying from a partial single layer of water molecules to those many molecules thick. The absorbed water is in a liquid-like state, having given up its heat of evaporation, and each molecular layer added to the film is less tightly held than its predecessor, until the molecules of the last layer are in a state of equilibrium with the ambient water-vapour condition. The film thickness is related to the relative vapour pressure, and increases with relative humidity. Most materials other than metals are porous, and so have internal and external surfaces that can be as high as 100 m²/g. Water can enter almost all connecting pores to form films in them (or even to fill them completely when the film thickness becomes greater than the pore radius) in response to high relative humidities. Because the process of adsorption and the reverse effect known as desorption involve a readjustment in the balance of molecular forces, there are accompanying small dimensional changes [**5.6–5.8**].

Curves of moisture content versus relative humidity at a given temperature are known as sorption isotherms. These are shown for three common materials in Figure 5.1, for both wetting and drying situations. Values are not greatly changed by moderate changes in temperature or by the presence of air, and are commonly determined in an atmosphere containing only water vapour to speed up the process. The difference between wetting and drying curves is known as hysteresis, and arises from differences in the ease with which water can enter and leave materials as a consequence of prior moisture conditions.

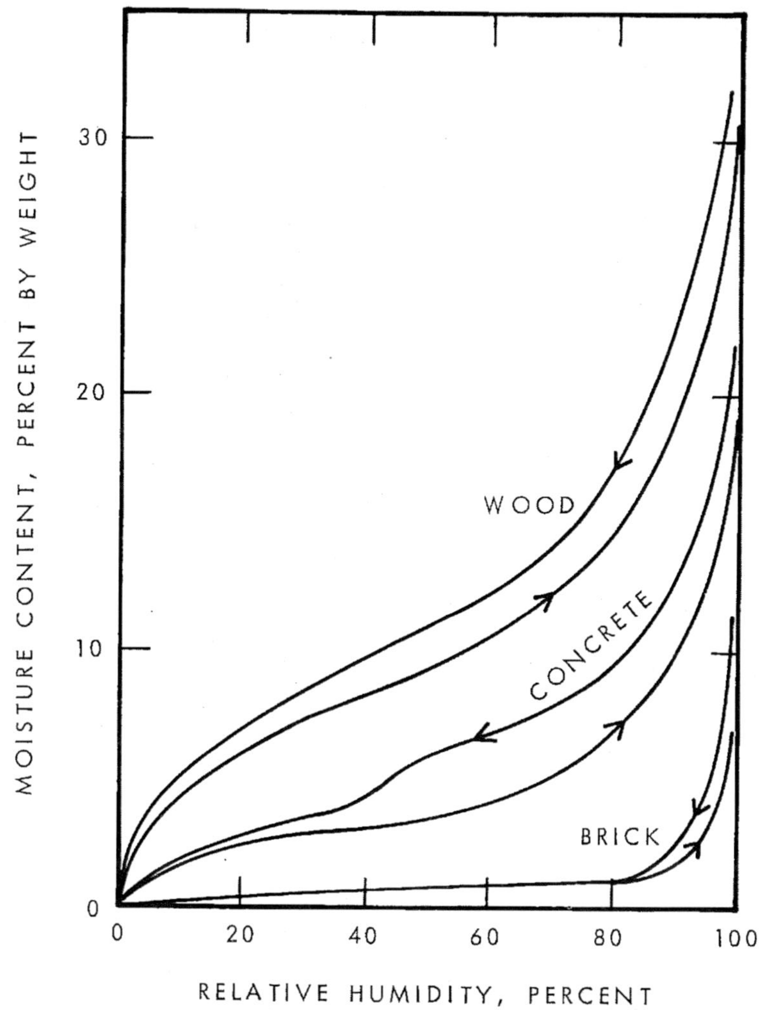

FIGURE 5.1 Water held in porous materials at various relative humidities.

The isotherms for wood, shown in Figure 5.1, indicate a relatively high sorption, characteristic of many natural organic products, with an orderly hysteresis effect amounting to about 14% between drying and wetting. Brick, at the other extreme, is an inorganic material which, when well burned, has relatively few very small pores and so sorbs moderate amounts of water at low and intermediate relative humidities. Concrete is more complex, since it is a combination of stone aggregate and hardened cement paste. The porosity of both can vary substantially. The hardened cement consists largely of material formed by hydration in the setting process, and so it is difficult to differentiate between water of hydration and sorbed water. In addition, water can enter into the platelike crystal structure of the cement, thus complicating further the sorption isotherm, including the hysteresis effect. The detailed study of these complications has provided insight into the nature of cement paste [**5.6, 5.8, 5.9**]. Sorption isotherms considered together with the corresponding length-change isotherms reveal much about the response of materials to changes in the moisture conditions to which they may be exposed [**5.10, 5.11**].

It will be noted that the sorption curves in Figure 5.1 rise at an increasing rate as the relative humidity approaches 100%. This is because more and larger pores, accounting for substantial amounts of water, are being filled at successively smaller incremental increases in relative humidity. In wood, the moisture content rises to as much as 200% of the dry weight in some cases by the time all the pores are completely filled. It may be that relative humidity can no longer be satisfactorily related to sorbed moisture at conditions near saturation unless it can be controlled and measured with great precision. Temperature must also be controlled since it determines the saturation vapour pressure.

Example 5.3

Calculate the change in relative humidity for an increase of 1 K in the temperature of an atmosphere initially saturated at 23°C.

From Table 5.1,
saturation vapour pressure at 23° = 2.809 kPa
and at 24° = 2.983 kPa

The relative humidity at 24°C will be

$$\frac{2.809}{2.983} \times 100 = 97\%.$$

The change in relative humidity is 100 − 97 = 3%.

5.4 Vapour Pressures over Curved Liquid Surfaces

When materials are exposed to water in its liquid state or when relative humidities are very high it is appropriate to consider the relationships between curvature of the liquid surface and the equilibrium vapour pressure above. This has direct relevance to the surfaces formed by water in sorbed layers within materials in pores and capillaries.

Molecules at the surface of any body, whether liquid or solid, experience different forces of interaction from their fellow molecules at some distance from the surface. This difference is reflected in the existence of interfacial tension (or surface tension, as it is called in the case of liquids). Forces between like molecules are cohesive; those that develop between unlike molecules are adhesive.

Surface tension can be measured by the use of a platinum ring of diameter D, usually 4 cm, which is withdrawn from the liquid surface so that the liquid still clings to it while the forces exerted on it are being measured [**5.1, p. 194**]. The arrangement is shown in Figure 5.2, which indicates that the force F is that which is required to rupture two concentric circumferential surfaces so that

$$F = 2 \times \pi D\gamma \tag{5.2}$$

where γ is the surface tension. Its dimensions are newtons per metre, corresponding to a contractile force per unit length, but could also be taken to be joules per square metre denoting surface energy per unit area. Both concepts are used in expressing the surface condition. The value of γ for water is about 0.072 N/m; for alcohol it is about 0.023 N/m.

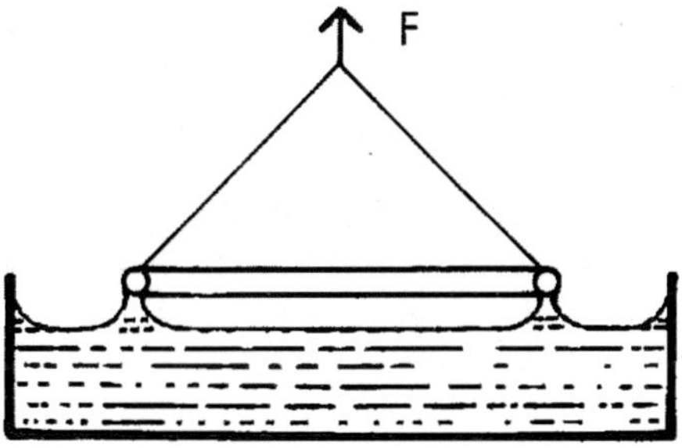

FIGURE 5.2 Surface tension.

The force given by Equation 5.2 is that of cohesion of the molecules in the liquid surfaces which were put under tension, but the measurement could not have been made if the liquid had not been held to the ring by strong forces of adhesion. A situation in which only cohesive forces are involved is presented by a falling drop of liquid which owes its spherical shape to them.

When a drop of liquid is placed on a horizontal surface, it may retain its shape, except for some flattening due to gravity (Figure 5.3(a)). In this case no adhesive forces exist and the liquid does not wet the surface. Mercury on glass is an example of such a situation. When the reverse is true, as in the case of water on glass shown in Figure 5.3(b) in which strong adhesive forces develop, wetting takes

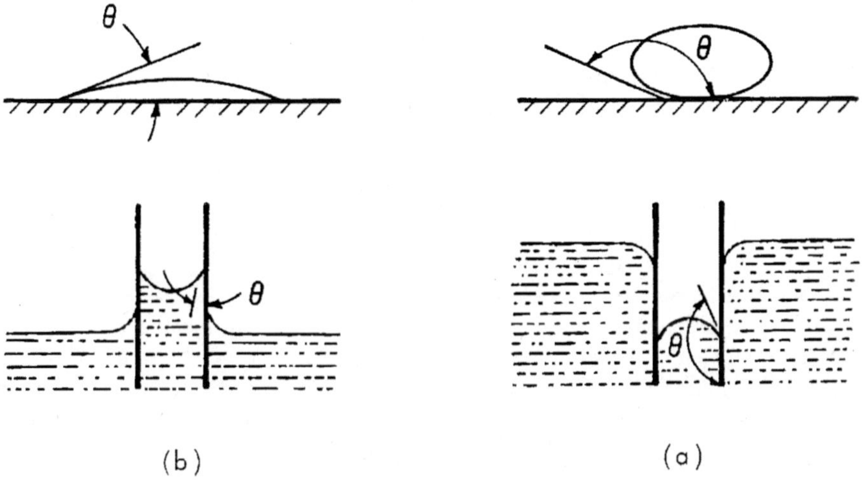

FIGURE 5.3 Contact angle for wetting and nonwetting liquids.

place. These tendencies can be measured in terms of a wetting angle or contact angle, θ, as shown. The corresponding meniscus and capillary-rise effects in each case are also shown. Surfaces may be coated with oil or substances such as silicones to form new surfaces which do not support adhesive forces and so are resistant to surface wetting as long as the added films remain in place.

5.5 Rise of Water in Capillaries

The rise of water in a glass capillary is due to the existence of strong adhesive forces. The balance of forces at the edge of a meniscus in a capillary is shown in Figure 5.4. The water surface must be perpendicular to the resultant force R, or flow will take place. The sharply curved concave surface can develop when the adhesion force A is large compared to the cohesive force C. When the capillary diameter is small, the surface approximates a hemisphere and the tangent to the edge of the water surface is nearly parallel to the glass surface. The capillary rise takes place in order to balance the upward component of the cohesive forces. The opposite effect (involving a negative rise) takes place in the case of mercury and other nonwetting liquid–surface combinations.

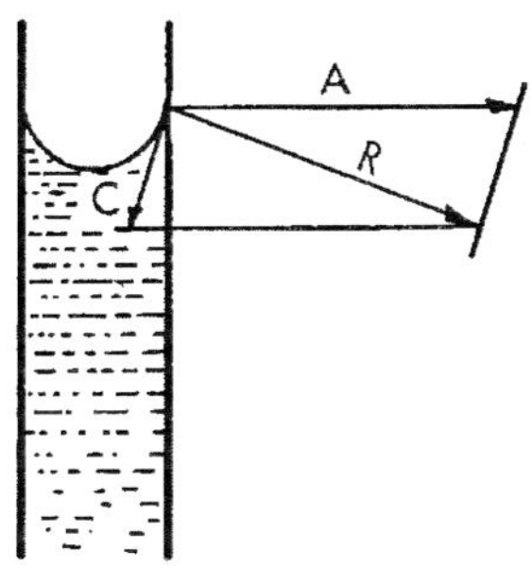

FIGURE 5.4 Adhesive and cohesive forces in a meniscus.

A rise h of water in a capillary of radius r is shown in Figure 5.5. When the surface tension is γ the total upward force is $2\pi r\gamma$. This force supports a water column of height h with a weight equal to $g\rho\pi r^2 h$.

$$g\rho\pi r^2 h = 2\pi r\gamma$$

and

$$h = \frac{2\gamma}{g\rho r} \tag{5.3}$$

where

γ = interfacial tension, water-to-air
ρ = density of water
g = gravitational constant.

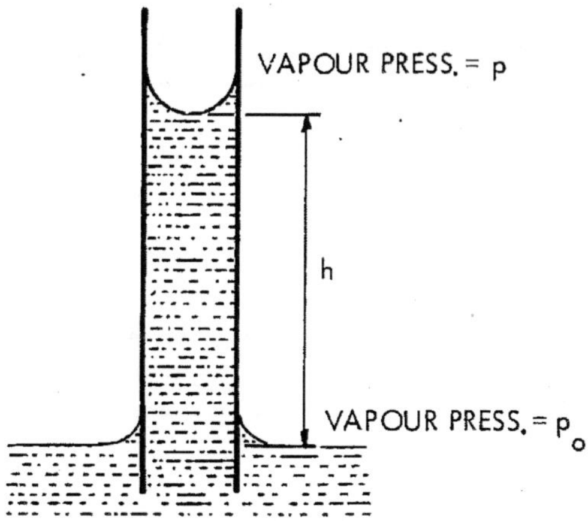

FIGURE 5.5 Capillary rise and equilibrium vapour pressures.

The vapour pressure, p, over the meniscus is related to the curvature of the surface and differs from the saturation vapour pressure p_o by an amount equivalent to the weight of the vapour column having a height equal to the capillary rise, h. The relationship can be derived using the equations of Chapter 3 for the weight of a column of gas compressed under its own weight, or may be taken from the widely recognized Kelvin equation:

$$\log_e p/p_o = -\frac{2\gamma M}{r\rho RT} \tag{5.4}$$

where

M = mole mass of water
R = gas constant
T = absolute temperature.

Combining equations 5.3 and 5.4,

$$h = -\frac{RT}{Mg}\log_e p/p_o. \tag{5.5}$$

This equation relates capillary rise to the relative vapour pressure. Equation 5.4 gives the relation between radius of curvature and relative vapour pressure. Thus, all three are related (see Table 5.3 [**5.12**]) and can be regarded as alternative ways of designating the moisture potential of water held in pores and capillaries.

The value of capillary rise calculated from Equation 5.5 can have large values when measured in centimetres. It has been the custom in soil science [**5.3, 5.12**] to deal with $\log_{10} h$ where h is in centimetres. Such values are given the designation *pF*, following Schofield [**5.13**]. Thus, a capillary rise of 100 cm becomes a *pF* of 2. Since increasing h means a lowering of the moisture potential, h is also called suction or negative hydraulic head and can be regarded as the potential for capillary flow. Under isothermal conditions h may also be a measure of the potential for vapour flow since p_o is then constant and p is proportional to p/p_o.

TABLE 5.3

Moisture Potential Relationships

(From Penner, E., *Proceedings, International Symposium on Humidity and Moisture, 1963*, Vol. 4, Chapter 29, Table 1, p. 247)
(NRCC 8601)[**5.12**]

p/p_o, %	Radius, r μm	Capillary rise cm	Pa	pF
49.0	0.001 47	10^6	9.79×10^7	6
93.0	0.014 7	10^5	9.79×10^6	5
99.3	0.147	10^4	9.79×10^5	4
99.9	1.47	10^3	9.79×10^4	3
99.999	147	10	9.79×10^2	1
100.0	∞	0	0	—

Example 5.4

Calculate the factors in Table 5.3 for

$$p/p_o = 0.993$$

Assume

$$T = 20 + 273 = 293 \text{ K} \qquad \rho = 998 \text{ kg/m}^3$$
$$\gamma = 0.072\,75 \text{ N/m} \qquad g = 9.807 \text{ m/s}^2$$

and find from Chapter 3, Example 3.2,

$$\frac{\text{R}}{\text{M}} = 461.5 \text{ J/(kg} \cdot \text{K)}$$

From Equation 5.4,

$$r = -\frac{2\gamma \text{M}}{\rho \text{R}T \log_e p/p_o}$$

$$= -\frac{2 \times 0.072\,75}{998 \times 461.5 \times 293 \times \log_e 0.993}$$

$$= 0.1535 \times 10^{-6} \text{ m} = 0.1535 \ \mu\text{m}.$$

From Equation 5.5,

$$h = -\frac{RT}{Mg} \log_e p/p_o$$

$$= -\frac{461.5 \times 293}{9.807} \log_e 0.993 = 96.85 \text{ m or } 9685 \text{ cm}$$

$$\text{Head} = 96.85 \times 998 \times 9.807 = 947\ 500 \text{ N/m}^2$$

$$pF = \log_{10} 9685 = 3.986.$$

From Table 5.1,
Vapour pressure at 20°C = 2.337 kPa = p_o

$$p = 0.993 \times 2.337 = 2.321 \text{ kPa}$$
$$p_o - p = 0.016 \text{ or } 16 \text{ Pa.}$$

Note that $p_o - p$ is the weight of 10^4 cm of vapour; h is for 10^4 cm of liquid. Note also that the values in the table are dependent on temperature, but not markedly so.

Capillaries in materials will seldom be uniform in diameter, will often be in the form of restricted passages between pores, and will be of limited length. This does not invalidate the concept, and capillary rise is often a useful way of thinking about moisture potential since it is a measure of the surface tension or surface energy. Thus, it may be appropriate to think of moisture potential at low moisture contents in terms of relative humidity and adsorbed films, and to think in terms of capillaries and curved surfaces at higher levels of moisture content when liquid-water movement in capillaries is possible. Soil scientists [**5.3**] find capillary rise a convenient concept for their purposes since they are mainly concerned with soil systems approaching saturation. Moisture situations in building may vary widely from dry to saturated and have an appreciable range in temperatures, but the capillary-rise concept is still useful.

It was noted earlier in connection with Figure 5.1 that large changes in sorbed moisture could take place near saturation for small corresponding changes in relative humidity. It can now be seen from Table 5.3 that 99.9% relative humidity corresponds to a water column of 1000 cm, which is a pF of 3. Thus, the capillary rise varies by 1000 cm while the humidity changes from 99.9 to 100%, providing a greatly expanded scale of measurement of moisture potential in an important part of the range. Other uses will be identified later.

The vapour pressures that have been discussed are equilibrium values determined by the curvature of the water surfaces to which they apply. They are independent of the presence of air apart from a small change in interfacial tensions when air is present, but the total pressures (the sum of the ambient air and water-vapour pressures) will act on the system. In Figure 5.6(a), the equilibrium vapour pressure over the meniscus is p. The total pressure p_t acts on top of the capillary column balanced by an equal pressure on the flat liquid surface. The actual partial pressure of the water vapour in the ambient air may not be equal to p, in which case there can be evaporation or condensation at the curved surface, at a rate dependent on the difference between p and the actual partial pressure as the system tries to achieve equilibrium of vapour pressures.

Figure 5.6(b) illustrates a short horizontal capillary filled until the surfaces at each end are flat. The equilibrium vapour pressure will be p_0. The total, or atmospheric, pressure must be the same on each side; otherwise the liquid moves as a piston until the difference in total pressures is balanced by different surface tension forces at the two ends. The actual partial pressure of the water vapour may be less than p_0, leading to evaporation from each end of the capillary until curvatures, and a balancing change in p, develop. If the actual partial pressure of the water vapour in the ambient air is less than the equilibrium pressure p for a fully developed curvature at the particular capillary diameter involved, it will be impossible for a balance to be achieved, and the capillary will be emptied by evaporation.

In Figure 5.6(c), balancing curvatures and surface tension forces are fully developed. The liquid in the capillary will now be in tension by an amount given by h, the theoretical capillary rise. The total pressures must be equal as before and the partial pressure of the water vapour in the ambient air may be different from the equilibrium vapour pressure p.

It can be concluded from further consideration of cases (b) and (c) that free water cannot be discharged at a surface as a result of capillary forces alone. Added forces of some kind must be brought into play to achieve this. Figure 5.6 (d) illustrates the effect of gravity on a column of height h, which is the capillary rise corresponding to the radius of the capillary. A flat water surface at the lower end will result, unless an additional head h' is developed to provide the forces necessary to form a free drop or a free film from the flat surface, thus delivering liquid from the capillary. A siphoning wick is an example of this. In Figure 5.6 (e) the force to deliver water is produced by a difference in total pressure which must overcome the surface tension, given by $h\rho g$, and provide the additional pressure p' necessary to deliver free water at the surface in the form of a film or drop.

The total pressure p_{t2} in Figure 5.6 (f) is maintained in a liquid reservoir so that there are no complicating surface-tension forces to consider at that end. The curvature of the liquid surface at the open end of the capillary and thus the equilibrium pressure p can now be varied at will by varying the total pressure p_{t1}. This can be done by providing an air chamber and some means of adjusting and controlling the air pressure. The vapour pressure in this chamber can now be allowed to come to equilibrium with the curved water surface. This principle can be used to generate known and controlled relative humidities in equilibrium with the corresponding curved surfaces, and is particularly useful in dealing with conditions near saturation.

It will be apparent that if $p_{t1} - p_{t2}$ in case (f) exceeds the surface tension forces as given by the capillary rise h, the meniscus will be broken and air will pass through the capillary into the liquid reservoir. The size of the capillary thus limits the extent to which p may be reduced below p_0. It is necessary to employ progressively finer capillaries in the production of lower relative humidities and correspondingly higher capillary suctions. Fine capillaries can be provided in the form of porous plates. Still finer pore structures can be obtained by using suitable membranes backed up by porous plates. The sample to be conditioned to a given moisture potential may either be exposed to the atmosphere in the chamber where it will respond to the equilibrium relative humidity p/p_o, or it can be placed in

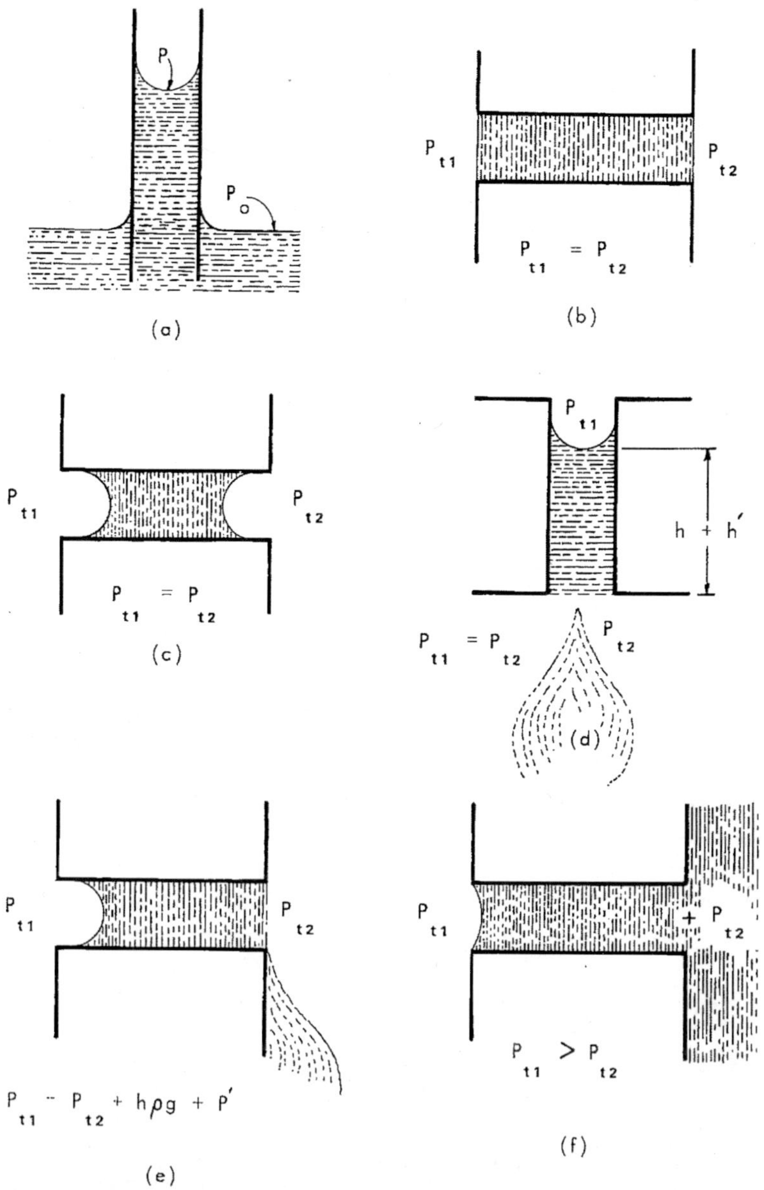

FIGURE 5.6 Some capillary and meniscus situations.

direct contact with the porous plate or membrane. A practical form of a pressure plate or pressure membrane apparatus is shown in Figure 5.7. Several different methods may be used to cover the complete range from saturation down to low relative humidities [**5.3, 5.12**].

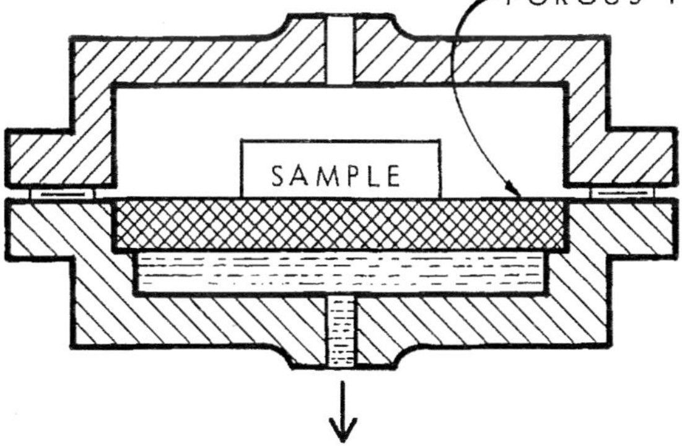

FIGURE 5.7 Schematic section of pressure plate apparatus.

Example 5.5

Find the suction and relative humidity that can be developed in a porous-plate apparatus with a plate having an air entry value of 5×10^5 Pa.

The largest pores in the plate begin to empty at an air pressure of 5×10^5 Pa. The corresponding suction as capillary rise is

$$h = \frac{\text{pressure}}{g\rho} \quad \text{so, using data from Example 5.4,}$$

$$h = \frac{5 \times 10^5 \ \text{N/m}^2}{(9.807 \ \text{m/s}^2)(998 \ \text{kg/m}^2)} = 51.09 \ \text{m or } 5109 \ \text{cm}$$

$$pF = \log_{10} 5109 = 3.7.$$

From Equation 5.5,

$$\log_e p/p_o = -\frac{h\,\text{Mg}}{\text{R}T} = \frac{51.09 \times 9.807}{461.5 \times 293} = -0.003\,705$$

and

$$p/p_o = 0.996 \text{ or } 99.6\%.$$

5.6 Measuring Moisture Potentials in Situ

Porous bodies are used for in-situ measurements of the moisture potentials in soils. The porous body with an internal chamber is filled with water and placed in contact with the soil. It gives up water to the soil until the reduced pressure within it is in balance with the soil suction. This pressure can then be read from the manometer or other pressure-measuring device that is part of the instrument. Such devices, called tensiometers, are useful for suctions up to about 0.8 atmospheres [**5.3, 5.12**].

Other types of moisture meters are made with parallel electrodes embedded in a porous body. The reading of electrical resistance of the porous material between them reflects the moisture content taken up in reaching equilibrium with the soil or other material. Changes in salt content and hysteresis effects, in both gauge and soil, limit the accuracy of such gauges, but for many purposes no better methods are available. When calibrated in a pressure-plate apparatus against known suction values, they become suction meters. The soil must then be tested to produce curves of moisture content versus suction so that suction values can be converted to moisture content for the particular soil involved [**5.12**].

5.7 Reduction of Vapour Pressure over Solutions

The attraction of surfaces for water molecules can be regarded as a suppressing influence on the attainment of saturation pressures. In a somewhat comparable way the addition of chemicals to water reduces the equilibrium vapour pressure over such solutions. Raoult's law states that

$$p = p_o x_1 = p_o (1 - x_2) \tag{5.6}$$

where x_1 is the mole fraction of solvent, x_2 is the mole fraction of solute given by $n_2/(n_1 + n_2)$ where n_1 and n_2 are moles of solvent and solute.

This relationship gives approximate results for dilute solutions and is not reliable for concentrated solutions.

Chemicals such as common salt and calcium chloride are used for snow melting and ice control on traffic surfaces. When spread as crystals or granules on snow or ice they cause partial thawing and degradation of slippery surfaces by forming a concentrated solution with a lowered freezing point. As temperatures are reduced progressively below the normal freezing point, the ability to hold water in solution decreases until a limiting eutectic temperature is reached, below which a solution cannot exist and salt will crystallize out [**5.14**].

A strong salt solution on the outside of a saturated porous material will cause water to migrate from the material to the solution, resulting in emptying of some coarse pores because of the lowered equilibrium vapour pressure. This effect is commonly experienced when salt is put on crisp vegetables or on meat that is being grilled or roasted. If the salt is already in the porous material, and not outside it, the reverse action can take place with water entering the material to saturate it at vapour pressures which are below those for saturation with no salt present.

This effect is thought to be a factor in damage to concrete in bridge decks when frozen with de-icing chemicals present.

Saturated solutions of various salts can be used on a laboratory scale to produce controlled relative humidity [**5.15**]. When extremely low relative humidities are required, materials known as desiccants (which are capable of adsorbing water strongly) are used. Any material with an affinity for water vapour will, when dry, take up water vapour within limits, and thus act as a drying agent. Calcium chloride is a well-known and relatively inexpensive desiccant with a substantial capacity for water, which it holds chemically in the form of hydrates. Other materials such as silica gel can be regenerated conveniently by heating; they are used for industrial dehumidification.

Cooling air–vapour mixtures below the condensation point on cold surfaces is by far the most widely used means of accomplishing dehumidification. Cold window surfaces act in this way to limit severely the relative humidities that can be maintained indoors in winter.

5.8 Transfer of Water between Materials

When two materials containing water are placed in contact with each other, moisture will migrate until the moisture potentials for the two materials are equal. It may be asked whether the transfer is by liquid or vapour or both. If the materials are not in contact but are together in a closed container, the transfer must then be entirely through the vapour phase. The Swedish results shown in Figure 5.8 [**5.16**] demonstrate that the end result is substantially the same if the bodies are separated or if they are in contact, given enough time. For certain materials and moisture levels the equalization may take place much more quickly when they are in contact, but in other cases, as the Swedish tests also showed, the rates are almost as great for vapour diffusion as for contact.

These results might have been expected in view of the relationship discovered earlier between relative humidity and capillary potential. When materials are separated, only vapour flow is possible; when they are fully saturated and in contact, only a liquid flow can take place, provided there is some external force to produce flow. For other cases involving partially saturated connected pore systems, both liquid and vapour flows are possible. This is the case for unsaturated flow within a material as well as for transfer between different materials in contact. Liquid-like flow can be expected along thick sorbed films and the smaller capillaries that have been filled by merged films, while film and vapour flow may take place in larger unfilled pores and capillaries.

The migration of water in sorbed films and capillaries will be determined by the gradient of p/p_0 or its close relative, h; vapour will be caused to flow under a vapour gradient. Under isothermal conditions these two kinds of flow will reinforce one another. It will be shown later that under a temperature gradient, the gradients of p and of p/p_0 may differ in direction and in relative magnitude so that liquid and vapour flows may be opposed. It should also be remembered that migrating liquid can carry dissolved salts, but vapour cannot.

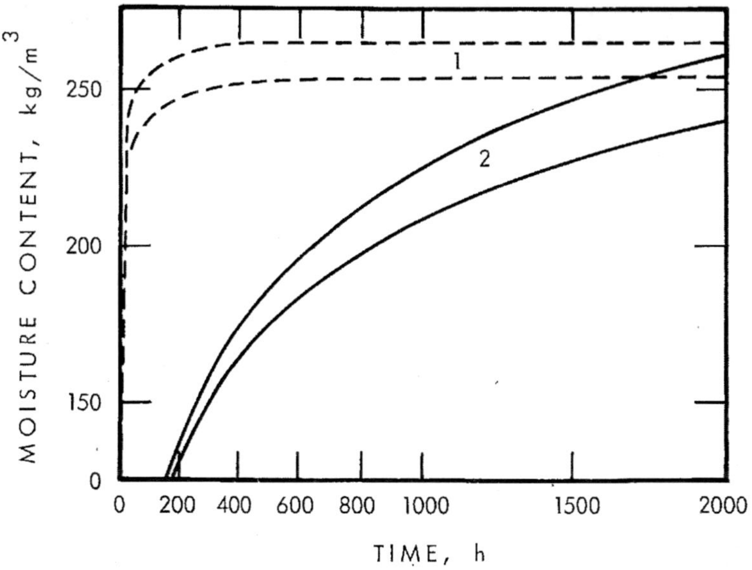

FIGURE 5.8 Moisture transfer from wet gypsum plate to aerated concrete sample at two moisture levels for: 1. samples in contact and 2. samples held 1 mm apart.
(From Bomberg, M., Report 52, Lund, Sweden: Lund Institute of Technology, 1974, Fig. 12, p. 42)[**5.16**]

The operation of suction plates, pressure plates, and several forms of moisture meters depends on the attainment of an equality of moisture potentials in materials in contact. The results will always be influenced by hysteresis, which leads to different moisture contents at a given applied potential, depending on whether the final condition is approached by wetting or drying [**5.12**].

The moisture contents resulting from an equilibrium established between materials in contact will depend on their respective moisture content versus relative vapour pressure relationships. These, in turn, are determined by porosity and pore-size distribution. It has been pointed out [**5.8**] that, for example, bricks may contain only 0.5% moisture, dry-weight basis, in contact with concrete at 5%. This can also be seen from Figure 5.1.

It is common practice to express moisture content as a percentage by weight of dry substance. This leads to very high values and, for many light materials, to percentages over 100. It has long been the practice in Europe to express moisture content for many situations as a percentage by volume. Thus, one cubic metre of a light cellular material having a dry density of 30 kg/m^3 and containing 900 kg of water would have a moisture content of 3000% by weight or by mass, dry basis. The percentage by volume, relating the volume of water present to the volume of material, which is also given by the mass of water present to the mass of water required to occupy the material volume, will be (900/1000) 100 = 90%. For some purposes, values are also given by expressing the moisture content simply in kilograms per cubic metre, or as a percentage of the moisture content at complete saturation.

5.9 Saturation of Materials

The rate and amount of water uptake by porous building materials have long been regarded as important properties for many purposes. For example, it has been widely held that the durability of bricks under freeze–thaw conditions is related to the ability of the brick to resist complete saturation when exposed to water. This can be measured in a standard way by comparing the amount of water taken up in 24 hours of immersion in cold water with that absorbed during a five-hour boiling test. When this cold/boiling ratio is 0.8 or less, depending on tensile strength, the brick is thought to be capable of resisting the expansive force of the freezing of the contained water.

One investigation [**5.17**] has shown that the boiling test is not a reliable measure of saturation in bricks compared to saturation in a vacuum to eliminate air. Values obtained for one brick selected as an example are given in Table 5.4. It was shown in the same study that the cold/boiling ratio is not a reliable indicator of freeze–thaw durability. This topic will be discussed later in this chapter.

The results of liquid-flow tests and water-uptake tests can be highly variable, owing in large part to variation in the amount of air trapped in pores. Such effects may account in part for the hysteresis noted between adsorption and desorption curves. Despite the variable nature of water uptake, however, the degree of saturation at the time of freezing is regarded as a highly important factor in the damage of materials caused by freeze–thaw cycling.

TABLE 5.4

Water Absorption of Brick Samples

(From Litvan, G.G., ASTM Special Technical Publication 589, 1975, Table 2, p. 126) (NRCC 14797 [**5.17**]. Copyright, ASTM, 1916 Race Street, Philadelphia, PA. 19103. Reprinted with permission.

	Cold-water immersion, days							
	1	2	3	4	5	6	7	8
Absorption, %	8.07	8.17	8.36	8.38	8.48	8.52	8.59	8.67
Saturation, %	58.35	59.07	—	—	—	—	—	62.69

	Repeated 5-hour boiling				Vacuum saturation
	1	2	3	4	
Absorption, %	11.04	11.17	11.29	11.44	13.83
Saturation, %	79.78	80.71	81.64	82.71	100
Saturation coeff.	0.73	0.72	0.71	0.70	0.58

5.10 Water-Vapour Transmission

Recognition of the effects of water in materials and constructions has led to the use of certain materials capable of restricting the flow of water vapour. Such materials may be used as packaging or incorporated as barrier membranes in constructions such as walls and ceilings. It becomes necessary to establish tests by which the permeance to water vapour of these and other materials can be measured and to develop methods of predicting vapour flow through building constructions. In the case of walls and ceilings, the objective is usually to restrict the flow of water vapour which might otherwise condense, producing undesirable amounts of water or ice within constructions.

The equation usually used to calculate water-vapour transmission through materials is based on a form of Fick's law and is as follows:

$$w = -\mu \frac{dp}{dx} \tag{5.7}$$

where

w = mass of vapour transmitted over unit time
p = vapour pressure
x = distance along the flow path
μ = permeability

so that

$\frac{dp}{dx}$ = vapour–pressure gradient.

The conditions producing flow are shown in Figure 5.9. Vapour pressures p_1 and p_2 are assumed to exist on either side of the specimen of thickness ℓ. It is assumed that the vapour-pressure gradient is not constant but varies because of a variation in μ, as shown.

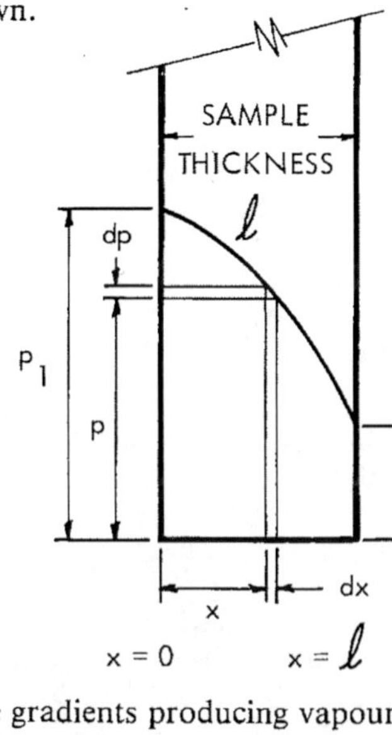

FIGURE 5.9 Pressure gradients producing vapour flow.

Integrating Equation 5.7 from $x = 0$ to $x = \ell$ and from p_1 to p_2 and rearranging

$$w = \frac{\int_{p_1}^{p_2} \mu dp}{p_1 - p_2} \times \frac{p_1 - p_2}{\ell}. \tag{5.8}$$

Let

$$\frac{\int_{p_1}^{p_2} \mu dp}{p_1 - p_2} = \bar{\mu}.$$

Then

$$w = \frac{\bar{\mu}(p_1 - p_2)}{\ell} \tag{5.9}$$

where

ℓ = length of flow path or thickness of material.

The integral expression for $\bar{\mu}$ may be seen to have the form of the area under a curve of μ versus p, between the limits p_1 and p_2, divided by $(p_1 - p_2)$, so that $\bar{\mu}$ is the average value of μ over the pressure range involved.

Equation 5.9 may be rewritten and units assigned:

$$W = \bar{\mu} A \theta \frac{(p_1 - p_2)}{\ell} \tag{5.10}$$

where

W = total mass of vapour transmitted, nanograms
A = cross-section area of the flow path, square metres
θ = time during which flow occurs, seconds
$(p_1 - p_2)$ = the vapour-pressure difference applied across the specimen, pascals
ℓ = length of flow path, or thickness, metres
$\bar{\mu}$ = average permeability, nanograms per second, square metre for one metre thickness and one pascal pressure difference which reduces to ng/(s • m • Pa).

The permeability $\bar{\mu}$ in Equation 5.10 refers to a specimen ℓ metres thick. When the material is commonly used in particular thicknesses, or when particular thicknesses are likely to have unique properties, it is appropriate and convenient to combine $\bar{\mu}$ and ℓ and to speak of $M = \frac{\bar{\mu}}{\ell}$, which then refers to the particular thickness of material stated or implied. The corresponding flow equation is

$$W = MA\theta\,(p_1 - p_2) \tag{5.11}$$

where

M = permeance coefficient, nanograms per second, square metre, pascal vapour-pressure difference which reduces to ng/(s • m^2 • Pa).

The units assigned above are those now recommended in the conversion to SI units in Canada. The term "perm" was formerly widely used to identify values of M stated in grains per square foot per hour per inch of mercury pressure difference. No short designation for unit permeance in SI units has yet been accepted in practice, and it is necessary to retain the basic units as a means of identification. Units in perms multiplied by 57.45 will give values of permeance M in ng/(s • m^2 • Pa). Units of permeability in perm-inches multiplied by 1.459 will give values of permeability $\bar{\mu}$ in ng/(s • m • Pa).

Equations 5.10 and 5.11 can be used to calculate water-vapour transmission through various sheet and board elements and combinations of them. It must be possible to assign appropriate values to p_1 and p_2 and to μ and M. It will be shown later how several elements can be combined.

5.11 Wet-Cup and Dry-Cup Methods

The simplest method of finding the permeance of a specimen is to seal it over the top of a cup containing either a desiccant or water, place it in a controlled atmosphere, and weigh it periodically. The steady rate of weight gain or loss will give the water-vapour transfer from which the permeance or the permeability can be calculated. There are standard methods for carrying out such tests [**5.19**]; the usual cabinet test conditions are 23°C and 50% rh. When the cup contains desiccant the vapour flows into the cup from the 50% condition to the dry condition assumed to be 0% rh inside the cup. This is known as the dry-cup test. When water is placed inside the cup the vapour flows from the condition near 100% rh inside the cup to the cabinet atmosphere at 50%; this is known as the wet-cup test. Although these are apparently simple procedures, care must be taken if reproducible results are to be obtained [**5.18**].

Wet-cup and dry-cup values of permeability or permeance often differ, with the wet-cup values usually higher. The relationship between them can be seen from Figure 5.10. If the coefficient of permeability μ is assumed to vary as shown, cup permeabilities will be the average values under the curve between appropriate vapour-pressure limits, which may also be expressed as relative humidity since these tests are carried out under isothermal conditions. The greater wet-cup value indicates a permeability curve which increases with increasing relative humidity. When such a curve is delineated, it is possible to find the average permeability for other limits of relative humidity by measuring the average height of the appropriate area under the curve.

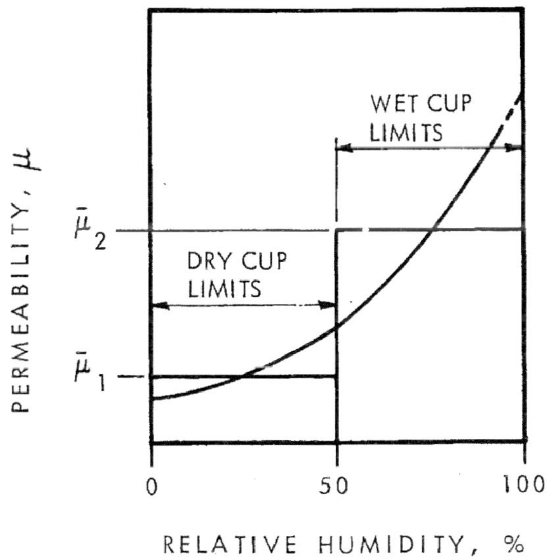

FIGURE 5.10 Relation between dry cup and wet cup permeabilities.

The experimental delineation of such permeability curves is time-consuming and is seldom considered justified. It is possible to judge from wet-cup and dry-cup tests whether permeability is likely to vary appreciably with relative humidity. In the wet-cup test the relative humidity at the surface of the specimen will be slightly less than 100%. This can have an effect, as might be anticipated from the rapidly increasing moisture contents near 100% rh shown in Figure 5.1. The wet-cup test can be carried out in the inverted position with water in contact with the specimen, and this often gives still higher values than the normal wet-cup test, thus providing a further indication of sensitivity of the permeability coefficient to moisture content or to relative humidity.

It is seldom considered justified to run cup tests at different temperatures, although changing the chamber temperature has a significant effect. It is necessary, however, to state the test conditions, and particularly the test humidities, when results are to be compared.

5.12 Coefficients of Vapour Transfer through Building Materials

Values of permeance for a range of building materials are given in Tables 5.5 and 5.6. A few values are given as permeabilities in Table 5.6. Table 5.5 provides both dry- and wet-cup values as well as a few inverted wet-cup results carried out on on the same or identical samples, so they can be compared with some confidence. The materials having low permeances that are commonly selected to provide resistance to vapour flow are naturally of greater interest than others, and so predominate among those listed. Additional values can be found elsewhere [**5.5, p. 21.6**].

TABLE 5.5

Water-Vapour Transmission Coefficients at 23°C
(From Joy, F.A., and Wilson, A.G., *Proceedings, International Symposium on Humidity and Moisture, 1963*, Vol. 4, Chapter 31, Table 1, p. 261) (NRCC 8838) [**5.18**]

Material	Permeance ng/(Pa • s • m^2)		
	dry cup 50–0%	wet cup 100–50%	inverted wet cup
Foamed polyurethane insulation 25 mm			
28 kg/m^3	75	75	—
31 kg/m^3	63	63	—
Foamed polystyrene insulation 25 mm			
Extruded 29 kg/m^3	92	92	—
Extruded 35 kg/m^3	44	42	—
Polyethylene film			
0.05 mm	9	8	—
0.10 mm	5	4	—
0.15 mm	3	2	—
Nylon film 0.025 mm	39	40	—
Vinyl film 0.05 mm	19	19	—
Cellulose acetate film 0.25 mm	270	640	—
Waxed building paper			
medium weight	5	9	—
heavy weight	6	51	—
Asphalt-saturated sheathing paper			
0.75 kg/m^2 (15 lb)	270	480	725
1.25 kg/m^2 (25 lb)	190	370	—
heavy weight	47	360	500
Asphalt-saturated roofing felt 0.75 kg/m^3	110	680	910
Tar-infused sheathing paper	375	1770	4050
Asphalt-infused sheathing paper	365	1080	2400
Asphalt-coated building paper	47	63	115
Perforated asphalt-coated sheathing paper	630	800	860
Structural clay tile 6 mm		660	—
Vitreous ceramic tile 9 mm	0.6	23	—
Fibreboard, untreated 12.5 mm	2470	2520	—
Fibreboard, sheathing grade 12.5 mm	1720	1780	—
Asbestos cement board	285	480	—

All values from [**5.18**] converted to SI units.
1 ng/(Pa • s • m^2) = 1×10^{12} kg/Ns = 0.0175 grains/(hr ft^2)(in. Hg)

TABLE 5.6

Water-Vapour Transmission Coefficients for Various Materials
(From *ASHRAE Handbook 1981 Fundamentals*, Table 2, p. 21.6, 21.7) [**5.5**]

Material	Permeance ng/(Pa • s • m²) dry cup	wet cup	other
Brick masonry, 10 cm	—	—	46
Concrete block, 20 cm, cored, limestone aggregate	—	—	138
Tile masonry, glazed, 10 cm	—	—	7
Asbestos cement board, 5 mm	31	—	—
Plaster on wood lath	—	630	—
Plaster on plain gypsum lath on studs	—	—	1150
Gypsum wallboard, 9.5 mm, plain	—	—	2870
Hardboard, 3 mm, tempered	—	—	290
Plywood, douglas fir, exterior glue, 6.5 mm	—	—	40
Enamels, 2 coats on smooth plaster	—	—	29–86
Primers, sealers, 2 coats on insulation board	—	—	52–120
Various primers, 2 coats + 1 coat flat oil paint on plaster	—	—	92–172
Flat paint, 2 coats on insulation board	—	—	230
Water emulsion, 2 coats on insulation board	—	—	1720 to 4900
Exterior paint, 3 coats white lead and oil on wood siding	17–57	—	—
Styrene butadiene latex coating 0.62 kg/m²	630	—	—
Polyvinyl acetate latex coating 1.25 kg/m²	320	—	—

Material (For unit thickness of 1 metre)	Permeability ng/(Pa • s • m) dry cup	wet cup	Other
Concrete 1:2:4 min		4.7	
Wood, sugar pine			0.58
Mineral wool, unprotected		170	1011.5
Expanded polystyrene—bead	3–8.5		

(All values from [**5.5**] converted to SI units)

The value of permeability for unprotected mineral wool of 170 ng(s • m^2 • Pa) should be noted. This is almost the same as the value for still air (not shown) of 175, indicating that the loose, low-density pack of mineral fibres does little to hinder vapour flow. Fibreboard, with a permeance of about 2500 in 12.5 mm thickness corresponding to a permeability of 30 ng/(s • m • Pa), is also relatively permeable, as are most other low-density porous products made from natural fibres. Papers heavily saturated or coated with bitumen are relatively impermeable, as are many plastic films and oil-paint films. Latex-paint films are relatively permeable. Metal foils should be impermeable, but the thin foils used in building often have pinholes in them. Some plastic films such as polyethylene are relatively impermeable, show little difference between wet, dry, and inverted wet-cup values, and show no anomalies in respect of thickness. The swelling of products made from natural fibres when exposed to high relative humidity may account for the high ratio of wet-cup to dry-cup values for some asphalt-treated papers. Some waxed papers show similar tendencies, with very high inverted wet-cup values.

In comparing materials it is useful to note that the Canadian standard for vapour-barrier membranes for use in construction above grade [**5.20**] requires that the permeance measured by the dry cup be not greater than 45 ng/(s • m^2 • Pa). The requirement for a higher grade of low-permeance materials is that they shall not exceed 15 ng/(s • m^2 • Pa) when tested wet-cup.

Example 5.6

The water-vapour transmission coefficient for foamed polyurethane in Table 5.4 could have been given as permeability, since it can be cut to various thicknesses. Convert the dry-cup permeance value for 25 mm thickness to permeability.

Permeance for 28 kg/m^3 material = 75 ng/(s • m^2 • Pa). This was for 1 in. thickness in the orginal table, now given as 25 mm. Permeability is the value for unit thickness, i.e., for 1 m thickness:

$$\bar{\mu} = \frac{75 \times 25}{1000} = 1.9 \text{ ng/(s} \cdot \text{m} \cdot \text{Pa)}$$

Example 5.7

In a wet-cup test the net sample area was 10 cm in diameter and the loss in weight in 30 days was 0.509 g. The cabinet condition was 50% rh at 23°C. Calculate the permeance.

From Table 5.1,

$$p_o \text{ at } 23°\text{C} = 2.809 \text{ kPa} = 2809 \text{ Pa}$$

$$\text{Pressure difference across specimen} = 0.50 \times 2809 = 1405 \text{ Pa.}$$

$$\text{Area of sample} = \left(\frac{10}{100}\right)^2 \times \frac{\pi}{4} = 0.007\ 85 \text{ m}^2$$

$$\theta = 30 \times 24 \times 3600 = 2.59 \times 10^6 \text{ seconds.}$$

From Equation 5.11,

$$M = W/A\,(p_1 - p_2)$$

$$= \frac{509 \times 10^6 \text{ ng}}{0.007\,85 \text{ m}^2 \times 2.59 \times 10^6 \text{s} \times 1405 \text{ Pa}}$$

$$= 17.8 \text{ ng/(s} \cdot \text{m}^2 \cdot \text{Pa).}$$

5.13 Vapour Flow through Composite Sections

Despite the difficulty of assigning appropriate permeance values to many materials under differing conditions, there is often much to be gained by way of a basis for improved judgement from calculations of the vapour flow through composite sections such as walls and ceilings. Simple theory can be used, with such permeance data as are available or can be assigned on the basis of judgement.

When a specimen such as that of Figure 5.9 is a composite of two elements in contact, the flow under steady conditions must be the same through each element and equal to that through the combination. Equation 5.10 can be written

$$\frac{W}{A\theta} = \frac{\bar{\mu}}{\ell}(p_1 - p_2).$$

Let the combined permeance coefficient be M_c and the vapour pressure at the interface of elements 1 and 2 be p'. Then

$$\frac{W}{A\theta} = \frac{\bar{\mu}_1}{\ell_1}(p_1 - p') = \frac{\bar{\mu}_2}{\ell_2}(p' - p_2) = M_c(p_1 - p_2) \tag{5.12}$$

From which

$$M_c\frac{\ell_1}{\bar{\mu}_1} = \frac{p_1 - p'}{p_1 - p_2} \text{ and } M_c\frac{\ell_2}{\bar{\mu}_2} = \frac{p' - p_2}{p_1 - p_2}.$$

Adding

$$M_c\left[\frac{\ell_1}{\bar{\mu}_1} + \frac{\ell_2}{\bar{\mu}_2}\right] = \frac{p_1 - p_2}{p_1 - p_2} = 1$$

and

$$M_c = \frac{1}{\dfrac{\ell_1}{\bar{\mu}_1} + \dfrac{\ell_2}{\bar{\mu}_2}} \tag{5.13}$$

The quantities $\ell_1/\bar{\mu}_1$ and $\ell_2/\bar{\mu}_2$ are the reciprocals of the permeances of the individual elements, which are resistances to vapour flow and can be added to find the composite resistance, as follows, from Equation 5.13:

$$M_c = \frac{1}{R_c} = \frac{1}{R_1 + R_2}. \tag{5.14}$$

The resistance of individual elements is given by $R = \frac{1}{M}$ or $\frac{\ell}{\bar{\mu}}$, whichever is appropriate, the thickness ℓ being in metres. Components of more than two elements can be handled by summing the resistances for the number of elements involved.

Example 5.8

An interior wall consists of wallboard on either side of steel studs 90 mm deep with the space filled with mineral wool. It is maintained at 20°C with a relative humidity of 70% on one side which has two coats of flat paint on insulation board, and 20% on the other side consisting of gypsum wallboard 9.5 mm thick. Find the composite permeance and the vapour flow in 24 hours through one square metre of wall area.

The vapour pressures are:

$$p_1 = 0.70 \times 2337 = 1640 \text{ Pa} \quad \text{and} \quad p_2 = 0.20 \times 2337 = 470 \text{ Pa}.$$

The permeances and permeabilities are given in Tables 5.5 and 5.6. The elements can be listed in order as follows:

Element	ℓ	$\bar{\mu}$ or M	R	Δp
Insulation board, two coats flat paint		$M =$ 230	0.004 35	973
Mineral wool	90 mm	$\bar{\mu} =$ 170	0.000 53	119
Gypsum wallboard	9.5 mm	$M =$ 2870	0.000 35	78
			0.005 23	1170

The overall permeance $M_c = \dfrac{1}{0.005\ 23} = 191.2$ ng/(s • m² • Pa)

The mass flow rate $= 191.2 \times 1(1640 - 470)$
$= 223\ 000$ ng/s • m²

The flow in 24 hours $= 24 \times 3600 \times 223\ 000 \times 10^{-9}$
$= 19.3$ g/m².

The last column in the table of Example 5.8 headed Δp shows the pressure drop in pascals across each element. This can be shown from equations 5.12 and 5.14 to be proportional to the resistance of the element in each case. The sum of these pressures should add up to the pressure difference across the composite sample.

5.14 Unsaturated Flow in Soils

Movement of water in soils is of great interest to the soil scientist [**5.3**]. Some water-transmission tests have been reported [**5.21**] for unsaturated soil samples with water at zero suction supplied at one face of the sample and an increasing suction, S, applied to the other. Results are shown in Figure 5.11. The flow equation used in such cases is of the form

$$v = k(S_2 - S_1). \tag{5.15}$$

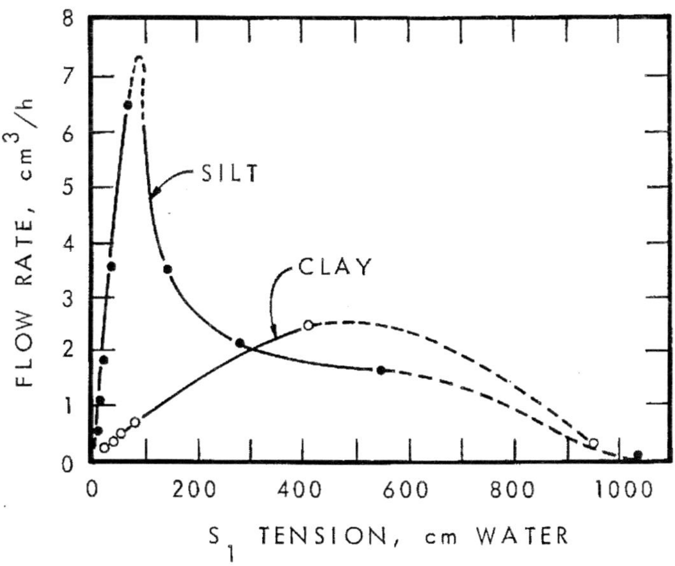

FIGURE 5.11 Flow rate in soil under a suction gradient as a function of S_1 with S_2 held at zero.
(From Penner, E., *Proceedings ASTM*, 1958, *58*, Fig. 6, p. 1214, NRCC 4994 [**5.21**].
Copyright, ASTM, 1916 Race Street, Philadelphia, PA. 19103. Reprinted with permission.)

It is evident from these results that k is not a constant, and that simple flow theory may be inadequate to describe unsaturated flow, even without the possible added complications of a temperature gradient. In this case the flow rate increases as the suction potential difference $(S_2 - S_1)$ increases. Meanwhile, increasing suctions lead to decreasing moisture contents, with large pores and capillaries being emptied and only smaller pores and capillaries remaining filled in support of water flow. The flow coefficient thus becomes dependent on moisture content or on suction; it decreases along the flow path and, on the average, as suction is increased. The net result is a flow maximum, as shown, at some intermediate suction. These results are of interest because they relate to the upper end of the moisture-content and relative-humidity scale and so provide some insight into the significance of some features of wet-cup and wet-cup inverted vapour-flow results. The relative humidity corresponding to a suction of 1000 cm is found from Table 5.3 to be 99.9%, so these results now shown are for the humidity range from 99.9 to 100%.

5.15 Moisture Flow with a Temperature Gradient

It has already been suggested that the imposition of a temperature gradient on a vapour-flow situation may lead to marked departures from the results for isothermal cases. It has not yet been possible, despite much research in many countries over the past 40 years, to reduce the functional relationships satisfactorily to equation form. The possibility of combined vapour and capillary flow was proposed as early as 1940 in Germany [**5.22**]. One early researcher described an interesting model, shown in Figure 5.12, in which opposing liquid and vapour flows were possible. Vapour flow occurs across the gap between plates maintained at temperatures T_1 and T_2 and returns in liquid form via the connecting capillary paths of blotting paper. Other researchers [**5.23**] reported an experiment with a soil column containing a salt solution and maintained under a temperature gradient along its axis. The salt was found to concentrate at the warm end, providing strong evidence that there was a movement of liquid from the cold end, which was also at higher moisture content, and a corresponding flow of vapour in the opposite direction from the warm end to the cold end.

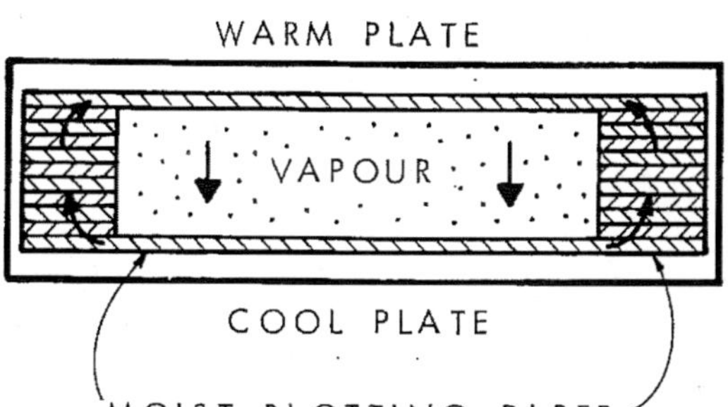

FIGURE 5.12 Model showing vapour and liquid flow in opposite directions under a temperature gradient.

Swedish research [**5.24**] paralleling that in Germany proposed a flow equation as follows:

$$w = -\mu' \frac{dp}{dx} - \mu'' \frac{dc}{dx} \tag{5.16}$$

where

$\dfrac{dp}{dx}$ = vapour-pressure gradient and

$\frac{dc}{dx}$ = concentration, or moisture-content gradient, μ' and μ'' being the corresponding flow coefficients.

The great difficulty seems to lie in the determination of these coefficients, which vary with changing size and proportion of unfilled pores as the moisture content changes. It is likely that liquid and vapour flows do not occur independently but rather are interrelated in a pattern of series/parallel situations determined by the particular pore structure of the material involved. It is possible also that the temperature gradients affecting vapour flow are those on a microscale which exist across unfilled pores, leading to a still more complicated situation, since they can differ markedly from the average or overall gradient as was shown in some experiments at DBR/NRCC [**5.26**].

Canadian work on the influence of moisture on thermal-conductivity measurements provided some interesting evidence of the complexities of moisture flow under a temperature gradient [**5.25**]. The material used was moist sawdust subjected to a temperature gradient in a closed system between 38°C and 16°C. The moisture was redistributed under the temperature gradient and assumed a steady-state condition with higher moisture contents at the cold side, as shown for three different average moisture contents in Figure 5.13. It was assumed that at the equilibrium condition the vapour pressure would be constant throughout the

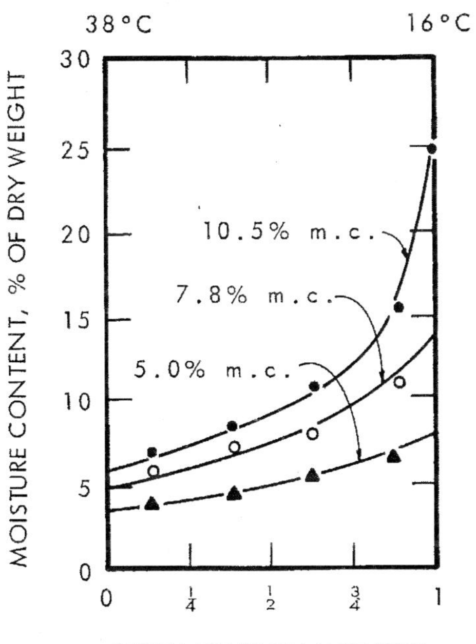

FIGURE 5.13 Moisture distribution in moist sawdust under a temperature gradient.
(From Paxton, J.A., and Hutcheon, N.B., *ASHVE Trans.*, *58*, 1952, Fig. 7, p. 307)[**5.25**]

sample. The relative humidities and the corresponding sorption values for the sawdust materials were used to predict the moisture distributions, which are shown by the curves in Figure 5.13. The experimentally measured values averaged over each quarter of the sample thickness and given by the plotted points indicate a satisfactory correspondence with the predicted values. The level of 10.5% moisture content was that for a vapour pressure throughout equal to that which would produce 100% rh at the cold face.

It may be argued from these results that no liquid flow was taking place from the cold side to the warm side despite the existence of substantial moisture-content gradients in all three cases. Because a closed system was involved, any

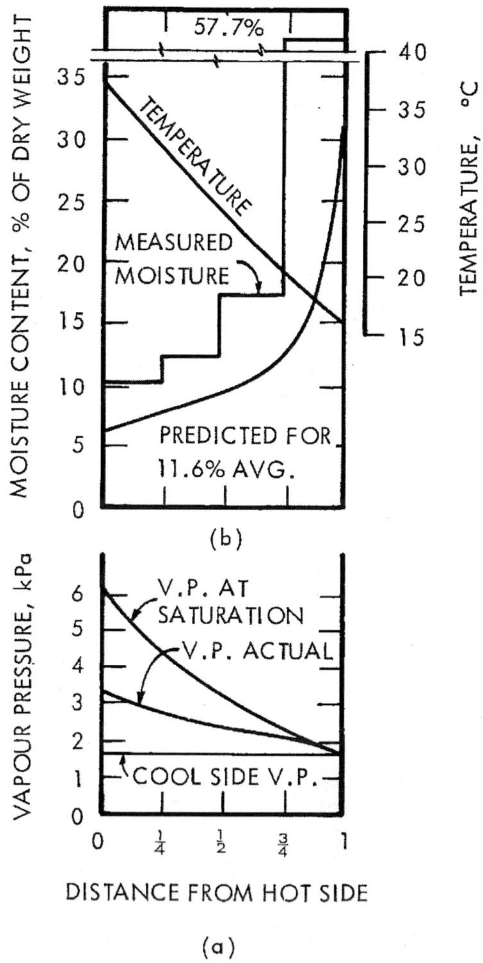

FIGURE 5.14 Moisture distribution in wet sawdust and calculated vapour gradient.
(From Paxton, J. A., and Hutcheon, N. B., *ASHVE Trans.*, 1952, *58*, Fig. 13(a) and (b), p. 316)[**5.25**]

flow across a transverse plane in the form of liquid must have been balanced by an equal and opposite flow of vapour; yet the evidence was in support of the assumption that the vapour pressure was constant so that no vapour flow could take place.

The introduction of higher amounts of water produced quite different results. Those for a moisture content of 24% are shown in Figure 5.14. Condensation at the cold face was expected; the average moisture content for the quarter-thickness was found to be 57.7%. The levels at all other points were also higher than those predicted from a cold-face vapour pressure equal to saturation at the cold-face temperature. When the calculation was made in reverse to find the vapour pressures corresponding to the measured moisture contents, the results shown in Figure 5.14(a) were obtained. The high moisture levels found in Figure 5.14(b) suggested a wicking of moisture from the cold side back to the warm side, but this could not occur without a corresponding vapour flow in the opposite direction. The calculations of vapour pressure based on measured moisture contents showed that such a vapour gradient did exist.

It can be recognized without further argument that with a given imposed vapour-pressure gradient the temperature gradient can be varied independently, giving rise to corresponding variations in relative humidity and thus in equilibrium moisture contents at various points along a flow path. There is reason to believe that different values of μ for use in Equation 5.7 may be found, depending on both temperature and vapour-pressure gradients. The evidence is that these complications are likely to be serious only under conditions close to saturation, at least for some materials. Although Equation 5.16 is scientifically more appropriate, its use may be of little practical value if two sets of coefficients varying with moisture content must be determined.

5.16 Combined Heat and Moisture Flow

The limitations in predicting moisture flow lead to corresponding difficulties in establishing the effects that the presence of moisture will have on heat flow. It has long been recognized that the effect of moisture is generally to increase the rate of heat flow compared with that in dry material under the same temperature gradient. It is unwise to assume, however, that new conductivities corresponding to, and depending only on, levels of moisture content can be assigned for use in conduction theory [**5.27**]. A completely saturated material is an exception because it is unlikely that liquid can circulate within the material. But when evaporation, vapour flow, and condensation within pores are possible, the added energy transfer due to moisture will be linked partly to vapour transfer and not directly and solely to moisture content.

Two sets of wet conductivities, one for organic and one for inorganic materials, were proposed many years ago. For the former the thermal conductivity is increased 1.25% for each percentage point increase in moisture, dry-weight basis. A series of values is to be applied in the case of inorganic materials for moisture content in percent by volume, as shown in Table 5.7.

TABLE 5.7

Increase in Thermal Conductivity with Moisture Content for Inorganic Materials

Moisture content, %	Increase in conductivity, %
1	32
2	48
3	75
4	108
5	144

These values were developed largely from the experiments of many researchers. They relate to one or more arbitrarily established test conditions and cannot rationally account for the range of possible moisture gradients that can exist. These or similar values have been widely used in Europe for want of a more rational method of calculating heat transfer through damp building constructions. The Canadian tests on moist sawdust [**5.25**] were not markedly in conflict with the 1.25% increase per percent of moisture proposed for organic materials, provided they were first allowed to adjust to steady-state conditions. The instantaneous heat flows following a sudden change of temperature while the moisture was changing from a uniform distribution to a new equilibrium condition were many times greater than the final steady-state values. Some similar experiments have been carried out in Ottawa on moist soils [**5.28**].

5.17 Effects of Moisture in Materials

Moisture content is usually determined by weighing before and after drying, the difference being taken as the moisture content. It must be asked whether the weight loss is loss of water or of some other material such as a volatile oil. If the drying is not severe enough, it may not remove all the water; complete drying may remove water which is chemically combined in the form of hydrates or in other ways so that it is really a part of the material under normal conditions. This difficulty has been encountered with portland cement paste, and it has been necessary to define a reference condition [**5.6**, **5.9**].

Drying is usually accompanied by shrinkage. Some materials are permanently changed by extreme drying and will show progressive damage under cycling conditions of wetting and drying, as evidenced by length changes, loss of strength, or loss of modulus of elasticity.

Length changes, whether due to moisture changes or factors other than loading, become important when they lead to discrepancies in dimensions of mating elements or are likely to produce rupture or distortion when constrained [**5.29**, **5.30**]. Dimensional changes in wood are commonly observed but are not always recognized as the inevitable consequence of moisture changes in the material [**5.10**, **5.11**].

Consideration of the strains in common structural materials at working-stress levels indicates that they are all about 0.001 or 0.1%, and for this reason they can be made to work in combination, as in the case of reinforcing steel in concrete. This value provides one practical reference point for judging the importance of changes due to moisture in materials that are to be combined to form building components. Concrete blocks, for example, can create difficulties in construction if they exhibit moisture movement from wet to dry greater than about 0.04%.

Many of the brittle construction materials, including most concrete products, are relatively weak in tension, showing tensile strengths about one-tenth of their compressive strengths. The tensile strains at which cracking may occur are correspondingly low, about 0.01%. As a consequence they often crack as a result of shrinkage; in many cases this must be regarded as normal.

For wood to meet Canadian lumber standards it must not have more than 19% moisture by weight. This limit is based partly on shrinkage and partly on the moisture level at which rotting can take place. Wood, when exposed to moisture and air, can support the growth of fungi which feed on it, resulting in rot [**5.31**]. The more common types of fungi will probably start to grow at moisture contents above 30% but, once started, can produce additional water from the breakdown of cellulose, the material on which they feed. Although wood will rot if subjected to conditions producing high moisture contents, it will remain sound almost indefinitely if it is kept reasonably dry or if it is immersed in water so that air is excluded.

Salts commonly appear on the exposed surfaces of brickwork. This is known as efflorescence, and is usually more unsightly than damaging [**5.32**]. Water repellents can be used to control it [**5.33**]. Salts dissolved in water in the pores of the brick are carried to the surface in solution, where they appear in powder form as the water evaporates. Under certain conditions the crystallization of salts may be damaging to porous materials; it is thought that water repellents may promote damage to the material if they prevent the movement of liquid to the surface and force evaporation and salt crystallization within the material. Efflorescence develops most extensively under moist cool conditions in late autumn and early spring, when moderate drying rates permit the moisture within the material to reach the surface before evaporating.

Corrosion of metals is also strongly moisture-dependent. It has been shown that atmospheric corrosion of exposed steel correlates with number of hours of wetness. Corrosion begins when the relative humidity exceeds about 75% at the surface [**5.34**].

Concrete and bricks in their newly manufactured state represent two extreme moisture conditions. More water is used in the making of concrete than is needed for chemical combination with the cement. For most uses where it is exposed to the atmosphere, the newly hardened concrete must lose substantial amounts of water and undergo appreciable initial shrinkage. Shrinkages from 0.3 to 0.95% on first drying have been reported [**5.35**] for a variety of cement pastes with and without additives. Bricks are made by high-temperature firing and, when taken from the kiln, contain various compounds which have not yet had the opportunity to take on water in equilibrium with the atmosphere. Initial moisture expansions of 0.03 to 0.20% (depending on raw materials and firing temperature) have been reported [**5.36**]. Much of this expansion takes place relatively quickly but can continue slowly for several years.

5.18 Freezing of Wet Materials

Damage to porous building materials can occur if they are exposed to freezing temperatures when wet. In extreme cases damage may occur with a single freezing, but more commonly the damage becomes noticeable only after a number of freeze–thaw cycles. Extensive investigations on this important subject have been carried on in Canada and many other countries. A comprehensive review paper [**5.37**] and a shorter summary paper [**5.38**] provide the basis for the discussion that follows. Experiments with materials such as porous glass and hardened cement paste have confirmed that only the water contained in large pores and cracks changes to ice, and that damage occurs only when the moisture content exceeds certain limits. These limits are related to the differences between the vapour pressures over water and over ice as given in Table 5.1 for various temperatures below 0°C.

The pressures given in Table 5.1 are called saturation pressures. They are the equilibrium vapour pressures developed over a flat surface of pure ice or of pure liquid water. A porous body brought to equilibrium with the saturation vapour pressure over a flat surface of pure water will be saturated, with all pores completely filled. As water does not freeze in the pores, the porous body must be exposed to saturation pressures over supercooled liquid in order to saturate it at temperatures below 0°C. This is not possible in practice because water outside the porous body will be frozen and capable only of generating the saturation vapour pressure over ice. For example, at −10°C the saturation pressure over ice is found from Table 5.1 to be 259.7 Pa, and over supercooled liquid 286.3 Pa. The moisture content in equilibrium with the saturation vapour pressure over ice will be that for a relative vapour pressure of 259.7/286.3 = 0.91. Menisci will form in the pores with a curvature appropriate to this relative pressure. Sorption isotherms drawn for various temperatures below freezing can be superimposed if they are plotted against relative vapour pressure with respect to liquid.

When salt solutions are involved, the situation can be much like that for ice. The vapour pressure over a salt solution outside the porous body is less than that for supercooled water and so cannot produce saturation. The situation is quite different if the porous body has a soluble salt in its pores, since it will then be fully saturated when in equilibrium with a vapour pressure equal to that over a salt solution of the same concentration as that in the pores.

A porous body saturated at some temperature above freezing and then cooled rapidly will be saturated, or nearly so, at the new freezing temperature. The vapour pressure inside the body will be that for supercooled liquid while that outside cannot exceed the saturation pressure over ice. The difference in thermodynamic potential as reflected in the difference in vapour pressures is available to promote the movement of water out of the sample in order to bring the pore system into equilibrium with its environment. With very slow cooling the imbalance may be minimized by the continuing rejection of water through migration to the surface.

Damage to the material of porous bodies is usually associated with temperature and moisture conditions that bring the material to subzero temperatures while it still contains an appreciable excess of water above the equilibrium moisture content over ice. Although this excess water does not form ice unless it moves to the surface of the material or to larger unfilled pores or cracks, there is evidence that with further cooling, water in the pores may change to a glassy noncrystalline state. Damage is most likely to occur in cases of high initial moisture, rapid freezing, and low permeability to water movement. The mechanisms leading to damage may include shrinkage and cracking due to desiccation from freezing, and the formation of ice in larger pores.

Air entrainment is effective in increasing the frost resistance of concrete because it forms uniformly spaced large pores into which water can pass without undue stress in the pore system. The degrading effect of de-icing salts on pavements is thought to arise because the salts lead to a higher degree of saturation at a given relative humidity, thus increasing the amount of water that must be rejected in order to maintain equilibrium as freezing takes place. Materials such as hardened cement paste with intermediate porosity are generally vulnerable, but concrete made with low water–cement ratios and with air entrainment is likely to be durable. Hard burned bricks having a predominance of large pores as reflected in surface areas of 0.8 m^2/g or less may be expected to be resistant to frost damage. Marble, which has low porosity, is generally durable. Surface coatings that effectively limit moisture contents will be beneficial, but they will be detrimental in applications where they serve mainly to block the escape of water that has entered elsewhere.

5.19 Water in the Ground

Two different conditions of moisture in the ground must be recognized: below the water table the condition will be saturation, and above the water table is normally one of partial saturation. Precipitation, flooding, and drainage are continually changing the surface and subsurface moisture conditions, and the water table itself may fluctuate with changing groundwater conditions. While the temperature of the ground at depths of 6 to 10 m remains at a value a few degrees above mean annual air temperature, the upper zone is affected by the daily and seasonal variations at the surface and varies widely in temperature, with frost penetration in winter up to 3 m in some regions of Canada. These variations in temperature and moisture give rise to many moisture-related effects of importance in building and construction.

Certain clays found in two regions of Canada exhibit marked shrinkage and swelling with changes in moisture content. Clays of the prairie region have been wetted and dried many times in their natural condition, and are of a type that shrinks on drying and then swells markedly on rewetting [**5.39**]. The Leda clay of eastern Canada shrinks markedly on initial drying but does not swell to the same degree on rewetting. When the groundwater is lowered by unusually dry weather, by underground construction, or by large trees [**5.40**], large settlements can take place.

Increases in moisture content of clay banks and slopes can lead to landslides [**5.41**]. Leda clay can, at critical high moisture contents, lose its structure suddenly and begin to flow like a viscous liquid.

Compressed sediments, or shales, are not free from difficulty. Some pyritic shales, when desaturated and exposed to air, begin to change chemically through a complex process involving bacterial action. The resulting end products bulk larger than the original shale, and heaving of buildings by 100 mm or more can take place [**5.42**].

Agricultural crops draw their water from the soil, which acts as a reservoir that is replenished periodically by precipitation or irrigation. The moisture content at which plants can no longer draw water freely from the soil can be defined in terms of suction or capillary tension, and tensiometers and moisture meters indicate when irrigation is needed.

5.20 Frost Heaving

The temperature gradients in the ground in winter cause some upward migration of water by vapour flow. When the ground freezes, another effect is introduced to produce upward migration. Ice crystals at a freezing plane can grow by drawing water from the unfrozen layers below. The water drawn to the ice interface releases its heat of fusion. When enough water is available, this heat of fusion can provide heat to balance the heat flow upward, thus preventing the further advance of the freezing plane until some change in conditions upsets the balance. Meanwhile, the growth of the ice lens, because of the migration and freezing of water at its lower surface, acts like an expanding pressure cell to lift the ice, soil, and any surface structure. A shortage of water will reduce the heat from freezing that is available to balance the upward heat loss, thus permitting the freezing plane to advance until conditions suitable for the formation of ice lenses are again encountered [**5.43**].

Frost heaving of this kind requires freezing conditions, water that can be drawn by suction to the freezing plane, and a fine-grained soil. When heaving occurs in fields it is seldom noticed, but when it occurs under roads and buildings or when it lifts posts or piles it can be very destructive. One of the most serious effects on roads and streets is caused by the upward movement of water and its storage in the form of ice in the frozen zone. In the spring, thawing from the surface releases this water, which adds greatly to the moisture content of the soil and reduces its bearing capacity. Heavy traffic may then rupture the pavement, which is no longer adequately supported. Posts and piles in the ground are lifted when the freezing soil bonds to the post. Once a sufficient bond has been developed, the post can be lifted along with the soil above the growing ice layers.

The foundations of heated buildings are not often affected by frost heaving. The heat from occupied structures usually prevents serious freezing adjacent to foundations, which in any event usually extend below the level of maximum undisturbed frost penetration. Adfreezing can occur with the walls of unheated basements; frost heaving can occur under unheated attached garages and under the entrances to basement garages [**5.44**]. Buildings are vulnerable during construction and at other times; if they are left unheated without protection, freezing may develop under basement floors and footings.

Substantial costs may be involved in dealing with frost action. Annual damage may be great if no steps are taken to prevent it, but the known preventive measures are also often costly: control of freezing by the use of insulation or (in special cases) by heating, limiting the groundwater available, or replacing soil with non-frost-susceptible material. There is a need for greater understanding of frost action to judge better the adequacy of preventive measures. The estimation of the heaving potential of the soil can be a difficult but highly important consideration in dcsign.

An understanding of frost action based on research done in Ottawa has been described [**5.45**]. A theoretical explanation has been given of the conditions of the ice–water–soil interface [**5.46**] as represented by the diagram in Figure 5.15. Others have contributed to the theoretical approach [**5.47, 5.48**].

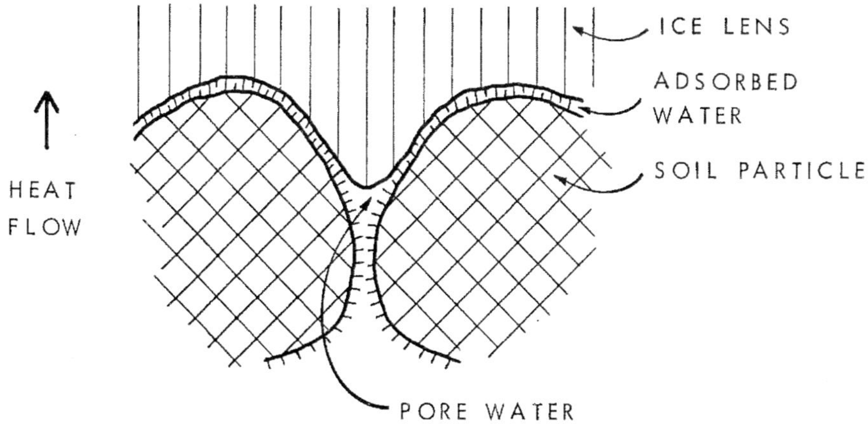

FIGURE 5.15 Schematic section showing ice advancing in a soil pore at the freezing point.
(From Penner, E., Highway Research Board, Special Report 40, Fig. 6, p. 196)(NRCC 5192)[**5.46**]

The essential features of the mechanism as understood at present are as follows. The attempted entry of the growing tongue of ice into a pore develops curvatures in the ice–water interface which bring into play interfacial tension forces. The net result as freezing progresses is a lowering of the capillary potential in the fine pores and a depression of the freezing point of the sharply curved front of the ice tongue. Water drawn to the pore by suction is able to move to the ice surface of lower curvature above the soil particle where it can freeze. The balance of forces is such that normally both a suction and a heaving pressure can be maintained. It has been shown theoretically and experimentally that heaving can be stopped by the development of a limiting suction or by a limiting value of soil loading for the particular conditions in a given case.

An equation has been produced [**5.47**] for the heaving pressure of a porous material consisting of uniformly sized spherical particles. The work in Ottawa has provided experimental evidence in support of the equation [**5.45**] as shown in Figure 5.16. These experiments also showed that although the fine particles determine the maximum heaving pressure, it is not necessary that they predominate. In dealing with actual soils, there may be difficulty in judging how the size and proportion of fine particles in relation to a given particle-size distribution for the soil as a whole will affect the result. The criterion suggested by Casagrande [**5.49**] in 1932 is still useful.

> *Under natural conditions and with sufficient water supply one should expect considerable ice segregation in uniform soils containing more than 3 per cent of grains smaller than 0.02 mm and in very uniform soils containing more than 10 per cent smaller than 0.02 mm. No ice segration was found in soils containing less than 1 per cent of grains smaller than 0.02 mm even if the ground-water level was as high as the frost line.*

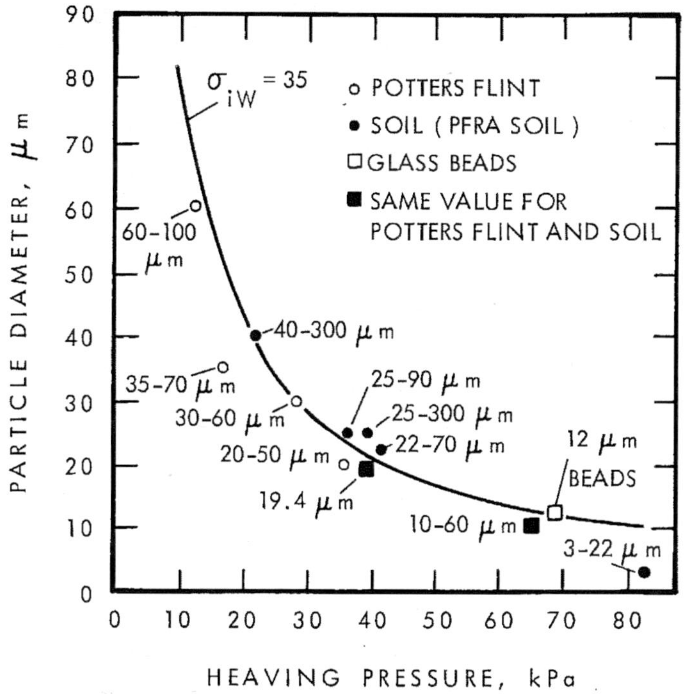

FIGURE 5.16 Relation between heaving pressure and smallest particle in fraction.
(From Penner, E., OECD Symposium, Paris 1973, Vol. 1, Fig. 1, p. 383)(NRCC 13674)[**5.45**]

It will be evident that a basic difficulty arises because it is the pores that determine the heaving potential; but these can only be determined indirectly in terms of the measurable particle sizes. It seems reasonable to expect that a small percentage of fine particles fitted into the spaces between larger particles can have a marked influence on the proportion of fine pore structure.

Combinations of suction and loading that would stop heaving have been reported [**5.46**] for a granular system of powdered quartz known as Potter's flint. The grain-size distribution varied from 0.001 mm (4%) to 0.04 mm. Changing the density by compaction was found to have a marked influence (Figure 5.17). For the intermediate density of 1.639 g/cm^3, heaving stopped when the soil-moisture tension reached 50 kPa without overburden pressure. When the water supply was unrestricted, the heaving pressure reached almost 100 kPa before heaving stopped. The marked influence of compaction is a measure of its effect on the development of pore structure, which is critical for heaving.

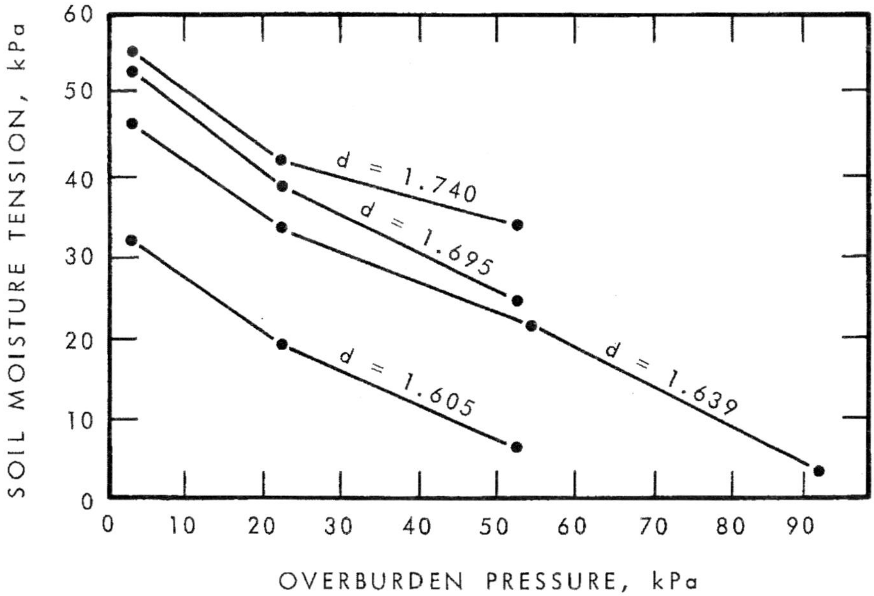

FIGURE 5.17 Relationship between soil moisture tension and over-burden pressure in Potter's flint when heaving ceased.
(From Penner, E., Highway Research Board, Special Report 40, Fig. 5, p. 195)(NRCC 5192)[**5.46**]

The pore structure essential for heaving need only exist at the freezing plane. This has very serious implications in construction, since all the soil likely to be swept by the advancing frost line must be non-frost-susceptible if all possibility of heaving is to be avoided (unless, of course, other measures are being taken). Some interesting results have been obtained in a laboratory test cell [**5.50**] showing that in a layered soil the limiting suction developed at the time heaving stopped was determined by the soil in which ice lensing was occurring; this suction was imposed on all layers below, whose water contents were reduced in accord with their respective suction/moisture content curves. Thus, a sand layer developed a pF of 2.13 (28% m.c.) when it was tested alone, but was drawn down to a pF of 3.50 (11.2% m.c.) when used as the lower, nonfrozen half of a composite sample. The upper half, in which the freezing plane was located, showed the same pF when the limiting suction was reached as when tested by itself. The pF-versus-moisture-content curves for these and other soils tested are given in

Figure 5.18. Samples 1, 2, 3, 4, and 6 were fine-grained. Sample 5 was a sand passing a 325 sieve, and sample 7 a sand passing 200 and retained on 325.

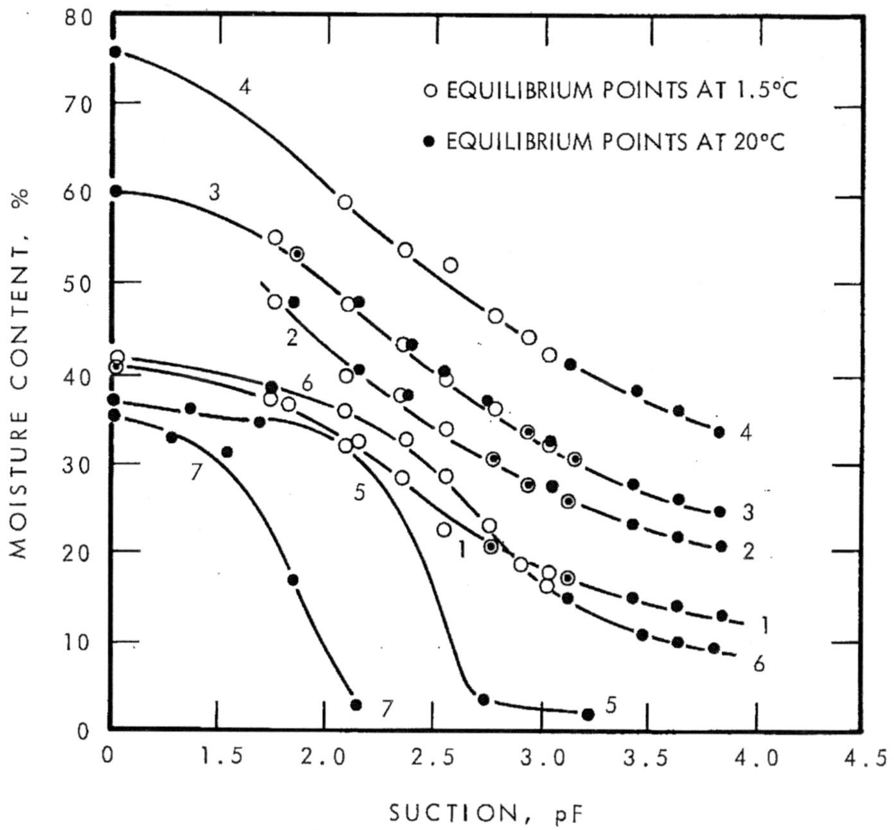

FIGURE 5.18 The pF-moisture content relationship for several soils. (From Penner, E., Highway Research Board, Bulletin 168, Fig. 4, p. 56)(NRCC 4738)[**5.50**]

A heaving pressure of 100 kPa is a stress commonly exceeded in many foundations. Thus, foundation loads would suppress heaving in many situations where freezing is allowed to occur under footings. However, it was also found that heaving pressures could be as high as 1400 kPa in clay, and these would not be overridden by normal foundation loads.

5.21 Heaving Pressures for Rigid Materials

The heaving-pressure apparatus developed for soils was also used to measure the heaving pressures developed by some cut-brick, stone, and mortar surfaces [**5.51**]. The maximum heaving pressures were from 90 kPa to 180 kPa for the bricks. The value found for the sandstone was 350 kPa, and for cement mortar

345 kPa. These values provide some evidence that the kinds of pressures developed in heaving soils may not be significant in porous materials exhibiting tensile strengths of about 1400 kPa, except perhaps in the case of weak mortars. A spalling type of fracture suggestive of ice lensing occurs in bricks and mortar under freeze–thaw conditions, and ice lenses develop. It may be that such fractures result from progressive damage caused by repeated cycles of freezing and thawing for the reasons discussed earlier in connection with rigid materials. It is possible that with further research the link between heaving in soils and frost damage in rigid materials will be understood. They are, after all, both related to the freezing of water in porous systems.

5.22 Conclusion

Much is known about the effects of water in materials. Extremes of wetting and drying and even moderate changes in moisture content can be avoided through proper design in order to ensure good service from the materials available. To accomplish this it is necessary to understand when and how much water will be present. It is necessary to be able to recognize and establish the potentials for moisture content and moisture flow, and to measure porosities and the coefficients describing the related permeabilities.

The movement of water in materials and some of its more important effects are closely related to temperature, temperature gradients, and heat flow. The imposition of a temperature gradient changes the isothermal relationship between vapour pressure and capillary potential, and makes the whole subject of combined heat and moisture flow a complex one. Temperatures below freezing introduce other considerations involving frost action in its several forms in both granular and solid materials.

The matters relating to water that have been discussed in this chapter provide important and useful background knowledge for the further study of building materials. In order to deal further with water, it is now necessary to probe the interrelated subjects of temperature and heat flow in building materials and components. The measurement and calculation of the properties of air–vapour mixtures merit discussion (Chapter 6). Finally, the possibilities and problems involving moisture must be considered along with those of air, heat, and radiation in relation to the characteristics of buildings and their use.

References

5.1 Miller, F., Jr. *College physics*. 2d Ed. New York: Harcourt Brace and World, 1972.

5.2 Vallentyne, J.R. *The algal bowl*. Department of the Environment. Ottawa: Information Canada. 1974.

5.3 Kohnke, H. *Soil physics*. New York: McGraw-Hill, 1968.

5.4 Byers, H.R. *General meteorology*. 4th Ed. New York: McGraw-Hill, 1974.

5.5 *ASHRAE Handbook 1981 Fundamentals*. Chaps. 5, 6, and 21.

5.6 Feldman, R.F., and Sereda, P.J. Moisture content—Its significance and interaction in a porous body. Vol. 4, *Proceedings, International Symposium on Humidity and Moisture*, Washington. 1963, Chap. 28, pp. 233-43. (NRCC 8600).

5.7 Sereda, P.J. *The structure of porous building materials.* Canadian Building Digest 127. DBR/NRCC, July 1970.

5.8 Sereda, P.J., and Feldman, R.F. *Wetting and drying of porous materials.* Canadian Building Digest 130. DBR/NRCC, Oct. 1970.

5.9 Feldman, R.F., and Sereda, P.J. A new model for hydrated portland cement and its practical implications. *Engineering Journal*, 1970, *53* (8/9), pp. 53-59. (NRCC 11604)

5.10 Hutcheon, N.B., and Jenkins, J.H. *Some basic characteristics of wood.* Canadian Building Digest 85. DBR/NRCC, Jan. 1967.

5.11 Hutcheon, N.B., and Jenkins, J.H. *Some implications of the properties of wood.* Canadian Building Digest 86. DBR/NRCC, Feb. 1967.

5.12 Penner, E. Suction and its use as a measure of moisture contents and potentials in porous materials. Vol. 4, *Proceedings, International Symposium on Humidity and Moisture.* Washington, 1963. Chap. 29, pp. 245-52. (NRCC 8601)

5.13 Schofield, R.K. The pF of water in the soil. *Transactions, Third International Congress on Soil Science*, 1935, *2*, pp. 37-48.

5.14 Williams, G.P. *Chemicals for snow and ice control around buildings.* Canadian Building Digest 191. DBR/NRCC, Oct. 1977.

5.15 Hedlin, C.P., and Trofimenkoff, F.N. Relative humidities over saturated solutions of nine salts in the temperature range from 0 to 90°F. Vol. 3, *Proceedings, International Symposium on Humidity and Moisture.* Washington, 1963. Chap. 31, pp. 519-20. (NRCC 8602)

5.16 Bomberg, Mark. *Moisture flow through porous building materials.* Report 52. Lund, Sweden: Lund Institute of Technology, 1974.

5.17 Litvan, G.G. *Testing the frost susceptibility of bricks.* Special Technical Publication 589, American Society for Testing and Materials, 1975, pp. 123-32. (NRCC 14797)

5.18 Joy, F.A., and Wilson, A.G. Standardization of the dish method for measuring water vapour transmission. Vol. 4, *Proceedings, International Symposium on Humidity and Moisture.* Washington, 1963. Chap. 31, pp. 259-70. (NRCC 8838)

5.19 *Standard test methods for water vapour transmission of materials.* ASTM E 96-80. Philadelphia: American Society for Testing and Materials, 1980.

5.20 *Vapour barrier, sheet, for use in above-grade building construction.* National Standard of Canada, CAN2-51.33-M80. Ottawa: Canadian General Standards Board, 1980.

5.21 Penner, E. Soil moisture suction—Its importance and measurement. *Proceedings, American Society of Testing and Materials*, 1958, *58*, pp. 1205-17. (NRCC 4994)

5.22 Krischer, O., Wissman W., and Kast, W. Feuchtigkeitseinwirkungen auf Baustoffe aus der umgebenden Luft. *Gesundheits-Ingenieur*, 1958, *79* (5), pp. 129-48.

5.23 Gurr, C.G., Marshall, T.J., and Hutton, J.T. Movement of water in soil due to a temperature gradient. *Soil Science*, 1952: *74* (5), Nov. 1952, pp. 335-45.

5.24 Johansson, C.H. Fuktgenomgång och Fuktfördelning i Byggnadsmaterial. (Moisture transmission and moisture distribution in building materials.) *VVS: Tidskrift för Värme-ventilations-sanitets och Kylteknik*, 1948, *19* (5), pp. 67-75. (NRCC Technical Translation 189)

5.25 Paxton, J.A., and Hutcheon, N.B. Moisture migration in a closed guarded hot plate. *ASHVE Transactions*, 1952, *58*, pp. 301-20.

5.26 Kuzmak, J.M., and Sereda, P.J. The mechanism by which water moves through a porous material subjected to a temperature gradient: I, Introduction of a vapour gap into a saturated system. *Soil Science*, 1957, *84* (4), pp. 291-99. (NRCC 4491)

5.27 Influence of moisture on thermal properties of materials, building products and building elements. Papers 4.0–4.11. *RILEM/CIB Symposium on Moisture Problems in Buildings*, Helsinki, 1965.

5.28 Woodside, W., and de Bruyn, C.M.A. Heat transfer in a moist clay. *Soil Science*, 1959, *87*, (3), pp. 166-73. (NRCC 5098)

5.29 Baker, M.C. *Thermal and moisture deformations in building materials*. Canadian Building Digest 56. DBR/NRCC, Aug. 1964.

5.30 *Effects of deformations in building components*. Commentary D, Chap. 4, Commentaries on Part 4 of the National Building Code of Canada 1980. The Supplement to the National Building Code of Canada 1980, pp. 178-80. Ottawa: National Research Council of Canada. (NRCC 17724)

5.31 Baker, M.C. *Decay of wood*. Canadian Building Digest 111. DBR/NRCC, March 1969.

5.32 Ritchie, T. *Efflorescence*. Canadian Building Digest 2. DBR/NRCC, Feb. 1960.

5.33 Ritchie, T. *Silicone water repellents for masonry*. Canadian Building Digest 162. DBR/NRCC, 1974.

5.34 Sereda, P.J. *Atmospheric corrosion of metals*. Canadian Building Digest 170. DBR/NRCC, 1975.

5.35 Feldman, R.F., and Swenson, E.G. Volume change on first drying of hydrated portland cement with and without admixtures. *Cement and Concrete Research*, 1975, *5* (1), pp. 25-35. (NRCC 14409)

5.36 Ritchie, T. *Moisture expansion of clay bricks and brickwork*. Building Research Note 103. DBR/NRCC, Oct. 1975.

5.37 Litvan, G.G. Adsorption systems at temperatures below the freezing point. *Advances in Colloid and Interface Science*, 1978, *9*, pp. 253-302. (NRCC 16810)

5.38 Litvan, G.G. Freeze–thaw durability of porous building materials. *Proceedings, First International Conference on Durability of Building Materials and Components*. Special Technical Publication 691, American Society for Testing and Materials, 1980, pp. 455-63. (NRCC 18638)

5.39 Hamilton, J.J. *Swelling and shrinking subsoils*. Canadian Building Digest 84. DBR/NRCC, Dec. 1966.

5.40 Legget, R.F., and Crawford, C.B. *Trees and buildings*. Canadian Building Digest 62. DBR/NRCC, Feb. 1965.

5.41 Eden, W.J. *Landslides in clays*. Canadian Building Digest 143. DBR/NRCC, Nov. 1971.

5.42 Penner, E., Eden, W.J., and Grattan-Bellew, P.E. *Expansion of pyritic shales*. Canadian Building Digest 152. DBR/NRCC, July 1973.

5.43 Penner, E. *Ground freezing and frost heaving*. Canadian Building Digest 26. DBR/NRCC, Feb. 1962.

5.44 Penner, E., and Burn, K.N. *Adfreezing and frost heaving of foundations*. Canadian Building Digest 128. DBR/NRCC, Aug. 1970.

5.45 Penner, E. Frost heaving pressures in particulate materials. Vol. 1, *OECD Symposium on Frost Action on Roads*, Paris, 1973, pp. 379-85. (NRCC 13674)

5.46 Penner, E. *Pressures developed in a porous granular system as a result of ice segregation*. Special Report No. 40. Washington: Highway Research Board, 1959. (NRCC 5192)

5.47 Everett, D.H., and Haynes, J.M. Capillary properties of some model pore systems with special reference to frost damage. *RILEM Bulletin*, No. 27, 1965, pp. 31-38.

5.48 Gold, L.W. *A possible force mechanism associated with the freezing of water in porous materials*. Bulletin 168. Washington: Highway Research Board, 1958, pp. 65-73. (NRCC 4737)

5.49 Casagrande, A. Discussion on frost heaving. *Proceedings, Highway Research Board*, 1932 *11* (1), 168-72.

5.50 Penner, E. *Soil moisture tension and ice segregation*. Bulletin 168. Washington: Highway Research Board, 1958, pp. 50-64. (NRCC 4738)

5.51 Penner, E. Pressures developed during the unidirectional freezing of water-saturated porous materials. Vol. 1, *Proceedings, International Conference on Low Temperature Science*, Sapporo, Japan, 1966. Part 2, 1967, pp. 1401-12. (NRCC 10383)

6

Air and Its Water Content

6.1 Introduction

The unusual importance of air as an intermediary in the exchange and transportation of heat, moisture, odours, and particulate matter in the indoor environment has been identified in previous chapters. The discussion in Chapter 5 has made it clear that the water-vapour content, though small in relation to the dry air, has a great influence on many matters, including several aspects of the thermal environment. The influence of the air and its water-vapour content is so great that air conditioning—the control of temperature, relative humidity, movement, and quality of the air in occupied spaces—has virtually become synonymous with control of the thermal environment. This means that the properties of air–vapour mixtures and the ability to calculate and to predict them are important in many aspects of building.

The relative humidity in the ambient air determines the equilibrium moisture content of materials; the vapour pressure determines the movement of vapour. These factors alone would justify a substantial interest in the moisture condition of the air. But there is a second important aspect: when the air is being conditioned, the water content may change from vapour to liquid to ice, or from ice to liquid to vapour. These changes have practical implications for handling water in its several forms, and they contribute greatly to the amount of heat that must be added or removed to bring about a given change in air temperature. Since these contributions are often of the same magnitude as those involved in changing the temperature of the dry air alone by the same amount, they cannot be disregarded.

It is difficult to avoid reference to moisture as part of the air, since it is thoroughly dispersed in the air mixture. The water vapour moves with the air gases under total pressure differences and experiences sensible heat changes along with them as the air is heated or cooled. It has already been shown in Chapter 5, however, that the water vapour can in many cases vary in concentration from point to point and can respond to potential differences substantially as if no dry air were present. It will generally be apparent from the context when water vapour must be treated on its own.

6.2 Vapour Pressure, Humidity, and Other Properties

The basic concept of equilibrium vapour pressures over a flat liquid-water surface has already been developed in chapters 3 and 5. Tables 5.1 and 5.2 show values of saturation pressures at various temperatures. Relative vapour pressure and relative humidity (rh) have been defined as the ratio of the actual partial pressure of water vapour to the value of the vapour pressure at saturation over a flat water surface at the same temperature. The calculation of some basic properties based on the assumption that the ideal gas laws and Dalton's law of partial pressures apply has already been described in chapters 3 and 5. These approaches neglect the intermolecular influences on equilibrium vapour pressures and any departures from the gas laws. There are two approaches to the thermodynamic properties of moist air: one is approximate but sufficient for many practical purposes, and the other is more rigorous and precise. The approximate method will be followed here; the more rigorous one, leading to some differences in definitions and in tabular values of properties and methods of calculation, is well documented elsewhere [**6.1, chaps. 5 and 6**].

The relationships already developed can be reviewed briefly as follows:

$$\text{For dry air } p_a V = n_a \mathrm{R} T, \text{ and}$$
$$\text{for water vapour } p_w V = n_w \mathrm{R} T$$

where

p_a = the partial pressure of the dry air
p_w = the partial pressure of the water vapour
V = the total mixture volume
R = the gas constant, and
n_a and n_w are moles of air and water respectively.

It was shown in Chapter 3 that when p is in pascals, V in cubic metres, and T in kelvins, the gas constant R has the value 8.314 J/(mol • K).

The equation of state can also be written:

$$pV = w \mathrm{R_g} T \tag{6.1}$$

where w now refers to mass of substance and $\mathrm{R_g}$ refers to unit mass of a particular substance and not to one mole, as in the discussion above.

The value of $\mathrm{R_g}$ for air is

$$\mathrm{R_a} = \frac{8.314}{0.028\,96} = 287.1 \text{ J/(kg} \cdot \text{K)}$$

and the value of $\mathrm{R_g}$ for water vapour is

$$\mathrm{R_w} = \frac{8.314}{0.018\,02} = 461.5 \text{ J/(kg} \cdot \text{K)}$$

and w in Equation 6.1 is the mass of gas in kilograms and 0.028 96 and 0.018 02 are the kilograms per mole of air and water vapour respectively.

The relative humidity as previously defined is given by

$$rh = \frac{p_w}{p_{ws}} \times 100 \tag{6.2}$$

where p_w is the actual partial pressure of the water vapour, and p_{ws} is the vapour pressure at saturation. (These were designated p and p_o respectively in previous discussions, following scientific usage.) Total or atmospheric pressure will be denoted by p_t.

The humidity ratio is a fundamental quantity defined as the ratio of the mass of water vapour to the mass of dry air contained in the sample:

$$W = \frac{w_w}{w_a}. \tag{6.3}$$

Applying Equation 6.1 to V cubic metres of moist air,

$$w_a = \frac{p_a V}{R_a T} \tag{6.4}$$

and

$$w_w = \frac{p_w V}{R_w T}. \tag{6.5}$$

Dividing Equation 6.4 by Equation 6.5,

$$\frac{w_w}{w_a} = \frac{p_w V}{R_w T} \times \frac{R_a T}{p_a V} = \frac{p_w}{p_a} \times \frac{R_a}{R_w}$$

but

$$R_a = \frac{R}{M_a} \text{ and } R_w = \frac{R}{M_w}$$

so

$$\frac{R_a}{R_w} = \frac{M_w}{M_a} = \frac{0.018\,02}{0.028\,96} = 0.622$$

and

$$W = \frac{w_w}{w_a} = \frac{p_w}{p_a} \times 0.622 = \frac{0.622\,p_w}{p_t - p_w}. \tag{6.6}$$

The mass density of water vapour is given by the value of w_w from Equation 6.5 for $V = 1\ m^3$ and, correspondingly, for dry air by Equation 6.4. The density of the mixture is given by the sum of the densities of air and water vapour.

The specific volume, v, of an air–vapour mixture is expressed not for unit mass of moist air, but for unit mass of dry air. Thus, from Equation 6.4,

$$v = \frac{R_a T}{p_a} = \frac{R_a T}{p_t - p_w}. \tag{6.7}$$

6.3 Enthalpy of Air–Vapour Mixtures

Enthalpy was formerly called total heat, or heat content. The enthalpy of moist air is given by the sensible heat of the air with reference to 0°C plus the enthalpy of the water vapour, which includes the heat of the liquid h_f, referred to 0°C, the heat of evaporation h_{fg}, and heat of superheat, if any, for unit mass of dry air. (See Table 5.2 for the values of the thermodynamic properties of saturated steam.)

When an air–vapour mixture is saturated at some temperature t', the enthalpy of the mixture per kilogram of dry air will be

$$h' = c_{pa}t' + W'(h_f' + h_{fg}') \tag{6.8}$$

where c_{pa} is the specific heat of the air, W' is the humidity ratio, h_f' is the enthalpy of the liquid, and h_{fg}' is the enthalpy of evaporation.

If this saturated mixture is now heated to some new temperature, t, it will no longer be saturated since the water required for saturation at the new temperature will be greater than that held in the mixture. The vapour will now be superheated and the enthalpy at the new condition will be

$$h = c_{pa}t + W[(h_f' + h_{fg}') + c_{ps}(t - t')] \tag{6.9}$$

where c_{ps} is the specific heat of water vapour, h is the enthalpy for 1 kg of dry air, and W is the mass of water vapour associated with 1 kg of dry air, which is the humidity ratio. W in this case will have the same value as W' in Equation 6.8, since the humidity ratio was not changed by heating. The expression inside the square brackets is for the enthalpy of 1 kg of water vapour saturated at t' and then heated to temperature t.

The enthalpy of saturated vapour by itself can be found from tables such as Table 5.2, but is also given with sufficient accuracy for present purposes by the empirical expression

$$h_g = 2500 + 1.86\,t. \qquad \text{(kJ/kg)} \tag{6.10}$$

The specific heat of water vapour is about 1.88 so that the expression for enthalpy of water vapour superheated to temperature t, given by the expression in the square brackets of Equation 6.9, becomes

$$h_g = (2500 + 1.86\,t') + 1.88(t - t') \tag{6.11}$$

and, since 1.86 t' is almost the same as 1.88 t', the expression could have been written

$$h_g = 2500 + 1.86\,t$$

which is the value given by Equation 6.10 for the enthalpy of vapour at temperature t. In other words, the enthalpy per kilogram of superheated vapour is almost the same as for saturated vapour at the same temperature and is given approximately by Equation 6.10.

The enthalpy of 1 kg of dry air with its associated water vapour, from Equation 6.9, now reduces to

$$h = c_{pa}\, t + W h_g \tag{6.12}$$

and, on substituting 1.005 for c_{pa} and the value of h_g from Equation 6.10, becomes

$$h = 1.005\, t + W(2500 + 1.86\, t). \qquad \text{(kJ/kg)} \tag{6.13}$$

Example 6.1

Given $p_t = 101\,300$ Pa, $t = 21$°C and rh = 45%, find the various properties of the air-vapour mixture.

From Table 5.1:
At 21°C $p_{ws} = 2486$ Pa, so $p_w = 0.45 \times 2486 = 1119$ Pa

$$W = \frac{0.622 \times 1119}{101\,300 - 1119} = 0.006\,95 \text{ kg water vapour per kg dry air}$$

$$\text{Density of air} = \frac{p_a}{R_a T} = \frac{101\,300 - 1119}{287.1(273 + 21)} = 1.187 \text{ kg/m}^3$$

$$\text{Density of water vapour} = \frac{p_w}{R_w T} = \frac{1119}{461.5(273 + 21)} = 0.0082 \text{ kg/m}^3$$

$$\text{Density of mixture} = 1.187 + 0.0082 = 1.195 \text{ kg/m}^3$$

$$\text{Specific volume, } v = \frac{1}{1.187} = 0.842 \text{ m}^3\text{/kg dry air}$$

$$\begin{aligned}\text{Enthalpy, } h &= 1.005 \times 21 + 0.006\,95(2500 + 1.86 \times 21)\\ &= 21.11 + 17.64\\ &= 38.75 \text{ kJ/kg.}\end{aligned}$$

6.4 The Psychrometric Chart

The values calculated in Example 6.1 and others can be found using a psychrometric chart such as that in Figure 6.1. The chart's one limitation is that it must be constructed for one atmospheric pressure; it gives results that are increasingly in error as the total pressure deviates from that for which the chart was constructed [**6.2**]. The principal change in pressure to be expected is that resulting from altitude above sea level. The chart shown is for the standard atmosphere of 101 325 Pa and will not be seriously in error for most purposes until the altitude exceeds 300 m above sea level. Charts for other altitudes are available.

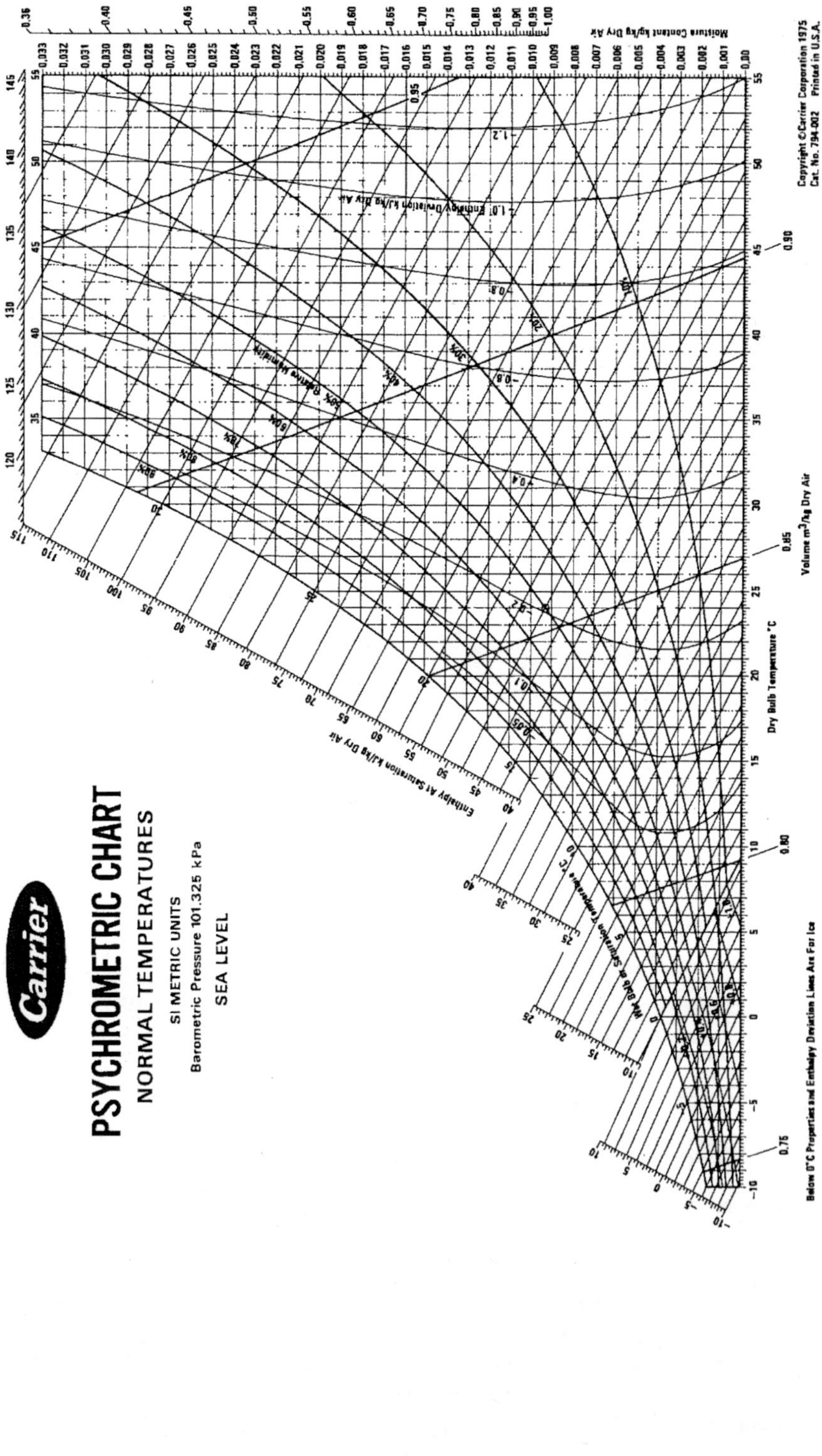

FIGURE 6.1 Psychrometric chart. (Courtesy Carrier Corporation)[6.2]

Note the basic construction of the chart, which has a horizontal scale of dry-bulb temperature and a vertical scale of humidity ratio. Against these two scales, the saturation line for 100% rh is plotted, curving from lower left to upper right. The intermediate curves for various relative humidities have been drawn, and inclined volume lines added. Note the wet-bulb temperature lines, which are approximately lines of equal enthalpy, or constant total heat, except for a small deviation (to be discussed later).

The psychrometric chart provides a complete plot of all the properties of air–vapour mixtures as previously found from equations. It has many uses in air-conditioning and other calculations, but initially it may be regarded as a quick and convenient way to find and to relate all the properties for a given moist air condition or "state point". When values of any two of the properties are known, the state point is found at the intersection of the lines defining these values, as shown in Figure 6.2. Values for all other properties can then be found.

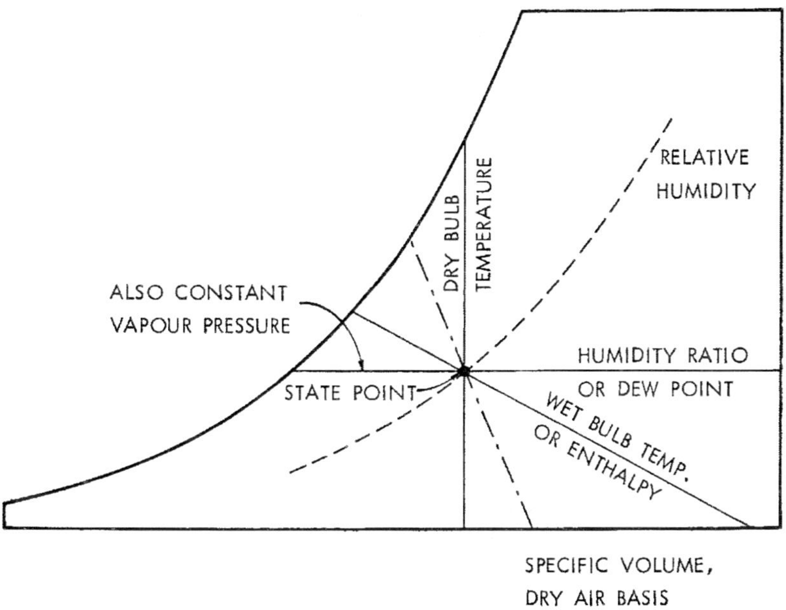

FIGURE 6.2 Outline of the psychrometric chart showing a state point and the lines defining the condition of the air–vapour mixture. Note that any two lines define the state point.

Example 6.2

Use the chart to find the quantities previously calculated in Example 6.1, as follows:

Locate the state point at the intersection of the lines for 21°C and 45% rh. Read horizontally to the right to find $W = 0.007$. Read from the state point relative to the volume lines the volume $v = 0.84$.

To find enthalpy, read from the state point along the inclined wet-bulb temperature lines to the enthalpy scale and find $h' = 38.7$. Read from the state point relative to the curved enthalpy deviation lines and find a value of -0.17 which, when added to h', gives the value of $h = 38.5$ kJ/kg.

The values found by calculation in Example 6.1 agree with those found from the chart. The value of the psychrometric chart lies in its ability to give values of properties for particular state points and a quick means of visualizing what is happening during heating, cooling, and other processes.

6.5 Heating and Cooling Processes

The horizontal lines in the psychrometric chart are lines of constant humidity ratio. Since the humidity ratio is given by $0.622\, p_w/(p_t - p_w)$ and total pressure p_t is constant, p_w must also be constant. If an air–vapour mixture is heated or cooled at constant total pressure, the humidity ratio and the partial pressure of the water vapour remain constant. The changing conditions can be depicted by successive points on the horizontal line in the chart between the initial and final state points.

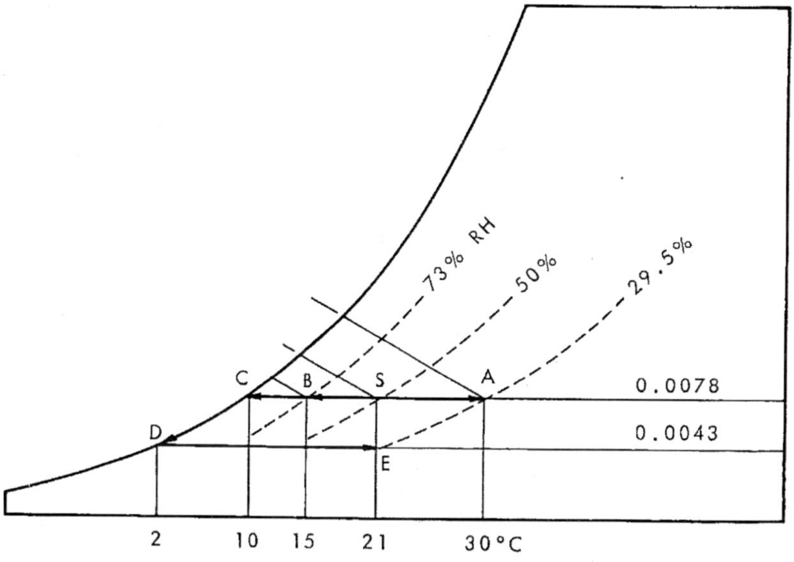

FIGURE 6.3 Outline of the psychrometric chart showing the heating and cooling processes of Examples 6.3 and 6.5 from original state point S for 21°C and 50% relative humidity. SA = heating, SB = cooling, SC = cooling to saturation, SCD = cooling and condensing, DE = heating to original temperature after condensing.

Example 6.3 (Figure 6.3)

Find the condition of an air–vapour mixture originally at 21°C and 50% rh when heated to 30°C.

From the chart, follow the horizontal line from 21°C and 50% rh to the intersection with 30°C and read rh = 29.5%.

Proceed in the opposite direction to find that on cooling to 15°C the rh is 73%.

Relative humidity drops with heating and increases with cooling. Although the explanation is simple, the implications for building are important. Both the vapour pressure and the vapour content required for saturation change with the changes in temperature produced by heating or cooling; but the actual partial pressure remains constant, so that the relative vapour pressure, or relative humidity, changes.

When, upon cooling, the saturation curve, or 100% rh line, is reached, the contained moisture is just sufficient for saturation. The temperature at which this occurs is known as the dewpoint temperature. A state point on the chart which has been defined by dry-bulb temperature and relative humidity can also be defined uniquely in terms of dewpoint and dry-bulb temperatures, or dewpoint and relative humidity. It will be seen that dewpoint temperature, water-vapour pressure, and humidity ratio define the same horizontal line. They cannot define the state point. Another intersecting line is required to locate it.

Example 6.4 (Figure 6.3)

Find from the chart the dewpoint temperature for the state point defined by 21°C and 50% rh.

Reading horizontally to the left to the saturation line, find the dry-bulb temperature of 10°C. This can also be found readily by calculation. The partial pressure of water vapour at 21°C is 2486 Pa × 0.50 = 1243 Pa. The temperature with a value of p_{ws} = 1243 is found by interpolation from Table 5.1 to be 10.2°C.

Further cooling below the dewpoint results in dehumidification, since the air can no longer hold the original moisture at these lower temperatures.

Example 6.5 (Figure 6.3)

Find the amount of water removed by cooling the air–vapour mixture with 50% rh at 21°C to saturation at 2°C.

From the chart, the original humidity ratio is 0.0078. The humidity ratio at the new condition is that for saturation at 2°C, found to be 0.0043 kg/kg of dry air. The water removed is 0.0078 − 0.0043 = 0.0035 kg/kg dry air. (This can also be found by calculation.)

If now this reduced mixture is rewarmed to its original temperature, it will have a reduced rh of 30%.

It should be noted that a cold surface such as that of a window or a wall in winter may dehumidify by condensation from the moist air in contact with it, whether or not this is desired, and may set a limit on the relative humidity or the vapour pressure that can be maintained in the space.

Using the unit mass of dry air as the reference base for calculating changes in heat and moisture is convenient since it does not change during a process, whereas the mass of the mixture may.

6.6 Mixing

When two masses of moist air are mixed, their heat and moisture contents are merged in the resulting total mass. The new enthalpies and humidity ratios can be found as follows:

$$(w_1 + w_2)h_3 = w_1 h_1 + w_2 h_2 \tag{6.14}$$

and

$$(w_1 + w_2)W_3 = w_1 W_1 + w_2 W_2 \tag{6.15}$$

where w_1 and w_2 are the masses at initial conditions 1 and 2, the final condition being designated 3.

Some thought will show that, since the scales on the chart for moisture and enthalpy are uniform and since the final enthalpy and moisture content following mixing are the weighted average of the values for the initial conditions, the state point for the mixture will be found on the straight line joining the initial state points and at a position dividing the distance between them in inverse proportion to the masses at each condition.

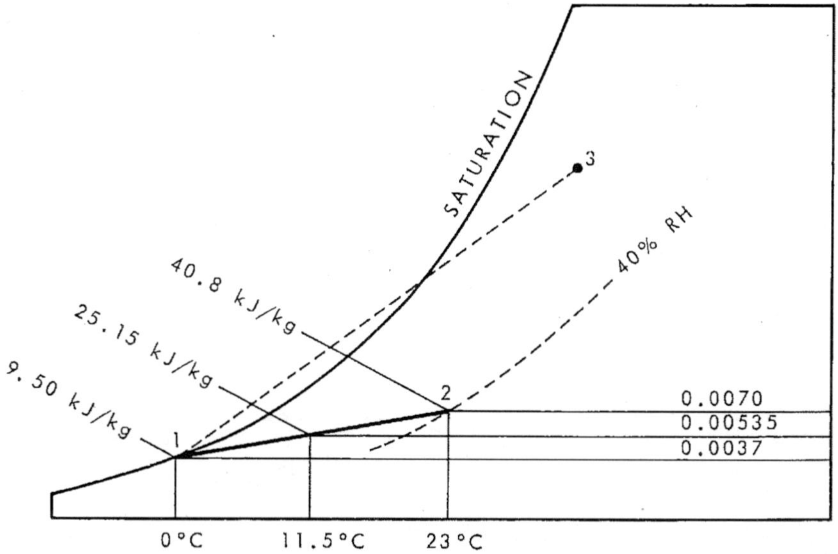

FIGURE 6.4 Outline of the psychrometric chart showing the mixing process of Example 6.6 involving equal quantities of air at conditions 1 and 2. Mixing of conditions 1 and 3 will produce fogging.

Example 6.6 (Figure 6.4)

Air drawn from a room maintained at 23°C and 40% rh is to be mixed with an equal quantity of cold outdoor air at 0°C and 100% rh, the total quantity then being returned to the room for purposes of ventilation. Find the condition of the mixture.

Since it is not otherwise specified, equal quantities can be taken to mean equal masses of dry air; this simplifies the calculation.

From the chart

$W_1 = 0.0037$ kg/kg dry air, $h_1 = 9.50$ kJ/kg dry air
$W_2 = 0.007$ kg/kg dry air, $h_2 = 41.0] 0.2$ kJ/kg dry air.

As equal amounts of dry air are to be mixed, it is only necessary to average the above quantities to find

$$W_3 = \frac{W_1 + W_2}{2} = \frac{0.0107}{2} = 0.005\ 35 \text{ kg/kg dry air}$$

and $$h_3 = \frac{h_1 + h_2}{2} = \frac{50.3}{2} = 25.15 \text{ kJ/kg dry air.}$$

These two quantities define the new state point on the chart which is at 11.5°C and 63% rh. This could have been found from equations 6.6 and 6.13, yielding values for p_w and t, or by bisecting the line joining the initial state points on the chart since, in this case, equal masses of dry air are involved. Had the amount of air at 23°C been much larger than that at 0°C, the final condition would have been much closer to condition 2.

It is possible to select two moist-air conditions for mixing so that the straight line joining them falls outside the saturation line as shown for conditions 1 and 3 in Figure 6.4. When the mixing proportion is such that the final condition falls on the mixing line but outside the saturation line, fogging is indicated, since the air cannot hold all the moisture as vapour.

Mixing in varying degrees takes place virtually everywhere throughout any space that is being heated or cooled. Mixing will be associated with convection streams around warm or cold objects and at walls, windows, ceilings, and floors, and will occur from air motion induced by fans or other moving objects in the space, by infiltration, and by air supply and exhaust.

It may be observed at this point that processes can be pencilled on the chart as a means of delineating them. Pads of charts can usually be obtained for this purpose. Once the chart becomes familiar, it is often sufficient to use a simple sketch outline on which the pertinent values, read from the chart, can be marked.

6.7 Adiabatic Saturation and Thermodynamic Wet Bulb

Another process of great significance in air conditioning involves spraying water into the air or, alternatively, passing the air over a wetted surface, with no heat added or extracted, until the air becomes saturated. Some of the sprayed water evaporates, finding its heat of evaporation from the air and original vapour, which are cooled in giving up sensible heat. The humidity ratio and dewpoint temperature are increased, and the dry-bulb temperature is decreased. The enthalpy remains almost but not quite constant during the process, since some heat of the liquid in the water is added by evaporation. The final temperature resulting from such a process when the added liquid is made available at the final temperature is known as the thermodynamic wet-bulb temperature, and the process is known as adiabatic saturation. The change from sensible to latent heat during adiabatic saturation carried to completion at the thermodynamic wet-bulb temperature, t', from an initial temperature of t_1 is

$$h'_{fg}(W' - W_1) = c_{pa}(t_1 - t') + W_1(t_1 - t')c_{ps}. \tag{6.16}$$

The energy balance is

$$h_1 + (W'_s - W_1)h'_f = h'_s \tag{6.17}$$

and the increase in enthalpy is

$$\Delta h = (W'_s - W_1)h'_f. \tag{6.18}$$

The specific heat of the liquid is 4.184 kJ/(kg • K) so that

$$h'_f = 4.18\,t'$$

and from Equation 6.13

$$h_1 = 1.005\,t_1 + W_1\,(2500 + 1.86\,t_1)$$

and

$$h'_s = 1.005\,t' + W'\,(2500 + 1.86\,t').$$

Substituting in Equation 6.17 and rearranging

$$W_1 = W' - \frac{1.005(t_1 - t')}{2500 - 2.3t'}. \tag{6.19}$$

Equation 6.19 is significant. It defines a basic air-conditioning process and it shows the relationship between thermodynamic wet-bulb temperature t', the humidity ratio W' at saturation at t', the dry-bulb temperature t_1, and the humidity ratio at the state point. Equation 6.19 thus provides another way of defining the state point, since if t_1 and t' are known, W_1 can be found. The actual partial pressure can then be calculated from Equation 6.6, which on rearrangement gives

$$p_w = \frac{p_t W}{0.622 + W} \tag{6.20}$$

where p_t is the total or barometric pressure. Knowing p_w and t_1, other properties can be calculated.

The values of t_1 and t' can also be used with the psychrometric chart, since the wet-bulb lines on it are lines of adiabatic saturation, and the state point can be found at the intersection of the lines for t_1 and t'.

A device known as a wet-bulb psychrometer (to be described later) gives approximate values of t' along with dry-bulb temperature t_1, and has been widely used for determining the moisture condition of the air.

Example 6.7

Given the thermodynamic wet bulb $t' = 15°C$ and the dry-bulb temperature 22°C at standard atmospheric pressure, find the actual partial pressure by calculation.

$$p'_{ws} \text{ at } 15°C = 1704 \text{ Pa} \qquad p_t = 101\,325 \text{ Pa}$$

$$W' = 0.622 \frac{p'_w}{p_t - p'_w} = \frac{0.622 \times 1704}{101\,325 - 1704} = 0.010\,64 \text{ kg/kg dry air.}$$

From Equation 6.19,

$$W = W' - \frac{1.005(t - t')}{2500 - 2.3\,t'}$$

$$= 0.010\,64 - \frac{1.005(22 - 15)}{2500 - 2.3 \times 15} = 0.007\,78 \text{ kg/kg dry air}$$

and from Equation 6.20,

$$p_w = \frac{101\,325 \times 0.007\,78}{0.622 + 0.007\,78} = 1252 \text{ Pa}$$

and

$$\text{rh} = 1252/2643 = 0.47 \text{ or } 47\%.$$

Example 6.8 (Figure 6.5)

Given the thermodynamic wet bulb = 15°C and the dry-bulb temperature = 22°C for standard atmospheric pressure, find from the chart the state point and enthalpy.

Reading the intersection of lines for t' and t gives

$$\text{rh} = 48\%.$$

Reading the enthalpy gives 42 kJ/kg at saturation at 15°C. This is not the exact enthalpy for the state point. The correction from the chart is –0.18, so enthalpy at the state point is

$$h = 42 - 0.2 = 41.8 \text{ kJ/kg dry air.}$$

The correction could also have been calculated from Equation 6.18:

$$\Delta h = (W' - W_1)h'_f$$

$$= 0.002\,86 \times 4.18 \times 15$$

$$= 0.179 \text{ kJ/kg dry air.}$$

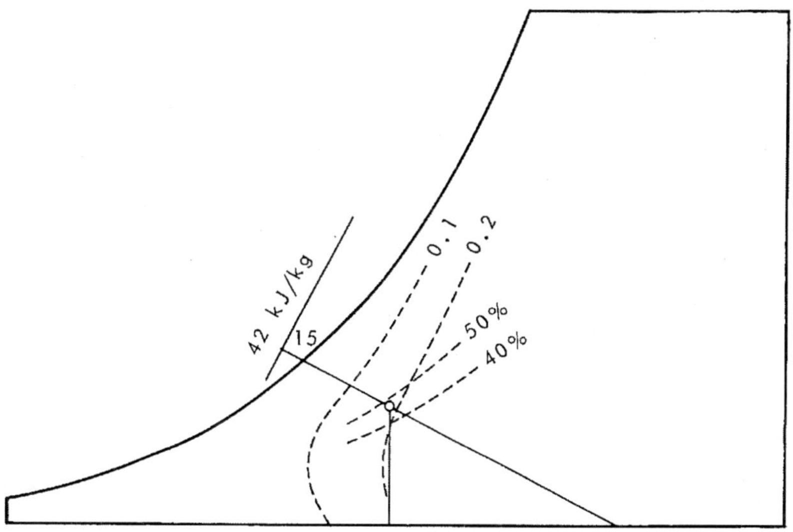

FIGURE 6.5 Outline of psychrometric chart showing the state point of 22°C dry-bulb temperature and 15°C wet-bulb temperature with corresponding rh, enthalpy, and enthalpy correction.

Several modern charts show lines for both wet bulb and enthalpy printed on the chart rather than wet-bulb lines plus enthalpy correction as used in Figure 6.1. One chart uses a nomograph approach, by which the enthalpy lines can be drawn in when needed. Only the sensible-heat factor scale on the right of Figure 6.1 remains to be explained.

6.8 Adiabatic and Evaporative Processes

The adiabatic process is commonly called an evaporative process. Evaporative processes are not always strictly adiabatic, but they can be represented generally by lines of constant wet bulb. They need not be carried to saturation.

An evaporative process is characterized by a decrease in the dry-bulb temperature and an increase in the moisture content and relative humidity up to the point of saturation at the wet-bulb temperature. This may be thought of as cooling the air, but no heat is actually removed. Evaporative methods can be effective in cases where the increased relative humidity and the limitation imposed by the state of the air are not serious disadvantages.

Water that is in contact with the air during the evaporative process is also brought closer to the wet-bulb temperature of the entering air. This is the operating principle of cooling towers and spray ponds, which can be used to cool large quantities of water almost down to the wet-bulb temperature of the air.

Devices called evaporative coolers are widely used in hot, dry climates to reduce the dry-bulb temperature of the air for purposes of improving comfort. The benefit in terms of comfort is partly offset by the increased relative humidities, which reduce the potential for cooling of the skin by the evaporation of perspiration. Evaporative coolers are most beneficial when the initial relative humidity is low and the dry-bulb temperature is high, so that substantial dry-bulb temperature reduction can occur before high relative humidities are reached.

The porous water bag or clay jug used to cool drinking water in areas of low humidities operates on the evaporative principle. Like other evaporative devices, the bags or jugs cannot cool below the wet-bulb temperature; they can only approach it. All evaporative devices can improve conditions, but because they are limited by the state of the air, a given end result cannot be guaranteed.

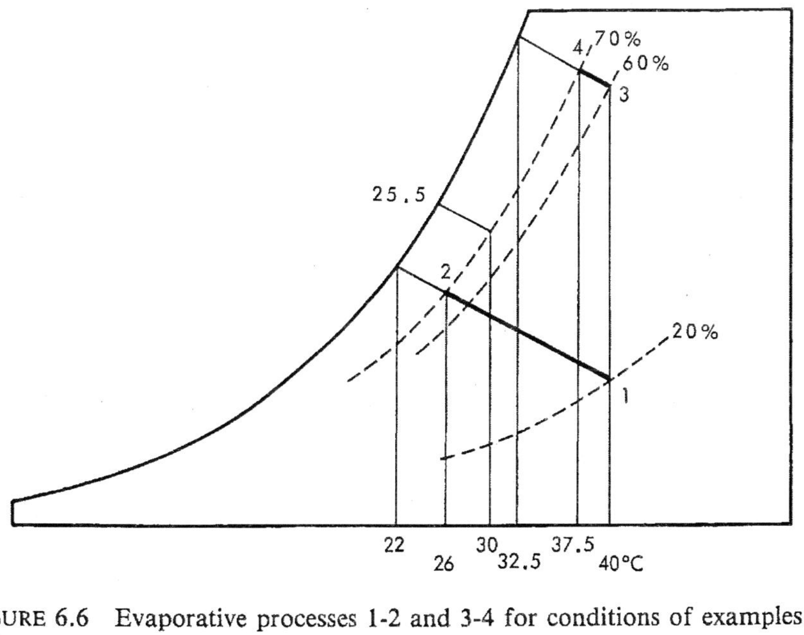

FIGURE 6.6 Evaporative processes 1-2 and 3-4 for conditions of examples 6.9 and 6.10.

Example 6.9 (Figure 6.6)

Desert air at 40°C and 20% rh is to be processed by evaporative cooling. Find the lowest temperature that can be produced if the relative humidity must be limited to 70%.

Find the state point on the chart at 40°C and 20% rh. Follow along the constant wet-bulb line to the point of intersection with the curve for 70% rh, and read the dry-bulb temperature for that state point as 26°C.

Note that the limit for cooling is the wet-bulb temperature of 22°C and that, if the initial rh had been 60% instead of 20%, the limit would have been 37.5°C at 70% rh and 32.5°C at 100% rh.

Example 6.10 (Figure 6.6)

Find the improvement in thermal comfort that could be made with evaporative cooling on typical hot, humid day in Eastern Canada at 30°C dry bulb and 70% rh. Find the wet-bulb temperature of 25.5°C from the chart. All possible conditions fall outside the comfort range (see figures 14.2 and 14.3).

Most room humidifiers operate as evaporative coolers, although their purpose is to evaporate water, not to reduce the temperature. They usually contain wetted wicks, porous plates, or other means of providing a large surface on which evaporation can take place. The process is essentially one of constant wet bulb. When the air is brought into contact with the wetted surface over a substantial path length, it will approach saturation. The reduction in actual dry-bulb temperature relative to the difference between initial dry-bulb and wet-bulb temperatures is called the humidifying efficiency of the evaporative device.

Example 6.11 (Figure 6.6)

A small evaporative-type humidifier has a humidifying efficiency of 50% when taking in air at 23°C and 40% rh. Find the leaving air temperature and humidity and the water added per kilogram of dry air.

From Figure 6.1 the wet-bulb temperature is 14.8° C (for convenience, use 15°C). The maximum possible reduction in dry-bulb temperature is $23 - 15 = 8$°C. The reduction at 50% humidifying efficiency is $0.50 \times 8 = 4$°C. The leaving air temperature will be $23 - 4 = 19$°C and the relative humidity will be 68%.

The humidity ratio at 23°C and 40% rh is 0.007 kg/kg dry air; at 19°C and 68% the humidity ratio is 0.009, so the moisture added is 0.002 kg/kg dry air.

When moist air is sprayed with a salt solution instead of water, dehumidification can result. The concentration of salt lowers the equilibrium vapour pressure of the solution below that of pure water. Lithium chloride is often used as the salt. In such cases the air and the salt spray increase in temperature, following a constant wet-bulb line, since latent heat of evaporation is now being converted to sensible heat. In other chemical dehumidifying devices, solid desiccant materials such as silica gel are used, and the air is caused to flow over and through them.

These, too, follow a constant wet-bulb process with an increasing dry-bulb temperature as moisture is removed. The desiccants, whether solid or solution, must be removed for processing by heating to drive off the absorbed water before they can be reused (Figure 6.7).

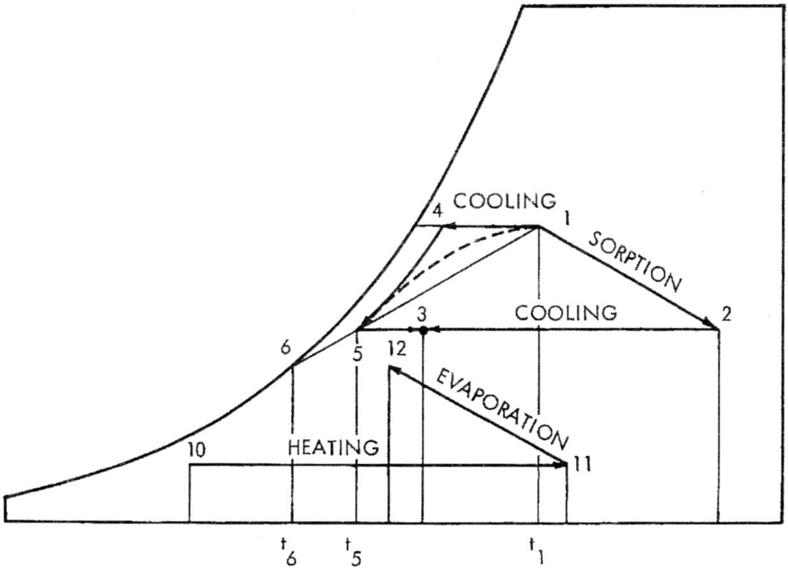

FIGURE 6.7 Use of various processes to condition air to a desired final condition. 1-2-3 involves chemical dehumidification followed by cooling, 10-11-12 involves heating to 11 followed by humidification through an evaporative process, 1-4-5-3 involves cooling followed by heating (also shows effect of cooling coil bypass).

6.9 Processes in Combination

Several processes can be used in combination to adjust the conditions of an air stream to some desired final state. Some combinations of the basic processes of heating–cooling and evaporation–condensation are illustrated in Figure 6.7. There is usually more than one way to arrive at the desired condition. The choice may depend on considerations such as cost, energy required, means of control, ease of operation, and the ultimate use (which is usually the maintenance of suitable conditions in one or more spaces). This is the business of the air-conditioning engineer, but individual processes and simple combinations of them may be of broader application and interest.

The mixing process discussed in Section 6.6 is often used to produce an air condition to meet changing requirements in a space. Two streams maintained at selected fixed conditions may be mixed to provide a mixed stream having the desired properties. The proportions can be varied to adjust the final conditions when requirements change. This flexibility also makes it possible to provide several streams at differing conditions simultaneously.

6.10 The Bypass Factor

The mixing principle and its representation on the psychrometric chart may be used to advantage in dealing with the characteristics of heating and cooling devices. The degree of saturation attained in the evaporative process in Example 6.11 was expressed as a humidifying efficiency (in that case, 50%). This factor reflects the extent and duration of contact of the air with the wetted surface at which evaporation is taking place. This may also be visualized and expressed as a mixing process with two streams of air flowing through the apparatus in parallel and then remixing. One part of the entering stream is brought to saturation at the water surface temperature, and the other part is considered to be unaffected. The relative flow rates for these two hypothetical streams will be in proportion to the changes in their dry-bulb temperatures.

The humidifying efficiency is given by the reduction in temperature divided by the maximum possible reduction, which also gives the ratio of the rate of flow of the fully humidified air to the total rate of flow. The proportion of air considered to be unaffected is found by difference to be $1\text{-}\eta_h/100$ where η_h is the humidifying efficiency expressed as a percentage. This is sometimes also called the bypass factor F for the apparatus involved.

A similar bypass situation is shown for the cooling process 145 of Figure 6.7 in which the air could not be cooled fully to saturation at the cooling surface condition. The proportion of air assumed to have passed through the coil without change is given by the bypass factor

$$F = (t_5 - t_6)/(t_1 - t_6).$$

The cooling process can now be viewed on the chart as a mixing process with the final condition at point 5 on the line joining points 1 and 6. When the bypass factor is known or can be estimated for a given apparatus and operating condition, the leaving air condition, or, more commonly, the cooling surface temperature to produce a given leaving air temperature, can be found.

A bypass may be arranged directly in parallel with a cooling coil or other apparatus to provide a further means of varying the final condition of the leaving air stream by adjusting the mixing proportions.

6.11 The Sensible-Heat Factor

The proportionality relations resulting from the linearity of the chart scales mean that all conditions at state points along any straight line on the chart will differ from one another by amounts of heat and moisture in a constant proportion as defined by the slope of the line. The slope of the line can also be expressed indirectly in terms of a sensible-heat factor (SHF), which is the ratio of the change in sensible heat to the change in total heat between any two points on the line. The SHF scale along the right-hand side of the chart used in conjunction with the reference point located at the state point for 24°C and 50% rh will give the SHF for any given slope of the line. Figure 6.8 shows the relationship between sensible, latent, and total heat and between changes in heat and moisture and the slope of a given line for a change from point 1 to point 3.

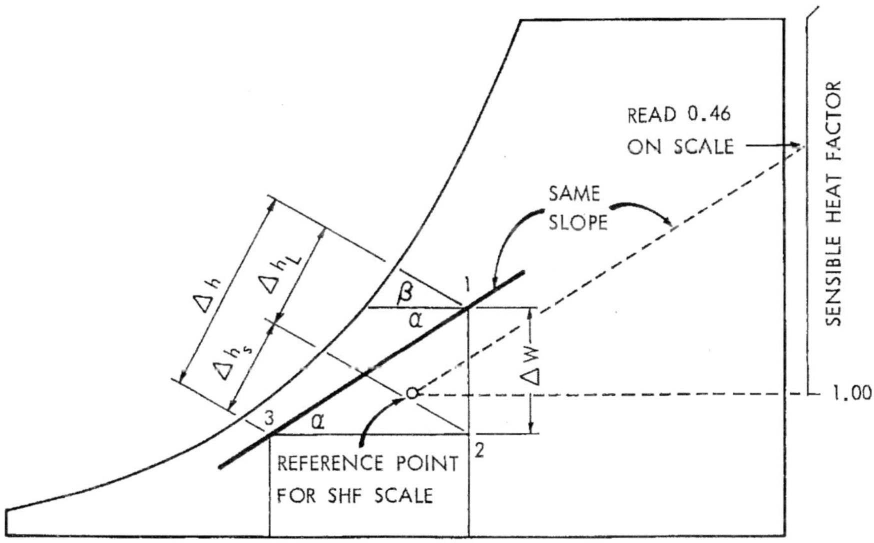

FIGURE 6.8 Outline of the psychrometric chart showing the definition of sensible and latent heat and the use of the sensible-heat factor scale.

The change in latent heat, Δh_L, between points 1 and 3 is found from the enthalpy change for a change from 1 to 2, removing moisture at constant temperature. The change in sensible heat, Δh_s, is found by changing temperature from 2 to 3 without change in moisture content. The ratio of $\Delta h_s/\Delta h$ can be shown from the trigonometric relationship on the chart to be given by (sin β cos α)/(sin ($\beta + \alpha$)). Since β has a fixed value, $\Delta h_s/\Delta h$, which is also the SHF, depends solely on α, which is the slope of the "condition" line. This feature of the chart is of great value in air-conditioning calculations.

It has been assumed in these discussions that the lines of adiabatic saturation on the psychrometric chart are also lines of constant enthalpy. This assumption is an approximation, since they differ by the amount of the enthalpy deviation arising from the heat of the liquid evaporated during an adiabatic saturation process. The condition line found from the chart for any given case may be suitably adjusted to take this small difference into account if desired. Charts showing lines of constant enthalpy [**6.1, p. 5.5**] will not be subject to this difficulty.

The slope of the line joining any two state points may also be expressed directly in terms of the ratio $\Delta h/\Delta W$ rather than as a sensible-heat ratio, where Δh is the difference in enthalpy and ΔW is the difference in humidity ratio between the two state points.

6.12 Measuring the Condition of Air–Vapour Mixtures

The most common instrument for measuring the state of an air–vapour mixture is the wet-bulb psychrometer, which consists of two identical thermometers: one has a clean wetted wick over the bulb, and the other is left bare. Evaporation from the wetted wick cools the thermometer bulb, giving a reading which approximates the thermodynamic wet bulb already discussed. The values of wet-bulb temperature and dry-bulb temperature are sufficient to define the state point. It was this instrument that gave rise to the terms "wet-bulb temperature" and "dry-bulb temperature".

The correspondence between the instrumental reading and the thermodynamic wet bulb is somewhat fortuitous. Radiation and speed of air flow over the wet bulb affect the readings. The influence of air speed is shown in Table 6.1 [**6.3**]. It is evident from the table that the unventilated or unaspirated psychrometer will have a large error of the order of 8% of the wet-bulb depression at room temperature with normal air movement, while at 5 m/s the error becomes relatively small. The common form of the instrument is a sling psychrometer, about 350 mm long, which can be whirled around a pivoted handle at one end at about 60 rpm to produce a speed over the wet bulb of about 1 m/s. The ventilation error at this speed is probably less than the accuracy with which the thermometers can be read. Low air speed and radiation both tend to make the wet-bulb reading too high, giving a wet-bulb depression that is too low, and an apparent relative humidity that is too high. Some types of wet-bulb psychrometers have shielded bulbs with the air drawn over them by a motor-driven fan.

TABLE 6.1

Correction for Air Flow
Over the Wet Bulb in a Psychrometer
(From Carrier, W.H., and Lindsay, D.C., *ASME Trans.*, Vol. 46, 1924) [**6.3**]

Wet Bulb Temp. t' °C	Error in percent of theoretical depression at various air speeds			
	0.125 m/s	1 m/s	5 m/s	10 m/s
0	15	5	1.5	1
10	12	4	1.2	0.8
20	8	3	1	0.6
30	6	2	0.7	0.4
40	4	1.5	0.5	0.3

The wet- and dry-bulb temperatures can be translated to relative humidity or to vapour pressure using the chart, provided that any barometric discrepancy can be disregarded. They can also be converted by calculation using Equation 6.19 for the thermodynamic wet bulb. The equation may be further transformed in combination with Equation 6.20 to obtain

$$p_w \times \frac{p_t - p'_w}{p_t - p_w} = p'_w - \frac{(p_t - p'_w)(t_1 - t')}{1547 - 1.4\,t'} \qquad (6.21)$$

and if $(p_t - p'_w)/(p_t - p_w)$ is neglected since it will, within the probable limits of error of the instrument, be equal to unity

$$p_w = p'_w - \frac{(p_t - p'_w)(t_1 - t')}{1547 - 1.4\,t'} \qquad (6.22)$$

where p_w is the actual partial pressure of the vapour, p_w' is the saturation pressure at the wet-bulb temperature, t_1 and t' are the dry- and wet-bulb temperatures, and p_t is the total or barometric pressure. Temperatures are in degrees Celsius but pressures may be in any consistent units. Equation 6.22 is of the general form proposed by Willis H. Carrier many years ago [6.3]. Some versions contain modified quantities to account in part for radiation effects when unshielded thermometers are used.

Example 6.12

Find the relative humidity for moist air with a dry-bulb temperature of 24°C and a wet-bulb temperature of 20°C at standard atmospheric pressure.

From the chart (Figure 6.1), at the intersection of the appropriate dry-bulb and wet-bulb lines, find the state point and read rh = 70%.

Using Equation 6.22, substitute $p_w' = 2337$ Pa (Table 5.1) so that

$$p_w = 2337 - \frac{(101\,325 - 2337)(24 - 20)}{1547 - 1.4 \times 20}$$

$$= 2337 - 261 = 2076 \text{ Pa}$$

and since p_{ws} at 24°C = 2983 Pa, rh $= \dfrac{2076}{2983} \times 100 = 69.6\%$.

Note that the value of $(p_t - p_w')/(p_t - p_w)$ which was assumed equal to 1 in the derivation of Equation 6.22 has a value in this case of 1.007. The value of p_w should have been more nearly $1.007 \times 2076 = 2091$ Pa and the relative humidity

$$\text{rh} = \frac{2091}{2983} \times 100 = 70.1\%.$$

Disregarding this correction has the effect of increasing the calculated wet-bulb depression, $(p_w - p_w')$, and this compensates in part for radiation effects in the psychrometer which tend to make the wet-bulb reading too high and the observed wet-bulb depression too low.

The wet-bulb psychrometer is still widely used. It gives reasonably accurate results without the need for calibration, provided some attention is given to the ventilation of the wet bulb.

Many hygrometers rely on the expansion produced in a sensitive material by the moisture taken up in equilibrium with the air to indicate relative humidity. Paper, animal membranes, human hair, and even wood can be used as sensitive elements. All require calibration and exhibit hysteresis effects. Various means are used to magnify the length change to produce movement of the indicating needle against a scale. The hair hygrometer has been one of the more reliable instruments of this type, and for many years hair was employed as the sensing element in hygrostats for humidity control.

There are several types of electrical-resistance hygrometers, capable of high sensitivity and adaptable to remote reading; they require care in use to protect them from wetting by condensation, and frequent calibration to ensure accuracy. A calibration method is available [**6.4**]. Other instruments operate on the dew-point principle, involving cooling a mirror until a visible or otherwise detectable moisture film is produced. An international symposium has been held on the important but difficult matter of measuring humidity and moisture [**6.5**]. A good description of current methods and a bibliography are available [**6.6**].

6.13 Conclusion

The instruments, calculation methods, and psychrometric charts described and discussed in this chapter are the essential tools by which the condition of moist air can be identified and its properties calculated. The psychrometric chart, introduced because of the contribution it makes to a more complete understanding of moist-air processes, is widely used in air-conditioning engineering. The visual images it provides are of great assistance in understanding many natural moist-air processes.

References

6.1 *ASHRAE Handbook 1981 Fundamentals.* Chaps. 5 and 6.

6.2 Bullock, C.E., and Carpenter, J.H. New psychrometric charts in S.I. metric units. *ASHRAE Journal*, 1975, *17*, pp. 30-33.

6.3 Carrier, W.H., and Lindsay, D.C. The temperatures of evaporation of water into air. *ASME Transactions,* 1924, *46*, pp. 739-80.

6.4 Hedlin, C.P. A device for calibrating electrical humidity sensors. *Materials Research and Standards*, 1966, *6* (1), pp. 25-29. (NRCC 8951)

6.5 Various papers in *Proceedings, International Symposium on Humidity and Moisture.* Washington, 1963. (several volumes)

6.6 *ASHRAE Handbook 1981 Fundamentals.* Chap. 13.

7

Fluids in Motion

7.1 Introduction

Fluids in motion are of concern in many aspects of the design and construction of buildings. The importance of this subject and the possibilities for its further development and application in building are now being recognized as an essential part of the study of buildings. In this chapter some of the essential features of fluids in motion will be reviewed and their influences identified and discussed. The effect of wind on buildings, which has now become a subject of considerable importance (particularly for tall buildings), and the subject of air leakage in buildings will be taken up in the following chapters.

It was emphasized earlier that since air is everywhere it is inevitably involved as an intermediary in exchanges of heat between every body and its surroundings whenever there are differences in temperature. Because air is less dense when warmed and increases in density when cooled in contact with warmer or colder surfaces, it moves under buoyancy forces, transporting heat, water vapour, and any other gaseous or airborne particulate matter it may contain. Such movement, known as natural convection or free convection, is of obvious importance in the environment, affecting individual objects and surfaces, entire buildings, and even the earth's atmosphere. Free convection in liquids takes place whenever density differences occur, and is principally of importance in underwater situations.

Air, water, steam, refrigerants, and other gases and liquids are delivered through pipes and ducts for use at various points in a building. The energy required to maintain flow is provided by pumps, fans, and other means.

7.2 Streamlines and Continuity

Streamline flow is the steady, nonturbulent motion of a fluid. Streamlines are defined as lines drawn tangential to the direction of flow and can be visualized as the boundaries of stream tubes through which a steady flow is taking place, as shown in Figure 7.1. Considerations of continuity of mass flow at planes 1 and 2 normal to the flow lead to the relationship

$$\rho_1 A_1 u_1 = \rho_2 A_2 u_2. \tag{7.1}$$

For conditions in which the flow can be considered incompressible,

$$Q = A_1 u_1 = A_2 u_2 \tag{7.2}$$

where Q = volume flow rate, A is the transverse area of the stream tube, ρ is the mass density of the fluid, and u is the velocity across the stream tube. The relationship between velocity and changing stream tube area is given by

$$\frac{u_1}{u_2} = \frac{A_2}{A_1}. \tag{7.3}$$

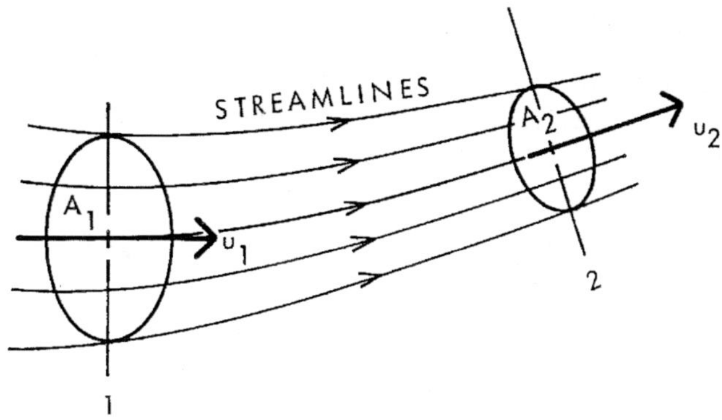

FIGURE 7.1 Continuity of flow along a stream tube.

7.3 Bernoulli's Equation

Bernoulli's equation, an important one in fluid mechanics, is essentially a statement of the conservation of energy along a stream tube in streamline flow. In its elementary form it takes into account kinetic energy of translation, pressure energy, and potential energy, and can be stated with reference to points 1 and 2 in Figure 7.1 as

$$p_1 + \tfrac{1}{2}\rho_1 u_1^2 + \rho_1 g z_1 = p_2 + \tfrac{1}{2}\rho_2 u_2^2 + \rho_2 g z_2 \tag{7.4}$$

where p is the pressure, ρ is the density, u is the velocity in the direction of flow, z is the height above some reference level, and g is the gravitational constant. A more general form can be written

$$p + \tfrac{1}{2}\rho u^2 + \rho g z = \text{a constant}. \tag{7.5}$$

Each of the terms has the dimension of pressure, which is the energy per unit volume of flowing fluid. There are no terms dealing with thermal energy, conversion of energy to heat by friction and eddy dissipation, or work done by the fluid, so that the equations as given here apply only to situations in which these effects are not present or are of such a nature that they can be neglected. Additional terms must be added when necessary to provide more general forms of the energy equation.

Another form of Bernoulli's equation can be obtained by dividing all terms by ρg, with results as follows:

$$\frac{p}{\rho g} + \frac{1}{2}\frac{u^2}{g} + z = \text{a constant.} \tag{7.6}$$

The dimensions are now of length, or of "head" of fluid, and the terms are the energies for unit weight of flowing fluid. This is the form commonly used in hydraulics.

The density, ρ, is independent of pressure, or nearly so, for liquids. Air can also be treated as an incompressible fluid in many cases when the pressure changes are relatively small.

7.4 The Venturi Meter and the Pitot Tube

Two of the instruments commonly used to measure flow velocities provide excellent examples of the application of Bernoulli's equation.

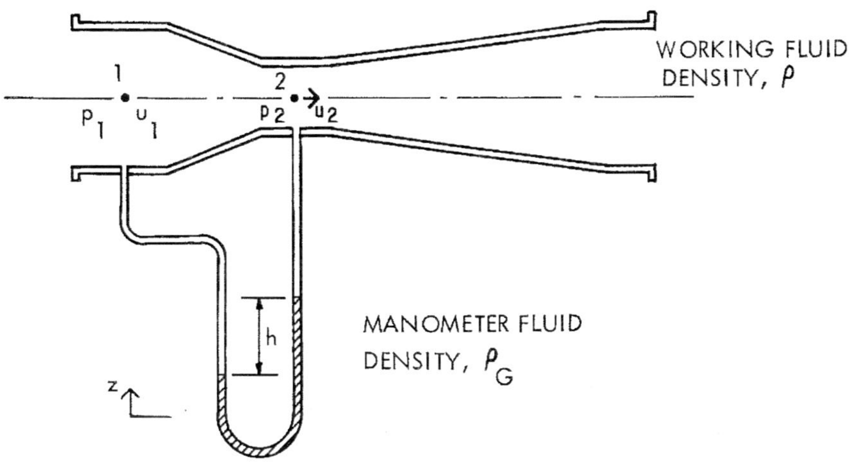

FIGURE 7.2 Venturi meter and manometer.
(From Aynsley, R.M., Melbourne, W., and Vickery, B.J., *Architectural aerodynamics*, London: Applied Science Publishers, 1977, Fig. 2.6, p. 49)[7.7]

The Venturi meter is inserted to form part of a pipe carrying fluid. The flow velocity is increased at the constricted throat that it provides. The accompanying decrease in pressure can be measured, and from it the flow in the pipe can be calculated. In the arrangement shown in Figure 7.2, a manometer containing a gauge fluid with a density ρ_G is used to measure the pressure change. Applying Bernoulli's equation to conditions at points 1 and 2,

$$p_1 + \frac{1}{2}\rho\, u_1^2 = p_2 + \frac{1}{2}\rho\, u_2^2 \tag{7.7}$$

and from continuity of flow

$$u_1 A_1 = u_2 A_2 \tag{7.8}$$

and from consideration of pressures at the manometer

$$p_1 + \rho g z_1 = p_2 + \rho g(z_1 - h) + \rho_G g h. \tag{7.9}$$

Combining all three equations yields the result

$$Q = u_2 A_2 = A_2 \left[\frac{2gh(\rho_G - \rho)}{\rho \left[1 - \left[\frac{A_2}{A_1}\right]^2\right]} \right]^{1/2}. \tag{7.10}$$

As the flow is not ideal, it is not given exactly by Equation 7.10; it will be from 1 to 5% lower, depending on the instrument, the flow rate, and the fluid properties. When the instrument is used for liquids, care must be taken to eliminate air from the manometer and connecting piping in order to avoid false readings of pressure on the manometer.

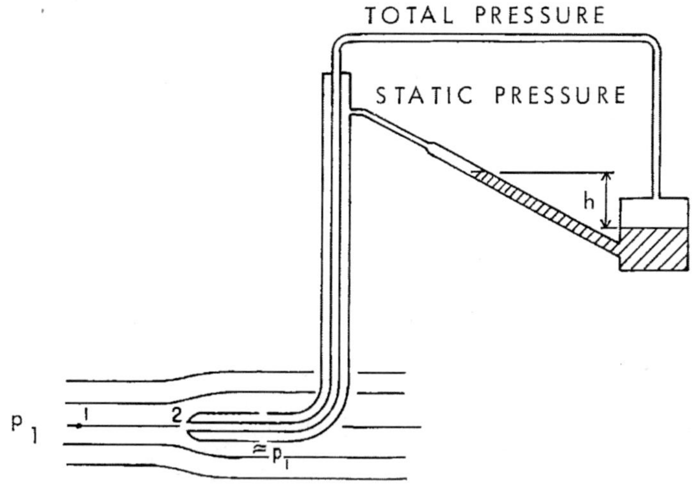

FIGURE 7.3 Pitot-static tube and inclined manometer.
(From Aynsley, R.M., Melbourne, W., and Vickery, B.J., *Architectural aerodynamics*, London: Applied Science Publishers, 1977, Fig. 2.7, p. 50)[**7.7**]

The pitot tube in its simplest form consists of a bent tube which can be inserted in a fluid stream with its open end facing directly into the fluid flow. When a manometer or some other suitable pressure-measuring device is connected to the other end of the tube, the pressure generated at the open end in the fluid stream can be measured. In another way the instrument is arranged so that it also measures the pressure of the undisturbed stream; this is known as a pitot-static tube. Figure 7.3 shows one such arrangement connected to a manometer. The flow along the streamline from 1 to 2 is brought to zero at the tip of the pitot tube. The resulting stagnation pressure can be found by applying Bernoulli's equation

$$p_1 + \tfrac{1}{2}\rho\, u_1^2 = p_2 + 0. \tag{7.11}$$

The pressure p_2 is the total pressure, since it is equal to the static pressure, p_1, plus the dynamic pressure, which is the pressure equivalent of the velocity of flow at 1. The dynamic pressure, $\tfrac{1}{2}\rho\, u_1^2$, commonly designated q, is given by the difference between the total and static pressures:

$$p_2 - p_1 = \tfrac{1}{2}\rho\, u_1^2 = q. \tag{7.12}$$

The inclined manometer in Figure 7.3 is connected to read the difference between the total pressure at 2 and the pressure at the static taps in the straight portion of the tube arranged parallel to the flow. The latter pressure is nearly equal to the static pressure p_1 when the instrument is carefully proportioned. The velocity in the undisturbed stream is given by

$$u_1 = \left[\frac{2(p_2 - p_1)}{\rho}\right]^{1/2} \tag{7.13}$$

and, considering the pressures at the manometer, which has a manometer fluid of density ρ_G,

$$u_1 = \left[\frac{2gh\,(\rho_G - \rho)}{\rho}\right]^{1/2}. \tag{7.14}$$

An elementary form of the pitot tube used with a separate static-pressure tap is shown in Figure 7.4. Such an arrangement can be used to measure total and static

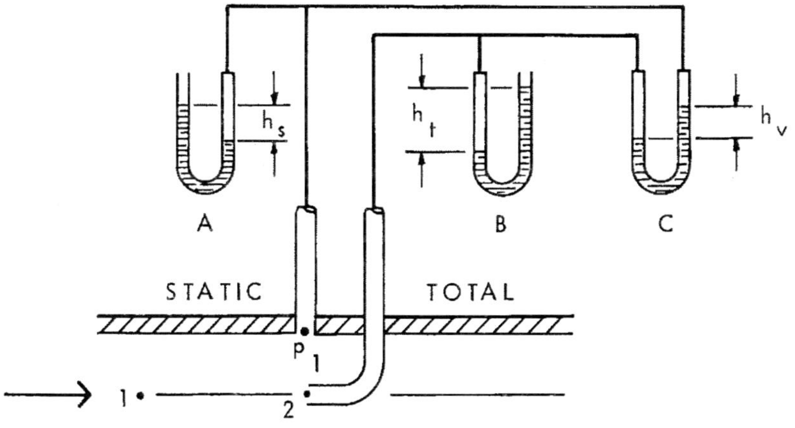

FIGURE 7.4 Pressure measurements.

pressures near bounding surfaces. For such a use, manometers measure the pressures separately. Manometer A connected to the static tap does not show the absolute static pressure p_1 directly, but gives the difference between it and the ambient pressure at the open limb of the manometer. Correspondingly, manometer B reads the impact pressure relative to the ambient pressure. The pressures shown are gauge pressures and must be treated accordingly. There is no problem in the case of dynamic pressure, since the manometer at C reads directly the head corresponding to the difference between total and static pressures. Pressure measurements made at the surfaces of buildings or on models in wind tunnels will usually, as a matter of convenience if not of necessity, be read as differences with respect to some reference pressure, and it is necessary in using any such results to keep this clearly in mind.

Example 7.1

Find the velocity of air flow as measured by a pitot-static tube when the manometer reading is 16 mm, the gauge fluid is water, the air and water are at 20°C, and the air pressure is 101.3 kPa.

The density of air for the conditions given is 1.205 kg/m³, the density of water is 998 kg/m³, and g = 9.807 m/s². Substituting in Equation 7.14 yields

$$u = \left[\frac{2 \times 9.807 \times h(998 - 1.205)}{1.205 \times 1000}\right]^{1/2}$$

and $u = 403\ h^{1/2}$.
For $h = 16$ mm, $u = 16$ m/s which corresponds to 58 km/h.

$$\text{The dynamic pressure} = \rho_G g h = \frac{998 \times 9.807 \times 16}{1000} = 156\ \text{Pa}.$$

Further consideration of Example 7.1 will show that the dynamic pressures encountered in many cases of air flow, which vary as the square of the velocity, will often be so small that they cannot readily be measured with manometers. Air flowing in ducts at about 10 m/s will involve dynamic pressures of 6 mm of water, while air issuing from supply grilles at 2 m/s will produce a head of only 0.24 mm of water, requiring an instrument that is sensitive to 0.002 mm of water to achieve an accuracy of 1%.

The visualization of pressures as a height of water column has some merit in air-flow work. This feature may be retained without prejudice to the use of the proper SI units by noting that a pressure in pascals divided by 10 gives the head in millimetres of water.

The velocity will usually vary over the cross-section of a fluid stream, and it becomes necessary to average the results of a number of readings taken at selected stations to find the flow rate. The pitot-static tube is well suited for this purpose. When the cross-section of a flowing stream is divided into equal areas, with readings taken at the centre of each area, the flow rate may be found from a simple average of the velocities. Averaging the observed dynamic pressures does not give the pressure equivalent of the average velocity.

7.5 Flow through Orifices

An orifice in a tank wall having a curved entrance as shown in Figure 7.5 will flow full of fluid and will provide a flow rate given by Bernoulli's equation.

$$p_1 + \tfrac{1}{2}\rho u_1^2 = p_2 + \tfrac{1}{2}\rho u_2^2$$

but $u_1 = 0$, so

$$p_1 - p_2 = \rho\frac{u_2^2}{2}. \qquad (7.15)$$

The theoretical flow rate is then given by

$$Q = u_2 A = A\left[\frac{2}{\rho}(p_1 - p_2)\right]^{1/2}. \qquad (7.16)$$

Since flow cannot take place without some loss, and since there may be contraction in the jet, it is necessary to include a coefficient C to compensate for this, so that

$$Q = CA\left[\frac{2}{\rho}(p_1 - p_2)\right]^{1/2}. \qquad (7.17)$$

For a well-rounded entrance, the contraction will be negligible and the coefficient C will be from 0.96 to 0.98 to account for velocity losses. For sharp-edged or square-edged orifices the effective jet cross-section is reduced by contraction to about 0.62 A and the coefficient for such cases is commonly taken as 0.6.

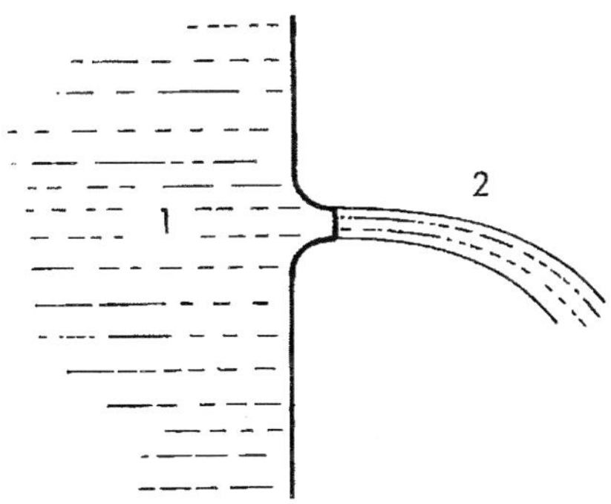

FIGURE 7.5 Orifice in a tank wall.

7.6 Streamline Flow around an Airfoil

Bernoulli's equation can be applied in the study of flow around an airfoil as shown in Figure 7.6. The streamlines become crowded above and separated below the air foil. It follows that for constant energy along any streamline there must be increased velocities with attendant reductions in pressure above, and increased pressures with attendant reductions in velocities below. This gives rise to the lifting force at right angles to the flow.

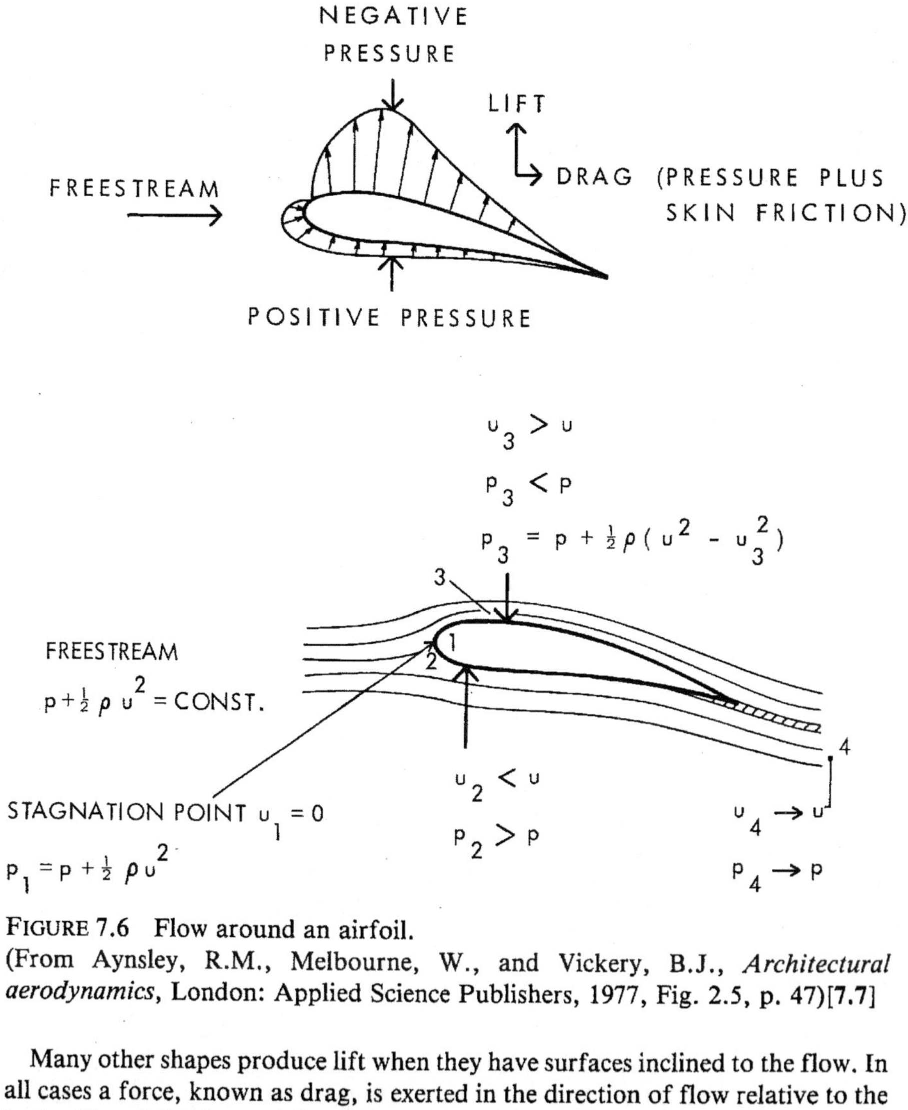

FIGURE 7.6 Flow around an airfoil.
(From Aynsley, R.M., Melbourne, W., and Vickery, B.J., *Architectural aerodynamics*, London: Applied Science Publishers, 1977, Fig. 2.5, p. 47)[7.7]

Many other shapes produce lift when they have surfaces inclined to the flow. In all cases a force, known as drag, is exerted in the direction of flow relative to the body. The airfoil is specially designed to produce high lift with low drag. Bernoulli's equation reveals something about the nature of the drag force, since it leads to some understanding of the pressures shown in Figure 7.6 which produce surface forces having components in the direction of motion and in the lift direction. It does not, however, explain the role of fluid viscosity, which leads to skin friction, turbulence, and other departures from streamline flow which must now be considered.

7.7 Viscosity in Fluids

Real fluids exhibit a resistance to being sheared which increases with the rate of shearing. In the situation shown in Figure 7.7, the fluid being sheared is between two parallel plates separated by a distance y. The velocity of the moving plate relative to the other, which is assumed to be fixed, is u. When the nature of the fluid is such that the resistance varies directly with the velocity gradient, the force F to move the plate of area A is

$$F = A\mu \frac{u}{y} \tag{7.18}$$

or in terms of the gradient expressed as a differential

$$\text{shearing stress} = \frac{F}{A} = \mu \frac{du}{dy} \tag{7.19}$$

where μ is the viscosity and has the units pascal seconds (Pa • s) or (N • s/m²).

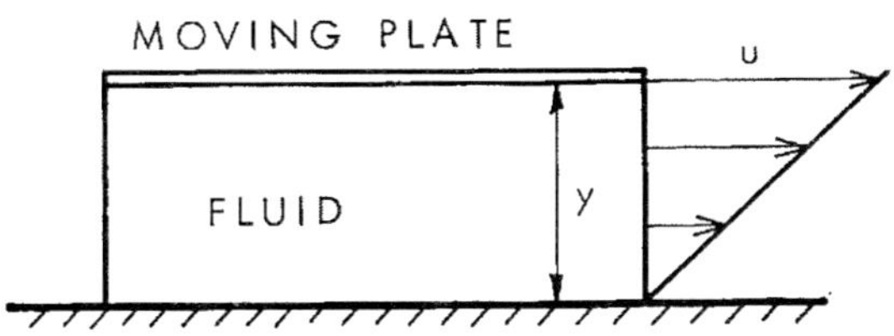

FIGURE 7.7 Shearing of a fluid.

TABLE 7.1

Properties of Water at Various Temperatures

Temperature, °C	Viscosity, Pa • s	Surface tension, N/m	Density, kg/m³	Heat capacity, kJ/kg
0	0.001 792	0.0756	999.87	4.215
5	0.001 519	0.0749	999.9	4.202
10	0.001 308	0.0742	999.7	4.189
15	0.001 140	0.0735	999.1	4.183
20	0.001 005	0.0727	998.2	4.179
25	0.000 894	0.0720	997.1	4.177
30	0.000 801	0.0712	995.7	4.176
40	0.000 656	0.0696	992.2	4.176
50	0.000 549	0.0679	988.1	4.178

Viscosity enters frequently into calculations in conjunction with density in the form μ/ρ, which is called kinematic viscosity. It has the units square metres per second (m²/s).

Values of the viscosities for water and for air can be found from Tables 7.1 and 7.2.

TABLE 7.2

Properties of Air at Various Temperatures

Temperature °C	μ Pa • s	k W/m • K	C_p kJ/kg • K	ρ* kg/m³	μ/ρ mm²/s
−46	14.7×10^{-6}	0.0204	1.000	1.555	9.45
−18	16.2×10^{-6}	0.0227	1.004	1.384	11.71
+10	17.6×10^{-6}	0.0247	1.005	1.247	14.11
+38	18.9×10^{-6}	0.0272	1.006	1.135	16.65
+66	20.0×10^{-6}	0.0289	1.008	1.041	19.21

*Dry air at standard atmospheric pressure

7.8 Laminar Flow in Pipes

The development of laminar flow in circular pipes with only shearing action taking place in the fluid, as shown in Figure 7.8, leads to a velocity distribution in the form of a paraboloid of revolution. The maximum velocity is

$$u_{max} = \frac{(p_1 - p_2)d}{16\,\mu\ell}. \tag{7.20}$$

The average velocity, which in the case of pipe flow will be designated U, is given by

$$U = \frac{(p_1 - p_2)d^2}{32\,\mu\ell} \tag{7.21}$$

the flow rate is

$$Q = \frac{\pi}{4} d^2 \times U = \frac{\pi(p_1 - p_2)d^4}{128\,\mu\ell}. \tag{7.22}$$

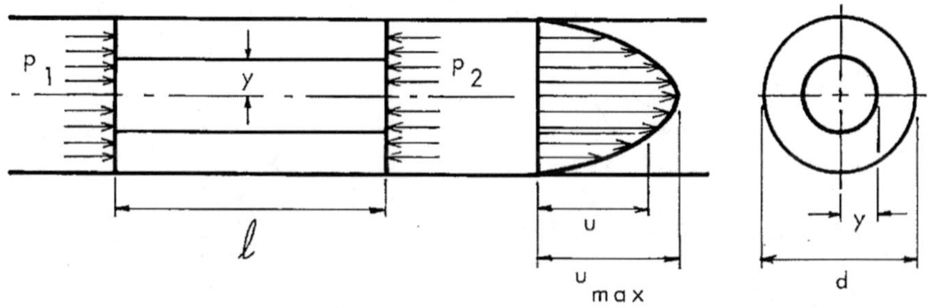

FIGURE 7.8 Laminar flow in a pipe.

Equation 7.22 is an expression of the Hagen-Poiseuille law, which states that discharge from a circular pipe with laminar flow is directly proportional to the pressure difference and to the fourth power of pipe diameter, and inversely proportional to viscosity and pipe length.

The resistance to flow measured as pressure loss in length ℓ can be found by rearranging Equation 7.22:

$$p_1 - p_2 = \frac{128 Q \mu \ell}{\pi d^4}. \tag{7.23}$$

The pressure drop can be expressed as loss in head of fluid by substituting $p = \rho g h$

$$h_1 - h_2 = \frac{128\, Q \mu \ell}{\rho g \pi d^4}. \tag{7.24}$$

The flow described by Equation 7.23 is obtained with viscous liquids flowing at low velocities in small pipes or tubes. The velocity distribution at the pipe entrance is not parabolic, but is determined by the entrance conditions. It changes slowly with distance from the entrance, first taking shape as a layer at the pipe wall which thickens to fill the pipe. The full parabolic form may only be reached at 50 to 100 pipe diameters downstream, and under certain conditions may require even greater distances. The characteristic steady laminar flow begins to change when velocity, density, or diameter is increased or viscosity decreased. Beyond certain limits the flow becomes turbulent over the whole of the pipe cross-section (except for a very thin laminar layer at the pipe wall), and Equation 7.23 no longer applies. The resistance to flow now varies as the square of the velocity.

The varied random motions of the turbulence introduce radial and tangential components into the flow. These have a mixing effect, transferring velocity energy in a radial direction. The velocity profile becomes flattened, as shown in Figure 7.9, with a ratio of centreline to mean velocity of about 1.25 instead of 2 as in the case of laminar flow. The velocity gradient at the pipe wall is sharply increased, since the reduction to zero velocity right at the pipe wall must now be accomplished over a much smaller radial distance. The flow develops a stable pattern at some distance downstream from any disturbing feature such as an entrance, an elbow, or a valve. The flow following a rounded entrance will begin as a laminar boundary layer at the pipe wall; it gradually thickens and then becomes turbulent under suitable conditions. The turbulent layer continues to grow in thickness until it fills the whole of the pipe cross-section and adjusts finally to a characteristic velocity profile for the conditions of flow at 10 to 30 or more pipe diameters from the entrance, as shown in exaggerated form in Figure 7.10.

It is clearly important to be able to determine whether the flow in a particular situation will be laminar or turbulent. A dimensionless ratio called Reynolds number, commonly designated *Re*, is widely used as an index of the flow patterns in geometrically similar situations with changing fluid properties and geometric scale. It provides the means of correlating and applying the results of experiments with any given geometry, and has relevance to pipe flow and to a wide range of other fluid-flow situations.

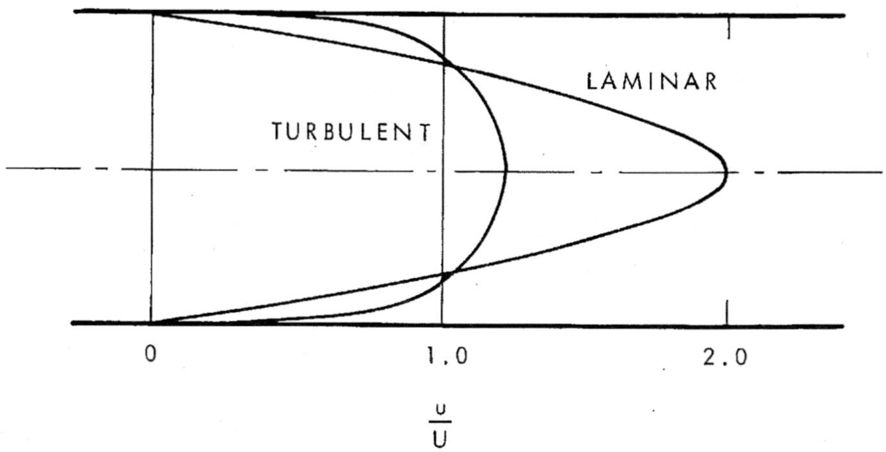

FIGURE 7.9 Velocity profiles of flow in pipes.

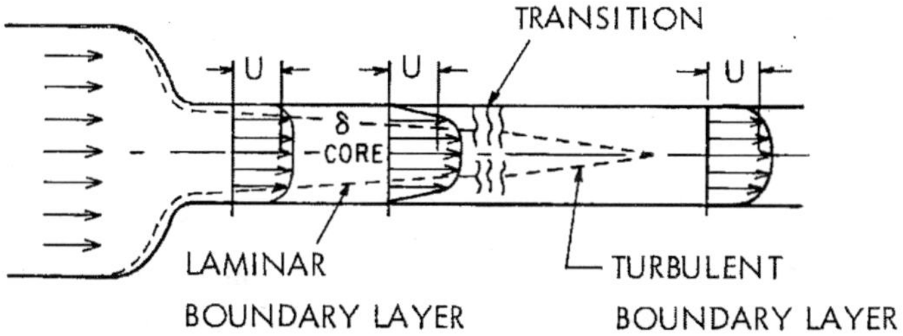

FIGURE 7.10 Flow in pipes following a rounded entrance.
(From *ASHRAE Handbook 1981 Fundamentals*, Fig. 5, p. 4.4)[**7.1**]

7.9 Reynolds Number

The changing nature of a flow pattern for differing geometric scale, fluid properties, and flow velocities can be attributed to changes in the ratio of inertia forces to viscous forces. Inertia forces are related to mass and acceleration. Mass has the dimensions $\rho \ell^3$, while acceleration can be taken as u/t, or, since time, t, is proportional to ℓ/u, as u^2/ℓ. Thus, inertia forces are proportional to

$$\rho \ell^2 \times \frac{u^2}{\ell} = \rho \ell^2 u^2.$$

Viscous forces are a result of shearing and are given by shear stress times area. The shear stress can be represented by $\mu u/\ell$. The viscous forces are thus proportional to

$$\frac{\mu u}{\ell} \times \ell^2 = \mu u \ell.$$

The ratio of inertia force to viscous force is given by

$$\frac{\rho\, \ell^2\, u^2}{\mu\, u\, \ell} = \frac{\rho u \ell}{\mu}$$

which is exactly the expression for Reynolds number,

$$Re = \frac{\rho u \ell}{\mu}. \tag{7.25}$$

Reynolds number includes not only velocity but also the kinematic viscosity μ/ρ and a characteristic length ℓ. It follows that if Re is an index of the flow pattern, then the same changes can be brought about by increasing the velocity, increasing the size or scale, or decreasing the kinematic viscosity. The characteristic length for pipe flow is the diameter, as it is for a cylinder lying across an airstream; for an airfoil section it is the chord length. These apparent inconsistencies are not as serious as they first appear, since Reynolds number is only used within geometrically similar situations. It is only necessary to remember what feature is used as the characteristic length in a particular family of cross-section geometries. In the case of pipes, there may be some doubt whether to use centreline or average velocities. Clearly it is much simpler to use average velocity of flow in a pipe, as long as it serves the purpose. The important thing is to know what the convention is.

Example 7.2

Calculate Re for a pipe of 5 cm inside diameter in which water at 20°C is flowing with a velocity of 5 m/s. From Table 7.1, μ = 0.001 005 N • s/m² and ρ = 998.2 kg/m³, so

$$Re = \frac{5 \times 0.05 \times 998.2}{0.001\,005} = 2.5 \times 10^5.$$

A check demonstrates that Re is dimensionless:

$$\frac{ud\rho}{\mu} = \frac{\text{m}}{\text{s}} \times \text{m} \times \frac{\text{kg}}{\text{m}^3} \times \frac{\text{m}^2}{\text{N} \cdot \text{s}}$$

$$= \frac{\text{kg} \cdot \text{m}}{\text{s}^2} \times \frac{1}{\text{N}}$$

but $$\mathrm{N} = \frac{\mathrm{kg} \cdot \mathrm{m}}{\mathrm{s}^2}, \text{ so } \frac{u\,d\,\rho}{\mu} \text{ is dimensionless.}$$

7.10 Flow in Pipes

In the case of pipe flow the principal coefficient of interest is the friction factor f, used in the Darcy-Weisbach equation

$$h_f = f\frac{\ell}{d}\frac{U^2}{2g} \tag{7.26}$$

where h_f is the loss due to friction measured as a height of fluid column, and U is the average velocity of flow over the pipe cross-section. Other factors are as previously defined. The friction factor f is dimensionless. Reynolds number provides the basis for relating friction factor f to flow situations, as shown in Figure 7.11 [**7.1, p. 4.10; 7.2**].

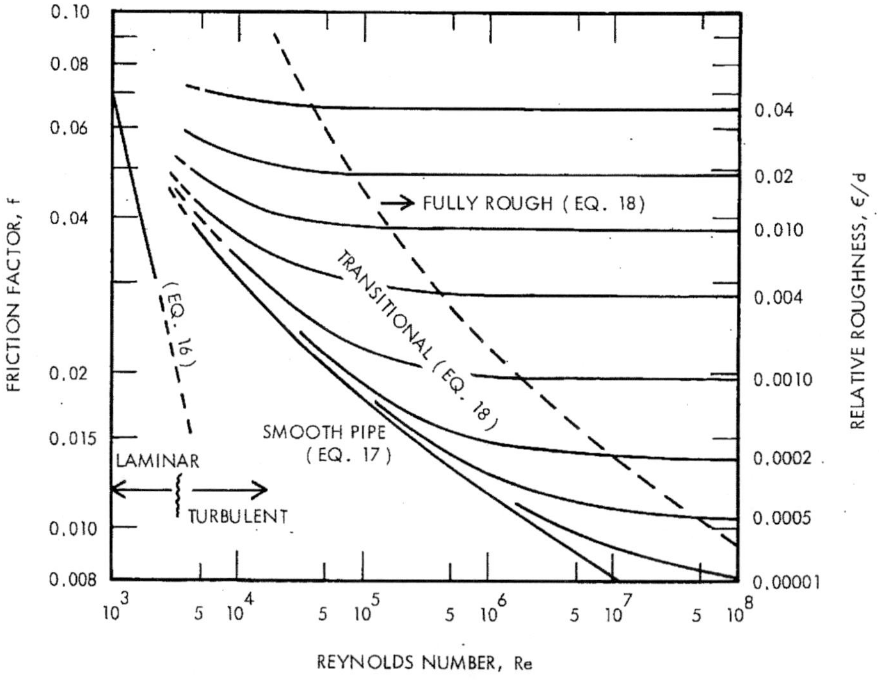

FIGURE 7.11 Friction factor and Reynolds number for pipe flow (equations 16, 17, and 18 given in Ref. **7.1**)
(From ASHRAE Handbook 1981 Fundamentals, Fig. 13, p. 4.10)[**7.1**]

Figure 7.11 shows that laminar flow exists for values of Reynolds number less than 3000. The friction factor f is also shown to vary, decreasing markedly with increasing Re. The reason for this is that the loss of head due to friction in laminar flow varies as the first power of velocity, while the Darcy equation to which f applies relates friction loss to the velocity head, which varies as the square of velocity. The friction factor must therefore vary inversely with velocity to compensate, as can now be shown. From the previous discussion of laminar flow and rearranging Equation 7.21,

$$p_1 - p_2 = \frac{32\mu \ell U}{d^2}$$

but

$$p = h\rho g$$

so

$$\rho g\, h_f = p_1 - p_2 = \frac{32\mu \ell U}{d^2}$$

and rearranging

$$h_f = \frac{32\mu \ell U}{\rho g d^2}.$$

But also

$$h_f = f \frac{\ell}{d} \frac{U^2}{2g}$$

and rearranging

$$f = h_f \frac{2gd}{U^2 \ell}$$

and

$$Re = \frac{Ud\rho}{\mu}$$

so

$$f = \frac{64\mu}{Ud\rho} = \frac{64}{Re}. \tag{7.27}$$

For laminar flow, then, f varies inversely with Reynolds number as already suggested and as shown by Figure 7.11.

In the turbulent-flow region of Figure 7.11, the friction factor for smooth pipes is shown to decrease with increasing Reynolds number but at a decreasing rate.

Pipe roughness becomes a significant factor in turbulent flow and determines a limiting value of f for any given roughness value, which is not changed by further increases in Reynolds number. Roughness must also be considered in relation to scale, and so is expressed as relative roughness ϵ/d. Values of ϵ in millimetres for some surfaces in new condition are given in Table 7.3 [**7.1, p. 4.10**].

When the kind and diameter of pipe are known, the value of ϵ/d in consistent units can be estimated for use with Figure 7.11.

TABLE 7.3

Values of Pipe Roughness
(From *ASHRAE Handbook 1981 Fundamentals*, Table 3, p. 4.11) [**7.1**]

Smooth brass, lead, copper, or transite	0.0015 mm
Steel and wrought iron	0.05 mm
Galvanized iron or steel	0.15 mm
Cast iron	0.25 mm

Example 7.3

A copper pipe with an internal diameter of 15 mm is installed to deliver hot water at 50°C at the rate of 0.25 litres per second. Find the friction loss in 100 metres of pipe.

For water at 50°C (Table 7.1), $\rho = 988.1 \text{ kg/m}^3$
$\mu = 0.000\,549 \text{ N} \cdot \text{s/m}^2$

$$\text{Area of cross-section} = \frac{\pi}{4} d^2 = \frac{\pi}{4} \times (0.015)^2 = 0.000\,177 \text{ m}^2.$$

$$\text{Average velocity } U = \frac{0.25 \times 10^{-3}}{0.000\,177} = 1.4 \text{ m/s}$$

$$\text{Reynolds number, } Re = \frac{1.4 \times 0.015 \times 988.1}{0.000\,549} = 37.800$$

From Table 7.3, Roughness, $\epsilon = 0.0015$ mm
and so $\epsilon/d = 0.0015/15 = 0.0001$.

From Figure 7.11, $f = 0.022$. Flow is transitional.

$$\text{Friction loss} = h_f = f\frac{L}{d}\frac{U_2}{2g}$$

$$= 0.022 \times \frac{100}{0.015} \times \frac{(1.4)^2}{2} \times \frac{1}{9.807}$$

$$= 14.6 \text{ m}.$$

Pressure equivalent $= h_f \times \rho g$
$= 14.6 \times 988.1 \times 9.807$
$= 141.5$ kPa.

The Darcy-Weisbach equation [**7.26**] or its equivalent is widely used as the basis for calculating losses for fluid flow in conduits as long as the volume changes due to the changes in pressure can be neglected and the flow thus regarded as incompressible. Various charts providing graphical solutions have been developed for particular purposes, such as the design of cold and hot water piping and the flow of air in ducts [**7.1**, chapters 33 and 34]. Some have been derived from specific test data, while others are adaptations of general correlations such as that shown in Figure 7.11.

It will be apparent that in many engineering calculations it is convenient to use U, the average velocity of flow in pipes and ducts. It should be appreciated, however, that in considering the energy in the flowing stream, the true velocity pressure is not given by $\rho\ U^2/2$ but must be found from integration over the cross-section, and can be expressed as $\alpha\ \rho U^2/2$ where the coefficient α has a value greater than 1. In the case of laminar flow in a circular conduit, the velocity head and the corresponding kinetic energy are exactly twice the value for a flow with a constant velocity U across the section. The corresponding value for turbulent flow is about 1.1. [**7.3, p. 200**]

7.11 Energy Losses in Conduit Flow

The Darcy-Weisbach equation gives the pressure head or pressure drop that is involved in overcoming the resistance to flow, which is also the energy required per unit weight of fluid to maintain the flow. This apparent loss in energy occurs in the shearing of the fluid and can also be regarded as the result of the drag of the pipe wall on the flowing stream. It is made up of the shearing in the longitudinal direction, which is the principal resistance in laminar flow, and of the shearing that occurs in a highly varied way when turbulence is also present. Eddies and related flow variations forming the structure of turbulence also involve shearing, are damped by the viscosity, and die out unless continuously generated. The energy that is dissipated in this way, forming heat which is taken up by the fluid, must finally be made up from the pressure energy of the fluid stream and so adds to the friction-pressure drop.

The uniform conversion of pressure energy to sensible heat that occurs owing to wall friction in steady flow can be augmented by disturbances arising from changes in conduit geometry and devices introduced for flow measurement and control. These include enlargement and contraction of the conduit, bends, elbows, and valves. Changes in pressure, velocity, and turbulence may be involved, with associated losses. The losses are commonly given as factors to be applied to the velocity head as shown in Table 7.4 [**7.1, p. 4.11**]. Although the actual losses might have been incurred partly as a result of effects persisting for many diameters downstream, they are taken in practice as having occurred at the fitting.

TABLE 7.4

Some Fitting Loss Coefficients for Turbulent Flow
(From *ASHRAE Handbook 1981 Fundamentals*, Table 4, p. 4.11) [**7.1**]

Fitting	Geometry	$K = \dfrac{\text{head loss}}{V^2/2g}$
Entrance	Sharp	0.50
	Well-rounded	0.05
Contraction	Sharp $\dfrac{D_2}{D_1} = 0.5$	0.38
90° elbow	Miter	1.3
	Short radius	0.90
	Long radius	0.60
	Miter with turning vanes	0.2
Globe valve	Open	10.0
Angle valve	Open	5.0
Gate valve	Open	0.19 to 0.22
	¾ open	1.10
	½ open	3.6
	¼ open	28.8
Any valve	Closed	∞
Tee	Straight thru flow	0.5
	Flow thru branch	1.8

Example 7.4

Find the increase in friction loss produced by incorporating three long-radius, 90-degree elbows in the copper pipe for Example 7.3.

From Table 7.4, the loss coefficient K = 0.6
From Example 7.3, U = 1.4 m/s

$$\text{Velocity head} = \frac{U^2}{2g} = \frac{(1.4)^2}{2 \times 9.807} = 0.100 \text{ m}$$

Friction head for 3 elbows = 0.6 × 0.100 × 3 = 0.18 m.

7.12 Flow Separation

The effect known as flow separation is of importance in many fluid-flow situations. A discussion of its role in conduit flow is appropriate here, and will serve also as an introduction to cases to be discussed later.

There will be little or no pressure variation over the cross-section of a fluid stream in steady straight flow, provided the differences in elevation are small enough to be neglected. When the flow lines must assume a curvature in order to follow a curving boundary surface, there must be pressure differences transverse to the flow. Quite simply, the flow can occur on a curved path only if there is an inward force to produce the necessary inward acceleration toward the centre of curvature. This can occur in accord with Bernoulli's equation if the velocity of flow at the outside of the curve decreases or if that at the inner side increases, allowing the necessary pressure gradient to develop.

When the flow is adjacent to a convex bounding surface, as in the case of the upper side of the airfoil in Figure 7.6, there can be situations when, with decelerating flows, it is no longer possible to maintain the flow parallel to the bounding surface. The flow breaks away at the boundary layer, following a tangential path from the separation point with an eddying back flow developing locally between it and the surface, as shown in Figure 7.12.

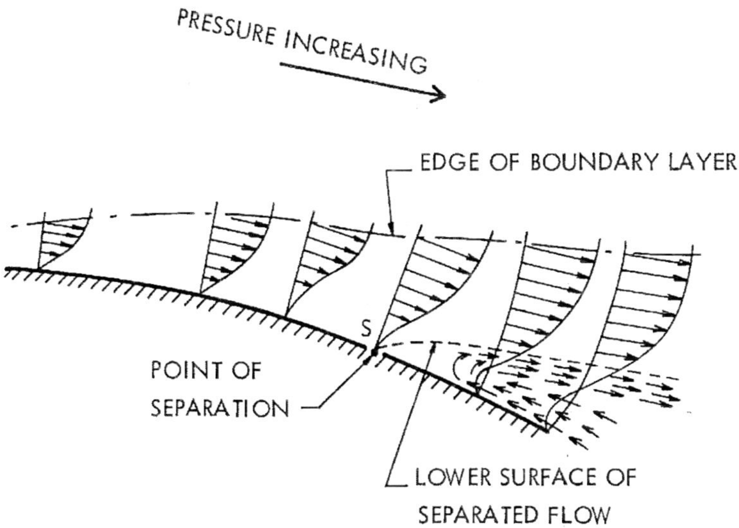

FIGURE 7.12 Boundary-layer separation.
(From Houghton, E.L., and Carruthers, N.B., *Wind forces on buildings and structures*, London: Edward Arnold Ltd., 1976, p. 6.6)[7.6]

An example of separation in conduit flow is provided by the diffuser shown in Figure 7.13. This simple device will be recognized as a component of the Venturi meter in Figure 7.2. It makes possible the recovery of kinetic energy by converting it to static pressure as the flow velocities decrease in accord with the increasing area of cross-section. When the slope of the wall is greater than about 1 in 10 the fluid is no longer able to adjust to maintain an orderly full flow, and a separation occurs. The attendant losses limit the static pressure regain that can be obtained. It is possible to achieve quite good recovery of velocity head if the cone angle is kept within proper limits. When the angle is small, separation is avoided, but the added length of path to achieve a given expansion introduces more friction losses which offset the gain from higher velocity conversion. The optimum cone angle is about 7°. Efficient diffusers of this type are of great interest for wind tunnels, where recovery of velocity head is important in order to reduce the power required.

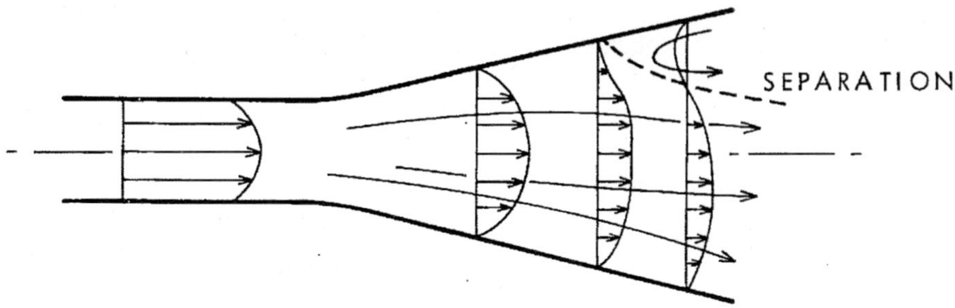

FIGURE 7.13 Flow separation in a diffuser.

Another example of separation is provided by the sharp entrance to a conduit as shown in Figure 7.14. Here the separation occurs at the sharp edge of the entrance, since the fluid is unable to change direction sharply to follow the conduit wall. Instead, it flows with a radially inward component past the sharp edge before expanding outward, creating the vena contracta. Eddying flow in the reverse direction occupies the space around the vena contracta, which will be at a reduced pressure as determined by the balance required to maintain stable flow. Such geometric separations are of interest in air flow around buildings and will be discussed in that context in another chapter.

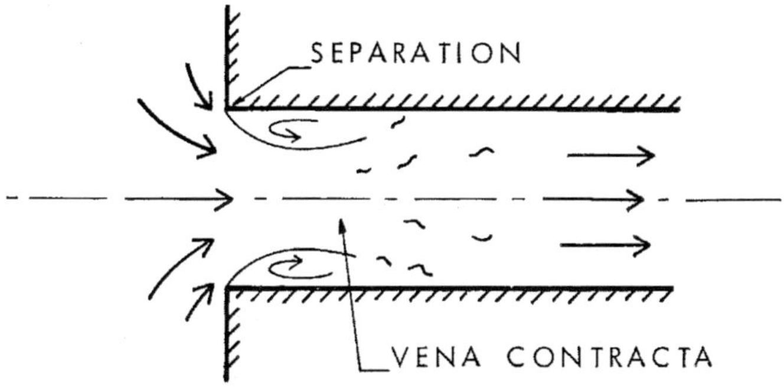

FIGURE 7.14 Geometric separation at a sharp entrance.

7.13 Flow around Immersed Bodies

The emphasis thus far has been on conduit flow, but it has many features in common with the flow around immersed bodies. Both are influenced by the boundary layers at bounding surfaces, and exhibit separation under certain conditions. The flowing stream is confined in conduit flow and it is convenient to think of the resistance to flow in terms of the drag of the pipe on the fluid and the loss in energy of the fluid stream. The volume flow of fluid involved in the case of immersed bodies is not readily identified, and in any event it becomes more convenient to deal with the force exerted by the flowing fluid on the immersed body.

The airfoil in Figure 7.6 is an example of a streamlined immersed body. It exhibits boundary-layer separation only when the angle between the chord dimension and the direction of the approaching flow (angle of attack) is increased appreciably. In the condition known as a stall there is a large turbulent wake and a large loss in lift. When operating in normal flight to produce high lift with low drag, the surface drag, which is governed by the boundary layers, is an important contributor to resistance to flight. A small turbulent wake results from the merging of the top and bottom boundary layers at the trailing edge.

Certain features of the flow around immersed bodies are well illustrated by the bluff body shown in Figure 7.15, which is a thick plate set at right angles to the approaching fluid. Boundary layers are not important in this case, since they have little chance to develop before the separation at the sharp forward edges of the plate takes over. The large turbulent wake is bounded by shear layers which separate it from the adjacent streamline flow in the freestream. These shear layers entrain fluid both from the wake and the freestream; this has a damping effect on the turbulence. The wake disturbance gradually becomes diffuse and tends to disappear as it is carried downstream in the moving fluid.

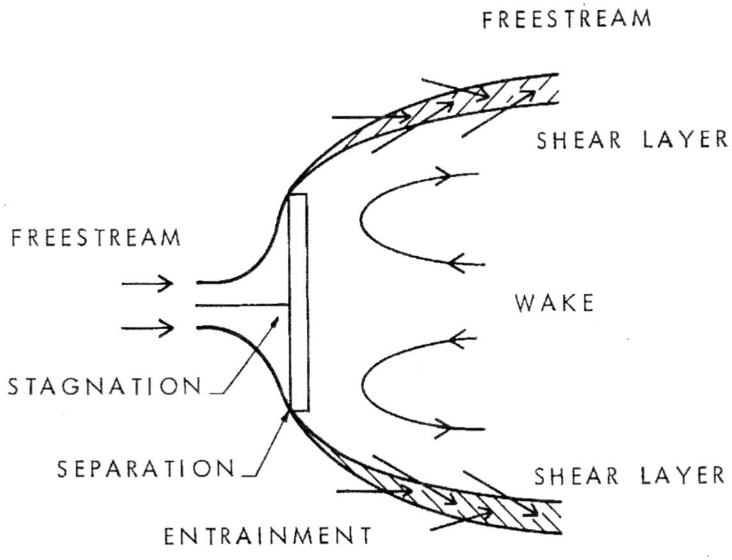

FIGURE 7.15 Principal flow features for immersed bluff body.
(From Aynsley, R.M., Melbourne, W., and Vickery, B.J., *Architectural aerodynamics*, London: Applied Science Publishers, 1977, Fig. 2.19, p. 68)[7.7]

The forces on the body in Figure 7.15 come from the pressure on its upstream face, which will be almost equal to the stagnation pressure, and from the suction or negative pressure of the wake. The large wake accounts for much of the drag force through its influence on the wake pressure. It does not change greatly with Reynolds number since the separation points are fixed by the sharp edges of the bluff body. Other shapes such as the sphere and the circular cylinder exhibit substantial boundary-layer effects which interact with and greatly influence the size of the wake, since shifting of the separation points can take place. The cylinder is of particular interest and merits more detailed consideration.

7.14 Flow around Circular Cylinders

The flow at right angles to a long circular cylinder changes markedly in character with changing Reynolds number [**7.4–7.6**]. At low values of *Re*, less than unity, the flow is entirely laminar and the fluid flows smoothly in behind the cylinder, leaving no appreciable wake. For values of *Re* from 1 to 10 the boundary layer is laminar and separation develops over a small area at the rear of the cylinder, producing a small turbulent wake. At values up to 60 the turbulent wake broadens as the separation points of the laminar boundary layer move outward and forward around the curved rear surface of the cylinder. With a further increase in *Re* the boundary layer remains laminar, but the separation points continue to move outward and reach a diametral position. A symmetrically disposed pair of vortices develops in the broad wake, but this changes at values of *Re* from 150 to 50 000 to a condition of alternate vortex shedding, first from one side and then from the other side of the cylinder. The boundary layer is still laminar, and the separation points shift forward and back alternately on each side in accord with the individual vortex generation. At this stage the wake, which is now quite broad, consists of a procession of equally spaced vortices alternating in the direction of rotation, shown in an elementary way in Figure 7.16. A similar wake pattern, known as a von Karman vortex street, can occur behind all long prisms of bluff cross-section at appropriate Reynolds numbers.

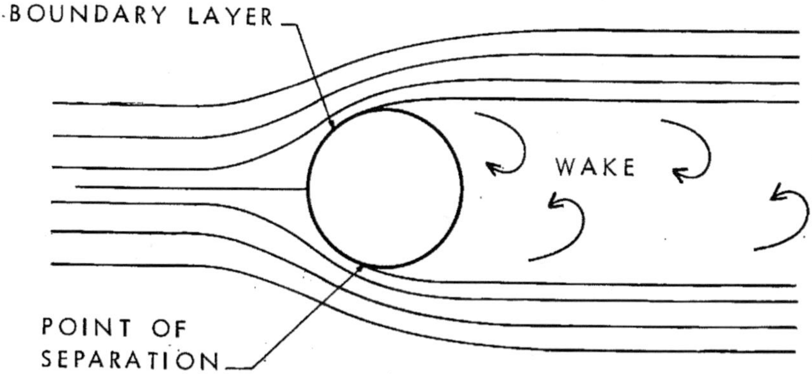

FIGURE 7.16 Separation and vortices behind a cylinder.

The generation of alternate vortices is accompanied by a loss in symmetry of the flow and gives rise to an oscillating transverse or lift force. The frequency is given by a dimensionless ratio known as the Strouhal number, equal to nd/u, which is uniquely related to Reynolds number [**7.6, p. 71**]. When the frequency coincides with the natural frequency of oscillation of a cylindrical structure, marked oscillations can be set up. This phenomenon produces singing of wires in the wind and has led to the destruction by wind of steel stacks and other flexible structures.

At a critical value of *Re* of about 4×10^5 the boundary layer develops a transition from laminar to turbulent in the region of the separation. The turbulent boundary layer is better able to resist separation, which now occurs more to the rear, reducing the wake and the total drag on the cylinder. The vortex pattern has now disappeared and the wake is turbulent and symmetrical. The pressures around the cylindrical surface for this condition are shown in Figure 7.17. Their components in the direction of flow summed over the circumferential area of a unit length of cylinder give the drag force per unit length.

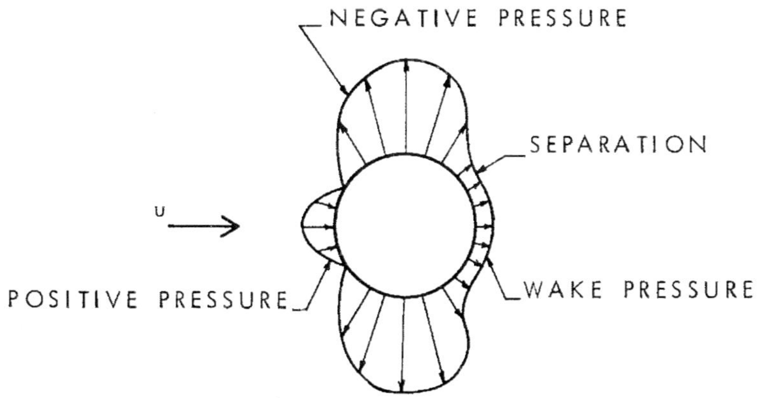

FIGURE 7.17 Pressure distribution on a circular cylinder.

7.15 Drag Coefficients for a Circular Cylinder

The drag force reflects directly the sequence of events in the changing flow pattern with increasing Reynolds number. It is a great convenience in such comparisons and in calculations to deal with a dimensionless drag coefficient, C_D, which is obtained by dividing the drag force per unit area by the velocity head, $\rho u^2/2$, where u is the freestream velocity. Thus,

$$C_D = \frac{2}{\rho u^2} \cdot \frac{\text{drag force}}{\text{area}}. \tag{7.28}$$

The drag may be taken for unit length of cylinder. The area will then be unit length times the characteristic dimension to which C_D and *Re* are to be referred—in this case the diameter d. Drag coefficients for other bodies must always be related to the proper characteristic dimension and area.

The change in drag coefficient for the circular cylinder with changing Reynolds number is shown in Figure 7.18 [**7.5, p. 8**]. The marked decrease in C_D as *Re* is increased from a low value reflects the initial true laminar flow for which the drag is proportional to u and not to u^2 as used in the calculation of C_D. Like the laminar pipe flow in Figure 7.11, the drag coefficient must decrease inversely with u or with *Re* to compensate.

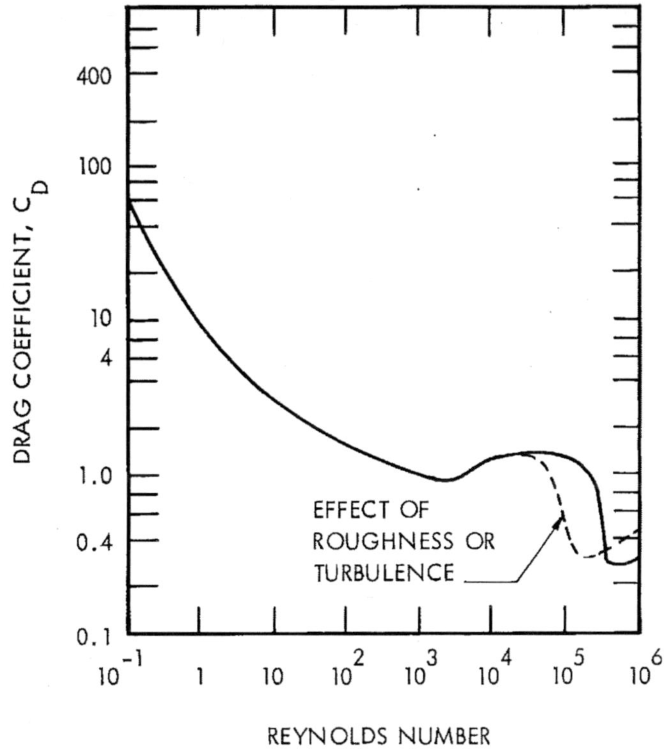

FIGURE 7.18 Drag coefficient variation for a cylinder.
(From White, F.M., *Viscous fluid flow*, New York: McGraw-Hill, 1974, Fig. 1.6, p. 8)[**7.5**]

The other outstanding feature is the marked reduction in drag and thus in C_D as the boundary layer becomes turbulent in advance of separation. The conditions at which this will occur are related mainly to Reynolds number but partly to surface roughness and to initial turbulence in the approaching freestream, both of which will promote the earlier development of the turbulent boundary layer. The dotted second curve shown in Figure 7.18 depicts such a condition.

7.16 Flow Past Spheres

The relationship between C_D and Re for spheres is much like that for circular cylinders. There is an inverse relationship for values of $Re < 1$ represented by $C_D = 24/Re$. According to Equation 7.28,

$$\text{drag} = C_D \cdot \rho \frac{u^2}{2} \cdot \text{area}$$

but substituting area $= \dfrac{\pi d^2}{4}$ and $C_D = \dfrac{24}{Re}$ and $Re = \rho \dfrac{ud}{\mu}$

$$\text{drag} = \frac{24\mu}{\rho u d} \cdot \rho \frac{u^2}{2} \cdot \frac{\pi d^2}{4}$$

$$= 3\,\pi\mu u d. \tag{7.29}$$

This is the Stokes equation for laminar flow past spheres. It can be applied to free-falling spheres used to measure viscosity and to the settling of fine particulate matter for conditions of $Re < 1$.

7.17 Circular Plate Normal to Flow

A circular plate provides a drag coefficient curve not unlike those for spheres and cylinders at Reynolds numbers up to 10^3. It shows a slope at $Re < 1$, which is very close to that for the sphere, indicating that in laminar flow the fluid flows in behind the plate almost as readily as in the sphere. At higher values of Re, above 10^3, the coefficient has been shown to be nearly constant at a value of about 1.1. For the reasons discussed in connection with Figure 7.15, the point of separation does not change and so the wake and the drag are not greatly influenced by increasing Reynolds number. More recent work has shown that freestream turbulence has an influence on the wake pressure and the drag [**7.7, pp. 67-71**].

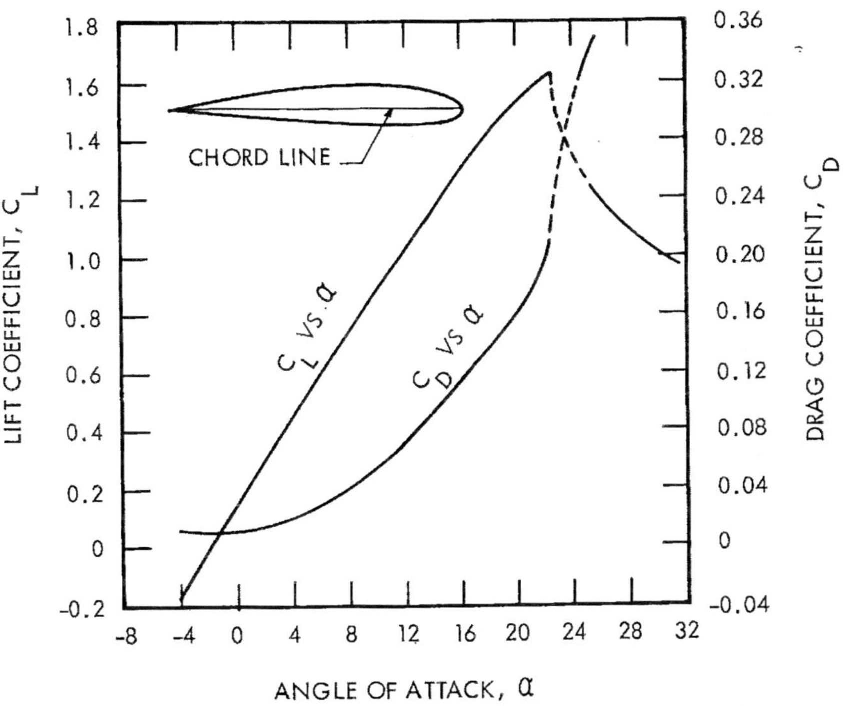

FIGURE 7.19 Lift and drag coefficients for an airfoil.
(From Jacobs et al., *The characteristics of 78 related airfoil sections from tests in the variable-density wind tunnel*, NACA Technical Report 460, National Advisory Committee on Aeronautics, 1939)[**7.8**]

7.18 Lift and Drag on an Airfoil

Lift and drag coefficients for a particular airfoil are shown in Figure 7.19 [7.8]. Both lift and drag increase with an increasing angle of attack, but not at the same rate, so that there is an optimum lift-to-drag ratio at some intermediate angle before the point of maximum lift is reached. Separation of the flow over the convex upper surface destroys the lift, and the drag coefficient increases rapidly at an angle of attack of 22 degrees, as shown in Figure 7.19.

The practice with airfoils is to use the chord line as the characteristic dimension. The wing area based on the chord times wing length is used as the basis for lift and drag coefficients. The lift force is given by the product of the lift coefficient C_L, the velocity pressure $\rho u^2/2$, and the area, in the same way as for drag in Equation 7.20.

7.19 Flow along a Flat Plate

The flow over a thin flat plate set parallel to the approaching stream provides the basis for further appreciation of the boundary layer and surface drag. Only the thin flat plate, with symmetrical flows on opposite sides, will be considered, to avoid the complications posed by the leading edges of plates of appreciable thickness and by arrangements confining the flow to one side of the plate. There can be no separation, so only skin friction is involved in the drag force in much the same way as in the straight circular conduit of Figure 7.10. The boundary layer grows in thickness with distance x from the leading edge and changes in character in accord with Reynolds number Re_x calculated using x as the characteristic length.

At low values of Re_x up to a critical value of about 500 000, for smooth plates the boundary layer will be laminar. With increasing Reynolds number the flow will become turbulent over that portion of the plate on which Re_x exceeds the critical value, giving rise to the situation shown for one side of the plate in Figure 7.20.

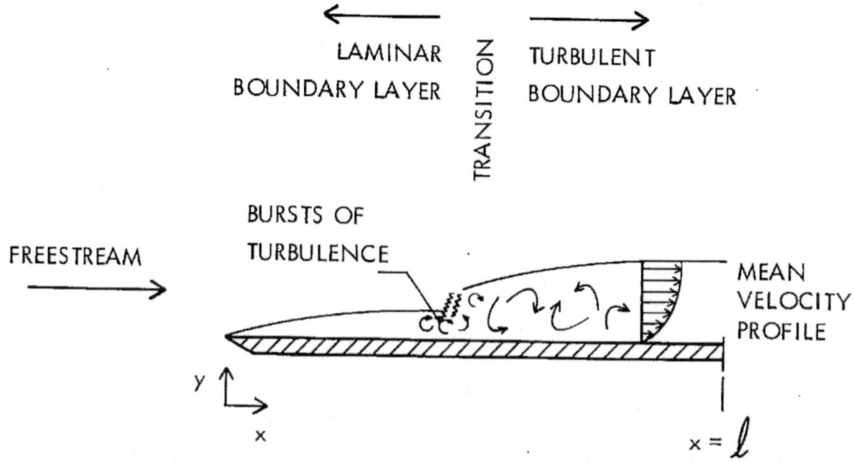

FIGURE 7.20 Transition from a laminar to a turbulent boundary layer.

Reynolds number, Re_ℓ, based on the plate length ℓ, refers to conditions at the trailing edge and can also be used as the index of flow over the plate length. A turbulent boundary layer is initiated at the trailing edge when Re_ℓ exceeds the critical value, and the transition point will shift gradually toward the leading edge until the boundary layer becomes fully turbulent at a value of Re_ℓ of about 15×10^6.

The overall drag coefficient C_D is referenced to Re_ℓ and is based on the integrated effect of the variations along the plate length. The drag coefficient for laminar flow (one side only) for a unit area of plate having a length ℓ in the direction of flow is

$$C_D = \frac{1.33}{\sqrt{Re_\ell}} \tag{7.30}$$

and for turbulent flow is

$$C_D = \frac{0.074}{(Re_\ell)^{1/5}}. \tag{7.31}$$

These values were developed from the work of Blasius, Prandtl, and others [**7.3**]. A more recent review [**7.5**] confirms that the Prandtl–Schlichting formula for turbulent flows gives more accurate values:

$$C_D = \frac{0.455}{(\log Re_\ell)^{2.58}}. \tag{7.32}$$

The above formulae are for smooth, or relatively smooth, plates. As in the case of pipe flow, surface roughness has little influence (within wide limits) on laminar flow, but can markedly increase the resistance for turbulent flow and can also promote the transition to turbulent flow at lower critical Reynolds numbers.

Von Karman has shown that for a turbulent boundary layer the velocity varies as the $\frac{1}{7}$ power of the distance from the surface, so that

$$u = U_1 \left(\frac{y}{\delta}\right)^{1/7} \tag{7.33}$$

where

u = the velocity at a distance y from the bounding surface
U_1 = the velocity at the outer edge of the boundary layer
δ = thickness of the boundary layer.

A similar relationship has been shown to apply in pipe flow with pipe radius substituted for boundary-layer thickness.

The $\frac{1}{7}$ power law has also been applied to the case of the earth's boundary layer to find the variation of wind speed with elevation above the ground. It has been found experimentally that the $\frac{1}{7}$ power closely fits the profiles of wind speed over flat open country, but for greater surface roughnesses higher exponents should be used, up to $\frac{1}{2}$ for the centres of large cities [**7.7**].

7.20 Wind Tunnels and Model Testing

Extensive use has been made of wind tunnels in studies of the aerodynamic effects of the flow of air over bodies of various shapes. Most wind tunnels have been constructed for aeronautical purposes. One such tunnel is shown in Figure 7.21; air streams of the desired velocity can be produced in the closed working section in which the model under test is suitably mounted. Pressures over the surface of the model and total forces and moments exerted by the air flow can be measured with pressure taps and with special balances. In aeronautical work the simulation of flight in still air requires that the velocity be constant over the test section. The turbulence in the test stream is characteristic of the design of the tunnel, being determined largely by the propellor and turning vanes unless other turbulence-generating devices are introduced. Many of the low-speed tunnels that were much in demand for aeronautical work before the days of high-speed flight are now used extensively for other kinds of studies, including wind effects on buildings as discussed in Chapter 10.

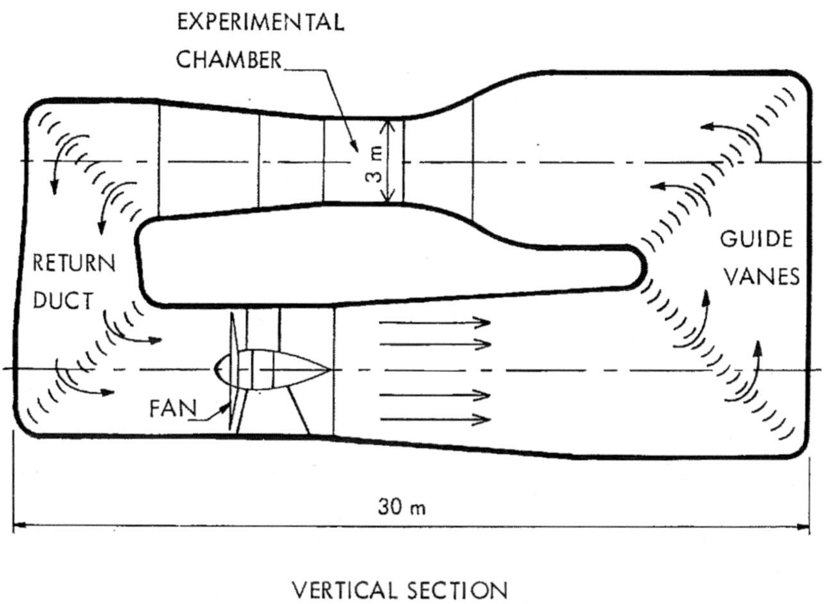

FIGURE 7.21 Closed-throat closed-return type of wind tunnel.

Wind tunnels are costly and can seldom be made large enough to accommodate airplanes to full scale. They were normally used to test small-scale models of aerodynamic shapes. For example, when tests were being run at $1/10$ geometric scale and $1/3$ real air speed, the test Reynolds number was only $1/30$ of that for full scale. A look at Figure 7.18 will serve as a reminder of the possible influence of Reynolds number, particularly when curved surfaces are involved, and will indicate the problem facing the designer in translating model results to full scale. There can be changes in the drag coefficient, particularly for rounded shapes with the possibility of marked flow transitions, as in the case of the sphere and the cylinder. Turbulence can also have an influence. The aeronautical designer who used these tunnels was primarily interested in flight of aerodynamically refined shapes through nonturbulent air and was concerned with the scale effects on skin friction and on flow over curved surfaces. Much ingenuity was brought to bear on these difficulties, even to the point of building a variable-density tunnel pressurized to 20 atmospheres or more to increase the test Reynolds number through the decreased kinematic viscosity resulting from such increased air densities.

Building designers have somewhat different problems [7.6, 7.7]. They are concerned in general with bluff, even sharp-edged, bodies. Since these shapes are seldom aerodynamically refined, the skin drag will be relatively small and, it is argued, points of flow separation will be determined by sharp edges and will not vary to influence greatly the size of the wake and the associated drag as Reynolds number is increased. Another problem in the model testing of buildings is the proper simulation of the turbulence and velocity-gradient conditions of the earth's boundary layer. These and other matters are discussed in Chapter 10.

7.21 Sound and Sonic Booms

The transmission of sound is a special kind of fluid-dynamics problem involving the transmission of sound waves through fluids. In air, the molecules are put into oscillatory motion leading to alternate rarefaction and compression; these pulsations, not the air itself, are transmitted. The velocity of sound in air is given by

$$c = \sqrt{kP/\rho} \tag{7.34}$$

where P is the atmospheric pressure, ρ is the density, and k is the ratio of specific heat at constant pressure to that at constant volume and has the value 1.41. The velocity of sound in air at standard conditions is 341 m/s.

Sound becomes involved in aerodynamics when flow velocities begin to approach the speed of sound. The sonic boom, which can have a dramatic destructive effect on buildings under certain conditions, is produced by aircraft flying at and above the speed of sound. The expanding spherical fronts of sound waves from successive aircraft positions cannot keep up with the aircraft, and merge like the bow waves from a boat to form a trailing conical surface of intensified pressure waves known as a shock front. The shock waves and associated compressibility effects become increasingly involved in the dynamics of fluid flow as velocities increase. An index similar to Reynolds number is used to characterize these kinds of flow patterns. It is known as the Mach number, and is given by

$$M = \frac{u}{c} \tag{7.35}$$

where u is the flow velocity and c is the speed of sound at the existing freestream conditions. It begins to be of significance at about $M = 0.5$, or half the speed of sound. Such acoustic effects have long been encountered in the flow of compressible fluids such as steam and air through orifices under substantial pressure differences. Today they are commonplace in the study of aerodynamics of high-speed flight up to and beyond the speed of sound.

7.22 Air Flow Induced by Jets

One of the problems encountered in manipulating the indoor environment is the need to ensure distribution of a conditioned air stream introduced into a room. The subject is of great practical importance and much attention has been directed to it [**7.1, Chap. 32**]. Reference is made to it here mainly to identify the flow-inducing action of jets as another significant aspect of fluids in motion.

A jet of air directed into a space entrains room air at its boundary and transfers energy to it. The volume of air in the stream increases and, if circumstances permit, will expand with an angle of divergence of about 22 degrees until it merges in the general room-air circulation at some considerable distance from the point of entry. Such a jet is readily affected by small forces and so is influenced by room-air turbulence and movement. It is also affected by buoyancy forces when it is warmer or cooler than the room air. A horizontal jet of cool air will fall, while a warm one will rise as it is projected into the room. In the case of jets discharged vertically, similar buoyancy forces reinforce or weaken the jet flow. The air at the jet boundary is caused to flow in the direction of the jet and becomes part of a much wider secondary circulation in the space.

7.23 Fluid Flow and Heat Transfer

When a fluid at one temperature is caused to flow over a surface at some other temperature, heat is transferred. The rate of transfer is dependent not only on the temperature difference but also on the characteristics of the fluid and of the flow which develops over the surface. When the flow is produced by jets, pumps, or fans, the process is called forced convection. Flow can also be induced by differences in density in a body of fluid. The circulation brought about by differences in density due to temperature changes is called natural convection. It has implications for the circulation and mixing of fluids as well as for heat transfer. Like forced convection, it is best discussed in the context of heat transfer (as will be done in the following chapter), but there are some fluid aspects that should be recognized here.

Natural convection in a fluid takes place as a result of heating or cooling at bounding or contacting surfaces provided the fluid is free to move in response to the changes in density. Circulation may be inhibited in the case of upward-facing, horizontal cool surfaces if the cooled fluid is retained against the surface by bounding walls in much the same way as water is held in a swimming pool. Heated fluid may be retained in a corresponding way under a hot, downward-facing surface. Thus, the transfer of heat by convection downward across a closed air space between horizontal parallel plates will be zero, while that for upward heat flow will be relatively great. The convective transfer across a closed air space between vertical parallel plates involves an upward flow at the warm side and a downward flow at the cold side. Convective circulation can also take place between heating and cooling surfaces that are more widely separated, as in the case of heated rooms with cold walls and windows.

The temperature in a fluid adjacent to a heated vertical plate varies from the original fluid temperature at some distance from the plate to the surface temperature at the plate. The changes in density of the fluid adjacent to the plate, varying in accord with the temperature profile at various levels, provide the force for the upward movement of fluid. The velocity profiles and the temperature profiles that give rise to them are closely linked; both change with distance above the lower or leading edge of the plate. Temperatures and velocities tend to increase at higher levels as more heat is transferred to the fluid while more of the adjacent fluid is drawn into the rising stream. A boundary layer characteristic of the conditions is thus developed, with properties that change along the flow path. The flow is frequently laminar, but turbulent conditions can also be created.

7.24 Convection Plumes

One interesting example of natural convection is the plume of heated air that rises from a hot source. It has been shown [**7.9**] that the plume grows with its diameter increasing linearly with distance above the source. Such information can be used to design hoods and vent openings to capture hot plumes containing unwanted heat, smoke, or vapours. There may be situations where it is important to recognize that such plumes do develop and often disturb existing conditions.

7.25 Conclusion

The motion of fluids, which is encompassed in the important and well-recognized field of study known as fluid mechanics, has been examined in relation to buildings. Wind forces on buildings will be examined in some detail later. Conduit flow has been reviewed, as have some other basic aspects in preparation for the identification and later development of many related topics. It is evident that the air circulation in a room is a function of the energy in the entering air stream and the buoyancy forces created by heating and cooling of air within the room in combination with the properties of air and the mechanics of its motion. When heat transfer is involved, fluid flow and heat transmission must be considered together, as will become more evident in the next chapter.

References

7.1 *ASHRAE Handbook 1981 Fundamentals.*

7.2 Moody, L.F. Friction factors for pipe flow. *ASME Transactions*, 1944, *66*, p. 672.

7.3 Dodge, R.A., and Thompson, M.J. *Fluid mechanics.* New York: McGraw-Hill, 1937.

7.4 Prandtl, L., and Tietjens, O.G. *Fundamentals of hydro and aeromechanics.* New York: McGraw-Hill, 1934, p. 259.

7.5 White, F.M. *Viscous fluid flow.* New York: McGraw-Hill, 1974.

7.6 Houghton, E.L., and Carruthers, N.B. *Wind forces on buildings and structures.* London: Edward Arnold Ltd., 1976.

7.7 Aynsley, R.M., Melbourne, W., and Vickery, B.J. *Architectural aerodynamics.* London: Applied Science Publishers Ltd., 1977.

7.8 Jacobs, E.N., Ward, K.E., and Pinkerton, R.M. *The characteristics of 78 related airfoil sections from tests in the variable-density wind tunnel.* NACA Report 460, National Advisory Committee on Aeronautics, 1939.

7.9 Hatch, T.F., and Barron-Oronzco, D. Air flow in free convection over heated bodies. *ASHAE Transactions,* 1957, *63,* pp. 275-90.

8

Heat Transfer

8.1 Introduction

Energy is a dominant factor in the thermal environment. Its movement or transfer, sometimes as radiation and sometimes as thermal or molecular energy, is a subject worthy of study in considerable detail. Heat transfer is strongly linked to radiation and radiative exchanges on the one hand, and involves conductive and convective exchanges on the other. It also has a strong link with changes in state of various substances of which water is the one most commonly involved. The topics of earlier chapters, although important in their own right, have been arranged with some care so that each one, building on the others, contributes to the background necessary for a proper understanding of heat transfer, which can now be approached directly.

8.2 The Heat-Flow Equation

The basic relationship for heat flow is given by Fourier's law, which states that

$$\frac{dq}{d\theta} = -kA\frac{dt}{dx} \tag{8.1}$$

where q is the quantity of heat in time θ, $dq/d\theta$ is the rate of heat flow, A is the area transverse to the flow, dt/dx is the temperature gradient, and k is the coefficient of thermal conductivity.

It is sufficient in many cases to deal with steady-state heat flow in one direction, so that the conduction equation can be written

$$q = A\frac{k}{\ell}(t_1 - t_2) \tag{8.2}$$

where q is the rate of heat flow, ℓ is the length of the flow path or thickness of the material, and t_1 and t_2 are the temperatures at either end of the flow path, so that $(t_1 - t_2)$ is the temperature difference producing flow. This equation and its variations will be used in the discussions to follow. Solutions to more complicated situations involving flow along two or three axes, unsteady flow, variable cross-section of flow path, or variation of conductivity with temperature must be sought in various texts on heat transfer and in their references to the pertinent literature [**8.1**, **8.2**].

Equation 8.2 describes simple steady-state flow in one direction along a path of constant cross-section in a material with thermal properties that are independent of temperature. The expression for k is found by rearrangement:

$$k = \frac{q}{A}\frac{\ell}{(t_1 - t_2)}. \tag{8.3}$$

The preferred SI units are

$$k = \frac{\text{(watts)(metres)}}{\text{(square metres)(kelvins)}}$$

$$= \frac{\text{watts}}{\text{metres} \times \text{kelvins}} = \text{W/m} \cdot \text{K}.$$

The coefficient of thermal conductivity k can be visualized as the heat flow in watts through a uniform one-metre cube of material between two opposite faces which differ in temperature by one kelvin. The cross-section of the flow path is one square metre and the length is one metre. The unit of time is one second, but this is implicit in the use of watts which have the dimensions of joules per second.

Real materials can be compared and rated on the basis of the one-metre cube as to k value. The length of the flow path, ℓ, for individual materials will be given by their thickness and will usually be much less than one metre, but it is still convenient to refer to k value and to calculate the heat flow for other thicknesses as required. When the product is only available in one or two discrete thicknesses and is not homogeneous but consists of several layers, as in the case of plasterboard, or has a particular geometry, like concrete block, it is more appropriate to refer to the conductance, C, having the units k/ℓ, referring to the thickness stated or implied, not to one metre in thickness. The coefficient of conductance C has the units W/(m^2 • K).

8.3 Coefficients for Building Elements

Values of the coefficients of conductivity or conductance as appropriate are given in Table 8.1 for a selected list of materials commonly used in building. Reference to the original sources [**8.1, p. 23.14**] will provide a broader coverage for purposes of design. All values are for a mean temperature of 24°C except those for concrete block with rectangular cores, which are for 7°C. Although most materials exhibit a variation with temperature, data are not always available for more than one temperature.

The wide range of conductivities for different classes of materials should be noted. They vary roughly in accord with density, and good correlations exist between thermal conductivity, k, and density, ρ, for certain groups of materials such as the concretes, as illustrated by the values assigned to them on the basis of density. The low conductivity for still air at 0.025 is equalled by a few of the very light foamed insulations. When air is not still but is circulating under buoyancy or other forces, the apparent conductivity increases greatly (as will be shown later). Most insulating materials are effective because they prevent air circulation and block radiative transfer in a space while, being of low density, they add little in the way of material through which conductive heat transfer can take place [**8.3**]. Materials used for thermal insulation in building will have k values up to 0.07 W/(m • K).

Many masonry products have conductivities grouped about k = 0.7 W/(m • K) but ranging quite widely, depending on density, from as little as 0.10 to 1.5 or more. Wood, by comparison, is both light and strong so that it is a relatively good insulator as well as a structural material. Metals are all relatively high in conductivity compared with nonmetals, but show a surprising range of values.

All values shown, with one exception, are for materials in the dry state. The presence of water can influence heat transfer, but the effect is not a simple one, as already discussed in Chapter 5. It is common practice to base calculations on the assumption of dry materials and to make adjustments for moisture only in cases where it is likely to have a substantial influence.

Most values for nonmetallic building materials ranging in thickness from 6 to 100 mm have been obtained from tests on sample sizes appropriate to the thickness in a form of thermal-conductivity apparatus known as the guarded hot plate [**8.4**]. Products such as hollow masonry units which are relatively thick or which are nonhomogeneous must be made into panels and tested in a hot-box apparatus [**8.5**]. This apparatus is also used to test complete wall, floor, and ceiling panels and windows and doors. It is difficult to devise suitable ways to meter heat transfer accurately and, at the same time, provide realistic conditions of radiative and convective heat exchange at the panel surfaces.

TABLE 8.1

Thermal Conductivities and Conductances of Building and Insulating Materials

(From *ASHRAE Handbook 1981 Fundamentals*, Table 3A, p. 23.14–23.17) [**8.1**]

Material	ρ kg/m³	k W/(m • K)	C W/(m² • K)
Building board			
Asbestos cement board	1920	0.58	—
Gypsum or plaster board, 9.5 mm	800	—	16.6
Gypsum or plaster board, 12.5 mm	800	—	12.5
Plywood	545	0.115	—
Insulating board, regular	290	0.055	—
Hardboard, medium density	800	0.105	—
Particle board, low density	590	0.078	—
Particle board, medium density	800	0.136	—
Particle board, high density	1000	0.170	—
Building paper			
Building paper	—	—	95
Vapour barrier, plastic film	—		negligible

TABLE 8.1 Continued (1)

Material	ρ kg/m³	k W/(m • K)	C W/(m² • K)
Insulating blankets and batts			
Mineral fibre 50–68 mm	5–32		0.81
(rock, slag, or glass) 75–88 mm	5–32		0.52
89–162 mm	5–32		0.30
Insulating boards and slabs			
Cellular glass	136	0.555	—
Glass fibre, organic bonded	65–145	0.036	—
Expanded polystyrene			
extruded, cut cell surface	29	0.036	—
extruded, smooth skin surface	35	0.029	—
extruded, smooth skin surface	56	0.027	—
Molded beads	16	0.040	—
Expanded polyurethane			
R-11 expanded	24	0.023	—
Mineral fibre, resin binder	240	0.042	—
Wood fibreboard, interior finish	240	0.050	—
Insulating materials loose fill			
Cellulose insulation (milled paper)	37–50	0.039–0.046	—
Sawdust or shavings	128–240	0.065	—
Mineral fibre (rock, slag, or glass)			
approximately 95–127 mm	10–32	—	0.52
approximately 165–222 mm	10–32	—	0.30
approximately 190–254 mm	10–32	—	0.26
approximately 260–350 mm	10–32	—	0.19
Vermiculite, expanded	110–130	0.068	—
	65–95	0.064	—
Roof insulation			
Various types supplied in thicknesses to provide the rated conductance			
Masonry materials			
Cement mortar	1860	0.72	—
Lightweight concretes	1920	0.75	—
(Various aggregates)	1600	0.52	—
	1280	0.36	—
	960	0.25	—
	640	0.17	—

Table 8.1 Continued (2)

Material	ρ kg/m³	k W/(m · K)	C W/(m² · K)
	480	0.13	—
	320	0.10	—
Dense concrete, dry	2250	1.32	—
Dense concrete, not dry	2250	1.82	—
Stucco	1860	0.72	—
Masonry units			
Brick, common	1920	0.72	—
Brick, face	2080	1.32	—
Clay tile, hollow:			
1 cell 100 mm	—	—	5.0
2 cells 200 mm	—	—	3.0
3 cells 300 mm	—	—	2.3
Concrete blocks, 3 oval core:			
Sand and gravel aggregate 200 mm	—	—	5.0
Lightweight aggregate 200 mm	—	—	2.9
Concrete blocks, rectangular core:			
Sand and gravel 2 core 200 mm	—	—	5.6
Same with insulation-filled cores	—	—	2.9
Lightweight aggregate 2 core 200 mm	—	—	2.6
Same with insulation-filled cores	—	—	1.1
Plastering materials			
Cement plaster, sand aggregate	1860	0.72	—
Gypsum plaster, light aggregate	720	0.23	—
Gypsum plaster, sand aggregate	1680	0.81	—
Gypsum plaster, vermiculite aggregate	720	0.25	—
Roofing			
Asphalt shingles	—	—	12.5
Built-up roofing 9.5 mm	—	—	17
Wood shingles	—	—	6.0
Siding materials			
Asbestos cement shingles	—	—	25
Wood shingles	—	—	6.7
Wood siding, bevel, 13 × 200 mm	—	—	7.0
Wood siding, bevel, 19 × 250 mm	—	—	5.4
Wood plywood 9.5 mm	—	—	10.0

Table 8.1 Continued (3)

Material	ρ kg/m³	k W/(m · K)	C W/(m² · K)
Metal siding, hollow backed over board sheathing	—	—	9.0
Architectural glass	—	—	50
Wood			
Maple, oak, and similar hardwoods	720	0.16	—
Fir, pine, and similar softwoods	510	0.12	—
Metals			
Aluminum	2 740	220	—
Brass, yellow	8 300	120	—
Copper	8 900	390	—
Lead	11 300	35	—
Nickel	8 890	60	—
Steel, mild	7 830	45	—
Miscellaneous			
Glass, soda lime	2470	1.0	—
Air, still	1.2	0.025	—
Water, still	1000	0.60	—

Metric conversions for heat transfer

U or C Btu/(ft² · hr · °F) × 5.678 = W/(m² · K)
k-ft Btu/(ft · hr · °F) × 1.730 = W/(m · K)
k-in. Btu · in./(ft² · hr · °F) × 0.1442 = W/(m · K)
Heat flow Btu/(ft² · hr) × 3.155 = W/m²
Specific weight, lb/ft³ × 16.02 = kg/m³ (mass density).

8.4 Conversion to the Metric System

At present, North America is in various stages of conversion to the metric system. Canada is fully committed to the system; conversion in the construction industry began in 1978 and was completed officially by 1980. In the United States no time commitment has been made but various preparatory steps have been taken. Various standards and guides for metric conversion and the establishment of uniform practices have been adopted in Canada [**8.6, 8.7**] and the United States [**8.1, Chap. 37**]. Although many details have been finalized, others will only become known as product and industry standards and practices change in response to the needs and opportunities presented by the adoption and use of the International System of Units (le Système international d'unités) (SI).

All the values given in Table 8.1 and the others to follow have been converted to SI from the original data in British thermal units and the foot–pound system. These data are, of necessity, based on practices and product dimensions in current use.

The dimensions of buildings in Canada are to be based on a module of 100 mm [**8.8**]. Boards that were formerly 4 by 8 ft (1219 × 2438 mm) will be reduced in size to fit into an area 1200 × 2400 mm. Similar changes can be expected in many product dimensions. Those that affect the thickness of walls, insulation, and air spaces are of particular significance for heat-flow calculations.

Thermal conductivity was formerly expressed in two ways. One method was based on a thickness of one inch and led to a mixture of units when an area was given in square feet. In the other less common but fundamentally more correct approach the coefficient was based on consistent units for a basic thickness of one foot, so that it related to a one-foot cube. The use of these two sets of units introduces the possibilities of error in conversion unless they are properly identified.

8.5 Heat Transmission through Building Sections

In most cases involving heat transmission through walls, windows, doors, floors, ceilings, and roofs it is necessary to establish the coefficients by calculation rather than by test. Tests may be called for when the arrangement is not amenable to calculation. Values for the individual elements making up the flow path must be combined appropriately. The procedure (already demonstrated in the case of vapour transmission in Chapter 5) involves adding the resistances of the various elements to find the resistance to heat flow of the combination.

The resistance for one square metre of material with a thickness of one metre is given by $1/k$ and has the units (m • K)/W. The conductance of an element one square metre in area, ℓ metres thick, with a coefficient of thermal conductivity k is given by

$$C = k/\ell \qquad (\mathrm{W/(m^2 \cdot K)}) \tag{8.4}$$

and its resistance is

$$R = \frac{1}{C} = \frac{\ell}{k}. \qquad (\mathrm{m^2 \cdot K/W}) \tag{8.5}$$

The heat transmission through one square metre of a composite section is

$$\frac{q}{A} = U(t_i - t_o) \qquad (\mathrm{W/m^2}) \tag{8.6}$$

where q/A is the heat-flow rate, U is the overall coefficient of heat transmission in W/(m² • K), t_i is the inside or warm-side air temperature, and t_o is the outside or cool-side air temperature. The U-value thus applies to the heat transfer air-to-air. The total resistance of the building section is

$$R_{\text{total}} = \frac{1}{U} = R_1 + R_2 + R_3 \ldots + R_n \tag{8.7}$$

where R_1, $R_2 \ldots R_n$ are the resistances of the individual elements. Since this includes air-to-surface energy exchange at each side, and surface-to-surface exchange across any spaces, it is necessary to include conductance coefficients f_i, f_o, and a to account for them, neglecting for the time being the complications of combined radiative and conductive-convective exchanges in establishing appropriate values for these coefficients. The U-value for a composite section consisting of sheathing on either side of an air space can be written

$$U = \frac{1}{\frac{1}{f_i} + \frac{\ell_1}{k_1} + \frac{1}{a} + \frac{\ell_2}{k_2} + \frac{1}{f_o}}. \tag{8.8}$$

It can also be shown by the method described in Chapter 5 for vapour transmission and accepted here without further argument that the temperature drop across each element will be in proportion to the individual resistances. It is now possible to calculate the heat transmission and associated temperatures for a real wall, given the appropriate coefficients. Values of conductances for surfaces and air spaces may be selected from Tables 8.2 and 8.4 with the help of Table 8.3 [**8.1, pp. 23.12 and 23.13**].

TABLE 8.2

Surface Conductances for Air W/(m² · K)
(From *ASHRAE Handbook 1981 Fundamentals*, Table 1, p. 23.12) [**8.1**]

Position of surface	Direction of flow	Surface emissivity		
		$\epsilon = 0.90$	$\epsilon = 0.20$	$\epsilon = 0.05$
Still air				
Horizontal	upward	9.3	5.2	4.3
Sloping 45°	upward	9.1	5.0	4.1
Vertical	horizontal	8.3	4.2	3.4
Sloping 45°	downward	7.5	3.4	2.6
Horizontal	downward	6.1	2.1	1.25
Moving air (any position)				
24 km/h, for winter	any	34	—	—
12 km/h, for summer	any	23	—	—

TABLE 8.3

Effective Emissivities of Air Spaces
(From *ASHRAE Handbook 1981 Fundamentals*, Table 2, Section B, p. 23.13) [**8.1**]

Surface	Reflectivity, %	ϵ average	E, effective: $\epsilon_1 = 0.90$, $\epsilon_2 = \epsilon$	E, effective: $\epsilon_1 = \epsilon_2 = \epsilon$
Aluminum foil, bright	92–97	0.05	0.05	0.03
Aluminum sheet	80–95	0.12	0.12	0.06
Polished aluminum paper	75–84	0.20	0.20	0.11
Steel, galvanized, bright	70–80	0.25	0.24	0.15
Aluminum paint	30–70	0.50	0.47	0.35
Building materials	5–15	0.90	0.82	0.82

TABLE 8.4

Thermal Conductances of Plane Air Spaces, W/(m² • K)
(From *ASHRAE Handbook 1981 Fundamentals*, Table 2, p. 23.12, 23.13) [**8.1**]

Position	Direction of flow	Air space: Mean temp., °C	Air space: Temp. diff., K	Thickness 19 mm: $E = 0.03$	Thickness 19 mm: $E = 0.82$	Thickness 92 mm: $E = 0.03$	Thickness 92 mm: $E = 0.82$
Horizontal	Up	30	5	2.4	7.5	1.9	7.0
		10	20	3.3	7.0	2.7	6.6
		−20	10	3.1	6.1	2.5	5.6
45°	Up	30	5	1.9	7.0	1.8	6.9
		10	20	2.8	6.8	2.4	6.5
		−20	10	2.6	5.6	2.3	5.3
Vertical	Horizontal	30	5	1.6	6.8	1.5	6.6
		10	20	1.9	5.9	2.0	6.1
		−20	10	1.8	4.8	1.9	4.9
45°	Down	30	5	1.6	6.8	1.2	6.3
		10	20	1.6	5.6	1.6	5.6
		−20	10	1.5	4.5	1.5	4.4
Horizontal	Down	30	—	1.6	6.8	0.56	5.7
		10	—	1.5	5.6	0.51	4.6
		−20	—	1.3	4.3	0.45	3.5

Example 8.1

A simple masonry wall consisting of hollow concrete blocks of dense concrete 200 mm thick is exposed to air temperatures of 23°C one side and −20°C on the other. Find the overall coefficient, U, and the block surface temperatures.

Find, from Table 8.1, the value 5.5 for two-core rectangular core block. Find, from Table 8.2 for horizontal heat flow and $\epsilon = 0.90$, the value for $f_i = 8.3$. Find from the same table the value for $f_o = 34$ for moving air in winter. The temperature difference is 23 to −20 = 43 K. It is convenient to arrange the elements as follows:

Element	Conductance	Resistance	Temperature difference, K	Temperature, °C
				−20
Outside film, f_o	34	0.029	3.8	
				−16.2
Concrete block	5.5	0.182	23.6	
				+7.4
Inside film, f_i	8.3	0.120	15.6	
				+23
		0.331	43.0	

$$\text{The overall coefficient is } \frac{1}{0.331} = 3.0 \text{ W/(m}^2 \cdot \text{K)}$$

$$\text{The heat flow } \frac{q}{A} = 3.0\ (23 \text{ to } -20) = 129 \text{ W/m}^2$$

The inside surface temperature is +7.4°C and that outside −16.2°C.

The relatively large contribution of the inside surface resistance, amounting in the case of example 8.1 to more than one-third of the total, should be noted. This means that the blocks provide only moderate resistance to heat flow. The surface temperature is correspondingly low, being at +7.4°C for the conditions given while the air inside the room is at +23°C. The outside surface resistance is relatively much smaller, because it is influenced by the wind which creates a kind of forced convection and promotes heat transfer.

Example 8.2

A window consisting of two layers of glass, each 3 mm thick, separated by an air space 92 mm wide is exposed to $t_i = 23$°C and $t_o = -20$°C. Find the overall coefficient, the rate of heat flow, and the surface temperatures.

From Table 8.1, k for glass = 1.0. From Table 8.2, $f_i = 8.3$ and $f_o = 34$. From Table 8.3, find for $\epsilon_1 = \epsilon_2 = 0.90$ that $E = 0.82$. From Table 8.4, find the appropriate air space conductance for $E = 0.82$, mean temperature 10°C, temperature difference 20 K, thickness = 92 mm, to be $a = 6.1$. Temperature difference = 43 K.

Element	Conductance	Resistance	Temperature difference, K	Temperature, °C
				−20
Outside surface, f_o	34	0.029	3.9	
				−16.1
Glass 3 mm $C = \frac{1.0}{0.003}$	330	0.003	0.4	
				−15.7
Air space	6.1	0.164	22.1	
				+6.4
Glass 3 mm	330	0.003	0.4	
				+6.8
Inside surface, f_i	8.3	0.120	16.2	+23
		0.319	43.0	

$$U = \frac{1}{0.319} = 3.1 \text{ W/(m}^2 \cdot \text{K)} \quad \text{and} \quad \frac{q}{A} = 3.1(23 \text{ to } -20) = 133 \text{ W/m}^2$$

The case of the double window in Example 8.2 demonstrates the relatively small contribution of the two layers of glass and the relatively large contributions of the air space and the inside surface film. It will be evident that the two surface coefficients can be substantially affected by air motion with corresponding large changes in surface temperature and heat transmission. The low inside glass-surface temperature of 6.8°C may be noted. Condensation will occur when the inside vapour pressure exceeds p_{ws} for 6.8°C = 0.987 kPa (Table 5.1), that is, when the relative humidity at 23°C is 0.987/2.81 = 0.35 or 35%. Limiting conditions at which condensation will occur on inside surfaces of windows are set out in detail elsewhere [**8.9**], and are discussed in Chapter 12.

8.6 Calculations for a Wood-Frame Wall

Wood-frame walls, which are widely used in Canada, introduce some further considerations. They are normally constructed with vertical framing members, or studs 42 × 92 mm in section and 400 or 600 mm on centres. The spaces between the studs are partially or completely filled with insulation. Part of the wall, 42 mm in every 400 mm (or 600 mm), has 92 mm of solid wood in the heat-flow path, while the remainder between studs will have a different value. A correction can be made when required by calculating the area-weighted average of the overall coefficients for the area over the studs and the area between studs. It is often ignored since the wood in the studs has substantial resistance to heat flow. This is not the case, however, with steel studs, which are relatively highly conductive.

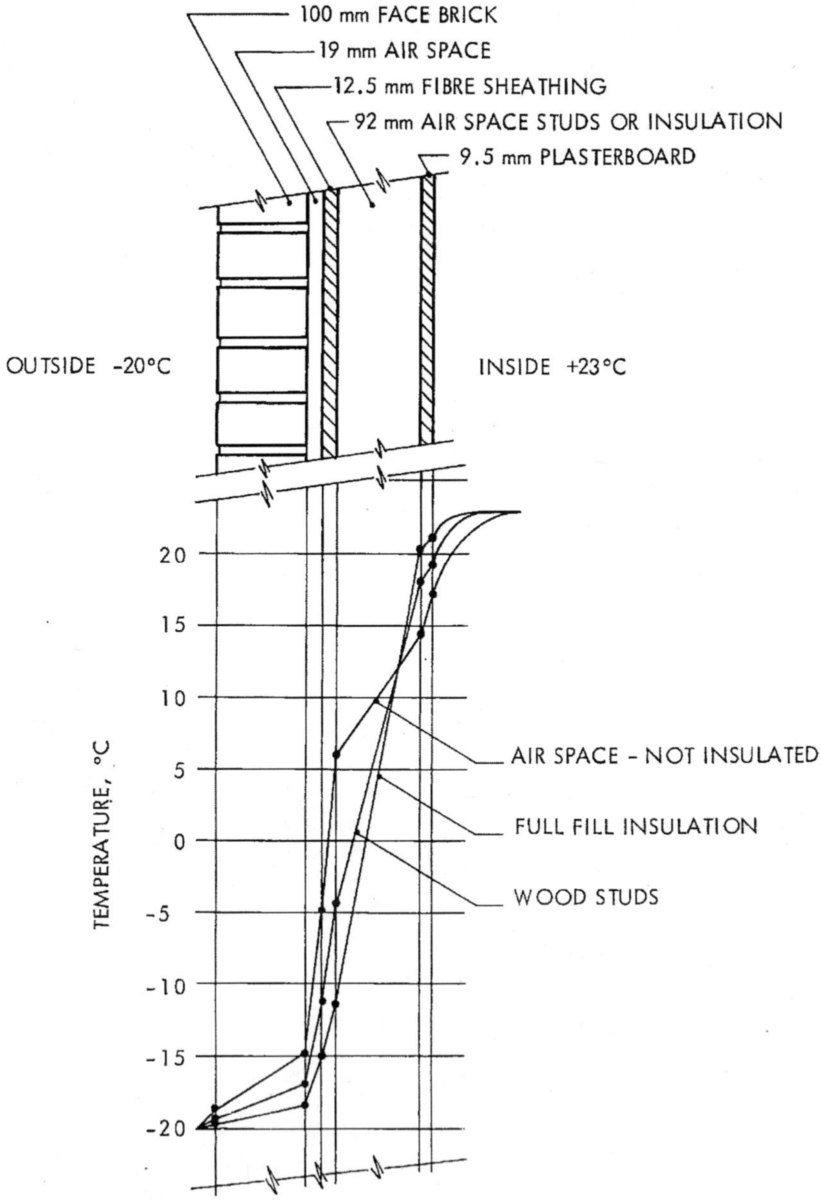

FIGURE 8.1 Temperature gradients in a wood-frame wall with brick veneer insulated and uninsulated.

Example 8.3

Consider the wood-frame wall shown in Figure 8.1, which is exposed to air at 23°C on the inside and -20°C on the outside. Calculate the coefficient and temperatures for an uninsulated wall and compare the result with the case where spaces between studs have been filled with mineral wool. Demonstrate the magnitude of the correction for framing members in each case.

Element	Conductance	Resistance	Temperature difference, K	Temperature, °C
				−20
Outside surface, f_o	34	0.029	1.4	
				−18.6
Face brick 100 mm 1.34/0.10	13	0.077	3.7	
				−14.9
Air space 19 mm	4.8	0.208	10.1	
				− 4.8
Fibre sheathing 12.5 mm 0.055/0.0125	4.4	0.227	11.1	
				+6.3
Air space 92 mm	6.1	0.164	8.0	
				+14.3
Plasterboard 9.5 mm	17.6	0.057	2.8	
				+17.1
Inside surface, f_i	8.3	0.120	5.9	+23
		0.882	43.0	

$$U = \frac{1}{0.882} = 1.13\ \mathrm{W/(m^2 \cdot K)} \quad \text{and} \quad \frac{q}{A} = 1.13(23 \text{ to } -20) = 48.8\ \mathrm{W/m^2}$$

The full calculation can be repeated for the area over the studs; the air space $R = 0.164$ will be replaced by 92 mm of wood, $k = 0.12$ and $R = 0.767$. Adjusting the total resistance accordingly yields $R_t = 1.485$, from which

$$U = \frac{1}{1.485} = 0.67.$$

The calculation for the area-weighted average for studs 42 mm wide, 400 mm on centres, assuming horizontal and vertical framing to cover 20% of the area, is

$$U_{AVGE} = 1.13 \times 0.80 + 0.67 \times 0.20 = 1.04 \text{ W/(m}^2 \cdot \text{K)}.$$

The calculations can be repeated, substituting for the initial air space ($R = 0.164$) the value of 92 mm of mineral wool, $k = 0.046$, $R = 0.092/0.046 = 2.00$. The total resistance becomes 2.72 and $U = 0.37$ for the insulated portion. The area-weighted average becomes

$$U_{AVGE} = 0.37 \times 0.80 + 0.67 \times 0.20 = 0.43 \text{ W/(m}^2 \cdot \text{K)}.$$

The wall in Example 8.3 is not complete. It should have a vapour barrier to control water-vapour migration into the insulated space from the warm side, where it can condense on the sheathing which is now, in the insulated wall, very cold. Installation of a vapour barrier will probably not change the heat-transfer calculation since its thermal contribution is negligible.

Another feature of this wall must be questioned. An air space of 19 mm is shown between the brick veneer and the fibre sheathing for which a resistance of 0.208 was included. There are several reasons why this value may not be realized in practice. The air space may not exist, because the bricks are placed directly against the sheathing, or the intended air space may be partially filled or bridged by mortar. If the cavity is vented and properly flashed at the bottom to provide drainage outward of any rain that penetrates the brick veneer, the circulation of cold air may enable the outward heat flow to bypass the air space and the brick veneer, reducing their value in the extreme case to that of a single-surface resistance at the sheathing. This is not of great importance when the wall is well insulated, since the outer air space and the brick contribute at best no more than 10% of the resistance for the wall. It should be realized that calling for an air space on a drawing or in a specification does not mean that it will necessarily exist in practice or provide thermal resistance as shown in tables. Judgement must be exercised about such matters, and corrections made accordingly.

The calculated temperatures for the additional cases of Example 8.2 are shown plotted in Figure 8.1. It should be noted that the resistance of the insulation is large compared with all other elements. Adding insulation raises all temperatures on the warm side of the insulation, while all temperatures on the cold side are lowered. Particular attention should be paid to the effect of insulation in lowering the temperature of the inside surface of the outside sheathing (in the case of Figure 8.1 from +6°C to −11°C). This will be of interest later when condensation within walls is discussed.

Windows and doors are incorporated in walls and must be allowed for in heat-transfer calculations. Values commonly used for windows are given in Table 8.5, values of doors in Table 8.6. Note that both are higher in overall heat transmission and therefore lower in resistance than the uninsulated wood-frame wall in Example 8.3. The values given are for heat transmission only and do not include solar heat gain through glass, a topic that will be discussed in a later chapter.

TABLE 8.5

Heat Transmission Coefficients (U) for Glass Elements, W/(m² • K)
(From *ASHRAE Handbook 1981 Fundamentals*, Table 81, p. 23.28) [**8.1**]

Description	Exterior Winter	Exterior Summer	Interior
Vertical			
Flat glass, single	6.4	6.0	4.1
Sealed glazing, double 12.5 mm air space, uncoated	3.3	3.2	2.6
Sealed glazing, triple 12.5 mm air space, uncoated	2.0	2.0	1.7
Storm windows 25–100 mm air space	3.2	3.1	2.5
Glass block 200 × 200 × 100 mm	3.2	3.1	2.5
Horizontal			
Flat glass, single	6.9	4.7	5.5
Sealed glazing, double 12.5 mm air space, uncoated	3.7	2.5	3.2

TABLE 8.6

Heat Transmission Coefficients for Doors, W/(m² • K)
(From *ASHRAE Handbook 1981 Fundamentals*, Tables 9A, 9B, p. 23-29) [**8.1**]

Thickness, mm	Winter: Solid wood, no storm door	Winter: Storm door, Wood	Winter: Storm door, Metal	Summer, no storm door
25	3.6	1.7	2.2	3.5
38	2.8	1.5	1.9	2.7
50	2.4	1.4	1.6	2.4
	Steel, no storm door			
45	(A) 3.4	—	—	3.3
45	(B) 1.1	—	—	1.0
45	(C) 2.7	—	—	2.6

(A) Mineral fibre core (32 kg/m³)
(B) Solid urethane foam core with thermal break
(C) Solid polystyrene core with thermal break

8.7 Heat Transfer in Ceilings, Attics, and Roofs

The calculation of heat transmission through horizontal components such as roofs is carried out in much the same way as for walls. It is necessary to select coefficients for surfaces and air spaces with due regard for the direction of heat flow, since convection is much greater for heat flow upward than for heat flow downward. Complications can arise when there are attic spaces. The area of the flow path is no longer constant and it becomes necessary to write equations for the heat flow into and out of the attic and to solve them for the attic temperature. Once this has been done, the losses or gains can be found readily and can, if desired, be expressed in terms of a U-value based on ceiling area and indoor–outdoor temperature difference.

Under winter conditions the attic gains heat through the ceiling and loses it through the roof and gable ends which may also have windows. Attic spaces often have additional openings for ventilation to control condensation. The air change represents a further heat loss from the attic which may be included in the heat balance in solving for attic temperature. Again, as in the case of the air space behind the brick veneer in Example 8.3, ventilation effectively reduces the resistance to heat flow of the roof and gable ends by providing an alternative path to the outside. Their contribution in the case of a well-insulated ceiling is relatively small in any event, and the effective U-value might be calculated on the basis of the ceiling alone, using an appropriate still-air surface coefficient for the attic side and assuming that the attic air is at outside temperature.

8.8 Basement Heat Losses

Basement walls pose two kinds of situations for heat flow. The above-ground portion (including windows) can be regarded in the same way as walls involved in air-to-air heat transfer. Below ground the situation is different: the earth becomes one of the elements in the flow path, which presents a two- or three-dimensional flow situation. There is also the probability that water will be present. These special problems are discussed later under the topic of ground-heat flow.

8.9 Amount of Insulation for Buildings

The problem of how much insulation to use in the enclosing walls, roofs, floors, and foundation walls of buildings is the general one of how to balance the capital investment in insulation and the return in annual benefits over the life of the building. Annual benefits have usually been in the form of annual savings in heating and cooling costs, but other considerations may be introduced in the face of a continuing energy scarcity. Other specific technical requirements can be identified in particular cases, such as the use of insulation to regulate or limit surface temperatures, but these must be considered independently and do not invalidate the general proposition: the economic benefit from a given amount of insulation decreases as the resistance to heat flow increases. The characteristic curves illustrating this situation are shown in Figure 8.2. When capital and operating costs are expressed on the basis of annual cost for various amounts of insulation used, the economic thickness is determined by the lowest annual cost. It is in the nature of things that the curve is rather flat, so that a thickness appreciably larger or smaller than the economic thickness does not greatly affect the annual cost. In wood-frame construction, the use of 50 mm of insulation where the economic thickness was 100 mm has led to cost increases of less than 10%, and it has been common practice to accept such increases in operating cost in order to reduce capital costs. The prospect that the cost of energy will increase markedly in the future makes it appropriate to consider installing amounts of insulation equal to or greater than the economic thickness in the interests of energy conservation. Calculation procedures have been described for the determination of economic thicknesses for various fuel and insulation costs [**8.10**]. There remains, however, the difficulty of judging what the installed cost of a given amount of insulation will be, and of taking into account the future changes in energy costs in what is now described as life-cycle costing.

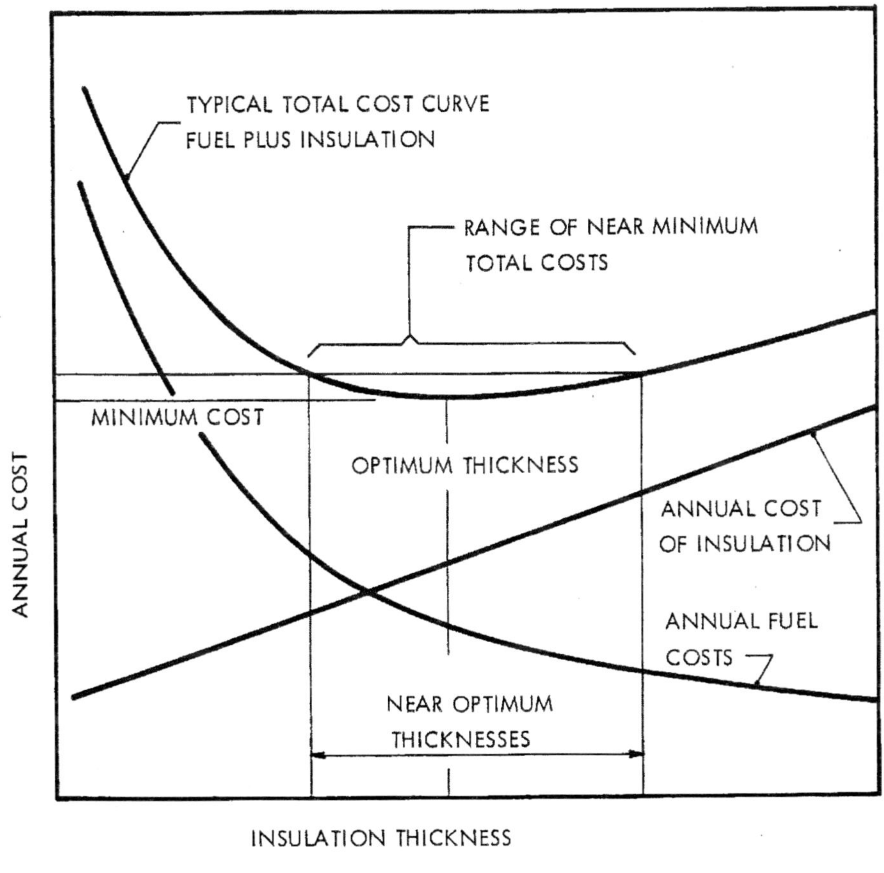

FIGURE 8.2 Typical cost thickness curves for house insulation.
(From Veale, A.C., *Insulation thickness for houses*, HN 21, DBR/NRCC, 1964)[**8.10**]

The curves of insulation cost in Figure 8.2 are linear with respect to insulation thickness. This is a valid assumption as long as a space already exists to accommodate the insulation, as in the case of wood-frame walls. Once the thickness exceeds the common depth of framing of 92 mm, costs increase. The insulation of metal and masonry walls has usually involved substantial increases in the basic wall costs, leading to lower economic insulation thicknesses than for wood-frame constructions, although it is usually possible to provide increased thicknesses in attics and roofs without such added costs.

The use of thickness as a measure of insulating value is an approximation justified by the fact that the conductivity of building insulations falls within a fairly narrow range. It is now common practice to speak of the thermal resistance, which takes into account both conductivity and thickness. Fifty millimetres of an insulation with a k-value of 0.04 will provide a resistance of $\ell/k = 0.050/0.04 = 1.25\ \mathrm{K \cdot m^2/W}$. Framing members 92 mm in depth will accommodate insulation to provide 2.5 or more units of thermal resistance.

It should be noted that the above discussion applies most directly to winter heating, the costs of which can be related directly to thermal resistance of the building enclosure. The relation for summer air conditioning is more difficult to establish, since the contribution of the heat transfer through walls and roofs to air-conditioning loads is usually much less than the contributions of the occupancy or of the solar heat gain through glass. Heat flow through roofs is usually more important than heat flow through walls. The summer heat gain is cyclical in nature, varying throughout each day, whereas the heat loss in winter periods of maximum heating demand may vary little throughout the day and from day to day.

The simple theory and methods by which heat transfer through walls, floors, and roofs is calculated can now be viewed as a whole. It is necessary to examine in more detail some of the assumptions that were made in establishing working coefficients, linking them to the earlier discussions of convection and radiation so that they can be better understood.

8.10 Convection in Building Heat Transfer

In building heat transfer, fluids are almost always set in motion by buoyancy effects as in natural convection, or by fans or pumps as in forced convection. The discussion of fluids in Chapter 7 identified the complex and changing nature of the flow pattern along the flow path resulting in laminar, transitional, and turbulent flow in the case of pipes and flat plates. When heat transfer is also taking place, further complications, including variations in density and viscosity due to temperature differences, are introduced. The forces producing flow are also related to the heat transfer in the case of natural convection. The net result is that the heat-transfer rate by convection may vary markedly from point to point along the path of the flowing fluid; the average coefficient to be used in calculations will also vary. Thus, length of surface in the fluid-flow direction becomes one of the variables.

Heat transfer by conduction involving a direct transfer of kinetic energy between molecules along a temperature gradient also takes place, but has significance mainly in situations where the fluid is relatively still or is in laminar flow and not subject to strong mixing by turbulence, which then dominates the heat-transfer situation. Some authors recognize the existence of conduction along with convection by using the term "conduction-convection".

Heat transfer by natural convection at a surface in contact with a fluid will vary with the properties of the fluid, the temperature and temperature difference, the orientation or inclination of the surface, the direction of heat transfer, and the dimensions of the heated or cooled surface. For example, convection upward from a horizontal heated plate takes place freely because the fluid is free to rise once it is heated. Heat transfer downward is different: the heated fluid is on top of the cooler, more dense fluid layers, and shows no tendency to circulate. In this case the fluid should be still, transferring heat by conduction only. In practice, however, there will always be some circulation because of edge effects, so the size of the plate becomes a factor. The case of downward heat flow to a cooled plate is similar; in horizontal flow at vertical surfaces there is usually a marked influence of surface dimension in the vertical direction.

The large number of variables to be considered in convective heat transfer has led to a great mass of experimental data. Correlations have been established using dimensionless ratios similar to Reynolds number and known as the Nusselt, Prandtl, and Grashof numbers. When air at ordinary conditions is involved in specific situations it is possible to eliminate some variables and substitute appropriate values for others in arriving at simplified equations such as those presented in Table 8.7 which have been converted to metric from their original form [**8.1**]. Three cases are of particular interest in building. One is the convective

TABLE 8.7

Convection Coefficients for Air at Various Surfaces

(From *ASHRAE Handbook 1981 Fundamentals*, Table 5, p. 2.12; Table 6, p. 2.15; [**8.1**] Min. T.C., Schutrum, L.F., Parmelee, G.V., and Vouris, J.D., *ASHRAE Transactions*, 1956, *62*, pp. 343–50 [**8.11**]

Natural convection [**8.1, p. 2.12**]	*Forced convection* (Simplified equations for air) [**8.1, p. 2.15**] (converted to SI)
Vertical plates, L = height, m	Vertical plane surfaces, room temperature
Small plates, laminar range	V = 5 to 30 m/s $h' = 7.34\,(V)^{0.8}$ W/(m² · K)
$h_c = 1.42\,(\Delta t/L)^{0.25}$ W/(m² · K)	V less than 5 m/s $h' = 5.62 + 3.91\,V$
Large plates, turbulent range	
$h_c = 1.31\,(\Delta t)^{0.33}$	
Horizontal plates, upward heat flow	*Convection at room surfaces* [**8.11**] (converted to SI)
L = length, m	Convection to walls, H = height, m
Small plates, laminar range	$h_c = 1.87\,(\Delta t)^{0.32} \times H^{-0.05}$ W/(m² · K)
$h_c = 1.32\,(\Delta t/L)^{0.25}$	Horizontal surface, warm ceiling or cold floor
Large plates, turbulent range	$h_c = 0.201\,(\Delta t)^{0.25} \times L^{-0.24}$ W/(m² · K)
$h_c = 1.52\,(\Delta t)^{0.33}$	Horizontal surface, warm floor or cold ceiling
Horizontal plates, downward heat flow, L = length, m	$h_c = 2.42\,(\Delta t)^{0.31} \times L^{-0.08}$ W/(m² · K)
Small plates	(L is the characteristic dimension in metres given by an equivalent diameter equal to 4 times the area divided by the perimeter for horizontal areas.)
$h_c = 0.59\,(\Delta t/L)^{0.25}$ W/(m² · K)	

transfer at the outside surfaces of a building, which is complicated by wind. The second is the natural convective heat transfer across air spaces. The third is the convective exchange between the air in rooms and the wall, floor, and ceiling surfaces.

The conditions conducive to convection, namely, a heated or cooled surface, will also produce radiation exchanges. Both convection and radiation become involved in determining the heat transfer at surfaces and across air spaces. The surface coefficients and air-space conductances presented earlier take both into account, and are always combined coefficients reflecting the characteristics of both kinds of heat transfer. When the appropriate coefficients are not known it becomes necessary to establish separate equations for the convection and radiation to be included in a heat-balance equation, which must then be solved for surface temperature. The general situation is as illustrated in Figure 4.9.

8.11 Coefficients for Building Surfaces

The radiative heat exchange between a surface and its surroundings is given by Equation 4.14:

$$\frac{q}{A} = \sigma F_A F_E (T_1^4 - T_2^4)$$

where $\sigma = 5.670 \times 10^{-8}$. For an exterior wall surface radiating to outdoors, assuming that all outdoor surroundings are at outdoor air temperature, and putting $F_A = 1$ and $F_E = \epsilon$,

$$\frac{q_r}{A} = \sigma \epsilon (T_s^4 - T_a^4). \tag{8.9}$$

This can be expressed in terms of a coefficient h_r for use in simple heat-flow theory as follows:

$$\frac{q_r}{A} = h_r(t_s - t_a) = \sigma \epsilon (T_s^4 - T_a^4)$$

so that

$$h_r = \frac{\epsilon\sigma(T_s^4 - T_a^4)}{T_s - T_a} = \epsilon\sigma(T_s^2 + T_a^2)(T_s + T_a). \tag{8.10}$$

Now considering convective heat transfer,

$$\frac{q_c}{A} = h_c(t_s - t_a) \tag{8.11}$$

so that the surface exchange is given by

$$f_o(t_s - t_a) = h_r(t_s - t_a) + h_c(t_s - t_a)$$

and

$$f_o = h_r + h_c. \tag{8.12}$$

Example 8.4

Derive the value of f_o for still air, $\epsilon = 0.90$, surface temperature −15°C, and outside air at −20°C.

From Table 8.7, $h_c = 1.31(\Delta t)^{1/3}$ for large plates, turbulent flow, so

$$h_c = 1.31(20 - 15)^{1/3} = 2.24 \text{ W/(m}^2 \cdot \text{K)}$$

and

$$h_r = \frac{0.90 \times 5.670}{(t_s - t_a)}\left[\left(\frac{273 - 15}{100}\right)^4 - \left(\frac{273 - 20}{100}\right)^4\right]$$

$$= \frac{0.90 \times 5.670\,[44.3 - 41.0]}{(-15 - (-20))} = 3.37 \text{ W/(m}^2 \cdot \text{K)}$$

$$f_o = h_c + h_r = 2.24 + 3.37 = 5.61 \text{ W/(m}^2 \cdot \text{K)}.$$

The value of f_o found in Example 8.4 can be compared with the value of 8.3 W/(m² • K) from Table 8.2. Note the large contribution of radiation. The value of f_o from Table 8.2 is not linked to any particular temperature, but the radiation component is markedly dependent on temperature. For an air temperature of 21°C and surface at 26°C, the value of h_r is 5.31 and the combined coefficient is 8.68. The value of h_c as shown by the equation is dependent on $t_s - t_a$, or Δt. Thus, the surface coefficients given in Table 8.2 with no correction for temperature or temperature difference are approximations only. There are, however, other uncertainties about emissivities and the correspondence of surrounding temperatures with air temperatures, which are assumed to be equal in the derivation and use of the combined coefficient. In most cases the outside surface contributes relatively little to the overall resistance, and these approximations can be accepted. When more exact values are required they can be found by calculation.

Example 8.5

Calculate the outside surface coefficient for the conditions of Example 8.4, but for a wind velocity of 24 km/h (6.7 m/s).

The radiation coefficient will remain unchanged. The convective coefficient must now be based on a kind of forced convection produced by the wind. It will not be known exactly what the effective flow rate parallel to the wall surface will be for a given wind speed, but it may be assumed equal to the wind speed.

The forced convection equation from Table 8.7 is

$$h' = 7.34\, V^{0.8}$$

and so

$$h_c = h' = 7.34(6.7)^{0.8} = 33.7 \text{ W/(m}^2 \cdot \text{K)}$$

adding

$$h_r = 3.37$$

gives

$$f_o = 37.07 \text{ W/(m}^2 \cdot \text{K)}.$$

This value can be compared with that of 34 from Table 8.2 for winter conditions. It will be seen that the convection coefficient is much greater (by 10 times) than the radiation coefficient when wind is present, and the resistance to flow at the outside surface is much less than for the still air condition.

8.12 Air-Space Coefficients

The radiation transfer across air spaces bounded by parallel planes is given by

$$\frac{q}{A} = \sigma F_A F_E (T_1^4 - T_2^4).$$

The configuration factor may be taken as unity, and F_E is given by

$$F_E = \frac{1}{\dfrac{1}{\epsilon_1} + \dfrac{1}{\epsilon_2} - 1}.$$

Values for various combinations of emissivities for the two bounding surfaces with emissivities ϵ_1 and ϵ_2 are shown in Table 8.3.

The convection coefficient for air spaces is shown in Figure 8.3 in the form of a correlation based on 96 separate tests carried out for a range of conditions [**8.12**]. When the temperature difference across the air space, θ, and the thickness, ℓ, are known, the value of $(\theta\ell^3)$ can be found and used as a point of entry to the graphs. Using the appropriate surface orientation curve, the corresponding value of $(h_c \times \ell)_{10}$ can be found and h_c calculated. The value thus found is for 10°C mean temperature. The value of h_c for other mean temperatures can be found from the following equations.

For $(h_c\ell)_{10} > 0.043$

$$(h_c)_t = (h_c)_{10} [1 - 0.0018 (t - 10)] \qquad (8.13)$$

and for $(h_c\ell)_{10} < 0.029$

$$(h_c)_t = (h_c)_{10} [1 + 0.003 (t - 10)]. \qquad (8.14)$$

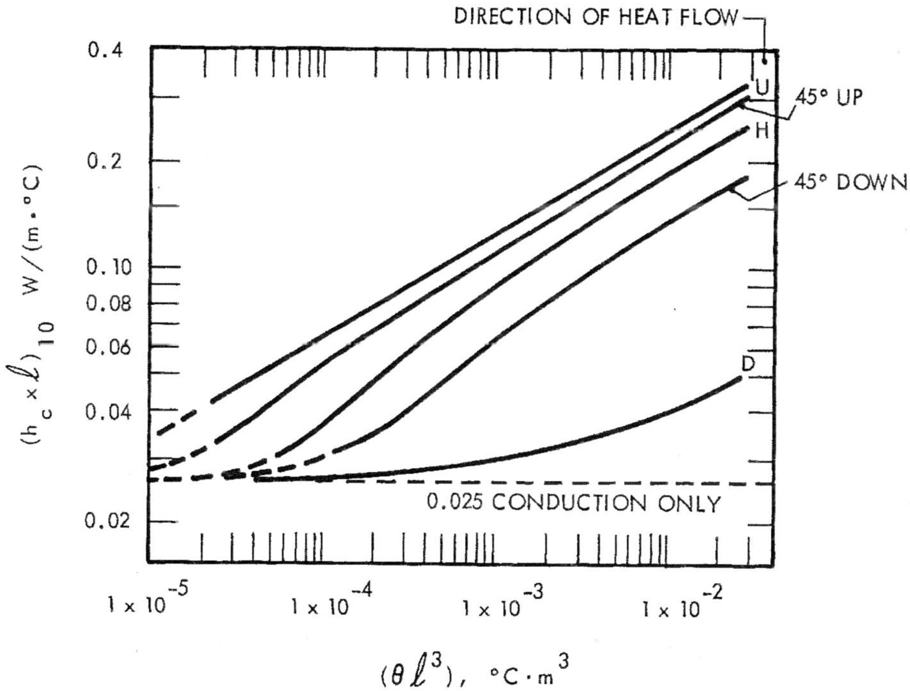

FIGURE 8.3 Correlation of conduction-convection in air spaces. (From Robinson, H.E., Cosgrove, H.A., and Powell, F.J., BMS Report 151, NBS, 1957. Courtesy of the National Bureau of Standards)[**8.12**]

No correction is necessary when the value of $(h_c\ell)_{10}$ is between 0.029 and 0.043.

A few cases giving air-space conductances were selected to make up Table 8.4. More complete tables are available [**8.1, p. 22.2; 8.13**] but it is not difficult to derive values, as required, from Figure 8.3.

Example 8.6

Using the data in Figure 8.3, calculate the conductance for a vertical air space at a mean temperature of 10°C and a temperature difference of 20 K if it is 92 mm thick and is between surfaces with emissivities $\epsilon_1 = \epsilon_2 = 0.90$. Compare with the value given in Table 8.4.

$$E = 0.82,\ t_1 = 10 + 10 = 20°\text{C},\ t_2 = 10 - 10 = 0°\text{C}.$$

The radiation exchange is (Equation 8.9):

$$\frac{q_r}{A} = 5.67 \times 0.82\ [(\frac{273 + 20}{100})^4 - (\frac{273 + 0}{100})^4]$$

$$= 84.6\ \text{W/m}^2$$

and $$h_r = \frac{q_r}{A(t_1 - t_2)} = \frac{84.6}{20 - 0} = 4.23 \text{ W/(m}^2 \cdot \text{K)}.$$

For convection

$$\theta\ell^3 = 20 \times 0.092^3 = 0.0156$$

From Figure 8.3 $(h_c\ell)_{10} = 0.21$

h_c is required for 10°C mean temperature so no correction is needed.

$$h_c = 0.21/0.092 = 2.28,$$

$$a = h_r + h_c = 4.23 + 2.28 = 6.51 \text{ W/(m}^2 \cdot \text{K)}.$$

From Table 8.4 the value of the conductance given is 6.1.

When air spaces are divided by framing members, the emissivity factor is no longer strictly equal to that for parallel planes. The discrepancy, however, is small enough that it can be neglected and the influence of framing members on the radiation transfer in the air space ignored. Table 8.4 shows that for Example 8.6, if both surfaces are made reflective, so that $E = 0.03$, the conductance will be reduced to 2.0. The change occurs in the radiation coefficient which is reduced from 4.2 to about 0.2, and the heat transfer is then almost entirely by convection. Using two reflecting surfaces instead of one for a single air space creates little change, since one reflecting surface will already have greatly reduced the transfer by radiation. It will be evident that high values of thermal resistance can be obtained by using bright metal-foil curtains to create several separate air spaces. The difficulties in ensuring that the air spaces are quite separate and that circulation does not take place from one to the other will be discussed later.

8.13 Thickness of Vertical Air Spaces

The resistance of a vertical air space bounded by ordinary materials varies as the thickness is increased, as shown in Figure 8.4. The radiative component changes not with thickness but with temperature, as evidenced in the difference between winter and summer curves. Increasing thickness above about 18 mm has little effect on the total resistance. This is so because the mechanism of convective transfer is one of transport and is therefore not affected by increasing the thickness above 18 mm. When the thickness is decreased below 18 mm, the upward and downward fluid flows begin to interfere, until at about 6 mm heat transfer is largely by conduction and radiation. The transfer by conduction increases as the thickness is decreased until at very small thicknesses the resistance to conduction approaches zero. These considerations are important in selecting air-space thicknesses for double windows.

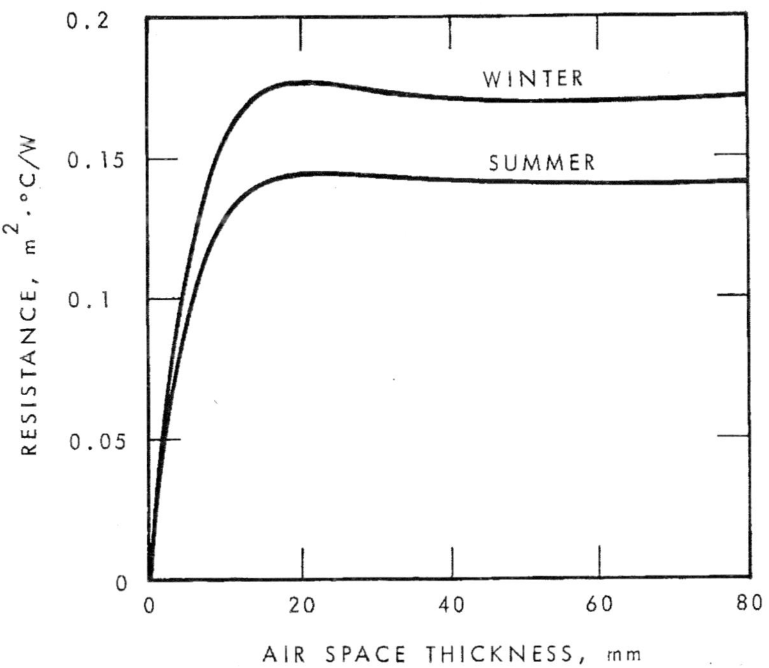

FIGURE 8.4 Air-space thermal resistance.

8.14 Air Spaces and Insulation in Frame Walls

Begininning in 1951, extensive experiments were carried out in Saskatoon to investigate some baffling aspects of field performance of reflective insulations [**8.14**]. Full-height wall test panels were placed between warm and cold chambers in which temperature and air flow could be controlled. The same basic wall construction was used throughout: 9.5 mm plasterboard, 38 × 92 mm studding, 20 × 250 mm spruce shiplap sheathing, building paper, and 150 × 12.5 mm bevel cedar siding. Heat flow was measured using heat meters. The following numbering system identifies the temperature curves in Figs. 8.5 to 8.8.

Curve Number	Location
1	Warm side, air 25 mm from wall surface
2	Inside surface of plasterboard
3	Outside surface of plasterboard
4	Air space, 19 mm from plasterboard
5	Air space, 19 mm from sheathing
6	Inside surface of sheathing
7	Cold side, air 50 mm from wall surface

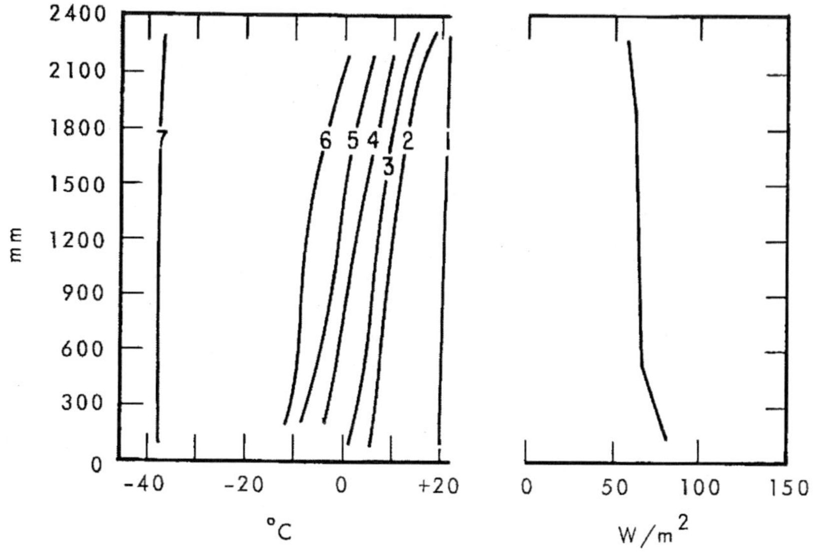

FIGURE 8.5 Temperature and heat flow in wood-frame wall uninsulated. (From Handegord, G.O., and Hutcheon, N.B., *ASHVE Trans.*, 1952, *58*, Fig. 3, p. 174)[**8.14**]

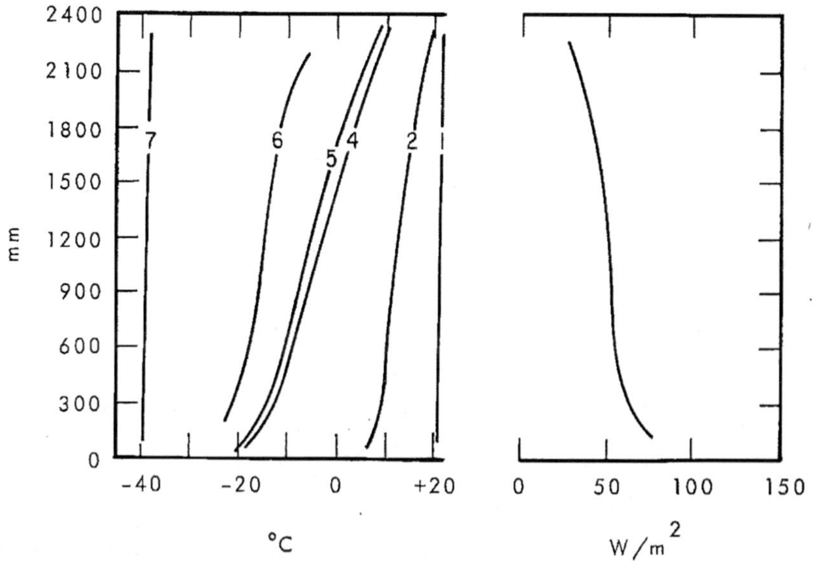

FIGURE 8.6 Temperature and heat flow in wood-frame wall with aluminum foil on warm side of air space.
(From Handegord, G.O., and Hutcheon, N.B., *ASHVE Trans.*, 1952, *58*, Fig. 4, p. 175)[**8.14**]

The first two tests, on an uninsulated wall and a wall with bright foil surface on one side of the air space, produced the results shown in Figures 8.5 and 8.6. The marked variation of temperature with height in the air space was unexpected, although it might have been anticipated on the basis of field observations which showed that frosting occurred first at the bottom of windows and sometimes on the lower portion of walls, indicating that they were lower in temperature there than at higher levels. The existence of such a temperature gradient vertically in the air space means that the temperature difference from the warm side to the air space is greater at the bottom than at the top of the wall. Thus, there is an increased heat flow in at the bottom of the wall. The reverse situation exists between the air space and the outside: the heat loss is greater at the top of the wall from the air space to outside. It follows that there must be a vertical upward movement of heat in the air space, which is, of course, the reason for the vertical temperature gradient. The higher heat flows entering at the bottom of the walls were confirmed by the heat-meter readings taken on the warm side only, between studs.

Results were more in accord with expectations when 50 mm of mineral-wool insulation were placed in the stud space. Temperatures were then reasonably uniform from the top to the bottom of the wall. The next two tests (reported in figures 8.7 and 8.8) were made with a foil curtain and an insulating board placed midway in the air space. Gaps of 9.5 mm were left at top and bottom to simulate common faults in installation. The high heat transfer, averaged over the height, was about the same as for the wall in Figure 8.6 with one foil-coated surface in

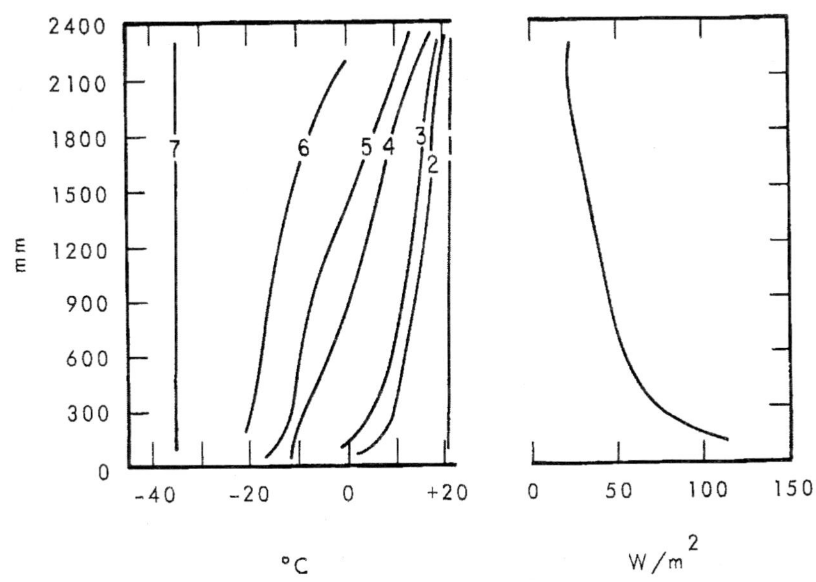

FIGURE 8.7 Temperature and heat flow in frame wall with single foil curtain forming two air spaces (9.5-mm gap top and bottom).
(From Handegord, G.O., and Hutcheon, N.B., *ASHVE Trans.*, 1952, *58*, Fig. 9, p. 179)[**8.14**]

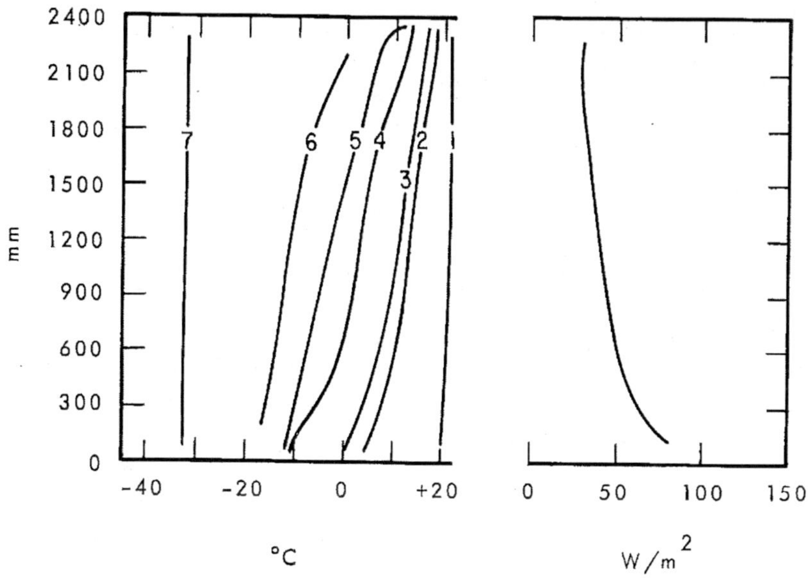

FIGURE 8.8 Temperature and heat flow in frame wall with 25-mm insulating board simulating blanket forming two air spaces (9.5-mm gap top and bottom). (From Handegord, G.O., and Hutcheon, N.B., *ASHVE Trans.*, 1952, *58*, Fig. 10, p. 180)[**8.14**]

the air space. The foil curtain and the insulating board intercepted the radiation and reduced the radiative heat transfer in much the same way as the single foil surface had done, but left the convection relatively undisturbed with air circulation from one space to the other almost as if the baffles were not there. The gap of 9.5 mm was apparently enough to permit almost full convective circulation. The net result was a single air space with convective heat transfer, but with the radiation exchange greatly reduced, and with a resistance about the same as a single air space faced one side with foil.

These experiments demonstrated conclusively that to develop the full insulating potential of multiple air spaces formed by foil curtains it is necessary to ensure that the air is not able to circulate from one space to the other. This is not easily accomplished in practice. The uncertainties of installation, coupled with some other characteristics of reflective insulation, led to a decision to restrict its use under the National Housing Act to regions in Canada having a severity of heating season not greater than 4440 degree days Celsius. The remarkable results that can be achieved with foil in downward heat-flow situations should not be overlooked, however.

8.15 Convection in Insulation

The tests that demonstrated the need to isolate air spaces to make them effective with reflective insulation also demonstrated that when any batt, blanket, or board insulation is installed with an air space on either side there must be no air circulation from one side to the other bypassing the insulation. A crack at top and bottom as small as 1 mm in width across a stud space 400 mm wide will produce 4 cm^2 of leakage area top and bottom, which is more than enough to reduce substantially the value of the insulation. The force to produce air flow from one space to the other results from the difference in weight of the two columns of air which are at different temperatures and have different densities. This same force can also produce flow horizontally through fibrous insulation having air spaces on either side unless the insulation has a paper backing or airtight film on one side to stop the air flow. This situation does not occur if the insulation is placed firmly against one side of the air space [**8.15**]. Convection does not take place to any appreciable extent within the insulation itself.

Much of the insulation being used today is only partially effective because of improper installation. Instructions for proper application are available. [**8.1, Chap. 20; 8.16**]

8.16 Wall-Surface Temperature Variations

Further experiments in Saskatoon [**8.17**] provided information on the variations of surface temperatures between framing members and opposite framing members on the warm side of wood-frame walls. The results are shown in Figure 8.9. Large variations result when the stud is left unprotected from the air space on the cold side of the insulation. The use of insulation makes this air space colder than it would be if there were no insulation, and leads to an increased cooling effect on the exposed portions of the studs. The smallest variation in surface temperature occurs when the insulation is located on the cold side of the air space so that the sides of the studs are exposed to warm rather than cold air. Filling the air space completely does not eliminate these temperature variations since the stud, having lower thermal resistance than the insulation, provides a path of higher conductivity for heat flow. It acts as a thermal bridge through the insulation, the effect of which will be in proportion to the area of the flow path it presents and its thermal conductivity. Wood framing is relatively low in conductivity but makes up about 20% of the wall area. Steel studs are high in conductivity, and unless arranged in special ways to minimize heat flow they produce a greater bridging effect than wood framing [**8.18**].

Heat flow and temperature are seldom constant in the plane of any wall or ceiling but may vary from point to point for a variety of reasons. Simple theory using averaging coefficients and simplifying assumptions can be remarkably accurate and effective for purposes of estimating and predicting many thermal situations. It is necessary, however, to recognize the limitations and to judge when more detailed analysis is needed.

8.17 Thermal Bridges

Thermal bridges through building enclosures can take many forms. Framing members in insulated construction, metal ties in cavity wall and panel constructions, structural members, and any other elements having relatively high conductivity that penetrate constructions that must provide resistance to heat transfer

1	2	3	4	5	6	7	8
WALL NO.	TEMPERATURE PATTERN	T_i	$T_i - T_o$	ΔT	$\frac{Q_{st}}{Q_{ss}}$	ΔT	$\frac{Q_{st}}{Q_{ss}}$
1 FOIL BLANKET	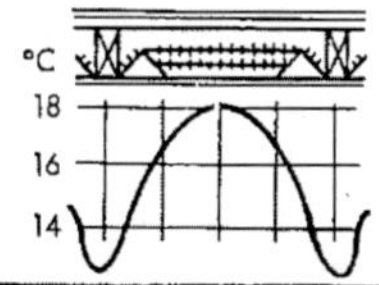	21.2	37.8	6.2	2.7	1.3	1.4
4 TWO FOIL CURTAINS	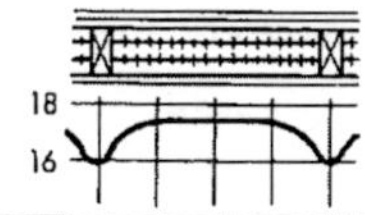	21.0	37.7	1.6	1.5	1.3	1.4
5 ONE FOIL CURTAIN	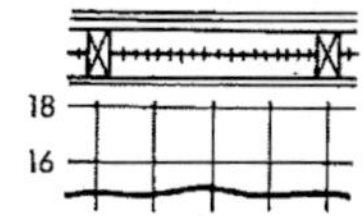	21.5	40.4	0	-	0.3	1.1
6 5 cm BATT	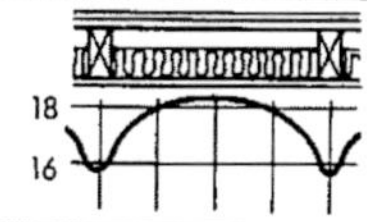	21.3	38.0	2.7	1.7	1.6	1.5
7 5 cm BATT	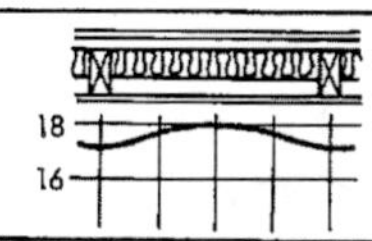	21.3	38.0	0.8	1.1	1.5	1.5
8 5 cm BOARD	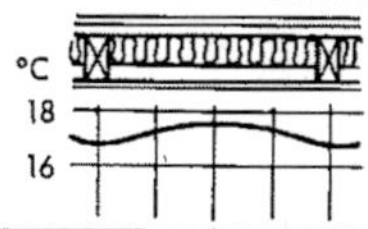	20.8	37.1	0.6	1.1	1.5	1.5
9 7.5 cm BOARD	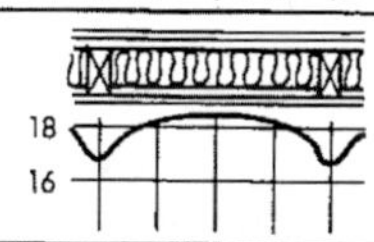	21.0	36.8	1.7	1.7	2.2	2.0
10 ROCK WOOL FILL	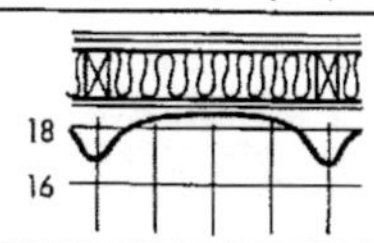	21.0	37.9	2.2	1.8	2.4	2.1
11 5 cm BOARD	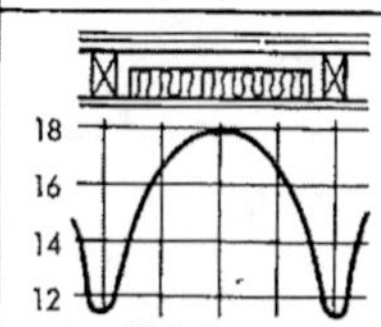	20.8	37.5	6.7	3.5	1.6	1.5

FIGURE 8.9 Surface temperatures on frame walls with various insulation arrangements. All values are for midheight of the wall.
Col. 5 - Measured surface-temperature difference over studs and between studs.
Col. 6 - Ratio of measured heat flows over studs and between studs.
Col. 7 - Calculated surface temperatures over studs and between studs.
Col. 8 - Ratio of calculated heat flows over studs and between studs.
(From Handegord, G.O., *ASHAE Trans.*, 1957, *63*, Fig. 1, p. 334-35)[**8.17**]

may have to be considered as thermal bridges. There may be undesirable effects owing to increased heat flow or decreased surface temperatures. Small differences in temperature may not have serious effects. When temperature differences on adjacent inside surface areas of walls and ceilings exceed about 2.5 K, accelerated soiling known as dust marking will occur on the colder surface. When temperature differences become large condensation may occur, creating a nuisance and promoting deterioration. In many cases the heat loss by way of thermal bridges may be large enough to require special attention.

It will be evident that the analysis of thermal bridges is not a simple matter; two- or even three-dimensional heat flow is involved in situations that are often complicated geometrically. Exact analysis may be possible in some cases; in others the means of extending the use of more simple analyses may be available. For example, there are ways of estimating the thermal effects of metal paths through masonry [**8.1, p. 22.3**]. Other results for insulated masonry situations have been presented as comparisons between calculated and measured differences in wall-surface temperatures [**8.19**]. These data, based on French [**8.20**] and Norwegian [**8.21**] studies, have been expressed as temperature differences for an overall temperature difference of 100°F. They may also be used for other temperatures without difficulty by noting that the temperature differences as expressed are really percentages of the overall temperature. These results will become more significant as increased insulation is used in masonry walls in response to anticipated higher energy costs. Some of the current practices such as extending concrete walls and floors beyond the outer walls to form balconies (but which also form large fins for enhanced heat transfer) will require careful consideration. The practice of insulating on the cold or outer side of masonry and metal constructions can resolve many thermal difficulties.

8.18 Thermal Aspects of Windows

Windows are often regarded, with good reason, as the most difficult of all design elements. They are often critical in respect of aesthetics as well as heat loss, solar heat gain, ventilation and air leakage, natural lighting, view, and condensation and indoor humidity. Many Canadian Building Digests have been written and published by DBR/NRCC with these varied but frequently interrelated considerations in mind.

Thermally, windows are weak spots in the building enclosure. Single glazing gives R = 0.15, double glazing R = 0.4, and reflective double glazing or triple glazing up to R = 0.6. It has already been shown that thermally the basic section of double glazing is essentially two air films plus an air space, since the two sheets of glass contribute only 2% of the resistance to heat flow. This means that heat flow through windows is highly sensitive to all the factors that influence surface film resistance. The proper simulation of surface air flow and radiation conditions is therefore a critical matter in the comparative thermal testing of windows. The air-space conditions are inherently more closely defined, but can be complicated by condensation between panes and exchange with indoor or outdoor air. The surface and air-space temperature variations [**8.22**] are remarkably similar to those for walls.

Windows almost always involve thermal bridges since they must ultimately be held structurally in the building enclosure. The edge temperature effects in double glazing [**8.22–8.24**] become important in determining surface condensation, since they define the coldest surfaces on which condensation will occur. They can also lead to serious glass breakage when a number of adverse factors, including high conductivity of sash, frames, and mullions, are combined to produce cold glass edges which then become stressed in tension. Glass is a brittle material and is sensitive to stress concentrations at the edges; this makes tensile loading critical.

Various mullion arrangements are shown in Figure 8.10. Successive improvements in thermal-bridge effects are provided by the drawings in order from left to right and from top to bottom [**8.25**].

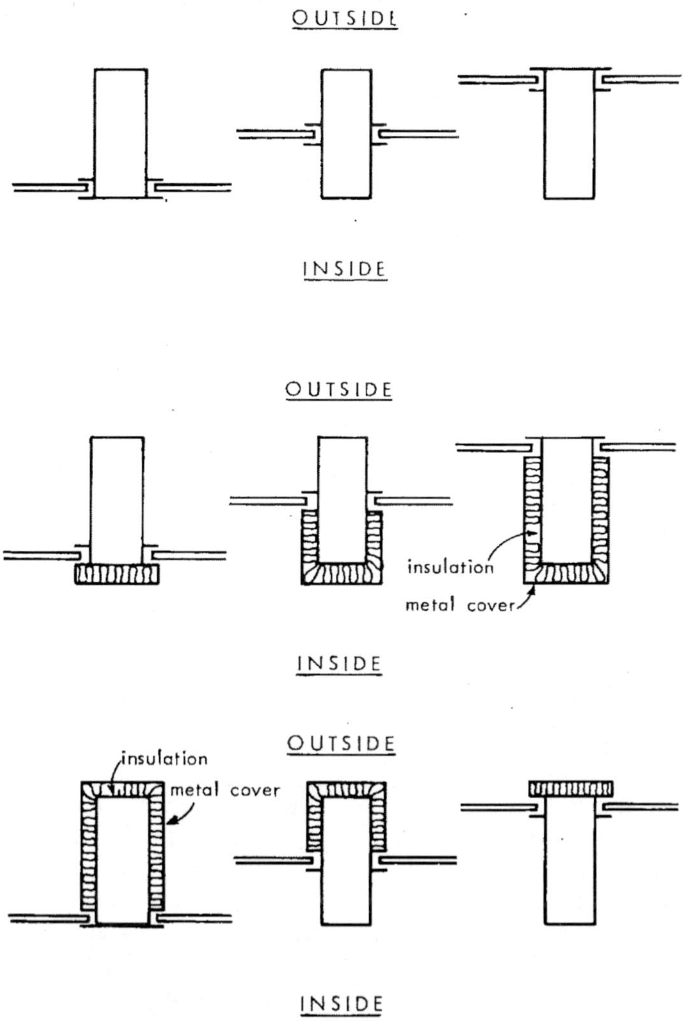

FIGURE 8.10 Thermal-bridge effect of a mullion with and without insulation. (From Latta, J.K., NRCC 13487, DBR/NRCC 1973)[**8.25**]

When a thermal bridge such as a mullion is highly conducting and of substantial cross-section, the air-to-surface heat transfer becomes the limiting factor to flow. In many situations the small temperature differences throughout the metal can be ignored and the calculations based only on the resistances of the surface air films. Consideration of the first three mullion arrangements will show that the extent and ratio of the areas exposed at the warm and cold sides then become of importance.

The arrangements shown in the third line are preferable to those in the second line because the mullion temperature is closer to indoor rather than outdoor conditions, thus minimizing both thermal movement and cooling of the edge of the window in contact with it. This is another example of the advantages of insulating on the outside whenever possible.

8.19 Two-dimensional Heat Flow

All of the heat-flow equations used so far have considered flow in only one dimension, although a number of the practical applications obviously involved flow in two or even three dimensions. The mathematical solutions become complex when the boundary conditions cannot be presented in simple expressions. Modern techniques of calculation can be used, provided the problem justifies the effort.

One analogue method that has been used recognizes the parallel between electrical flow and simple heat flow and uses a sheet-type resistor of conductive paper cut to conform to the geometrical shape involved in the plane of the two dimensions of flow. Electrical potentials are then applied at the boundary in conformity with the known boundary temperatures. The enclosed area is then probed with a voltmeter to find the voltages corresponding to the desired temperatures.

A graphical method uses the superposition of flow-network temperature patterns for elementary situations by which the real case can finally be approximated [**8.26**]. It has been applied in ground-heat problems, including the calculation of basement heat losses [**8.25, p. 7**]. Thermal conductivities for soils are discussed in Section 8.24.

8.20 Transient Heat Flow

Modern calculation methods may also be used, when justified, to deal with unsteady flow conditions. One example is the transfer function method now widely promoted [**8.1, Chap. 25**] to deal with air-conditioning loads and equipment operation. This subject cannot be covered in detail here, but some basic characteristics and several typical cases of unsteady flow can be usefully discussed.

Analysis of the steady-state condition is based on the assumption that heat in equals heat out, whereas in the unsteady-state situation the basic equation for flow involving an elementary volume of material is

$$\text{heat in} = \text{heat out} + \text{heat stored}$$

Consider an elementary volume $dx \cdot dy \cdot dz$, as shown in Figure 8.11. The basic differential equation for the unsteady state can be arrived at by equating heat flow into the elementary volume to heat flow out plus storage.

The heat flow in along the x-axis becomes

$$k_x\,(dy \times dz)\,(\frac{\partial t}{\partial x})\,d\theta. \tag{8.15}$$

The heat stored in the elementary volume is

$$dx \times dy \times dz\,(\rho C_p \frac{\partial t}{\partial \theta})\,d\theta. \tag{8.16}$$

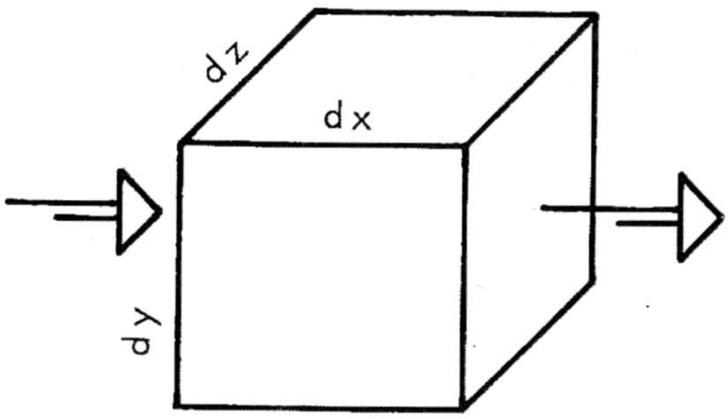

FIGURE 8.11 Elementary block with unidirectional heat flow.

When all these quantities are properly summed for the three-dimensional case, a complicated equation or set of equations is obtained which can only be solved for a limited number of actual cases. A considerable simplication results if flow takes place only along one axis.

In the case of steady-state heat flow, the only coefficient involved is that of the conductivity, k. For a given set of conditions all materials having the same k-value yield the same results thermally. Where thermal storage is involved, the corresponding coefficient is given by the ratio of the conductivity to the specific heat capacity of the unit volume of the material. This coefficient, usually designated as α, is called thermal diffusivity and $\alpha = \frac{k}{\rho C_p}$.

Referring now to Figure 8.12, consider an elementary flow path extending through a wall thickness to be divided into four elementary volumes. The centres of each cube are taken as the reference points for storage temperature. The material lying between these centre points constitutes a resistance to flow, and the whole assembly can be seen as a combination of resistances and capacities. The analogous electrical and hydraulic circuits can be quickly appreciated. They offer alternative means of visualizing the response to temperature changes. They also provide useful analogue solutions in particular cases. It is necessary in the hydraulic analogue to have resistance proportional to the first power of the flow rate; this requires the use of viscous fluids and fine tubing in which the flow will be laminar.

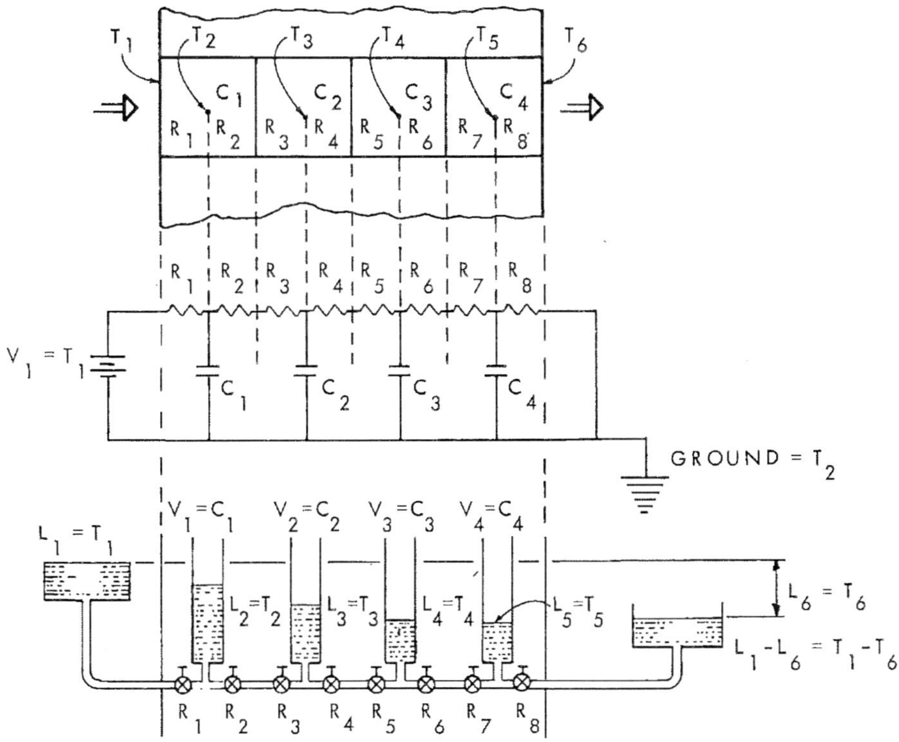

FIGURE 8.12 Electric and hydraulic analogues of an elementary thermal circuit.

8.21 Sudden Change in Temperature of a Slab

One of the simplest cases of transient heat flow is provided by a slab, initially at a temperature t_1 throughout, when the temperature of one face is quickly raised to and maintained at a temperature t_2 so as to provide a one-dimensional heat flow into the slab. Figure 8.13 shows in a simple way the nature of the successive temperature distributions with time. Solutions are available, usually in graph form, for many of the more common situations of this kind [**8.1, p. 2.4**].

The Fourier equation applies to any point in the slab so that at any time the heat flow at any point is given by the product of k and the temperature gradient. It will be recognized that at any point in the slab a change in temperature gradient cannot take place instantly; it must be accompanied by changes in heat storage. On an overall basis it takes some time for the change from temperature t_1 to t_2 at the slab face at time $\theta = \theta_0$ to be felt at the other side of the slab. Until this occurs the case can be regarded as that of a semi-infinite slab.

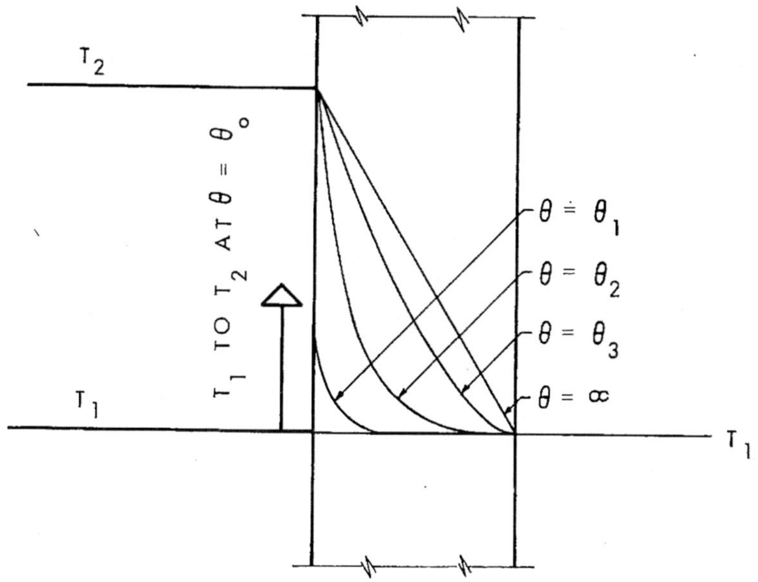

FIGURE 8.13 Sudden change of temperature at the face of a slab.

8.22 Cyclical Heat Flow

If in the case of the slab the temperature was increased as stated and then decreased to t_1 at time $\theta = \theta_2$, heat flow would take place from the inner portion of the slab toward both faces at the same time until the heat stored and the temperature reached a uniform value throughout the slab.

If the temperature at one face is cycled regularly, another typical case is obtained, which for a finite slab is approximately that obtained in a building wall in summer exposed to solar radiation during the day, or which for a semi-infinite slab represents approximately the condition in the ground under the influence of the cyclical variation in heat-flow conditions at the surface from summer to winter. The general characteristics of such situations are as follows.

1. The average heat flow can be calculated using average temperatures in the steady-state equation.
2. If the average temperatures at two points in a slab do not differ, there will be no net heat flow.
3. If an overall temperature difference does exist, it is usually possible to reduce the case to two simpler cases: a steady-state flow at the mean temperature plus a periodic flow about a mean temperature. The temperature patterns can then be combined.
4. The wave of temperature and heat flow within the slab will always be reduced in amplitude and will lag behind the applied temperature wave. Temperature lags in walls under daily cycles vary from one hour to ten hours, depending on the mass of the wall. Temperature lags in the earth may be months or even years, as a result of seasonal changes, depending on depth.

8.23 Lag and Amplitude Change in Temperature Cycles

The time lag or the time required for the temperature at a given point in a body to be influenced by a sine-wave temperature change at the surface may be found from the relation

$$\Delta\theta = \frac{x}{2}\sqrt{1/\alpha\pi n} \tag{8.17}$$

where x = distance from the surface
n = number of complete changes per unit of time
α = thermal diffusivity of the material.

In Figure 8.14 the surface temperature curve is shown in full lines, and the temperature at a distance x from the surface is shown by the dotted curve.

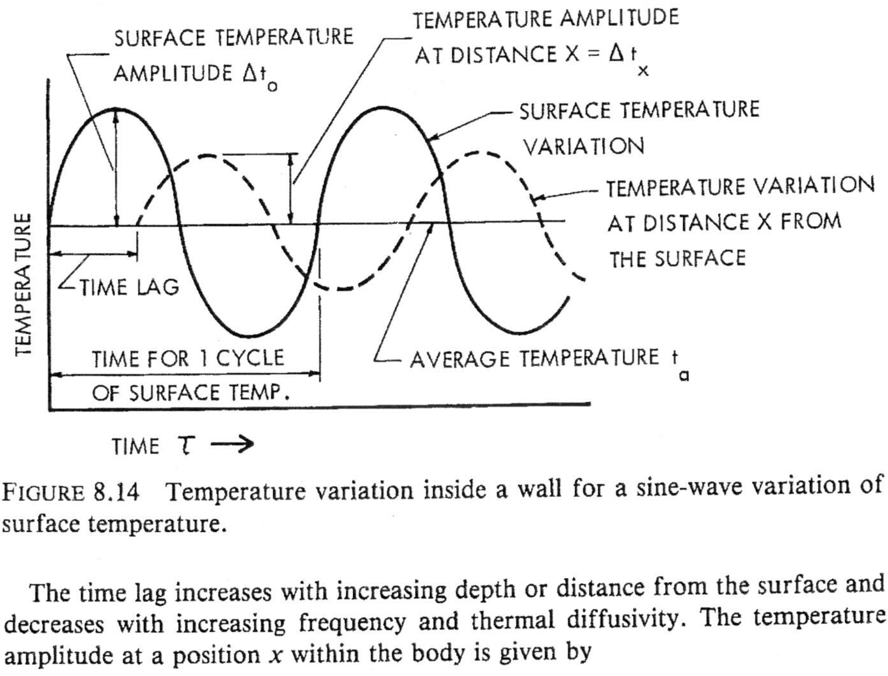

FIGURE 8.14 Temperature variation inside a wall for a sine-wave variation of surface temperature.

The time lag increases with increasing depth or distance from the surface and decreases with increasing frequency and thermal diffusivity. The temperature amplitude at a position x within the body is given by

$$\Delta t_x = \Delta t_o \times e^{-x\sqrt{\pi n/\alpha}} \tag{8.18}$$

where

Δt_o = surface temperature amplitude.

In more descriptive terms, the temperature varies between $t_a + \Delta t_x$ and $t_a - \Delta t_x$ where t_a is its average value. The amplitude of oscillation decreases with increasing distance x from the surface.

The calculation of the air-conditioning load is greatly complicated by these characteristics of periodic flow, but they must be taken into account. A large periodic heat exchange at the surface of a wall or roof gives rise to a periodic heat flow into the space, which is greatly reduced in amplitude and occurs several hours after the peak flow at the surface.

The ground presents another important example of periodic heat flow. Measured ground temperatures for Ottawa are shown in Figure 8.15 [**8.25, p. 6**]. The ground temperature in May is still decreasing at a depth of 3 to 4 m, long after the ground closer to the surface has begun to warm up. There is, of course, a corresponding lag six months later when the ground is still increasing in temperature at the 3-m depth.

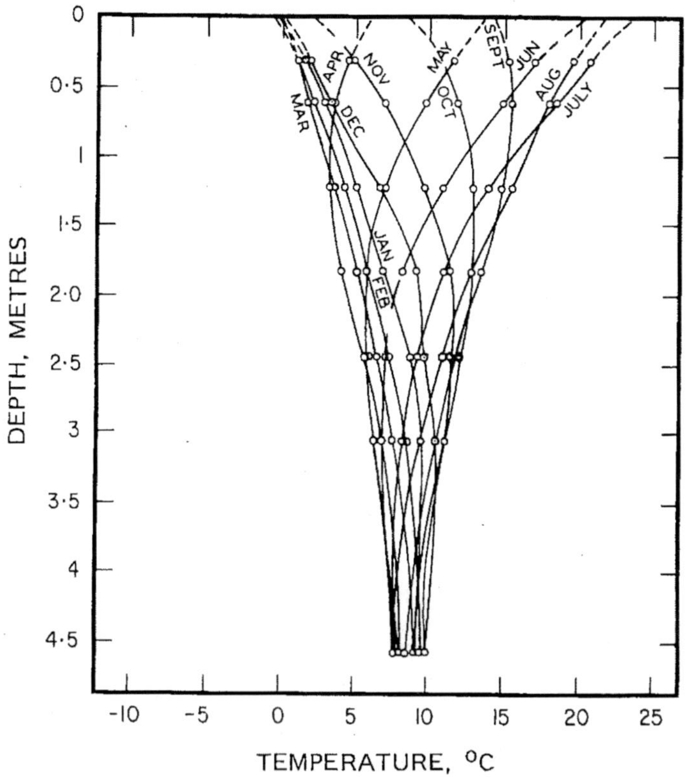

FIGURE 8.15 Monthly average ground temperatures measured in clay soil under natural surface cover in Ottawa.

The mean ground temperature is shown as a straight vertical line. It should show a gradient of about 2 K per 100 m, but this is so small as to be invisible in the illustration. These "geothermal gradients" are important in deep drilling and mining, and in the northern regions where they are related to the depth of permafrost.

Because geothermal gradients are small, the steady upward heat flow from the earth at depth is relatively small for many purposes. Mean ground temperatures are related to, but are not exclusively dependent on, air temperatures, and run from 2 to 6 K above mean annual air temperature. When the mean annual air temperature is well below freezing, it may be expected that the mean ground temperature will be below freezing and that there will be a permafrost condition [**8.27**]. If it is known that the mean ground temperature near the surface is, for example, −10°C, then for a geothermal gradient of 2 K per 100 m, a depth of permafrost of up to 500 m may be expected. The annual depth of frost penetration is of interest in construction [**8.28**].

8.24 Thermal Conductivity of Soils

The problem of heat transmission through moist materials was discussed in Chapter 5, where it was shown that a combined mass and energy flow is involved and that it may not be adequate to deal with it using conduction theory alone. Despite this complication, k-values for moist soils are widely used with conduction theory, and those of Kersten [**8.29**] are well known. Values for specific soils are given in Reference **8.1, p. 22.28**. Kersten's work has been discussed in relation to some Canadian studies [**8.30, 8.31**] which also demonstrate the complications of heat transfer in moist soils.

Kersten summarized his results in several empirical relationships (which have been converted to SI) as follows:

For sandy soils above 0°C

$$k = (0.1 \log_{10} M + 0.06)10^{0.000\,62D} \tag{8.19}$$

For sandy soils below 0°C

$$k = 0.001\,44 \times 10^{0.001\,33D} + M(0.0037 \times 10^{0.000\,91D}) \tag{8.20}$$

For silt and clay soils above 0°C

$$k = (0.13 \log_{10} M - 0.029)10^{0.000\,62D} \tag{8.21}$$

For silt and clay soils below 0°C

$$k = 0.001\,44 \times 10^{0.001\,33D} + M(0.0036 \times 10^{0.0087D}) \tag{8.22}$$

when

M = moisture content, percent, by weight, dry basis
D = mass density of dry soil, kg/m³
k = thermal conductivity, W/(m • K)

These relationships were considered to provide values of soil conductivity correct to within 25% when applied to sandy soils with $M \geq 1$ and to silt and clay soils with $M \geq 7$.

8.25 Wind Chill

Weather forecasters often express severe cold-weather conditions in terms of wind chill. The concept was introduced by Paul Siple, a member of the U.S. Army's Antarctic expedition. It describes the heat loss from a litre cylinder of water at 33°C as a function of ambient temperature and wind velocity. This cooling effect is a measure of the severity of conditions for a surface maintained at 33°C. It does not apply rigorously to exposed human flesh which, before it freezes, must drop from the normal skin temperature of about 29°C to below 0°C. Nevertheless, the wind-chill index has been useful as a means of identifying conditions where there is a high risk of frostbite on exposed skin. The index relates only indirectly to conditions for clothed or insulated bodies or buildings, since it is concerned primarily with the influence of velocity on the surface-heat transfer. This effect is small when the resistance of the added insulation becomes large in relation to the surface film resistance [**8.1, p. 8.17**].

8.26 Clothing

Clothing is, for many purposes, simply body insulation. The insulating value of clothing becomes an important factor when protection from low temperatures is required. Clothing affects energy exchange principally through conduction, but also through convection, radiation, evaporation, vapour diffusion, and air circulation. As in the case of building heat transfer, the simplest exchange is dry-heat conduction; there are increasing difficulties in handling the other factors. Further complications arise because the clothed human body does not remain stationary, but presents constantly changing radiation, convection, and ventilation effects.

The insulation value of clothing is expressed in resistance units of heat transfer, the reference unit being 1 clo. This unit was selected for convenience; it is approximately the thermal resistance of a business suit and its related garments, or a substantial overcoat.

$$1 \text{ clo} = 0.155 \text{ (m}^2 \cdot \text{K)/W}. \qquad (8.23)$$

Values for a variety of clothing ensembles are given in Chapter 14. The protective clothing required for different locations in Canada at different times has been set out [**8.32**]. The clothing must, of necessity, be related to the level of physical activity, which determines the rate of the clothed body's heat generation.

References

8.1 *ASHRAE Handbook 1981 Fundamentals*. Chaps. 2 and 23.

8.2 Gebhart, B. *Heat transfer*. 2d Ed. New York: McGraw-Hill, 1971.

8.3 Shirtliffe, C.J. *Thermal resistance of building insulation*. Canadian Building Digest 149. DBR/NRCC, May 1972.

8.4 *Standard test method for steady-state thermal transmission properties by means of the guarded hot plate*. ASTM C177-76. Philadelphia: American Society for Testing and Materials, 1976.

8.5 Solvason, K.R. Large-scale wall heat-flow measuring apparatus. *ASHRAE Transactions*, 1959, *65*, pp. 541-50. (NRCC 5421)

8.6 *The international system of units (SI)*. National Standard of Canada CAN3-Z234.2-76. Rexdale: Canadian Standards Association, 1966.

8.7 *Canadian metric practice guide*. National Standard of Canada CAN3-Z234.1-76. Rexdale: Canadian Standards Association, 1976.

8.8 Strelka, C.S., Loshak, L., and Torrance, J.S. *Manual on metric building drawing practice*. DBR/NRCC, June 1977. (NRCC 15234)

8.9 Wilson, A.G. *Condensation on inside window surfaces*. Canadian Building Digest 4. DBR/NRCC, April 1960.

8.10 Veale, A.C. Insulation thicknesses for houses. *Canadian Builder*, 1964, *14*, pp. 48–50. (Also available as DBR/NRCC Housing Note 21.)

8.11 Min, T.C., Schutrum, L.F., Parmelee, G.V., and Vouris, J.D. Natural convection and radiation in a panel-heated room. *ASHAE Transactions*, 1956, *62*, pp. 337–58.

8.12 Robinson, H.E., Cosgrove, H.A., and Powell, F.J. *Thermal resistance of air spaces and fibrous insulations bounded by reflective surfaces*. Building Materials and Structures Report 151. Washington: National Bureau of Standards, 1957.

8.13 Adams, L. Thermal conductance of air spaces. *ASHRAE Journal*, 1976, *18*, pp. 37–38.

8.14 Handegord, G.O., and Hutcheon, N.B. Thermal performance of frame walls. *ASHVE Transactions*, 1952, *58*, pp. 171-88. (DBR 30)

8.15 Wolf, S., Solvason, K.R. and Wilson, A.G. Convective air flow effects with mineral wool insulation in wood-frame walls. *ASHRAE Transactions*, 1966, *72* (Part II), pp. II.3.1–II.3.8. (NRCC 9649)

8.16 Ball, W.H. *Thermal insulation in dwellings*. Canadian Building Digest 16. DBR/NRCC, April 1961.

8.17 Handegord, G.O. Thermal performance of frame walls. Part III—Wall surface temperatures. *ASHAE Transactions*, 1957, *64*, pp. 331–46. (NRCC 4342)

8.18 Sasaki, J.R. Thermal performance of exterior steel-stud frame walls. *ASHRAE Transactions*, 1972, *78* (Part I), pp. 192–98. (NRCC 13858)

8.19 Brown, W.P., and Wilson, A.G. *Thermal bridges in buildings*. Canadian Building Digest 44. DBR/NRCC, Aug. 1963.

8.20 Berthier, J. Les points faibles thermiques ou ponts thermiques. *Cahiers de Centre Scientifique et Technique du Bâtiment*, 1960, No. 334, pp. 1–20. (Available in part in Technical Translation TT-1143, DBR/NRCC.)

8.21 Birkeland, Ø. Energy losses through thermal bridges. *Building Research and Practice*, 1979, *7* (5), pp. 284–91.

8.22 Wilson, A.G., and Brown, W.P. *Thermal characteristics of double windows*. Canadian Building Digest 58. DBR/NRCC, Oct. 1964.

8.23 Solvason, K.R. and Wilson, A.G. *Factory-sealed double-glazing units*. Canadian Building Digest 46. DBR/NRCC, Oct. 1963.

8.24 Sasaki, J.R. *Potential for thermal breakage of sealed double-glazing units.* Canadian Building Digest 129. DBR/NRCC, Sept. 1970.

8.25 Latta, J.K. *Walls, windows and roofs for the Canadian climate.* DBR/NRCC, Oct. 1973. (NRCC 13487)

8.26 Brown, W.G. *Graphical determination of temperature under heated or cooled areas on the ground surface.* DBR/NRCC, Oct. 1963. Ottawa. (NRCC 7660)

8.27 Johnston, G.H. *Permafrost and foundations.* Canadian Building Digest 64. DBR/NRCC, April 1965.

8.28 Penner, E. *Ground freezing and frost heaving.* Canadian Building Digest 26. DBR/NRCC, Feb. 1962.

8.29 Kersten, M.S. *Thermal properties of soils.* Bulletin No. 28, Engineering Experiment Station, University of Minnesota, 1952.

8.30 Woodside, W., and Cliffe, J.B. Heat and moisture transfer in closed systems of two granular materials. *Soil Science,* 1959, *87,* pp. 75–82. (NRCC 5030)

8.31 Woodside, W., and de Bruyn, C.M.A. Heat transfer in a moist clay. *Soil Science,* 1959, *87,* pp. 166–73. (NRCC 5098)

8.32 Auliciems, A., Freitas, R. de F., and Hare, F.K. *Winter clothing requirements for Canada.* Climatological Studies No. 22. Toronto: Environment Canada, 1973.

9

Solar Radiation and Buildings

9.1 Introduction

The discussions in earlier chapters (chapters 2, 4, and 8 in particular) have provided a basis for understanding the general nature of solar radiation. Its intensity outside the earth's atmosphere varies significantly with time of year as the earth-to-sun distance changes. Reflection and scattering in the atmosphere produce diffuse, short-wave sky radiation, while absorption and reradiation produce incoming long-wave radiation. The intensity of the direct radiation is reduced accordingly as it passes through the atmosphere.

At the earth's surface the direct solar radiation may be reflected, producing further diffuse short-wave radiation, or it may be absorbed. Receiving surfaces, while absorbing direct and diffuse short-wave and long-wave radiation, also radiate long-wave to their surroundings and to the sky. They are involved in heat transfer by convection, by conduction, and sometimes by evaporation. These components of the energy balance at a surface have already been shown in Figure 4.9, and all play a part in determining the effect of solar radiation on buildings.

The effect of solar radiation on buildings is complex, since the rate at which energy is delivered is highly variable with latitude, time of year, time of day, weather, surroundings, orientation of receiving surfaces and transmitting areas, and absorptance and emittance of receiving surfaces. The resulting transient and cyclical heat flows are not easy to deal with. The essential features of all these matters must be well understood so that the relevance of many of the simplifying assumptions introduced in dealing with solar energy can be judged in relation to a particular situation.

Various aspects of solar radiation, including the means by which solar heat gains and cooling loads can be estimated, are dealt with at length in the ASHRAE 1981 Handbook [**9.1, Chaps. 26, 27**]. Solar energy utilization is also discusssed in the 1978 Handbook [**9.2**].

9.2 Sun Direction

The sun's position in the sky can be expressed in terms of the solar altitude, β, above the horizontal and the solar azimuth, ϕ, measured from the south. These angles are determined by the local latitude, L, the solar declination, δ, and the apparent solar time (AST). The solar declination, which is the angle between the earth-sun line and the equatorial plane as shown in Figure 2.3, varies from

+23.45° (23° 27′) at the summer solstice to −23.45° at the winter solstice. Values for the twenty-first of each month are given in Table 9.1. There are small variations over the years which can be disregarded for most building purposes. The solar altitude at solar noon is given by

$$\beta_N = 90 - L + \delta. \qquad (9.1)$$

TABLE 9.1

Extraterrestrial Solar Radiation and Related Data For the 21st day of each month
(From *ASHRAE Handbook 1981 Fundamentals*, Table 1, p. 27.2) [**9.1**]

	I_o W/m²	Equation of time, min.	Declination, δ, deg.	A W/m²	B Dimensionless	C Dimensionless
Jan	1396	−11.2	−20	1229	0.142	0.058
Feb	1384	−13.9	−10.8	1214	0.144	0.060
Mar	1363	−7.5	0.0	1185	0.156	0.071
Apr	1341	+1.1	+11.6	1135	0.180	0.097
May	1321	+3.3	+20.0	1103	0.196	0.121
June	1310	−1.4	+23.45	1088	0.205	0.134
July	1311	−6.2	+20.6	1084	0.207	0.136
Aug	1324	−2.4	+12.3	1107	0.201	0.122
Sept	1345	+7.5	0.0	1151	0.177	0.092
Oct	1367	+15.4	−10.5	1192	0.160	0.073
Nov	1388	+13.8	−19.8	1220	0.149	0.063
Dec	1398	+1.6	−23.45	1233	0.142	0.057

The solar altitude for other times of day depends on the hour angle, H, which gives the rotation of the earth in the interval from noon to a particular time. Since the earth rotates 360° in 24 hours the hour angle changes 0.25° for each minute. Time must be measured from noon AST and H = 0.25 times the number of minutes.

AST differs from local standard time (LST) by four minutes for each degree between the local standard time meridian (LSM) and the local longitude (LON). A further adjustment is known as the equation of time (ET), since the earth does not move in its orbit around the sun at a constant angular velocity. Thus,

$$\text{AST} = \text{LST} + \text{ET} + 4(\text{LSM} - \text{LON}). \qquad (9.2)$$

Values of the equation of time are given in Table 9.1. The standard time meridians for North America are 60° for Atlantic, 75° for eastern, 90° for central, 105° for mountain, 120° for Pacific, and 135° for Yukon standard time.

Example 9.1

Find the hour angle, H, for 08:30 eastern standard time on July 21 at a place where the longitude is 79°W.

From Table 9.1, ET = 6.2 min.

From Equation 9.2,

$$\text{AST} = 08{:}30 - 6.2 \text{ min} + 4\,(75 - 79) \text{ min.}$$
$$= 08{:}08.$$

Time from solar noon = 232 minutes.

Hour angle $H = 232 \times 0.25 = 58°$.

The solar altitude β and the solar azimuth ϕ can be found from

$$\sin\beta = \cos L \cos\delta \cos H + \sin L \sin\delta \tag{9.3}$$
$$\sin\phi = (\cos\delta \sin H)/\cos\beta. \tag{9.4}$$

The direction of incoming solar radiation for a given latitude L and time of day can now be defined in terms of β and ϕ from available data. It is also possible to find the angle of incidence θ which gives the angle between the incoming solar rays and a line normal to any receiving surface. The relevant angles for horizontal and vertical surfaces are shown in Figure 9.1. The earth–sun line is OQ. NOS is the north–south line; OV is perpendicular to the horizontal plane. The vertical surface is oriented at a surface azimuth ψ. The angle POQ is the angle of incidence θ_v for the vertical surface and can be found from the relationship

$$\cos\theta_v = \cos\beta \cos\gamma \tag{9.5}$$

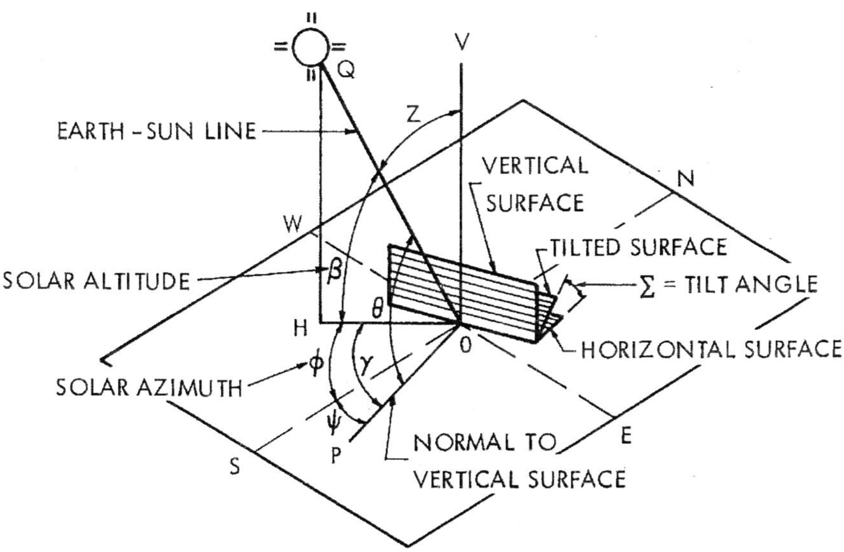

FIGURE 9.1 Solar angles for vertical and horizontal surfaces. (From *ASHRAE Handbook 1981 Fundamentals*, Fig. 2, p. 27.3)[**9.1**]

where γ = difference between the solar azimuth ϕ and the azimuth ψ of a normal to the surface. Values of γ greater than 90° mean that the surface is in the shade.

QOV is the angle of incidence, θ_h, for the horizontal surface and is given by

$$\cos\theta_h = \sin\beta. \tag{9.6}$$

When a surface is tilted from the horizontal by a tilt angle Σ, the angle of incidence, θ, is given by

$$\cos\theta = \cos\beta \cos\gamma \sin\Sigma + \sin\beta \cos\Sigma. \tag{9.7}$$

Example 9.2

Find for the time and location of Example 9.1 and for a latitude of 43°N the altitude and azimuth angles.

From Example 9.1, $H = 58°$
From Table 9.1, $\delta = 20.6°$
$L = 43°$

and from Equation 9.3,

$$\begin{aligned} \sin\beta &= \cos(43°)\cos(20.6°)\cos(58°) + \sin(43°)\sin(20.6°) \\ &= 0.73 \times 0.94 \times 0.53 + 0.68 \times 0.53 \\ &= 0.60 \text{ and } \beta = 36.9°. \end{aligned}$$

From Equation 9.4,

$$\begin{aligned} \sin\phi &= [\cos(20.6°)\sin(58°)]/\cos(36.9°) \\ &= 0.936 \times 0.848/0.800 = 0.992 \\ \text{and } \phi &= 83°. \end{aligned}$$

Example 9.3

Find the angle of incidence for a vertical surface facing southeast for the conditions of Example 9.2.

$$\psi = 45°, \quad \phi = 83°, \quad \Sigma = 90°, \quad \beta = 36.9°, \quad \gamma = 83 - 45 = 38°,$$

and from Equation 9.7, noting that $\cos(90°) = 0$,

$$\begin{aligned} \cos\theta &= \cos(36.9°)\cos(38°)\sin(90°) \\ &= 0.80 \times 0.79 \times 1.0 = 0.63 \\ \text{and } \theta &= 51°. \end{aligned}$$

Once the angle of incidence is known for a given surface orientation, it is possible to establish by calculation or by graphical means the shadows cast on surfaces so oriented by various projections such as window reveals, eaves, overhangs, and even adjacent constructions.

9.3 Intensity of Solar Radiation

The intensity of solar radiation, I_o, on a plane normal to the sun's rays outside the earth's atmosphere varies from 1398 W/m² on December 21 when the earth is closest to the sun to 1310 W/m² on June 21. The value at the mean earth–sun distance is known as the solar constant, and has the value 1353 W/m². Values for the other months are given in Table 9.1.

The total short-wave radiation, I_t, reaching a surface on the earth is given by

$$I_t = I_{DN} \times \cos\theta + I_d + I_r \qquad (9.8)$$

where I_{DN} is the direct normal radiation, I_d is the diffuse sky radiation, and I_r is the short-wave radiation reflected from surroundings. These factors have been discussed in a general way in Chapter 4; means of evaluating them will now be considered. The calculation of θ, the angle of incidence the direct radiation makes with the receiving surface, has just been dealt with.

The intensity of the direct normal radiation depends on the clarity of the atmosphere and the length of the path of the solar beam through it. The length of path depends on the altitude angle, β. The intensity at the earth's surface can be found from the equation

$$I_{DN} = A/e^{B/\sin\beta}. \qquad (9.9)$$

The values of A and B for use in this equation proposed by Stephenson [**9.3**] were adjusted to provide values of I_{DN}, which were in close agreement with those of Threlkeld and Jordan [**9.4**]. They are given in Table 9.1. They represent the conditions on average cloudless days, and they vary throughout the year owing to changing dust and water-vapour content of the air. The value of I_{DN} for clear days can be multiplied by a clearness number which may vary from 0.8 for industrial areas to as much as 1.2 for clear skies. Values of I_{DN} for average clear days can be found from various tables [**9.1, Chap. 27; 9.3**]. Values for 45°N latitude for January 21 and July 21 with corresponding solar altitude and azimuth are given in Table 9.2.

The diffuse solar radiation, I_{ds}, from a clear sky varies with sun position and surface orientation. That reaching a horizontal surface is given by

$$I_{ds} = CI_{DN} \qquad (\text{W/m}^2) \qquad (9.10)$$

where C, the diffuse radiation factor, is given in Table 9.1.

The diffuse radiation from the sky that is incident on a vertical surface can be found by multiplying I_{ds} by an empirical function, Y, developed by Threlkeld, which varies from 0.45 to more than 1.0. A simplified approximation for the diffuse solar radiation I_{ds} from a clear sky falling on any surface is given by

$$I_{ds} = CI_{DN}F_{ss} \qquad (\text{W/m}^2) \qquad (9.11)$$

where F_{ss} is the angle factor between surface and sky giving the fraction of short-wave radiation emitted by the sky that reaches the surface [**9.1, p. 27.9**]. F_{ss} is 0.5

TABLE 9.2

Solar Position and Intensity for 45°N Latitude for January 21 and July 21

(From Stephenson, D.G., *Tables of solar altitude, azimuth intensity and heat gain factors*, NRCC 9528, 1967) [9.3]

Date	Solar time	Solar position		Direct normal W/m²
		Alt.	Azimuth	
Jan. 21	08:00	5.2	54.8	255
	09:00	13.2	43.0	659
	10:00	19.5	29.9	803
	11:00	23.6	15.4	862
	12:00	25.0	0.0	878
	13:00	23.6	15.4	862
	14:00	19.5	29.9	803
	15:00	13.2	43.0	659
	16:00	5.2	54.8	255
July 21	05:00	4.4	114.9	75
	06:00	14.4	104.9	472
	07:00	24.8	94.9	663
	08:00	35.4	84.2	759
	09:00	45.8	71.7	813
	10:00	55.3	55.3	843
	11:00	62.6	31.8	859
	12:00	65.6	0.0	864
	13:00	62.6	31.8	859
	14:00	55.3	55.3	843
	15:00	45.8	71.7	813
	16:00	35.4	84.2	759
	17:00	24.8	94.9	663
	18:00	14.4	104.9	472
	19:00	4.4	114.9	75

for vertical surfaces and 1.0 for horizontal surfaces. For tilted surfaces, $F_{ss} = (1.0 + \cos\Sigma)/2$ where Σ is the tilt angle measured from the horizontal as in Figure 9.1.

Solar radiation to a surface by diffuse reflection from an adjacent surface is given by the intensity falling on the reflecting surface times its reflectance times the angle factor between the receiving and reflecting surfaces. Radiation reflected from the ground arises from diffuse sky radiation and direct solar radiation on a horizontal surface. The radiation reaching the ground is

$$I_{th} = I_{DN}(C + \sin\beta) \qquad (\mathrm{W/m^2}) \qquad (9.12)$$

where $I_{DN}\sin\beta$ is the direct radiation falling on the ground.

The ground-reflected radiation incident on any surface can be found from

$$I_{dg} = I_{th} \times \rho_g \times F_{sg} \tag{9.13}$$

where ρ_g is the reflectance of the ground and F_{sg} is the angle factor between the surface and the ground. If the surface sees only ground and sky, the sky-angle factor is $F_{ss} = 1 - F_{sg}$ since angle factors must sum to unity. Since $F_{sg} = (1 - \cos\Sigma)/2$, F_{ss} can be found.

For a vertical surface, when $F_{sg} = 0.5$ the incident diffuse plus direct radiation reflected from the ground is

$$0.5\rho_g I_{DN}(C + \sin\beta). \tag{9.14}$$

The total diffuse radiation incident on a vertical surface is given by the sum of that reaching the surface directly from the sky as given by Equation 9.11 and that reflected from the ground as given by Equation 9.14. The value of F_{ss} for a vertical surface is 0.5. A more nearly correct value would be given by Threlkeld's function Y [**9.5**], in place of F_{ss} in Equation 9.11. The total diffuse radiation incident on a vertical surface would then be given by

$$I_d = I_{DN}[CY + 0.5\rho_g(C + \sin\beta)]. \tag{9.15}$$

Example 9.4

Find the short-wave radiation falling directly on a vertical surface facing south-east at 10:00 solar time on July 21 at 45°N.

From Table 9.2, $\beta = 55.3$ and $\phi = 55.3$
From Table 9.1, $A = 1084$ W/m². $B = 0.207$.
From Equation 9.9,

$$\begin{aligned} I_{DN} &= A/e^{B/\sin\beta} \\ &= 1084/e^{0.207/\sin 55.3^\circ} \\ &= 843 \text{ W/m}^2. \end{aligned}$$

This value might also have been read from Table 9.2.
From Equation 9.5,

$$\begin{aligned} \cos\theta_v &= \cos\beta\cos\gamma \\ &= \cos(55.3^\circ)\cos(55.3 - 45) \\ \theta_v &= 55.9^\circ. \end{aligned}$$

the direct short-wave radiation on the vertical surface is given by

$$\begin{aligned} I_{DN}\cos\theta_v &= 843 \times 0.56 \\ &= 471 \text{ W/m}^2. \end{aligned}$$

Example 9.5

Find the diffuse short-wave radiation on a vertical surface for the conditions of Example 9.3 and for a ground reflectance of 0.20.

From Equation 9.14, the diffuse radiation reflected from the ground is

$$0.5\rho_g \cdot I_{DN}(C + \sin\beta) = 0.5 \times 0.20 \times 843\,(0.136 + 0.822)$$
$$= 80.8 \text{ W/m}^2.$$

From Equation 9.11, the diffuse radiation received directly from the sky is

$$I_{ds} = CI_{DN}F_{ss} \quad \text{and} \quad F_{ss} = 0.5$$
$$= 0.136 \times 843 \times 0.5$$
$$= 57.3 \quad \text{W/m}^2.$$

The diffuse and reflected short wave $= 80.8 + 57.3$
$= 138 \text{ W/m}^2$.

The total, I_t, of direct radiation as found in Example 9.4 and the diffuse radiation as now calculated is

$$138 + 471 = 609 \text{ W/m}^2.$$

It can now be appreciated that the values of total solar radiation falling on various surfaces as given in equations and tables and as used in many computations are often based on typical or average conditions, and may be subject to substantial adjustment for particular situations. Other simplifying assumptions may also have been introduced. In the discussions just completed, the incoming radiation for clear days was calculated for average conditions and is subject to adjustment through the clearness factor. The calculations of diffuse and reflected radiation were based on the assumption of angle factors for a flat, unobstructed ground surface. Surroundings projecting above the horizontal plane introduce the need for more elaborate analysis. Adjustments may also be indicated for ground reflectances other than those assumed and for changing reflectance with the seasons—for example, when there is snow on the ground.

9.4 Solar Heat Gain

Solar radiation on buildings may be incident on opaque surfaces such as roofs or transparent surfaces such as windows. When the receiving surfaces are opaque, the radiation is partially reflected; the remainder is absorbed, producing a rise in surface temperature. Since the solar radiation is not constant, the temperature and the associated heat flow are also variable. The thermal-storage characteristics as well as the heat-transmission properties become involved in determining the rate at which heat will be delivered through the construction under the combined effects of the solar irradiation of the outer surface and the outdoor-to-indoor temperature difference.

In the case of windows and other transparent constructions (known as fenestration), the transmitted radiation is usually large; the thermal storage of the glass is small and can be disregarded. These differences between transparent and opaque constructions lead to quite different procedures in the estimation of the solar heat gain.

9.5 Solar Heat Gain through Fenestration

The solar heat gain through single glazing is illustrated in Figure 9.2. The incident solar radiation intensity is I_t. The reflected radiation is ρI_t, the absorbed radiation is αI_t, and the transmitted radiation is τI_t, where ρ, α, and τ are the reflectance, absorptance, and transmittance respectively and

$$\rho + \alpha + \tau = 1 \tag{9.16}$$

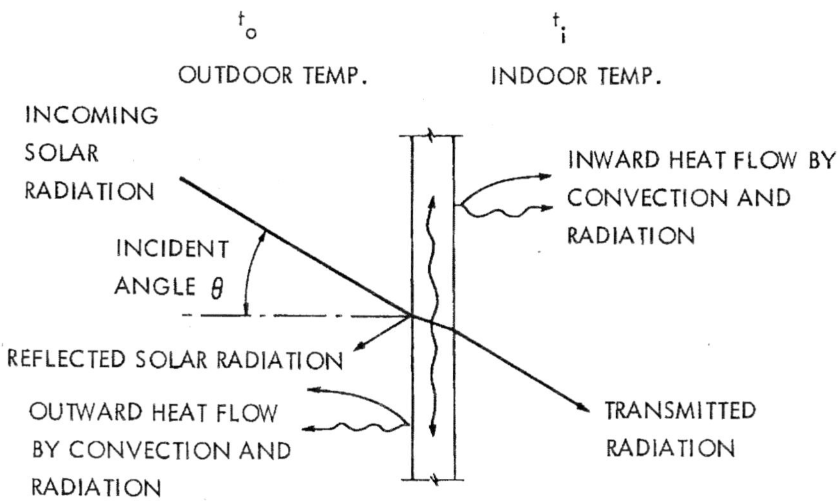

FIGURE 9.2 Instantaneous heat balance for sunlit glazing material. (From *ASHRAE Handbook 1981 Fundamentals*, Fig. 13, p. 27.17)[**9.1**]

The reflectance varies with the angle of incidence, and may be different for the direct short-wave radiation meeting the surface at an angle and for the diffuse and reflected short wave. This point can be ignored for our present purposes and the solar heat gain given as

$$q_s = \tau I_t + N\alpha I_t \qquad (\text{W/m}^2) \tag{9.17}$$

where N is the fraction of absorbed radiation which reaches the inside by reradiation and convection as determined by the indoor and outdoor surface conductances, with a value of about 0.3.

TABLE 9.3

Transmittance and U-Values for Some Glazing Units
(From Stephenson, D.G., Canadian Building Digest 101, Table 1, DBR/NRCC, 1968) [9.6]

	Transmittance without shades		*U*-values W/(m² · K)	
Type of Window	Light	Solar heat	No shade	Curtain or blind
Single glazing				
Clear sheet glass	0.90	0.80	5.7	4.5
Regular plate glass	0.87	0.77	5.7	4.5
Heat-absorbing plate	0.50	0.45	5.7	4.5
Double glazing				
Regular plate / Air space / Regular plate	0.77	0.60	3.4	2.8
Heat-absorbing plate / Air space / Regular plate	0.45	0.35	3.4	2.8
Regular plate-reflective / Air space / Regular plate	0.35	0.16	1.7	—

Note: Clear sheet 1/8 in. (3 mm) thick. Plate glass 1/4 in. (6 mm) thick.
Air space 1/2 in. (12.5 mm) thick.

The window will also transmit heat because of the difference in temperature between outside air at t_o and inside air at t_i.

The total energy gain becomes

$$q_t = \tau I_t + N\alpha I_t + U(t_o - t_i) \tag{9.18}$$

where U is the overall coefficient of heat transmission, or U-value, for the single glazing. Since thermal storage in the glass is negligible, q is regarded as the instantaneous heat gain. Actually, the gain is not all in the form of heat. The transmitted short-wave component τI_t must still be absorbed at interior surfaces.

The problem becomes increasingly complex when glazing with different properties, multiple glazing, and the addition of blinds and drapes are considered. It is now common practice to calculate the solar heat gain, as defined by Equation 9.17, for double-strength clear glass (approximately 3 mm thick) for all solar intensities and incidences. This is known as the solar heat-gain factor (SHGF). All other glazing and shading arrangements are described in terms of individual

TABLE 9.4

Shading Coefficients for Some Glazing Units
(From Stephenson, D.G., Canadian Building Digest 101, Table 1, DBR/NRCC, 1968) [**9.6**]

Type of Window	Shading coefficient: No shade	With curtain: Min.	With curtain: Max.	With venetian blind
Single glazing				
Clear sheet glass	1.00	0.45	0.65	0.55
Regular plate glass	0.95	0.45	0.65	0.55
Heat-absorbing plate	0.70	0.40	0.50	0.47
Double glazing				
Regular plate / Air space / Regular plate	0.83	0.40	0.60	0.50
Heat-absorbing plate / Air space / Regular plate	0.55	0.33	0.43	0.36
Regular plate-reflective / Air space / Regular plate	0.25	—	—	—

Note: These are the same glazing units as in Table 9.3.

shading coefficients (SC) by which the SHGF found for double-strength clear glass must be multiplied to give the appropriate solar heat gain.

U-values and shading coefficients for selected glazing arrangements are given in Tables 9.3 and 9.4. The total instantaneous heat gain is given by

$$q_t = \text{SC} \times \text{SHGF} + U(t_o - t_i). \tag{9.19}$$

Tables of solar heat-gain factors for various latitudes, times of year, times of day, and orientation of receiving surface are available [**9.1, Chap. 27, 9.3**]. Hourly values for morning and afternoon may be compared by recognizing the symmetry in the sun position with respect to solar noon. The radiation on a vertical surface facing east in the afternoon will be the same as that for a surface facing west in the morning for equal times from solar noon. The half-day totals [**9.1, 9.3**] in watt-hours per square metre are integrated values based on a ten-minute time interval and are the summation of hourly values. Values for January 21 and July 21 for 45°N latitude [**9.3**] are listed in Table 9.5 in a whole-day format.

TABLE 9.5

Solar Heat-Gain Factors, W/m², for 45°N Latitude
(From Stephenson, D.G., *Tables of solar altitude, azimuth intensity and heat gain factors,* NRCC 9528, 1967) [**9.3**]

January 21	North	East	South	West	Horizontal
08:00	8	199	137	8	17
09:00	29	422	454	29	111
10:00	42	358	646	42	222
11:00	50	177	756	50	299
12:00	53	57	792	57	326
13:00	50	50	756	177	299
14:00	42	42	646	358	222
15:00	29	29	454	422	111
16:00	8	8	137	199	17
Daily totals, W • h/m²	312	1346	4798	1346	1628
July 21	**North**	**East**	**South**	**West**	**Horizontal**
05:00	32	71	5	5	9
06:00	117	472	38	38	119
07:00	83	651	68	63	286
08:00	87	679	107	82	454
09:00	97	606	209	97	595
10:00	107	457	318	107	704
11:00	114	252	394	114	772
12:00	116	126	420	126	795
13:00	114	114	394	252	772
14:00	107	107	318	457	704
15:00	97	97	209	606	595
16:00	87	82	107	679	454
17:00	83	63	68	651	286
18:00	117	38	38	472	119
19:00	32	5	5	71	9
Daily totals, W • h/m²	1360	3785	2700	3785	6664

Example 9.6

Find the daily total of solar heat-gain factors for January 21 on an east wall at 45°N.

From Table 9.5,

$$\text{SHGF day total for east} = 1346.$$

(Note that the units are W • h/m².)
The 24-hour average will be 1346/24 = 56 W/m².

Example 9.7

Find the solar heat gain at 10:00 solar time for January 21 for a window on a south wall at 45°N. The window is double glazing, regular plate with venetian blind.

From Table 9.5, SHGF = 646 W/m^2
From Table 9.4, SC = 0.50
Solar heat gain = 646 × 0.50 = 323 W/m^2

Example 9.8

Find the heat-transmission loss and total heat gain for the window in Example 9.7 if the indoor-to-outdoor temperature difference is 45°C.

From Table 9.3, U = 2.8 $W/(m^2 \cdot °C)$
Heat-transmission loss = 2.8 × 45 = 126 W/m^2
Heat gain = 323 − 126 = 197 W/m^2.

Compare the rate of heat loss by transmission, 126 W/m^2, with the 24-hour average solar heat gain, 56 W/m^2, from Example 9.6, assuming that the temperature difference applies over the 24-hour period.

The calculations by which shading coefficients are established will not be given here. Extensive information and many further references are available [**9.1, Chap. 27**]. In many cases solar-calorimeter tests are required for accurate evaluation. Some properties of glass and glazing and shading arrangements will be discussed here, however.

Only the unshaded glass area of a window transmits at the rate given by the value of solar heat gain. This area may be reduced by shading owing to a significant depth of reveal. The encroachment of shadows cast by mullions and transoms can be calculated and the effective glass area reduced accordingly. The shaded area will still receive diffuse radiation, however, and a further adjustment can be made [**9.1, p. 27.47**]

9.6 **Properties of Glass**

The normal-incidence spectral transmittance of three typical glasses is shown in Figure 9.3. Values are presented in relation to the spectra of solar radiation and of blackbody long-wave radiation at 35°C. The change in scale between the solar and blackbody spectra of 50× should be noted. The solar spectra are given for no atmosphere, m = 0, and for an atmospheric path double that for the sun at the zenith, m = 2. The clear glass transmits 85% or more of the radiation between 0.3 and 3.0 μm, while the others transmit about 47% of the solar spectrum. All transmit highly in the visible range from 0.4 to 0.8 μm. Note that glass is opaque to almost all long-wave radiation.

The variation with angle of incidence of the solar-optical properties for clear and heat-absorbing glass is shown in Figure 9.4. The decrease in transmittance at high angles of incidence is due to the increased reflectance, but all properties are reasonably constant from 0° to 60°, which covers the majority of cases of high solar heat gain. Diffuse radiation properties are close to the direct radiation properties for 60° [**9.1, p. 27.14**].

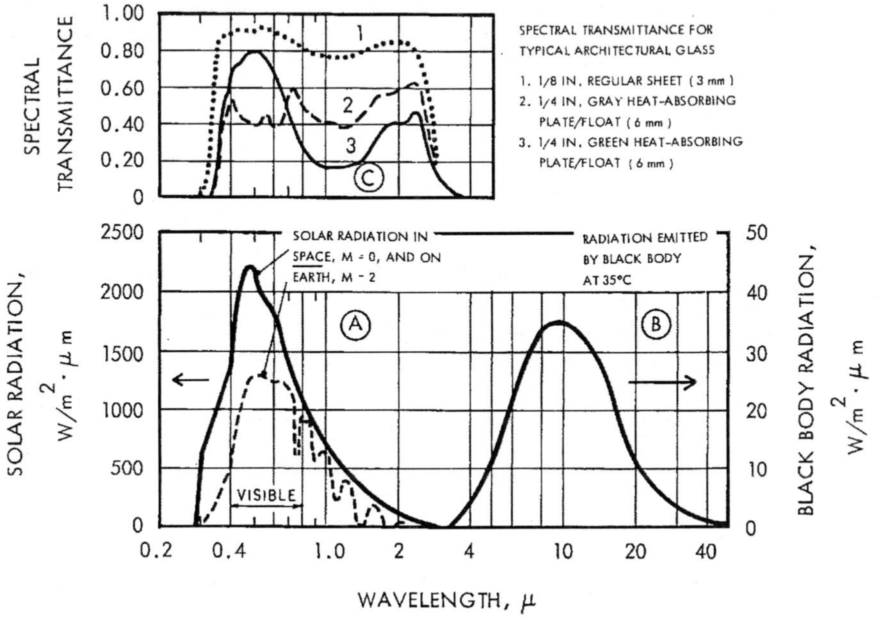

FIGURE 9.3 Spectral transmittances and solar radiation.
(From *ASHRAE Handbook 1981 Fundamentals*, Fig. 11, p. 27.14)[**9.1**]

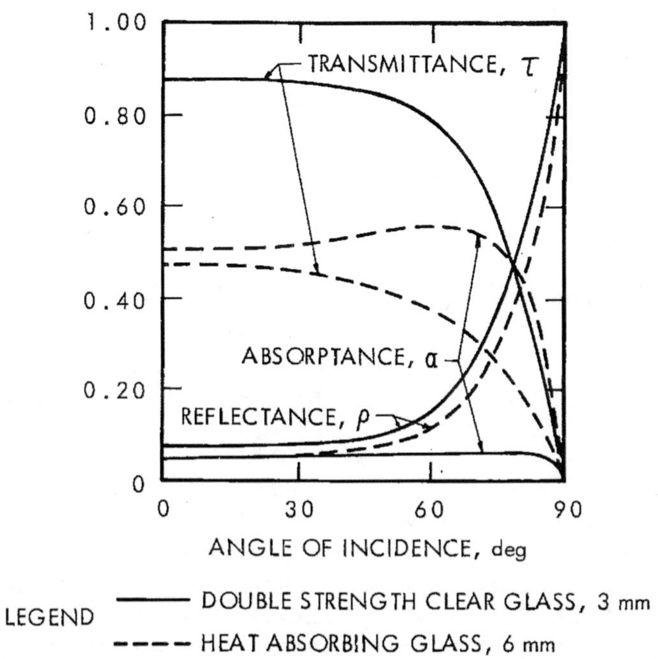

FIGURE 9.4 Solar-optical properties for double-strength clear sheet glass and heat-absorbing glass.
(From *ASHRAE Handbook 1981 Fundamentals*, Fig. 12, p. 27.14)[**9.1**]

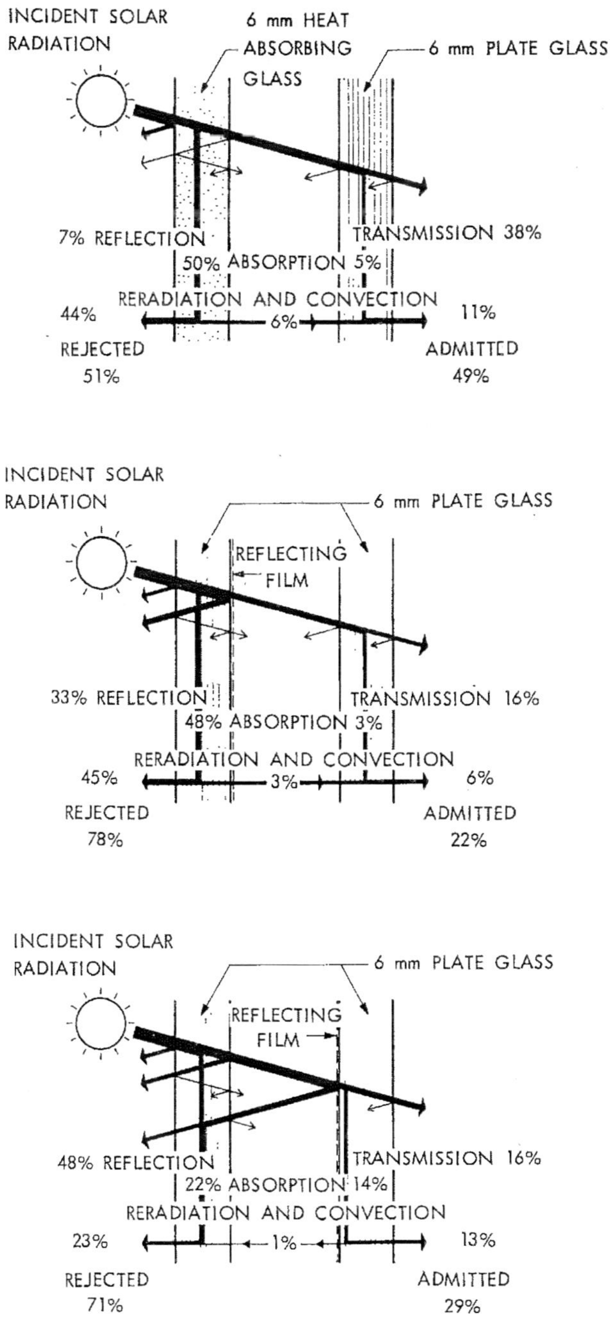

FIGURE 9.5 Components of solar heat admission and rejection for heat-absorbing and heat-reflecting double-glazing units.
(From Stephenson, D.G., Canadian Building Digest 101, Fig. 1, DBR/NRCC, 1968)[**9.6**]

Values of transmittance and *U*-values are given in Table 9.3 for some single- and double-glazing arrangements. The corresponding shading coefficients are given in Table 9.4 [**9.6**]. These values are subject to variation with different glasses and shading. *U*-values vary with surface and air space coefficients, as discussed in Chapter 8. More appropriate values for particular cases can be found from the information in Ref. **9.1, Chap. 27.**

The performance of double glazing and the use of heat-absorbing glass and glass with reflective coating are shown in Figure 9.5. Heat-absorbing glass is characterized by low transmittance and correspondingly high absorptance, as shown in Figure 9.4. Heat-reflecting glass is coated on one surface with a thin metallic film adjusted to limit by reflection the transfer of radiant energy. Because these films are easily damaged they can only be used on the interior surfaces of sealed glazing units. They reflect long- as well as short-wave and so reduce the heat transmission coefficient by reducing the air-space conductance. Reflective double glazing is more effective in excluding solar energy when the film is used on the inside of the outer pane rather than the air-space side of the inner pane. The transmittance is unchanged but more of the absorbed energy is returned to the outside. The outer pane, however, becomes hotter, and the risk of breakage increases.

Heat-absorbing glass can be used in single glazing, but the heat that it absorbs must be lost by long-wave reradiation and by convection from both sides, thus delivering part of the absorbed energy to the inside. It is much more advantageously used as the outer pane of double glazing so that a major portion of the absorbed heat escapes to the outside. It could be used as the inner pane to reduce the light intensity while maintaining a relatively high solar heat gain.

9.7 Control of Solar Radiation

Control of solar radiation on buildings may be wanted for many reasons, ranging from the need to trap and conserve as much energy as possible, to the other extreme of excluding as much as possible. Fenestration is always associated with appearance and view and is selected to have a light-transmitting capability. It also permits the entry of much additional radiant energy when exposed to the sun, and has relatively high *U*-values leading to substantial heat losses at times of low outdoor temperature. In some cases, there may be a need for vision and daylighting with the smallest possible solar heat gain: in other cases the solar energy may be wanted for heating; the need will vary from winter to summer. The highly variable nature of solar radiation from sunny to cloudy periods and with time and season must always be taken into account.

The building designer must decide whether fenestration is wanted and if so, how much, where, and what kind. The designer may wish to have generous areas of glazing with low solar heat gain, or smaller areas with higher transmittance. There may be a choice, within limits, of the shape and orientation of the building. The maximum values of solar heat-gain factors for 45°N from Table 9.6 suggest some possibilities in orientation. Northern exposures offer vision and light with small solar heat gain. Eastern and western exposures have the highest solar heat gain during the months when cooling may be required rather than heating. Southern exposures, because of the seasonal change in angle of incidence, have the highest solar heat gain in midwinter when the sun's altitude is reduced.

TABLE 9.6

Maximum Solar Heat-Gain Factors for 45°N, W/m²
(From Stephenson, D.G., *Tables of solar altitude, azimuth intensity and heat gain factors*, NRCC 9528, 1967) [**9.3**]

Date	Direction		
	N	E and W	S
Jan 21	53	422	791
Feb	68	554	784
Mar	85	661	698
April	102	699	554
May	114	692	432
June	118	679	378
July	116	679	419
Aug	107	672	534
Sept	88	621	675
Oct	70	533	758
Nov	54	412	777
Dec	46	342	769

Overhangs and other fixed outdoor shading devices can be used to advantage on southern exposures to provide shade in summer and sun in winter. These are not effective on eastern and western exposures, where the sun direction is more nearly normal to the building surface.

In general, outside shades are more effective than indoor shades in reducing solar heat gain, since the incoming energy can be more readily rejected to the outdoors by reflection, reradiation, and convection before it is allowed to pass through the glazing. Heat-reflecting and heat-absorbing double glazing can be more effective than other kinds of glazing with shades, as shown by Table 9.4, but cannot be adjusted as desired. Interior drapes and shades are often used as a means of control for indoor locations that are in a position to receive transmitted solar energy directly. They may also be wanted for privacy and for some measure of control over solar heat gains and heat losses generally.

9.8 Exterior Shading

In selecting solar shading it is necessary to be able to predict the shadows that will be cast on the surface to be shaded. The relationship between the sun's altitude and azimuth, the wall orientation, and the shadow angles are shown in Figure 9.6. The sun altitude is β and the surface-solar azimuth is $\gamma = \phi - \psi$, where ϕ is the solar azimuth and ψ is the surface azimuth (as previously presented in Figure 9.1). The incident angle, θ, for sun on the wall is angle *aOc*.

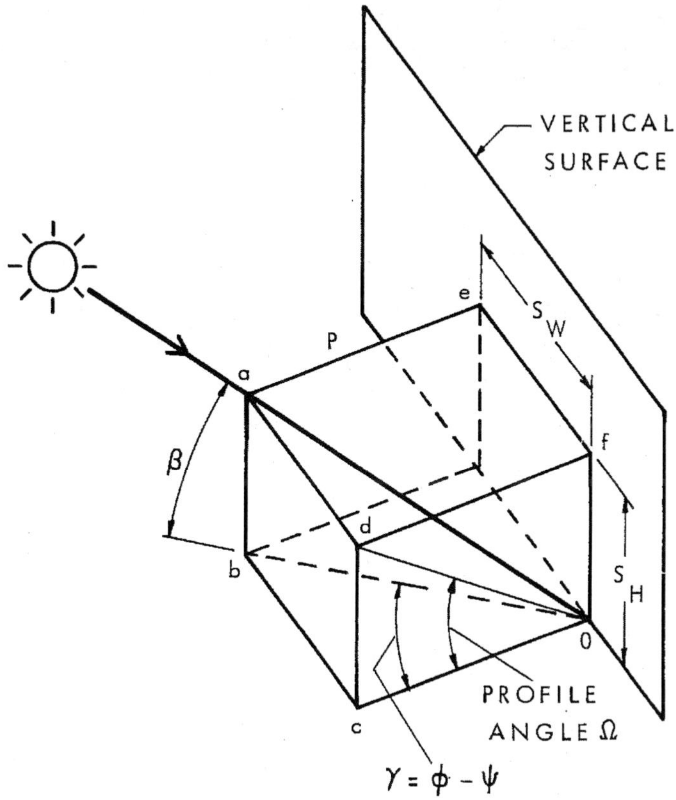

FIGURE 9.6 Geometry of exterior shading.

A vertical shading projection extending a distance p from the wall with its outstanding vertical edge coincident with ab in Figure 9.6 will cast a shadow on the wall horizontally as far as point O with a width S_W, so that

$$S_W = p \tan\gamma. \tag{9.20}$$

A horizontal shading projection extending a distance p with its outstanding horizontal edge coincident with ad will cast a shadow on the wall having a height S_H so that

$$S_H = p \tan\Omega \tag{9.21}$$

where Ω is the profile angle (sometimes called the vertical shadow angle) given by

$$\tan\Omega = \tan\beta/\cos\gamma. \tag{9.22}$$

The argument can be simplified somewhat by regarding ea as a rod of length p extending at right angles to the wall. Its outer end will cast a shadow at O, displaced horizontally by a distance S_W and vertically by a distance S_H from point e. Values of the profile angle are given for various latitudes in **Ref. 9.1, pp. 27.4 to 27.8.**

Example 9.9

Calculate for noon on December 21 and June 21 at 45°N the length of the shadow cast by an eave that overhangs the wall by 600 mm. Also find the shadow height for 09:00 solar time on June 21.

On December 21 the declination is $\delta = -23.5°$
The altitude of the sun at noon is

$$\beta = 90 - L + \delta = 90 - 45 - 23.5 = 21.5°$$

The shadow height is 600 tan 21.5° = 236 mm.
On June 21 the declination is $\delta = 23.5°$
The altitude of the sun at noon is

$$\beta = 90 - 45 + 23.5 = 68.5°.$$

The shadow height is 600 tan 68.5° = 1520 mm.
From [**9.3**] at 09:00 solar time, 45°N, June 21

$$\beta = 47.7° \qquad \phi = 74.7°$$

For a south wall $\psi = 0$ and the solar-surface azimuth is 74.7°. From Equation 9.22

$$\tan\Omega = \tan\beta/\cos\gamma = \tan 47.7°/\cos 74.7° = 4.18.$$

From Equation 9.21

$$S_H = p \tan\Omega = 600 \times 4.18 = 2510 \text{ mm}.$$

The shading of a window in a south wall by an overhanging eave, as calculated in Example 9.9, is shown in Figure 9.7. The overhang shades the window completely when the sun is at its highest in June, and allows full solar exposure on the window in December. Example 9.9 also demonstrated that in June the shadow heights would be greater for times other than noon.

Horizontal slat-type shades, either fixed or adjustable (such as the venetian blind), can be arranged to provide a wide range of shading geometries. Even when they are arranged so that direct sunlight is cut off they may still transmit appreciable diffuse and reflected energy [**9.7**]. Venetian blinds can be used to provide shading against the direct solar beam while transmitting diffuse and reflected radiation in a useful way for daylighting.

In some cases a designer is able to take advantage of the high reflectance of glass at large angles of incidence. Increasing the angle of incidence by tilting the glass can provide significant reductions in solar heat gain under some conditions [**9.8**].

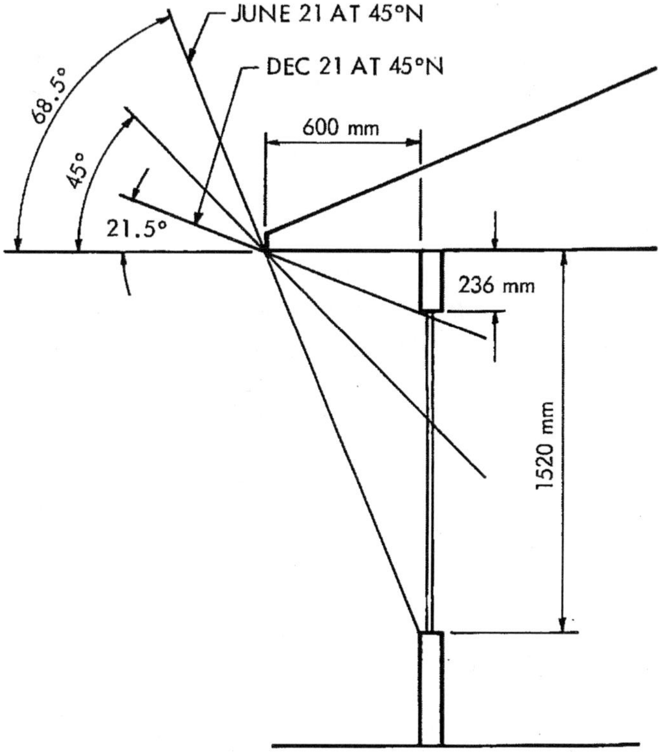

FIGURE 9.7 Shading angles for a south wall at 45°N

9.9 Solar Calculation Aids

In addition to extensive tables such as those given in [**9.1**], there are a number of aids useful in the rapid determination of solar data. They vary from slide-rule calculators to graphs from which it is possible to determine directly or by simple procedures solar-direction shadow patterns and shadow angles [**9.7, 9.9, 9.10**]. Some of them also give values of solar intensity or solar heat-gain factors. It is also possible to devise rotating table devices called heliodons which permit a model of a building under study to be oriented and rotated in sunlight or in the beam of a spotlight to reproduce the daily and seasonal patterns of shadows [**9.11**].

9.10 Heat Gain and Cooling Load

It has been common practice to calculate air-conditioning loads in two stages. The first step involves the determination of the energy gains of the space being conditioned. The various contributions to the heat gain from walls, windows, roofs, people, lights, and equipment when summed for each hour of the day provide a profile of the rate at which the space is gaining energy. The second stage involves the calculation of the cooling load, which is the rate at which the air-conditioning system must remove heat from the space in order to keep the space-air temperature constant. The cooling load at any given time usually differs from the corresponding heat gain. In the case of windows, for example, a substantial proportion of the heat gain is in the form of direct radiation which falls on various surfaces and is absorbed as heat. The associated rise in temperature results in a heat loss to the air, but it takes time for all the stored energy to be released and the temperature of the surface reduced accordingly. Peak cooling loads are usually less than maximum heat gains, but the principal difference is the delay in time, often up to several hours, before the entering energy finally passes to the room air and contributes to the cooling load. Such delays are associated with cyclical heat transfer and do not enter into steady-state situations.

The calculation of cooling loads is now commonly carried out by computer, using special methods of computation [**9.1, Chap. 26**]. Simple manual procedures have been devised, using the data from computer analysis. In the case of fenestration, the cooling loads at each hour can be found by multiplying the appropriate value of the maximum solar heat gain by cooling load factors (CLF). The cooling load due to solar heat gain is given by

$$q = A \times \text{SC} \times \text{MSHGF} \times \text{CLF} \tag{9.23}$$

where A = area, SC = shading coefficient, and MSHGF = maximum solar heat-gain factor from sources (such as Table 9.6) for the appropriate date, orientation, and latitude. Values of CLF for glazing with and without interior shading

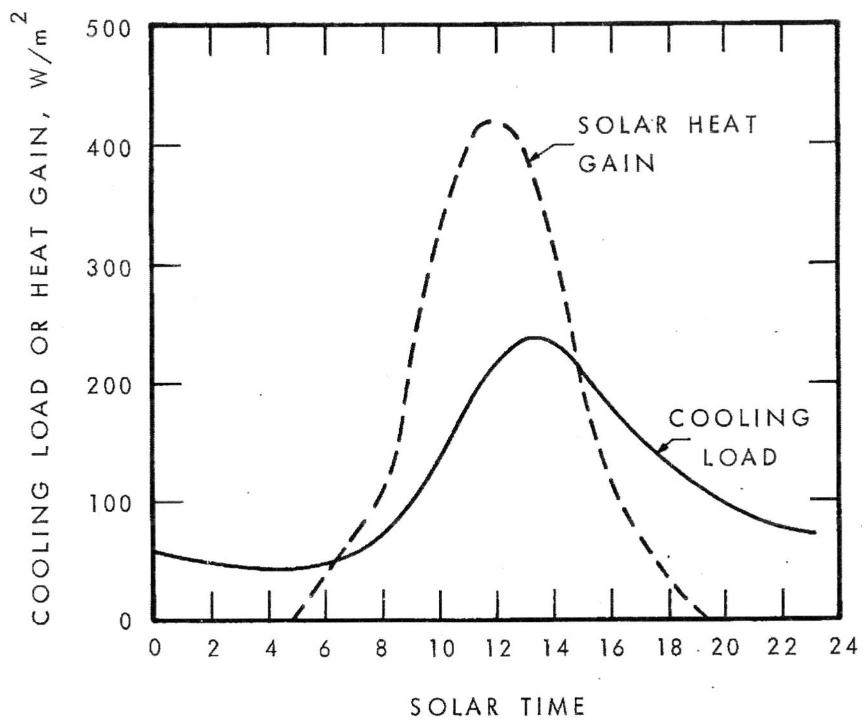

FIGURE 9.8 Solar heat gain and cooling load for single glazing facing south on 21 July at 45°N with heavy construction.

are given in [**9.1, p. 26.21**]. These are related to time of day, orientation, and type of construction, which is classed as light, medium, or heavy. Light wall and floor constructions with low thermal-storage capacity allow absorbed radiation to pass to the air relatively quickly; heavy constructions retain the energy longer.

The heat gain and cooling load for single glazing of clear glass on a south wall on July 21 at 45°N with heavy construction are shown in Figure 9.8. Since there are no shades and only single clear glass, the shading coefficient is unity and the values of SHGF from Table 9.5 apply. The maximum solar heat gain, at solar noon, is 420 W/m². The cooling load is found using the method described above. The peak cooling load for unit area of glass occurs about 14:00 and has a value 240 W/m². It can be deduced from this that the CLF was 240/420 = 0.56 at 14:00. Although there is a marked reduction from peak solar heat gain to peak cooling load, the cooling load is increased at later times and the cooling load summed over 24 hours is roughly the same as the 24-hour heat gain. Shades tend to increase the CLF but this is offset by the reduced shading coefficient, which results in a reduction in the 24-hour cooling load.

The cooling load resulting from the conduction heat gain through fenestration is not included in the cooling load as found above. It may be calculated using adjusted temperature differences to account for the difference between cooling load and heat gain [**9.1, p. 26.16, Table 10**].

9.11 Solar Radiation on Opaque Walls and Roofs

The solar radiation falling on opaque surfaces can be calculated directly as has already been done for windows, but it can also be found with reasonable accuracy from the solar heat-gain factors (SHGF) available for windows as in Table 9.5 and in [**9.1**]. Because the SHGF is calculated for single glass having a transmittance of 0.87, the solar radiation incident at the surface is given by

$$I_t = (1/0.87) \times \text{SHGF} = 1.15 \times \text{SHGF W/m}^2. \tag{9.24}$$

The values of SHGF were calculated for a ground reflectance of 0.2. Corrections can be made for other situations [**9.1, p. 27.37**].

Part of the incident solar energy is reflected; the rest is absorbed at the outer surface of the wall or roof and is converted to heat. The surface temperature is raised, temperature gradients are established, and nonsteady heat transfer takes place under the influence of the solar radiation input, which is variable with time.

The components of the energy exchange at an opaque surface that is receiving radiation are shown in Figure 9.9. This is identical to Figure 4.9, except that evaporation has been omitted. The symbols are as follows:

I_t = incident short-wave radiation
L = incident long-wave radiation
α = absorptance for solar radiation (see Table 4.1)
ϵ = emittance for long-wave radiation
h_c = surface coefficient for heat transfer by convection
t_o = outside air temperature, °C

t_s = surface temperature, °C
T_s = surface temperature, K
σ = Stefan-Boltzmann constant
q/A = heat conducted into the wall, W/m².

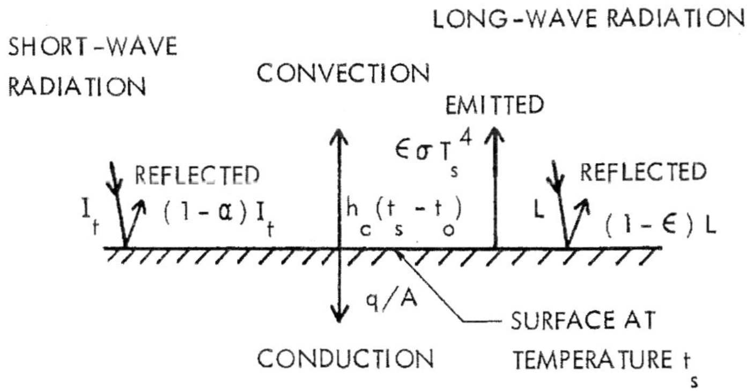

FIGURE 9.9 Components of heat balance at an opaque surface.

The incident solar radiation received at the surface is αI_t. The net long-wave radiation is $\epsilon L - \epsilon\sigma T_s^4$. The convective transfer is $h_c (t_s - t_o)$. The heat balance may be written as

$$q/A = \alpha I_t - h_c (t_s - t_o) + (\epsilon L - \epsilon\sigma T_s^4). \tag{9.25}$$

A procedure has been developed to simplify the handling of the surface-energy exchange in calculations of heat transmission through various constructions. The net exchange is expressed in terms of a surface conductance and a temperature difference so that it can be dealt with by conduction theory. The temperature difference is then added to the air temperature to give a fictitious outdoor air temperature, called the sol-air temperature. The net surface exchange is then expressed as

$$q/A = h_o (t_e - t_o) \qquad (\mathrm{W/m^2}) \tag{9.26}$$

where t_e is the sol-air temperature and h_o is the assigned surface conductance. The combined coefficient h_o is equal to the sum of the convection coefficient h_c and a radiation coefficient h_r, as developed in Chapter 8.

The sol-air temperature, t_e, can be envisaged as the temperature reached by an irradiated surface which is not losing heat by conduction, i.e., $q/A = 0$. This condition is approached if a good thermal insulating material is used as the exterior cladding. This definition of t_e was given by Mackey in his discussion of a paper by Parmelee and Aubele [**9.12**].

It will be recalled that the radiation component of the surface conductance, h_r, was developed on the assumption that the long-wave radiation exchange takes place between the surface and sky and surroundings considered as blackbodies at air temperature, so that the incoming long-wave radiation, L, equals σT_o^4 and

$$h_r = \epsilon\sigma (T_o^4 - T_s^4)/(t_o - t_s). \qquad (9.27)$$

The surface-heat-balance equation can be rewritten as

$$q/A = \alpha I_t + h_o (t_o - t_s) - \epsilon \Delta R \qquad (\text{W/m}^2) \qquad (9.28)$$

where $h_o (t_o - t_s)$ is the combined long-wave and convection exchange at the surface with blackbody surroundings at air temperature, t_o, and ΔR is a quantity included to account for the difference, if any, between the long-wave radiation incident on the surface from the sky and surroundings and the radiation emitted by a blackbody at outdoor air temperature. Equations 9.26 and 9.28 can be combined, yielding the equation for sol-air temperature as follows:

$$t_e = t_o + \alpha I_t/h_o - \epsilon \Delta R/h_o. \qquad (9.29)$$

The value of the emittance, ϵ, is commonly taken as 1. The value of h_o is subject to substantial variation with both temperature and air movement over the surface, but a value of 17 W/(m² • °C) is commonly used. The absorptance, α, for solar radiation was given for a number of surfaces in Table 4.1. It varies from about 0.4 for light-coloured surfaces to 0.9 or more for dark surfaces, including most masonry.

Long-wave radiation exchange with the sky was studied by Parmelee and Aubele [**9.12**] and more recently by Bliss and others [**9.2, p. 58.10**]. Approximately 90% of the incoming long-wave radiation from the atmosphere comes from the lowest 90 m. Water vapour in the air has a great influence. When skies are overcast with low-level cloud cover, the effective temperature is usually close to air temperature at ground level. On clear, dry winter nights the effective sky temperature may be as much as 20 K below air temperature, and surfaces at ground level may be cooled below air temperature. Dew or frost forms on those surfaces if they are cooled below the dewpoint temperature of the air. These radiation exchanges with the atmosphere occur in the daytime as well as at night, but they are masked during the day by the short-wave radiation. An approximate value of ΔR in Equation 9.29 for surfaces that "see" only the sky is about 60 W/m². When h_o = 17 W/(m² · K) and ϵ = 1, the adjustment to sol-air temperature represented by $\epsilon \Delta R/h_o$ is about −4 K. When the sun is shining, surrounding objects may be at higher temperatures than the air, thus compensating for the lower effective sky temperature. For vertical surfaces it is commonly assumed that $\Delta R = 0$.

Values of $t_e - t_o$ are given in Table 9.7 for January 21 and July 21 at 45°N. These must be added to air temperature in order to obtain sol-air temperature, t_e. They have been calculated using values of I_t derived from the solar heat-gain factors of Table 9.5 using Equation 9.24. Values of I_t and of $\alpha/h_o = 0.053$ were used in Equation 9.29 to find the values given in the table for $(t_e - t_o)$. A value of

$\Delta R = 0$ was used for vertical walls and a correction of −4°C was made for horizontal surfaces. A much larger correction of −10°C or more may be justified for clear, dry air conditions in January.

TABLE 9.7

Sol-Air Temperature Minus Air Temperature for 45°N Latitude for January 21 and July 21, degrees Celsius

January 21	N	NE	E	SE	S	SW	W	NW	Horizontal
08:00	0	2	12	15	8	0	0	0	−3
09:00	2	2	26	38	28	2	2	2	3
10:00	3	3	22	44	39	10	3	3	9
11:00	3	3	11	41	46	23	3	3	14
12:00	3	3	3	34	48	34	3	3	16
13:00	3	3	3	23	46	41	11	3	14
14:00	3	3	3	10	39	44	22	3	9
15:00	2	2	2	2	28	38	26	2	3
16:00	0	2	0	0	8	15	12	2	−3
July 21	**N**	**NE**	**E**	**SE**	**S**	**SW**	**W**	**NW**	**Horizontal**
05:00	2	4	4	2	0	0	0	0	−3
06:00	7	26	29	15	2	2	2	2	3
07:00	5	30	40	26	4	4	4	4	13
08:00	5	27	41	32	6	5	5	5	24
09:00	6	18	37	35	13	6	6	6	32
10:00	6	9	28	33	19	7	6	6	39
11:00	7	7	15	27	24	9	7	7	43
12:00	7	7	8	18	26	18	8	7	44
13:00	7	7	7	9	24	27	15	7	1
14:00	6	6	6	7	19	33	28	9	2
15:00	6	6	6	6	13	35	37	18	3
16:00	5	5	5	5	6	32	41	27	4
17:00	5	4	4	4	4	26	40	30	5
18:00	7	2	2	2	2	15	29	26	6
19:00	2	0	0	0	0	2	4	4	7

Note: These values are for dark-coloured walls for which $\alpha/h_o = 0.053$.
For light-coloured surfaces, divide values by 2.

Sol-air temperatures can be used with conduction theory in the calculation of the cyclical heat flow conditions produced by the solar radiation. They must then be associated with a particular daily air-temperature cycle. Figure 9.10 shows the daily air temperature used in the derivation of many of the tables of data given in [**9.1**], and the corresponding sol-air temperatures shown for July 21 at 45°N.

The computations of cyclical heat flow through various wall and roof constructions are now carried out by computer using special methods [**9.1, Chap. 26**] as already discussed in connection with cooling loads for fenestration. Manual methods based on these computations have also been developed. The cooling loads for various constructions have been calculated for each hour of the day and for the conditions applicable on July 21 at 40°N using 26°C as an indoor temperature and a daily air-temperature cycle similar to that shown in Figure 9.10. These values have been divided by the *U*-values of the constructions to produce a series of cooling-load temperature differences (CLTD) which have been set out in tables for walls and roofs [**9.1, pp. 26.7 to 26.12**]. They are applicable directly to a range of latitudes and times of year and can be extended to conditions other than those given by suitable adjustments. Values must be provided for various orientations in the case of walls, but not in the case of flat roofs.

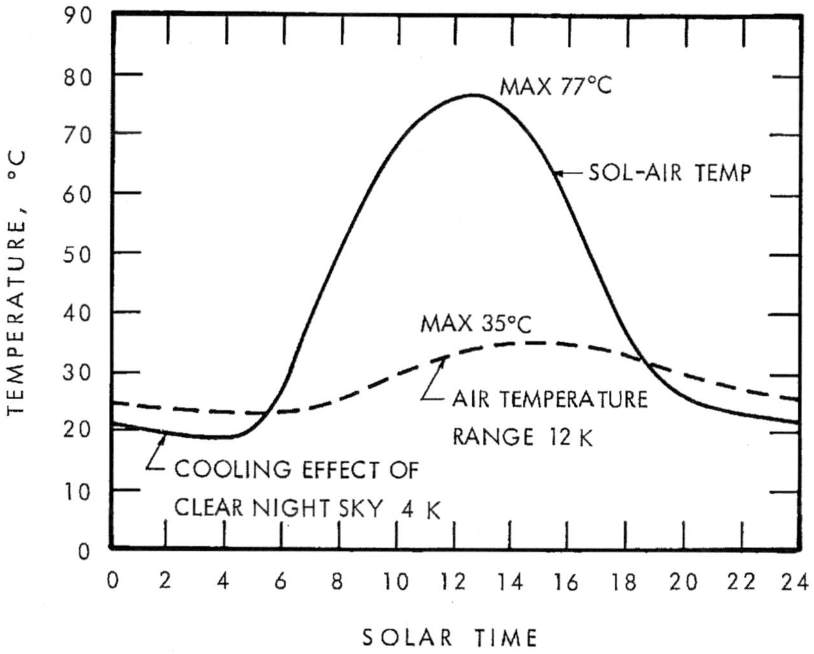

FIGURE 9.10 Typical hot day temperatures and sol-air temperatures for a dark horizontal surface on 21 July at 45°N.

9.12 Cooling Loads for Walls and Roofs

Cooling-load temperature differences are shown in Figure 9.11 for three roof constructions on July 21 at 45°N. The corresponding *U*-values, which when multiplied by the values of CLTD give the cooling load, are shown in each case. One roof is of sheet steel with insulation and roofing, weighing 39 kg/m^2. The cooling-load temperature differences may be compared with the sol-air temperature curve of Figure 9.10. It will be seen that the rise in sol-air temperature is 44 K above air temperature at noon, and the peak CLTD is also 44 K. The light roof provides little thermal capacity, and so the cooling load does not depart greatly from instantaneous heat gains calculated on the basis of steady-state conditions at any given hour. The peak CLTD occurs at 14:00.

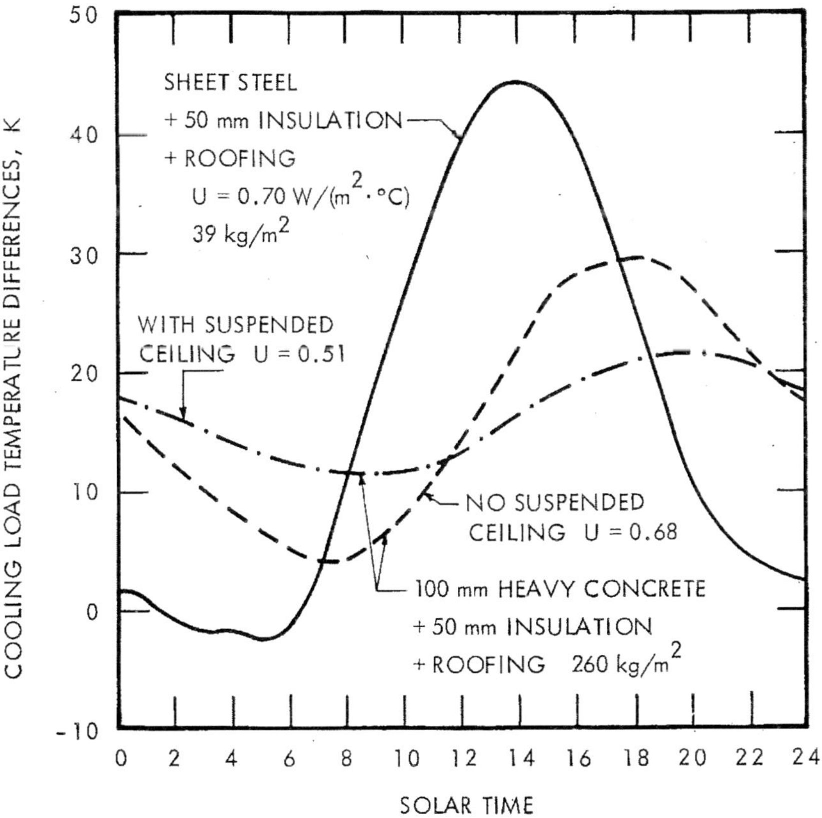

FIGURE 9.11 Cooling load temperature differences for three flat roofs on 21 July at 45°N.
(From *ASHRAE Handbook 1981 Fundamentals*, Table 5, p. 26.7)[**9.1**]

The second roof is of relatively heavy construction with a concrete slab 100 mm thick, but without a suspended ceiling. The *U*-value is almost the same as for the steel deck roof, but the peak CLTD is markedly reduced, from 44° to 29°, and delayed until 18:00. The cooling load from 18:00 to 20:00 is greater, and the 24-hour cooling load is not markedly different. The curve for the same heavy roof with a suspended ceiling has a still more marked redistribution of cooling load. The peak CLTD is reduced to 21°C and occurs at 20:00. The curve of CLTD does not show the further decrease in cooling load resulting from the decrease in *U*-value from 0.68 to 0.51 W/(m^2 • K) by the addition of the suspended ceiling. The substantial influence of the suspended ceiling can be attributed to the delay in the heat transfer from the slab to the ceiling, which must be mainly by radiation since convective transfer in downward heat flow is very small. The thermal-storage capacity of the ceiling, although relatively much smaller than the concrete slab, is very effective in reducing the peak CLTD.

The addition of increased insulation—from 25 to 50 mm, for example—has negligible influence on the CLTDs. It reduces the *U*-value and thus decreases the cooling load at all hours in proportion to the reduction in *U*-value.

Example 9.10

Find the peak cooling loads for the three flat roofs in Figure 9.11.

U-value for the steel roof is 0.70. The peak CLTD is 44°, so peak cooling load = 0.70 × 44 = 31 W/m^2.

U-value for the concrete slab roof without ceiling is 0.68 and the peak cooling load = 0.68 × 29 = 20 W/m^2.

U-value for the concrete slab roof with ceiling is 0.51 and the peak cooling load is 0.51 × 21 = 11 W/m^2.

Cooling loads for walls can be found in much the same way as for roofs. It is necessary to take into account the orientation of the receiving surfaces, since this determines the intensities of the incident solar radiation at various times of the day.

9.13 Temperatures at Building Surfaces

The computation of temperatures produced at building surfaces and through the building enclosure has not received the same attention as the calculation of cooling loads, and manual methods of computation have not yet been developed. Estimates based on steady-state heat flow can be made readily and are adequate for many purposes. It is often useful to know the limiting temperatures to which materials will be subjected in service. Some materials may soften and flow at elevated temperatures; others may become brittle at reduced temperatures. All materials expand and contract with changes in temperature. The structural significance of these dimensional changes has not always been recognized, but they merit consideration in the design of many features of the building enclosure. Another consequence of sun-induced daily temperature changes is the increase in freeze–thaw cycles, which may occur even on days when the outdoor air temperature remains well below freezing.

The calculations already made of sol-air temperature give some indication of the substantial increases in surface temperature that can be produced by solar radiation. Refer to Table 9.7, which shows that the sol-air temperature rise above air temperature in summer is 26°C for a south wall at noon for 45°N; it is even greater, 44°C, in January, since the sun's direction is then more nearly normal to the wall. The sol-air temperature is thus above the freezing point for every clear winter day, even with air temperatures of −20°C or lower. The surface temperatures that are of interest will be only a few degrees below the sol-air temperature. Some observations of wall-surface temperatures in Ottawa as shown in Figure 9.12 confirm this. The surface temperature on a south wall was as high as +20°C when the air temperature was only −17°C, a difference of 37 K. Snow on window sills thaws in the sun, producing wetting followed by freezing when the sun goes down. The effects of this can often be seen on masonry buildings in the form of staining, efflorescence, and spalling.

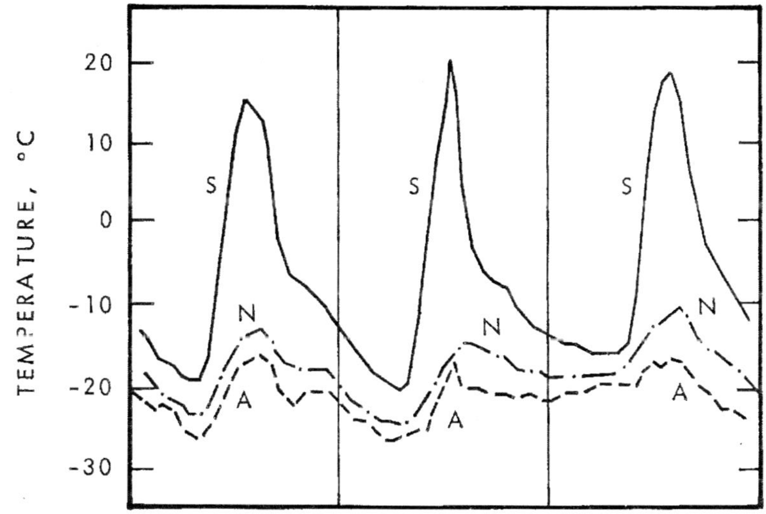

FIGURE 9.12 Effect of solar radiation on the surface temperature of a south-facing brick wall as compared with a north-facing wall and air temperature.

The daily and annual ranges of temperature are illustrated in Figure 9.13 in which air and sol-air temperatures are shown, along with temperature gradients based on steady-state conduction theory, for sunny days in July and January at 45°N. The winter temperature assumed is typical of Ottawa. The temperature profile through the wall for the July noon sol-air temperature of 59°C may now be considered in light of what has been learned about cyclical heat flow. The actual temperatures will never quite reach the levels of the steady-state profile. The surface temperature may be expected to rise to within a few degrees of the sol-air temperature. The face wythe of brick, providing high thermal storage, will rise slowly in temperature and will not follow the sol-air temperature as closely as would a lightweight cladding. The heavy inner wythe, which experiences reduced heat flow because of its position inward of the insulation, will lag markedly in temperature. Despite such discrepancies between cyclical and steady-state conditions, useful estimates of the probable daily and annual ranges of temperatures in walls and roofs can be based on simple steady-state calculations such as are shown in Figure 9.13.

The daily air temperature for the conditions of Figure 9.13 is shown to range from a night temperature of 23°C to a sol-air temperature of 59°C in July, a difference of 37 K. The corresponding range in winter for a moderately severe winter climate may be from −30°C to +24°C, a difference of 54 K. It may be assumed that these are also the approximate ranges of wall-surface temperature and of the mean temperature of the outer brick wythe. The extreme range from winter to summer is from −30°C to +59°C, a total of 89 K. It is significant that daily temperature excursions may be as great as 60% of the maximum annual ranges even for conditions of low winter temperatures.

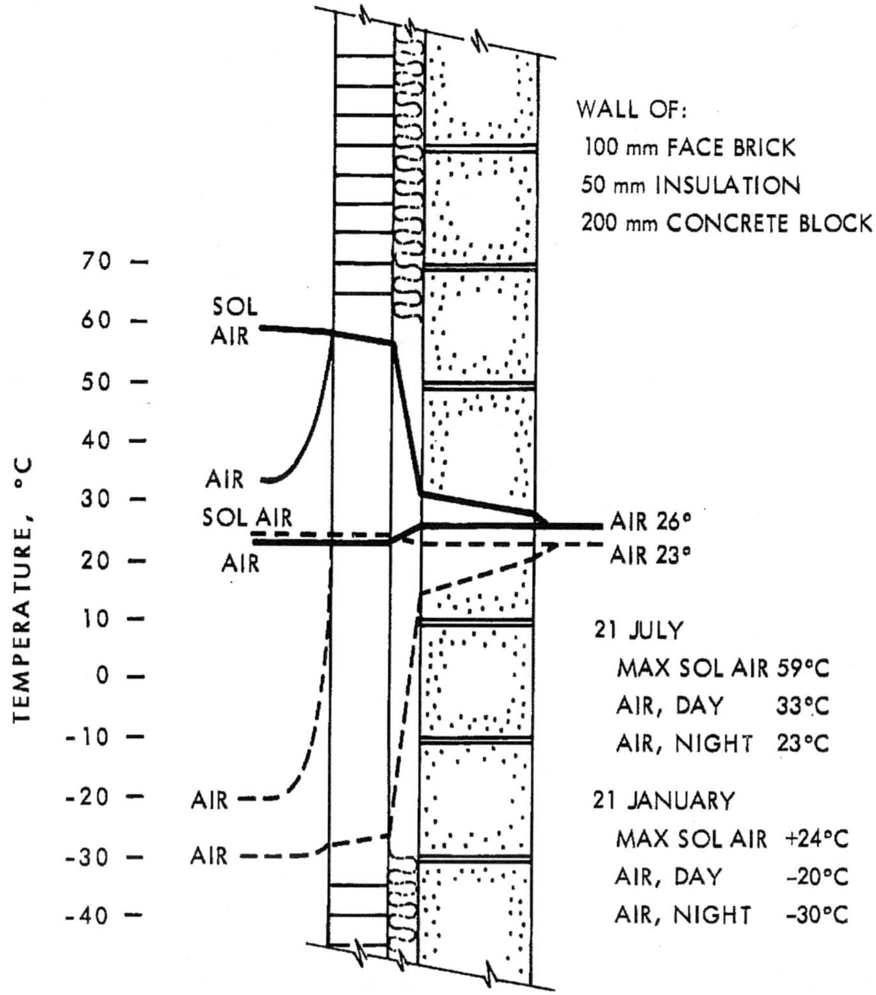

FIGURE 9.13 Daily and annual range of temperatures for an insulated masonry wall at 45°N facing south (gradients calculated from steady-state theory).

Conditions may differ from those illustrated. Wind reduces surface temperatures under sun conditions by increasing the heat loss from the surface by convection. Light-coloured surfaces with relatively low absorptances for solar radiation will markedly reduce the sol-air temperature. Incident radiation may be increased by reflection and reradiation from adjacent surfaces. Walls in particular may contribute significantly to the radiation received by adjacent low roofs. Estimates can be made of the extent of these contributions using the methods discussed in this chapter.

The thermal-storage capacity and the conductance of the exposed layer of a wall or roof will also influence the temperature rise of the surface receiving radiation. An unprotected black membrane over insulation will tend to follow the sol-air temperature closely, while the temperature of one placed directly on a concrete slab may be substantially lower owing to the higher rate of conduction of heat away from the surface. Simple rules have been proposed for estimating surface temperatures in such cases [**9.12–9.14**]. Roof membranes on an insulating substrate may rise by as much as 55 K, or even 70 K, if they receive radiation reflected from a light-coloured wall, while a roof with a grey surface can be expected to show a rise of at least 40 K. Similar effects can be expected for walls, but with the added complication of a greater range of constructions and the influence of orientation on the amount of solar energy received. Note the values shown in Figure 9.13.

9.14 **Conclusion**

The sun provides radiation at building surfaces at a rate that varies with time of year, time of day, orientation of the surface, clarity of the sky, surroundings, and surface absorptance. This energy input falls in rotation each sunny day on east-, south-, and west-facing surfaces in turn, raising surface temperatures, producing heat flow, and passing directly through fenestration. The substantial amount of energy involved has a marked influence on the heating and cooling of indoor space. The proportion of the incident energy that is transmitted in the case of fenestration is relatively large compared with that for opaque walls and roofs. The increased thicknesses of insulation currently being called for will reduce even more the proportion of the solar energy transmitted through walls and roofs.

The energy from the sun that reaches the interior of a building has an important influence on the temperatures of interior surfaces. These temperatures in turn along with corresponding changes in air temperature affect the indoor thermal environment and influence the rate at which energy must be added or removed in order to maintain a given comfort condition. These important matters will be explored in later chapters. The daylighting provided by the sun is also of considerable importance, although it has been supplanted in some cases by electric lighting. There is now a renewed interest in natural lighting because of the need for energy conservation.

Altho... ...n can have an important influence on the indoor environment, it ... in large part the radiation and thermal environment for the ...e as well as for many of its parts. The building may be shaded ..., and in turn may shade its surroundings. There may also be a ...on by one building of the radiation directed toward another. ...se due to radiation on buildings means that they expand on the ...; on a sunny day the top of a tall building goes through a cir... sun direction changes from east to west. The daily tempera... ...note expansion and contraction in many elements of the ...ture, or, if these changes are resisted, cause corresponding ... Ultraviolet radiation, extremes of temperature, and freeze–thaw cycles induced by the cyclical energy input of the sun are environmental factors affecting the performance of the many materials making up the building enclosure.

There is now great interest in manipulating the solar energy falling on buildings so that the energy required from other sources for lighting, heating, and cooling will be minimized. The variable nature of solar energy coupled with the difficulties in storing excess thermal energy for release when it is wanted makes this a difficult task, one which requires a thorough understanding of all aspects of energy in buildings.

References

9.1 *ASHRAE Handbook 1981 Fundamentals*. Chaps. 26 and 27.

9.2 *ASHRAE Handbook 1978 Applications*. Chap. 58.

9.3 Stephenson, D.G. *Tables of solar altitude, azimuth, intensity and heat gain factors for latitudes from 43 to 55 degrees north*. DBR/NRCC, April 1967. (NRC 9528)

9.4 Threlkeld, J.L., and Jordan, R.C. Direct solar radiation available on clear days. *ASHAE Transactions*, 1958, *64*, pp. 45–68.

9.5 Threlkeld, J.L. *Thermal environmental engineering*. New York: Prentice-Hall, 1962.

9.6 Stephenson, D.G. *Reflective glazing units*. Canadian Building Digest 101. DBR/NRCC, May 1968.

9.7 Stephenson, D.G. *Principles of solar shading*. Canadian Building Digest 59. DBR/NRCC, Nov. 1964.

9.8 Stephenson, D.G. *Solar heat gain through glass walls*. Canadian Building Digest 39. DBR/NRCC, March 1963.

9.9 Stephenson, D.G. *CSTB solar diagrams*. Building Research Note 48. DBR/NRCC, Dec. 1964.

9.10 White, R.W. Pseudoshadows for site planning. *ASHRAE Transactions*, 1982, *88*, (Part 2), pp. 368–87.

9.11 Mitalas, G.P. *The heliodon*. Building Research Note 47. DBR/NRCC, Dec. 1964.

9.12 Parmelee, G.V., and Aubele, W.W. Radiant energy emission of atmosphere and ground. *ASHVE Transactions*, 1952, *58*, pp. 85–106.

9.13 Garden, G.K. *Thermal considerations in roof design*. Canadian Building Digest 70. DBR/NRCC, Oct. 1965.

9.14 Baker, M.C. *Roofs: Design, application and maintenance*. Montreal: Multi-Science Publications Ltd., 1980.

10

Wind on Buildings

10.1 Introduction

Wind exerts an influence on buildings and their operation through the pressures generated by the flow of air over them. The resulting forces are often important determinants of the structural loads that the building and its parts must withstand. The pressures caused by wind can also lead to air leakage, which is a major consideration in heating and cooling and in the movement of dust and smoke. The deposition of pollution products, snow, and rain is markedly influenced by wind, as is the serious matter of rain penetration. The air pressures generated by wind can be useful for purposes of natural ventilation, but they can also affect adversely the operation of air-handling equipment, natural draft stacks, and chimneys.

Many practical considerations arise from the way in which buildings affect the wind, thus changing the effects of wind on people and objects. A group of buildings upwind may contribute to turbulence, or a single building may affect the wind in its immediate vicinity. Thus, the nature of wind velocities in the flows around buildings and the resulting pressures are of obvious importance.

A basis for some appreciation of the relation between fluid velocity, boundary geometry, and pressures was established in Chapter 7. Many complications arise, however, in dealing with wind over buildings, since the fluid stream involved includes the turbulent boundary layer of the earth and departs considerably from the uniform freestream conditions assumed in much of the fluid-mechanics theory that has been developed for other purposes. The study of wind on buildings and structures has been developing rapidly since 1960 as a discipline in its own right, based on existing theory with the aid of testing and experience as well as research in the development of new theories and new concepts better suited to predicting wind effects. The proceedings of international symposia and other important publications, including a number of textbooks, provide a record of these advances and confirm the widespread interest in the subject.

Wind loads are of importance in structural engineering, and the continuing need for safety and economy in building structures (particularly in tall buildings) has been the principal reason for the promotion of wind studies, which has led to an emphasis on measurements and methods related to strong winds and associated critical loading conditions. For other situations, average wind conditions may be more significant. What is wanted, in any case, is the ability to predict the pertinent characteristics and effects of the wind for a given situation.

10.2 The Nature of the Wind

The wind layer at the surface of the earth on which buildings are located can be regarded, at least for strong wind conditions, as a turbulent boundary layer. Its thickness, which can be as much as 600 m, will be dependent on the freestream velocity and the upstream surface roughness. The freestream velocity at the upper surface of the boundary layer is determined by global wind movements modified by weather forces and by topography on a regional scale (as discussed briefly in Chapter 2). It is subject to marked variations in speed and direction with location and time, and it may be gusty or turbulent.

This inconstant basic flow is further modified by the roughness of the earth's surface, which may vary from the relative smoothness of cultivated fields and waves on water to the relative unevenness of tall buildings in urban cores. Roughness is the cause of turbulence arising from flow separation, shear layers, wakes, and vortices generated by individual roughness elements. Since turbulence has components of velocity transverse to the main flow, it is propagated upward, thus generating the turbulent boundary layer with a velocity gradient and a scale of turbulence varying appropriately with height, roughness, and length of flow path. The turbulence generated by roughness gradually dies out over distances in the direction of flow up to 100 or more times the height of the disturbing element, so it must be continuously generated to maintain the turbulent layer. The added complications of variations in roughness with upstream distance and direction can readily be visualized.

It is this disordered, variable flow whose speed, direction, and turbulence or gustiness must somehow be described. It is necessary to make the best possible use of both theory and wind records. Wind records provide evidence of the variations with time at a particular location and height, while boundary-layer theory must be used to adjust available records to other locations and heights above ground. Statistical theory provides a means of describing such highly variable conditions.

10.3 Measuring and Describing the Wind

Instruments for routine measurement of the wind vary in type and in ability to respond to and record detailed variations in wind speed and direction. One common type is the cup anemometer (Figure 10.1), which rotates around a vertical spindle as determined by the wind speed in the horizontal plane, without regard for wind direction. The revolutions over a fixed time or, alternatively, the times for a given number of revolutions, are recorded. This input can be sorted as to direction and recorded in different ways using additional information relayed from a wind vane. This type of instrument is usually set up to provide information about the mean wind speed over a measurement period, which may be as long as one hour.

An instrument known as the Dines pressure tube provides information about wind fluctuations. It is essentially a form of pitot tube, arranged to pivot around a vertical axis so that it is turned into the approaching wind by the attached wind vane. The dynamic pressure is recorded remotely, along with information on the wind direction obtained from the electrical signals from a direction rheostat attached to the vertical shaft.

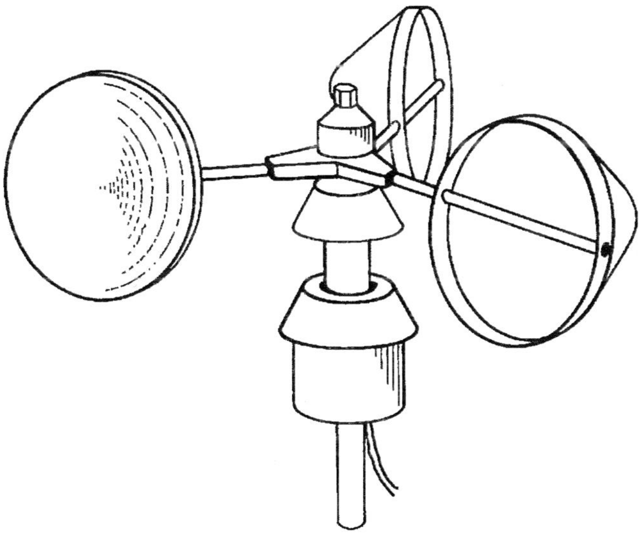

FIGURE 10.1 A cup anemometer.

A record from a Dines pressure tube (Figure 10.2) shows the marked fluctuations in wind speed that can occur. These fluctuations must be taken into account in any proper description of the wind, and this can be done in terms of a mean wind speed and deviations from it. The discussion of this approach that follows is based on the presentation by Aynsley et al. [**10.1, p. 86**].

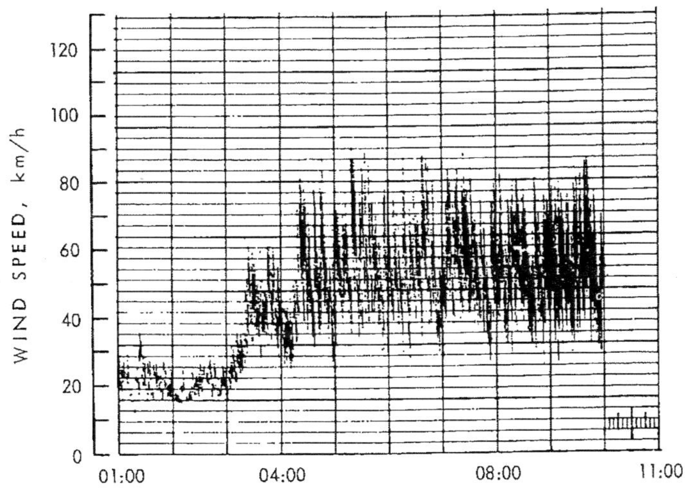

FIGURE 10.2 Typical pressure tube anemometer record.
(From Dalgliesh, W.A., and Boyd, D.W., Canadian Building Digest 28, DBR/NRCC, 1962, Fig. 2)

The mean wind speed can be defined mathematically as

$$V = \frac{1}{T} \int_{t_o - T/2}^{t_o + T/2} V_t \cdot dt \qquad (10.1)$$

where V_t = the instantaneous wind speed at time t
V = the mean speed averaged over some suitable time interval T, centred on t_o.
T may vary in practice from a few minutes to as much as one hour.

Figure 10.3 shows a section of a wind record to an enlarged time scale on which the mean speed V and the peak gust speed G have been drawn.

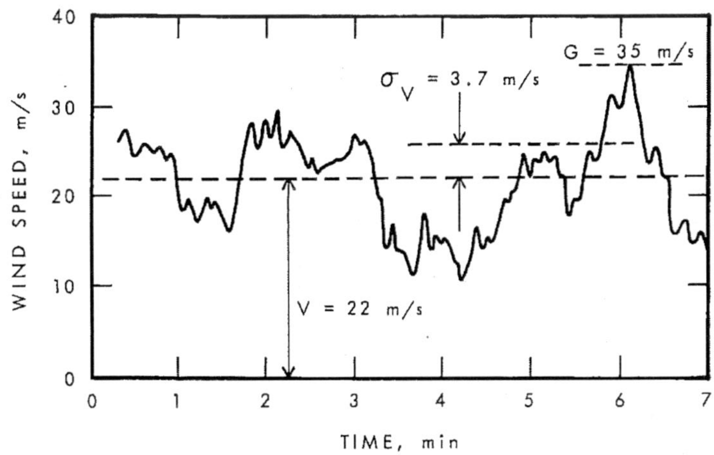

FIGURE 10.3 Expanded plot of a wind record.
(From Aynsley, R.M., Melbourne, W., and Vickery, B.J., *Architectural aerodynamics*, London: Applied Science Publishers Ltd., 1977, Fig. 3.5, p. 87)[**10.1**]

A further measure of the gustiness or turbulence is needed. This can be found from a statistical measure, the standard deviation σ_v, where

$$\sigma_v^2 = \frac{1}{T} \int_{t_o - T/2}^{t_o + T/2} (V_t - V)^2 \, dt. \qquad (10.2)$$

The relative gustiness or intensity of turbulence can be taken as

$$I_v = \frac{\sigma_v}{V}. \qquad (10.3)$$

As the distribution of wind speed follows approximately a normal or Gaussian distribution, it is possible to draw a probability curve for the record of Figure 10.3, as shown in Figure 10.4. This gives additional useful information about the time for which any given wind speed will be exceeded. It is stated [**10.1**] that the peak gust as recorded by most meteorological anemometers will exceed the mean by about 3.5 σ_V. Clearly, these ways of describing the strength and turbulence of the wind are important in the estimation of the critical wind loads on structures. As will be discussed later, a further measure, one which gives information on the spacing or frequency of gusts, is required when considering the dynamic response of structures.

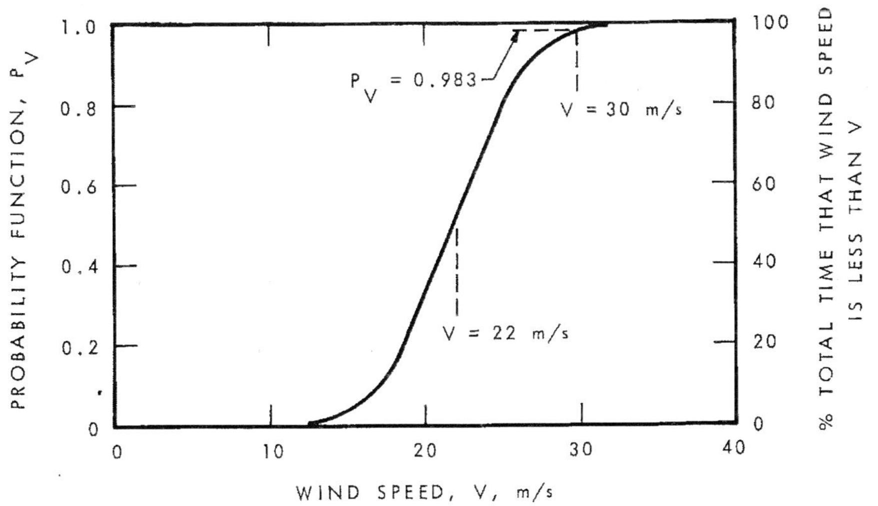

FIGURE 10.4 Probability distribution of wind speed.
(From Aynsley, R.M., Melbourne, W., and Vickery, B.J., *Architectural aerodynamics*, London: Applied Science Publishers Ltd., 1977, Fig. 3.6, p. 88)[**10.1**]

10.4 The Estimation of Wind Speeds

Wind records should be kept over as long a period as possible, preferably 20 years or more, in order to establish a statistically adequate measure of the wind at a particular location. In most countries, a national meteorological service forecasts the weather and keeps suitable weather records for as many selected locations as is practicable. Since the roughness of the earth's surface has a marked effect on both speed and turbulence, the topography in the vicinity of the local weather station and the type, arrangement, and height above ground of the wind-measuring instruments must be taken into consideration in interpreting the records. Because of the close association of weather forecasting with aviation, many weather stations are located at airports, which provide large expanses of relatively flat smooth terrain that does not change greatly over the life of the airport. Practical considerations, however, often rule out the locations of instruments at the standard height of 10 m and in positions that provide an unobstructed wind approach over terrain of similar roughness for one or two kilometres in all directions.

The wind at a building site will also be influenced by the upstream surface roughness; in city cores, buildings of various heights create a topography that differs appreciably from the airport site. In these cases it is necessary to make corrections to the observed wind speeds to take into account the effects of surface roughness and height above ground.

The 1/7 power law for the variation of velocity with distance from a surface in boundary layers on flat plates and in pipes is a reasonably good representation of the variation of mean wind speed with height for strong winds over smooth terrain [**10.2**]. The power law with appropriate exponents is also found to be reasonably applicable to cases of rough terrain, and is widely used. The basic relationship is based on the concept of a gradient mean wind speed V_g at the gradient height Z_g above ground, at which ground roughness no longer has an effect. The mean gradient wind speed V_g corresponds to the freestream velocity at the edge of a boundary layer with a thickness Z_g. The mean wind speed V_z at any height is then given by

$$\frac{V_z}{V_g} = \left[\frac{Z}{Z_g}\right]^{\alpha}. \qquad (10.4)$$

The exponent α varies with roughness from as low as 0.1 for open sea and other smooth terrain to as high as 0.4 for city centres with many tall buildings. The disturbed surface layer increases in thickness with increasing roughness, as in boundary-layer theory, producing variations in the gradient height from about 250 to 500 m.

The variation of gust speed with height can be represented by a power law

$$\frac{G_z}{G_g} = \left[\frac{Z}{Z_g}\right]^{\beta} \qquad (10.5)$$

in which G_z = peak or gust speed at height Z; G_g = peak or gust speed at gradient height Z_g; β = an exponent related to terrain roughness with a value of about 0.6 α.

An approximate relationship between mean wind speed and gust speed has been proposed [**10.1, p. 93**], in which

$$G_g = 1.35\ V_g. \qquad (10.6)$$

Terrain categories with appropriate gradient heights and exponents were proposed by Davenport following his review of the literature on wind loads in 1960 [**10.2**]. Since then, various authorities in different countries have adopted values that are quite similar. Typical values given by Aynsley et al. [**10.1, p. 89**] as shown in Table 10.1 have been used in plotting the mean wind speeds at various heights for a mean gradient wind of 100 km/h, presented in Figure 10.5.

TABLE 10.1

Values of Gradient Height and Power Law Exponents for Wind Profiles
(From Aynsley, R.M., Melbourne, W., and Vickery, B.J., *Architectural aerodynamics*, London: Applied Science Publishers Ltd., 1977, Table 3.1, p. 89) [**10.1**]

Terrain category and description	Gradient height Z_g m	Mean speed exponent α	Gust-speed exponent β
1. Open sea, ice, tundra, desert	250	0.11	0.07
2. Open country with low scrub or scattered trees	300	0.15	0.09
3. Suburban areas, small towns, well-wooded areas	400	0.25	0.14
4. Numerous tall buildings, city centres, well-developed industrial areas	500	0.36	0.20

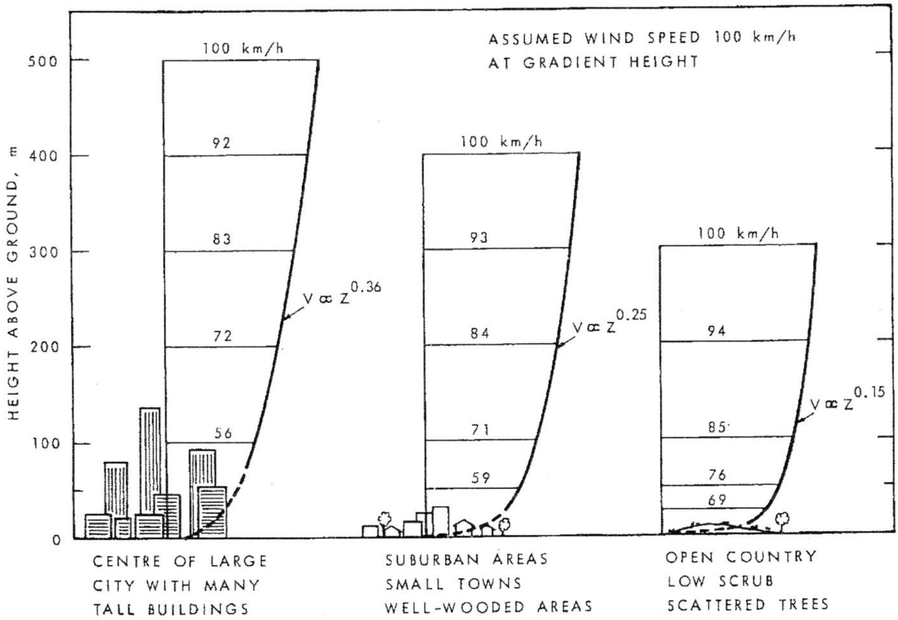

FIGURE 10.5 Mean wind-speed profiles over terrain with different roughness characteristics for assumed gradient wind speed of 100 km/h.
(From Dalgliesh, W.A. and Boyd, D.W., Canadian Building Digest 28, DBR/NRCC, 1962, Fig. 1)

Example 10.1

Find the mean wind speed at a height of 100 m for the three terrain categories of Figure 10.5 and for a mean gradient wind speed of 100 km/h.

From Table 10.1, for terrain category 2,

$$Z_g = 300, \alpha = 0.15.$$

From Equation 10.4,

$$V_z = V_g \left[\frac{Z}{Z_g}\right]^{\alpha}$$

substituting

$$V_{100} = 100 \left[\frac{100}{300}\right]^{0.15} = 85 \text{ km/h}.$$

Similarly, for terrain category 3,

$$Z_g = 400 \qquad \alpha = 0.25$$

$$V_{100} = 100 \left[\frac{100}{400}\right]^{0.25} = 71 \text{ km/h}$$

and for terrain category 4,

$$Z_g = 500 \qquad \alpha = 0.36,$$

$$V_{100} = 100 \left[\frac{100}{500}\right]^{0.36} = 56 \text{ km/h}.$$

The marked reduction in mean wind speed with increasing terrain roughness but for the same gradient wind speed is shown by the results of Example 10.1 as well as by Figure 10.5. The mean wind speed at a height of 100 m in the city centre will be only two-thirds of that at the same height in open country. Even greater differences are indicated at lower levels if the power law is assumed to apply down to ground level. Clearly some judgement must be applied in such applications, since most city centres contain many buildings in the four-to-ten-storey range (16 to 40 m) interspersed with streets and parks. The downtowns of larger cities are likely to contain buildings whose heights range from 20 storeys (80 m) to 50 storeys (200 m) or even more.

Equations describing strong wind profiles have been developed from measurements in the boundary layer at various levels above the roughness height; the buildings for which the predictions of wind are needed are part of the roughness that is generating the boundary layer. The equations can be expected to apply strictly only to the prediction of winds approaching isolated buildings and those portions of buildings in built-up areas that are above the level of any upstream roughness. The prediction of wind effects on buildings at levels below that of the upstream roughness requires that account be taken of the individual roughness elements in the vicinity.

Example 10.2

Calculate the mean wind speed and the peak gust speed to be expected at the top of a building 100 m high located in the city centre, when the mean speed being recorded at the anemometer at 10 m height above ground at the city airport is 25 m/s.

From Table 10.1, terrain category 2, $\alpha = 0.15$, $Z_g = 300$ m.

From Equation 10.4, the mean gradient wind is given by

$$V_g = V_z \left[\frac{Z_g}{Z}\right]^{\alpha} = 25 \left[\frac{300}{10}\right]^{0.15} = 41.6 \text{ m/s.}$$

Using $V_g = 41.6$ m/s for the gradient wind at 500 m over the city centre, and for $\alpha = 0.36$, the mean wind speed at 100 m is given by

$$V_{100} = 41.6 \left[\frac{100}{500}\right]^{0.36} = 23.3 \text{ m/s}$$

and from Equation 10.6

$$G_g = 1.35\, V_g = 1.35 \times 41.6 = 56.2 \text{ m/s.}$$

From Table 10.1, $\beta = 0.20$ and from Equation 10.5, the gust speed is

$$G_{100} = G_g \left[\frac{Z}{Z_g}\right]^{\beta} = 56.2 \left[\frac{100}{500}\right]^{0.20} = 40.7 \text{ m/s.}$$

10.5 Determination of Wind Pressures on Buildings

It is much more difficult to relate velocities and pressures over building surfaces to the approaching wind than in the case of the uniform flow conditions on which much present knowledge of fluid mechanics is based. In the case of uniform flow, it is only necessary to identify the fluid and its velocity which is the same everywhere in the approaching stream. In the case of wind on buildings it may be necessary to take into account not only the particular wind-speed gradient but also the gustiness or turbulence, and both may be subject to variation from one case to another, and from one wind direction to another.

Wind pressures on buildings have been determined largely by experiments either on actual buildings or on models in wind tunnels. The extreme variability in the wind makes field measurement both tedious and costly, and much of the available information has been obtained from models tested in aeronautical wind tunnels with uniform velocity profiles. This practice was continued long after the existence of wind-speed gradients and their effects were well recognized [**10.2**]. Information from such experiments is still in use for some situations for want of better data. Turbulence characteristics of the wind tunnels used were not always measured or even recognized.

Interest in the use of wind tunnels arranged to simulate wind speed profiles grew rapidly following the first International Conference on Wind Effects on Buildings and Structures held in Teddington, U.K., in 1963. Over the next few years a number of tunnels in several countries were either adapted or developed for wind studies, and papers describing the results began to appear with increasing frequency. A short historical review with a list of references is available [**10.1**].

Some investigators have concentrated on reproducing some particular wind-speed profile. Screens, horizontal bars, and spires have been introduced upstream of wind-tunnel working sections for this purpose. In Denmark, Jensen and Franck [**10.3**] emphasized the need to reproduce the turbulence to scale as well as the wind-speed gradient. They advocated the use of wind tunnels with long approach sections immediately upstream of the working section; upstream roughness to the proper scale is introduced to produce an appropriate boundary layer. The tunnel operated at the University of Western Ontario by Davenport and colleagues is of this type [**10.4**]. The tunnel in use at the Building Research Establishment in the U.K. employs appropriately spaced horizontal bars located immediately upstream of the working section [**10.5**].

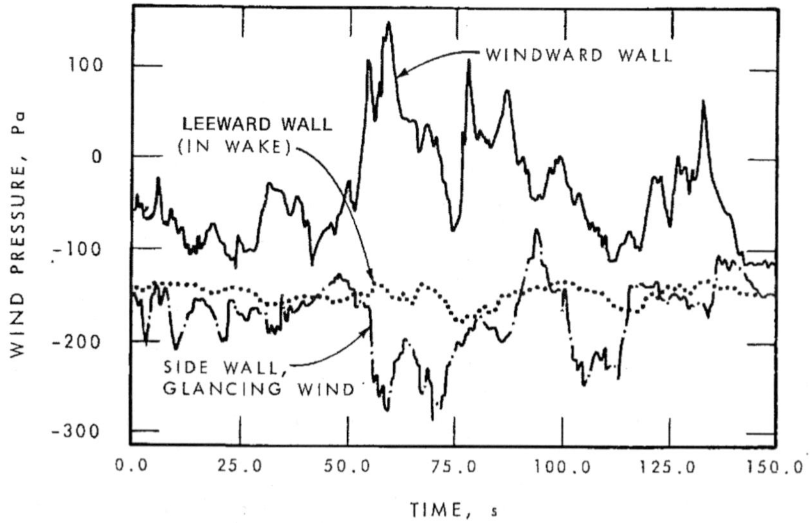

FIGURE 10.6 Characteristic surface-pressure fluctuations in different regions of flow.
(From Dalgliesh, W.A., *Journal of Industrial Aerodynamics*, 1975, *1*, pp. 55-66, NRCC 15130, Fig. 2, p. 59)[**10.6**]

A limited number of measurements on full-scale buildings has been made since 1963. Because wind is highly variable in its turbulence, gradient, speed, and direction, and being affected markedly by adjacent and upstream buildings, much data and careful analysis are required to form a basis for generalizations. Full-scale measurements cannot be made before a building is built; unless an identical building and neighbourhood situation has already been tested, the wind tunnel provides the only rational way of predicting any unusual wind effects in advance of construction. The great value of full-scale measurements lies in the validation of wind-tunnel techniques and results so that they can be used with confidence. Such studies have been made, for example by Dalgliesh [**10.6**], who has been able to compare full-scale measurements on a tall building with model results from the wind tunnel at the University of Western Ontario. The representative measurements shown in Figure 10.6 illustrate the variability of wind pressures; Figure 10.7 shows the kind of agreement that was achieved between wind-tunnel and

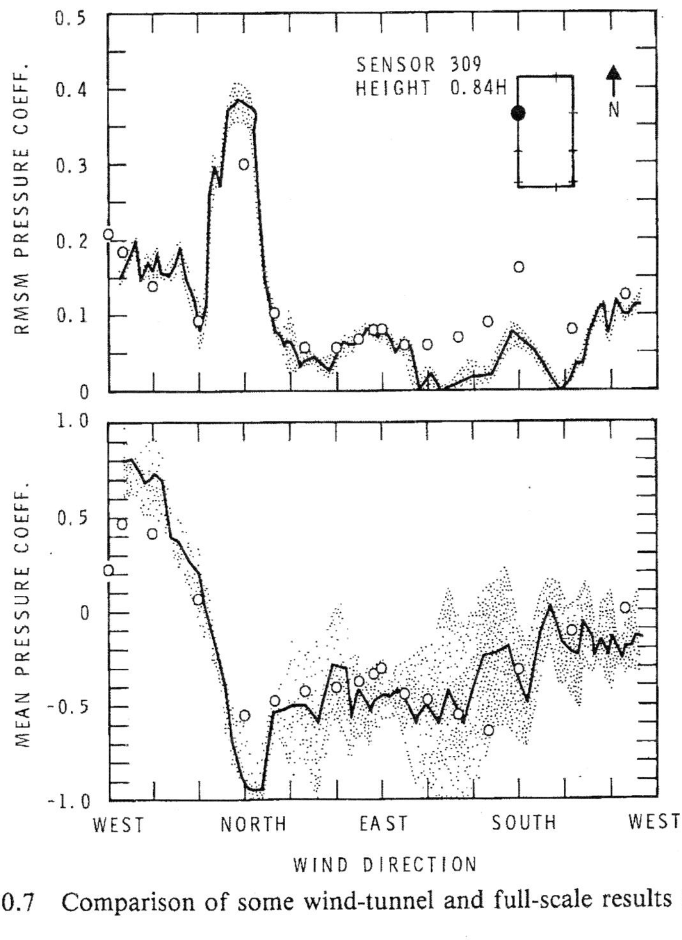

FIGURE 10.7 Comparison of some wind-tunnel and full-scale results for a tall building.
(From Dalgliesh, W.A., *Journal of Industrial Aerodynamics*, 1975, *1*, pp. 55-66, NRCC 15130, Fig. 5, p. 61)[**10.6**]

full-scale tests. The extent of the modeling of terrain and surrounding buildings that may be necessary in achieving such agreement is evident from Figure 10.8.

FIGURE 10.8 Model of downtown Montreal in the boundary-layer wind tunnel of the University of Western Ontario.
(Ron Nelson Photography, London, Ontario)

Because available information on wind pressures on buildings has been obtained from a variety of real or simulated wind situations, it is always desirable to consider carefully the conditions under which it was obtained. When the information available is inadequate, it may be desirable to commission wind-tunnel tests. As is now apparent, however, these can vary in the extent to which the general wind-speed profile and the buildings in the immediate and upstream vicinity are modeled. As a further complication, modeling may have to be carried out for several different wind directions. Finally, when the data are wanted for structural analysis of tall or slender structures, the building under examination may be modeled dynamically as well as geometrically so that the dynamic response to the variable loading provided by the gustiness in the wind can be examined experimentally.

Dynamic testing will not be justified in all cases; less elaborate modeling may be sufficient for a variety of structural and other purposes. One may choose whether the wind-speed gradient and turbulence are to be modeled with the building model standing free of any immediate obstructions as if in an isolated situation, or whether the upstream terrain and buildings in the vicinity are to be modeled more realistically. The latter approach is the only way of establishing the pressures on portions of a proposed building below the level of upstream obstructions, and is used to examine the effects of any new obstructions that may subsequently be built. The former approach can yield results with general application to buildings of the same geometry that are shielded to varying degrees. It is based, in effect, on the application down to ground level of the equation for the boundary-layer wind-speed profile appropriate to the general terrain roughness.

Considerations arising in the application of wind data to problems such as air leakage, natural ventilation, and winds at street level may justify attention to conditions other than those of strong winds and to modeling and testing techniques other than those developed to suit structural engineering needs.

10.6 The Flow of Wind around Buildings

The flow of wind over a bluff body on the ground has much the same features as the flow around fully immersed bluff objects, as discussed in Chapter 7. The wind flow over a long wall is illustrated in Figure 10.9. The essential characteristics are the separation at the sharp leading edge; the windward vortex region, which will be at a positive pressure with respect to atmospheric; the leeward vortex region, which will be at a negative pressure, or suction; and the shear layer between the vortex flows and the outer streamline flow. Bernoulli's equation can be applied to the streamline portion (with some reservation) despite the turbulence that is known to be present in the approaching wind.

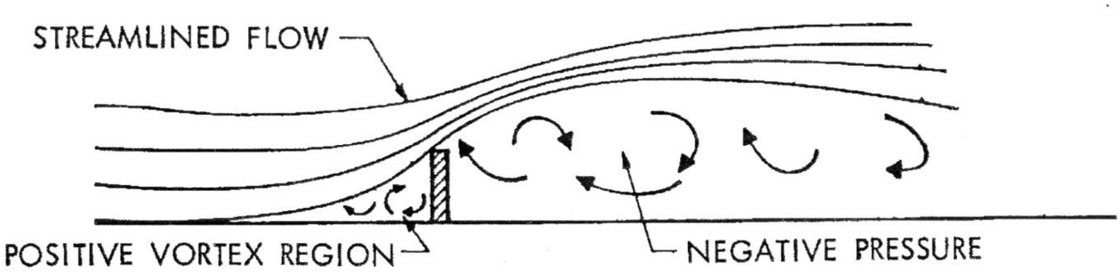

FIGURE 10.9 Wind flow over a long wall.

When the bluff body has limited length, as in the case of a square prism, which is representative of a low building, the basic elements of the flow are still present, but flow takes place in three dimensions. In particular, the large pressure difference between upstream and downstream regions produces vigorous flow around the ends of the simulated building. These flows separate at the sharp vertical edges, and they combine with the flow over the top to produce a complex three-dimensional wake.

10.7 Wind-Pressure Coefficients

The pressures over a building, either model or full-scale, are almost always presented as dimensionless coefficients obtained by dividing the measured values by the dynamic pressure corresponding to the wind speed, V, so that

$$C_p = \frac{p}{(1/2)\,\rho V^2} \tag{10.7}$$

in which p is the amount by which the measured surface pressure is found to be above or below the static, or barometric, pressure in the approaching stream. A positive pressure difference equal to the dynamic pressure is represented by a value of $C_p = 1$. Negative pressures or suctions give rise to negative coefficients and may in some cases be several times greater numerically than the dynamic pressure.

In uniform flows in which velocity is constant over the approaching stream there is no difficulty in establishing the reference dynamic pressure, $(1/2)\,\rho V^2$. When there is a wind-speed gradient, the dynamic pressure varies with height also, and it is necessary to make a choice as to the relative height in the wind-speed profile on which will be based the value of V and therefore of $(1/2)\,\rho V^2$ for a particular set of pressure coefficients. This is usually taken as the dynamic pressure of the approaching wind at the height of the building. In interpreting and using the pressure coefficients, it is always necessary to know what this reference height was.

Some thought will show that the selection of reference heights for full-scale testing may involve practical difficulties, since it is not a simple matter to arrange for the necessary measurements in an approaching wind that is changing in direction as well as speed. One solution is to use the wind speed measured at a mast on top of the building that is high enough to be free of the influence of the building. Pressure coefficients can be adjusted to another reference height by multiplying them by the inverse ratio of the dynamic pressures at the two heights.

Example 10.3

Given the wind speed at building height of 30 m to be 35 m/s and the wind-pressure coefficient for a point on the upwind face at a height of 20 m to be 0.7, find the pressure. The pressure coefficient is based on the dynamic pressure at building height. The density of air is 1.293 kg/m³ at a temperature of 0°C and standard barometric pressure.

$$\begin{aligned}\text{The dynamic pressure at 30 m} &= (1/2)\,\rho V^2 \\ &= 1/2 \times 1.293 \times 35^2 \\ &= 792\ \text{Pa} = 0.79\ \text{kPa}.\end{aligned}$$

The pressure on the building face at 20 m is

$$p = 0.7 \times 0.79 = 0.55\ \text{kPa}.$$

Pressure coefficients may also be based on or adjusted to the dynamic pressures of the wind-speed profile for the particular heights to which they apply. This can be done for purposes of correlating results from various shapes and wind profiles or for comparing boundary-layer data with constant-velocity data. This procedure may also be required for use in certain calculations of wind pressures or wind loads.

10.8 Power Law for Wind Pressures

The dynamic pressures at various heights in a wind-speed profile can be found from the wind speed for the height in question. When the power law is used for wind speed, it can also be applied to adjust dynamic pressures directly for height. It is only necessary to double the exponent being used, since pressures vary as the square of the wind speed. Thus, it follows from Equation 10.4 that

$$\frac{(1/2)\,\rho V_z^2}{(1/2)\,\rho V_g^2} = \frac{q_z}{q_g} = \left[\frac{Z}{Z_g}\right]^{2\alpha} \qquad (10.8)$$

Correspondingly, following Equation 10.5, a gust pressure relationship may be derived. Thus, for a value of $\beta = 0.1$ ($1/10$ power law) the gust pressures follow the $1/5$ power law.

Example 10.4

The dynamic pressure, q, at an elevation of 200 m in the centre of a city (terrain category 4), is 0.8 kPa. Find the dynamic pressure at a building height of 170 m.

From Table 10.1, $\alpha = 0.36$.
From Equation 10.8, it follows that

$$\frac{q_1}{q_2} = \left[\frac{Z_1}{Z_2}\right]^{2\alpha}$$

so

$$q_{170} = 0.8\left[\frac{170}{200}\right]^{0.72} = 0.71 \text{ kPa.}$$

10.9 Pressure Coefficients for Various Building Shapes

Pressure coefficients over a transverse midsection parallel to the wind flow for a building with a steep roof are shown in Figure 10.10 [**10.3**]. The roof slope is steep enough to force the streamlines upward, generating positive pressures on the windward slope. Separation takes place at the ridge, giving rise to negative pressures, or suctions, on the leeward wall and roof slope. Suctions are high immediately following a sharp edge at which separation is occurring. Reducing the roof slope reduces the pressures on the windward roof slope progressively to zero. Further reduction produces suctions instead of pressures on the windward slope until separation develops at the windward eave, as shown in Figure 10.11. A reattachment of the separated flow occurring between the eave and the peak results in a third small vortex region with high suctions, but this disappears on a further decrease of roof slope, and suctions are reduced.

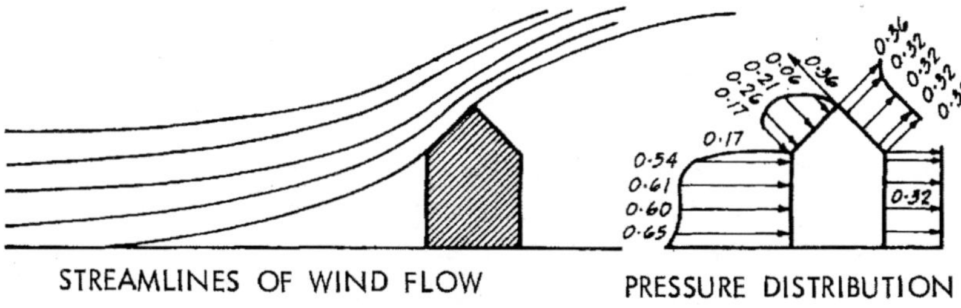

FIGURE 10.10 Two-dimensional wind flow over a building.
(From Jensen, M., and Franck, D., *Model-scale tests in turbulent wind, Part II*, Copenhagen: Danish Technical Press, 1965, Fig. 36b, p. 36)[**10.3**]

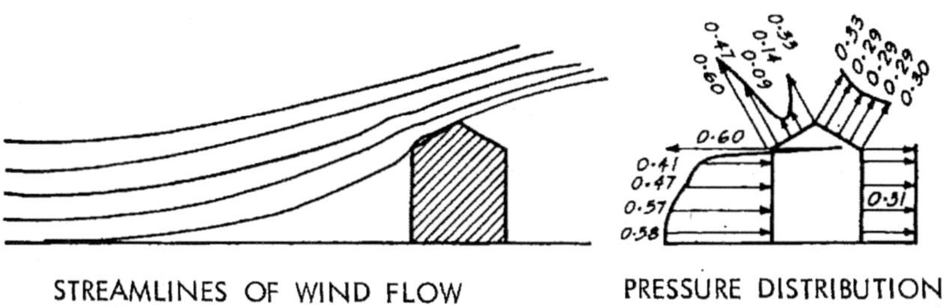

FIGURE 10.11 Separation of wind flow at eave causing high suction.
(From Jensen, M., and Franck, D., *Model-scale tests in turbulent wind, Part II*, Copenhagen: Danish Technical Press, 1965, Fig. 37b, p. 37)[**10.3**]

The leeward wall and roof slope are exposed to the same turbulent wake and experience similar negative pressures relatively unaffected by the changes in roof slope. There is also flow around the ends of the building with attendant separations at the sharp corners, and the possibility of reattachments as the wind direction changes. Changes in the ratios of length to width and height to width of the building affect the flows and, in particular, the critical slope at which maximum suctions will occur.

High local suctions are generated on long, low-slope roofs at the windward corner when the wind is at an angle of about 45° to the eave. Coefficients obtained from model tests in a wind tunnel with a uniform wind-speed profile are shown in Figure 10.12 [**10.7**]. Peak suctions as great as –4.0 times the dynamic pressure occurred as a result of the combined effects of the separations at the tops of the two windward walls. Suctions are likely to be greatest when reattachment of the flow takes place on a long roof. Parapets can influence the flow by lifting it to prevent reattachment, but if they are not high enough they may produce even higher suctions. These critical high suctions can lead to problems such as roofing failures in high winds.

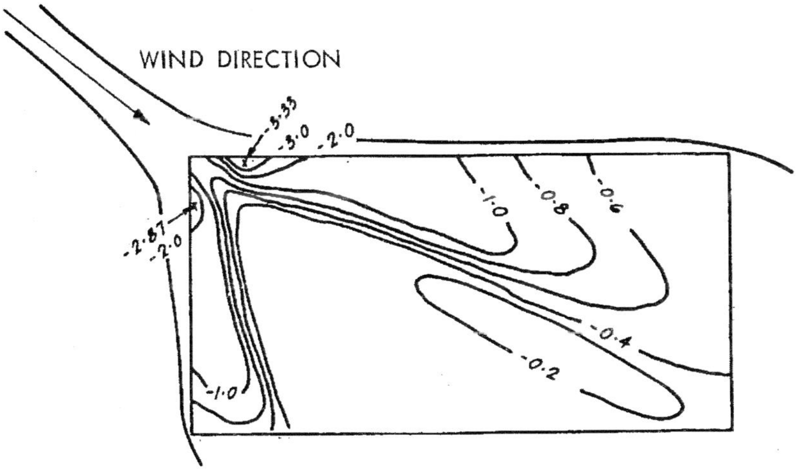

FIGURE 10.12 Plan view of roof with contours showing negative pressure distribution.
(From Leutheusser, H.J., University of Toronto, Department of Mechanical Engineering, TP 604, April 1964, Fig. 15)[**10.7**]

Pressure coefficients for a cube are shown in Figure 10.13 [**10.8**]. These were obtained in a wind tunnel for a constant velocity and for a boundary-layer profile produced by means of screens. The coefficients for the boundary-layer profile were based on the dynamic pressure at the height of the model. The maximum pressure coefficient on the windward face for constant velocity was 0.99, very close to the expected stagnation pressure coefficient of 1.0. The stagnation point was raised by changing the boundary-layer profile. Since the pressure coefficient is based on the dynamic pressure at cube height rather than stagnation-point height, the maximum value of the coefficient was less than 0.8. The reduction in pressure from the stagnation point downward for the boundary layer case is attributable to the decreasing velocities as the ground is approached. It was found that there was a marked downward flow ending in a strong standing vortex or "roller" at the base of the windward wall. This feature of the flows in boundary-layer profiles has been the subject of much study; it is associated with strong winds at street level around tall buildings and, as will be discussed later, creates discomfort and sometimes danger to pedestrians.

The general nature of the flow over a building in a boundary layer is shown in Figure 10.14. The downward flow over the windward face produces the standing vortex, which extends around the ends of the building. It is reinforced by the flows around the building at higher levels, which have also acquired a downward-velocity component to produce high speeds at the front, sides, and rear of the building at street level. These effects result mainly from the high-speed winds that are intercepted and directed downward by the upper part of the building face lying below the stagnation point. This means that they will develop only if the building extends well above the wakes of upstream obstructions, which will have a reduced speed. High speeds can be generated around lower buildings on unobstructed sites or on open terrain.

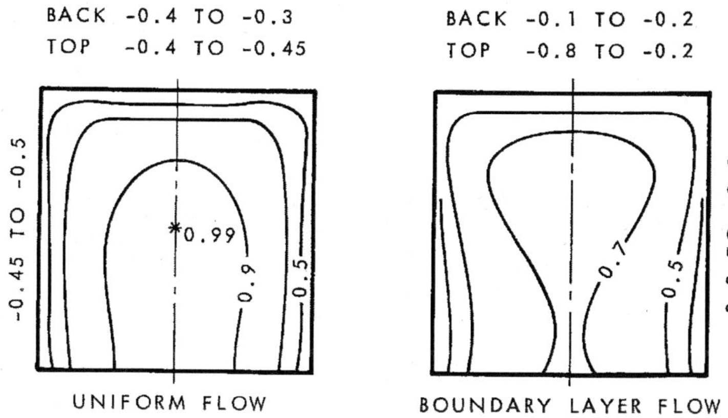

FIGURE 10.13 Pressure coefficients for a cube in uniform flow and typical boundary-layer flow.
(From Baines, W.D., Paper 6, International Conference on Wind Effects on Buildings and Structures, 1963, Published by Her Majesty's Stationery Office, London, 1965. Figs. 10, 11)[**10.8**]

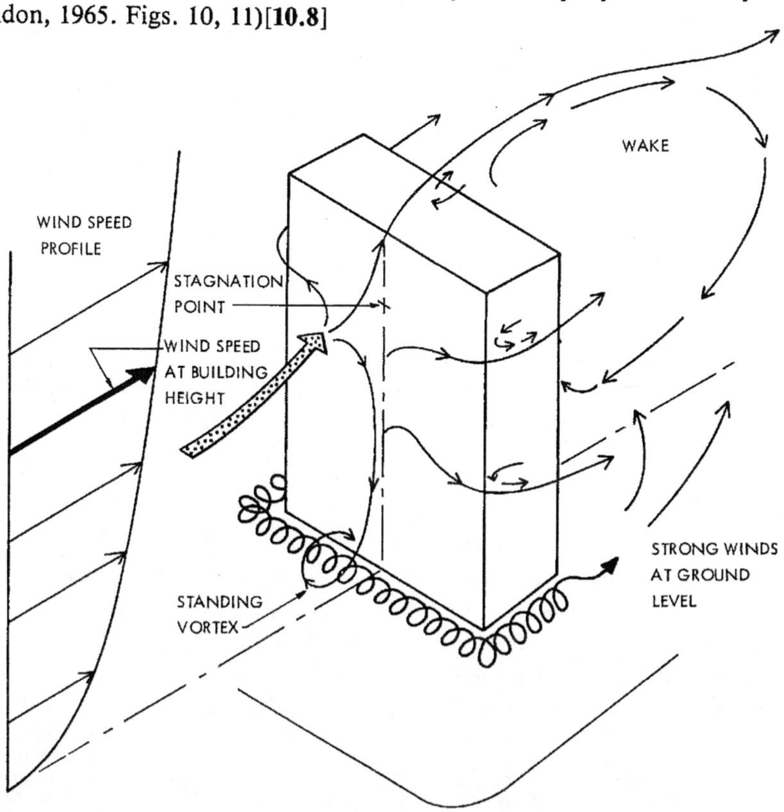

FIGURE 10.14 Flow around a building in a boundary layer.

The downward diversion of high-speed flows to lower levels complicates further the prediction of wind effects in the immediate vicinity of tall buildings. Buildings, signs, trees, lamp posts, vehicles, and people in the immediate neighbourhood may be adversely affected. When it is important to know what these effects are likely to be, a properly modeled wind-tunnel test may be called for.

10.10 Wind Loads from Pressure Coefficients

Pressure coefficients have frequently been determined on the basis of mean wind speed, V, and measured mean pressures, p. The pressure coefficient is correctly expressed by

$$C_p = \frac{p}{(1/2)\,\rho V^2}. \tag{10.9}$$

Values of C_p found in this way are commonly used in conjunction with estimated maximum gust speeds in the calculation of maximum pressures from the relationship

$$p_{max} = C_p \cdot (1/2)\,\rho G^2. \tag{10.10}$$

These maximum pressures are used in the calculation of equivalent static loads for design purposes. This "simple" procedure, also called a quasi-static approach, may underestimate the peak pressures and fails to take into account the increases in peak stresses that occur in buildings having a significant dynamic response to the gust loading [**10.1, p. 121**]. It is widely used, however, in the design of buildings that are unlikely to respond dynamically to wind loading.

Pressure coefficients for use in calculating structural loads are often adjusted and presented as average values over appropriately designated areas of wall and roof surfaces. They may also have been adjusted to correspond to the dynamic pressures from the wind-speed profile at various heights rather than to a fixed reference pressure such as that at building height. Such design values must always be judged in relation to the design procedures in which they will be used, and in particular to the rules by which wind speeds are selected and allowances made for terrain roughness and for gusts.

Load calculations begin with the selection from the available records of suitable values describing the maximum wind speeds to be expected for the locality. These will usually be values of mean hourly wind speed and should represent strong winds that will occur only once in a suitably long "return" period, such as 10, 30, or 100 years. A long or short period may be selected, depending on the risk that may be accepted in connection with the particular feature under consideration. Once this decision has been made, the value can be adjusted for terrain roughness and height to provide an estimated value of maximum gust speed and dynamic pressure at building height in accord with the methods discussed in this chapter. In an alternative procedure, these adjustments are incorporated in "exposure" factors, C_e, and gust factors, C_g, to be applied to the dynamic pressure for the reference wind speed selected from the records at the weather station to give values of maximum wind pressures.

When the pressure coefficients for the building shape under study are based on the dynamic pressure at building height, they may be multiplied directly by the design dynamic pressure at building height to find the pressures on various surfaces. If they are based on a uniform-velocity wind, or if they have already been suitably adjusted, they may be applied to the dynamic pressures for various heights corresponding to the wind-speed profile. The pressures over the building surfaces found in this way can then be used in the computation of structural wind loads as required.

Loads producing moments, torsion, and shear on a structure as a whole result from the pressures on the exterior surfaces of the building. Detailed procedures for the resolution of exterior pressures into loads for various planforms (and particularly for curved surfaces) have been devised and described [**10.9**]. When the building planform is rectangular and the wind direction is normal to one face, the effective load in the wind direction may in some cases be calculated conveniently by using the algebraic difference in the pressure coefficients for the windward and leeward faces.

10.11 Internal Pressures and Cladding Loads

Loads for the design of cladding can be found from the algebraic difference between the exterior and interior pressures. Interior pressures can be stated in terms of interior pressure coefficients, C_{pi}, referenced to some specified dynamic pressure such as that at the midheight of the building.

The interior pressure under steady state in the case of a simple building that has no interior floors or partitions creating barriers to flow will be that at which the flows of air in and out are balanced. Natural air leakage is determined by the pressure difference across any openings, including holes and cracks. To this may be added the supply or exhaust of air under pressure by fans. The flow through a wall having a total area of openings, A, for which the discharge coefficient is C_d with exterior and interior pressures p_e and p_i, will be

$$Q = C_d A \left[\frac{2}{\rho}(p_e - p_i)\right]^{1/2}. \tag{10.11}$$

Equations describing the flows through the various enclosure elements such as walls (including windows and doors) and roofs may be added, set equal to zero, and solved for p_i. The problem becomes more complicated when the building is divided into compartments, with restricted flows taking place between compartments as well as through the building enclosure. Equations must be solved simultaneously for the several compartments.

The values of p_i can be used to find values of $(p_e - p_i)$ for the calculation of cladding loads. They can also be substituted in the original expressions to find values of Q describing air exchanges between various compartments and with the outside, information that may be wanted for other purposes.

When there are major openings in one wall, the pressure drop $(p_e - p_i)$ for that face will be small, and p_i will be nearly equal to p_e. Thus, for a large open door in a windward wall the interior pressure will be substantially equal to the dynamic pressure in the approaching wind, while for a leeward location the interior pressure will be equal to that of the wake. The combination of high interior pressures and high suctions on the exterior can produce high lifting forces on roofs. If the roof deck is not airtight, high pressure differences may develop across the roof covering.

10.12 Design of Buildings to Resist Wind Loads

Many texts, papers, and reports contain information on the design of building structures, a field that has long been a large and important area of specialized study. Several textbooks are devoted entirely to the consideration of wind loads, and deal with the complex matter of the dynamic response of buildings to the variable pressures produced on them by wind [**10.1, 10.9–10.11**]. This subject has also received adequate treatment elsewhere, but a brief discussion of the topic, linked to the rules and requirements set out in the National Building Code of Canada, will illustrate the use of wind data as well as some of the design options for wind loads.

10.13 The National Building Code of Canada

Building codes are technical documents used to regulate construction in the interests of structural safety, fire protection, and health. The codes set out minimum requirements and specify procedures by which compliance can be determined. The National Buidling Code of Canada (NBCC) sets out a simple procedure for the determination of wind loading for general use and a detailed procedure for use with buildings of a height more than four times their width or greater than 120 m, or with other buildings having characteristics that are likely to lead to dynamic effects [**10.12–10.14**]. Buildings for which the detailed procedure is appropriate may also be designed on the basis of experimental wind-tunnel testing using dynamic models; this becomes the preferred method when application of the detailed calculation procedure indicates that significant dynamic effects can be expected.

The reference wind speeds for various locations in Canada for use with the NBCC have been derived from hourly mean wind speeds chosen in most cases to be representative of a height of 10 m in an open exposure. The corresponding dynamic pressure, q, is set out for three levels of probability (1 in 10, 1 in 30, and 1 in 100) that will be exceeded in any one year [**10.15**].

Design pressures, p, are given by

$$p = q\, C_g\, C_e\, C_p \tag{10.12}$$

in which q is the reference mean dynamic pressure at 10 m at the weather station, C_g is a gust-effect factor, C_p is a mean pressure coefficient, and C_e is an exposure factor.

The exposure factor C_e takes into account height and terrain roughness. The gust-effect factor C_g takes into account not only the increased wind speeds and wind pressures due to gusts but is also adjusted (in the case of the detailed procedure) to take account of the complex fluctuating response of the structure to the gusts. In this way the format based on Equation 10.12 is used in both simple and detailed procedures to arrive at design pressures which, when applied as static loads, will produce the same peak stresses as real wind.

In the simple procedure the exposure factor C_e is adjusted so that it varies over the height of the building. It is assigned a value of 1 for heights up to 12.7 m and thereafter varies in accord with the ⅕ power law. No adjustment is made for terrain roughness. The gust-effect factor is taken as 2 for the building as a whole and as 2.5 for cladding.

The exposure factor in the detailed procedure is also varied over the building height and is adjusted for height above ground and for terrain roughness in accord with the following:

Exposure A: open or standard exposure

$$C_e = \left[\frac{Z}{10}\right]^{0.28}, \; C_e \geq 1.0. \tag{10.13}$$

Exposure B: suburban and urban areas, wooded terrain or centres of large towns

$$C_e = 0.5\left[\frac{Z}{12.7}\right]^{0.50}, \; C_e \geq 0.5. \tag{10.14}$$

Exposure C: centres of large cities with heavy concentrations of tall buildings

$$C_e = 0.4\left[\frac{Z}{30}\right]^{0.72}, \; C_e \geq 0.4. \tag{10.15}$$

Since the exposure factor C_e is varied appropriately over the height of the building, the recommended average pressure coefficients C_p have also been adjusted to conform. They are referenced to actual dynamic pressure at various heights in the assumed wind-speed profile, not to that at a fixed level such as building height.

Example 10.5

Demonstrate that the exposure factors from equations 10.13 to 10.15 are consistent with the adjustments for height and terrain roughness discussed in sections 10.4 and 10.8.

Note that Section 10.4 relates to wind speeds, while the exposure factor C_e is applied to the reference dynamic pressure at 10 m over flat terrain. Note also from Section 10.8 that the exponent in the power law for pressures is double that in the equation relating wind speeds. Compare exponents 0.28, 0.50, and 0.72 with values for α in Table 10.1 of 0.15, 0.25 and 0.36. Compare also terrain categories 2 and 3 in Table 10.1 with exposures A and B.

Find the ratio of pressures at 12.7 m over terrain category 3 to those at 10 m over terrain category 2 using the methods of Section 10.4.

From Equation 10.4

$$\frac{V_z}{V_g} = \left[\frac{Z}{Z_g}\right]^{\alpha}$$

and for terrain category 2 from Table 10.1

$$Z_g = 300 \text{ m}, \alpha = 0.15$$

so that the gradient wind speed

$$V_g = V_{10}\left[\frac{300}{10}\right]^{0.15}$$
$$= 1.66\ V_{10}$$

where V_{10} is the reference wind speed at 10 m at the open site.

For terrain category 3, from Table 10.1

$$Z_g = 400 \text{ m}, \alpha = 0.25.$$

The gradient wind as found above

$$= 1.66\ V_{10},$$

and

$$\frac{V_{12.7}}{V_{10}} = 1.66\left[\frac{12.7}{400}\right]^{0.25} = 0.70$$

but the ratio of dynamic pressures

$$= \left[\frac{V_{12.7}}{V_{10}}\right]^2 = (0.70)^2 = 0.49.$$

Now C_e also gives the ratio of dynamic pressures, and for 12.7 m, from Equation 10.14, $C_e = 0.5$ which is nearly equal to 0.49; the discrepancy is due to differences in exponents and gradient heights used, not to differences in method.

The gust-effect factors for use in the detailed method must take into account the dynamic response of the building to the gustiness of the wind. The method has not been developed to the point of accounting for directional effects of the wind, the interaction of neighbouring structures, and certain other effects such as transverse vibrations and fatigue [**10.14**].

10.14 The Nature of Dynamic Response

All structures deform under loads. This deformation is proportional (or nearly so) to the load for framed structures of steel and reinforced concrete under steady or static loads. When such structures are subjected to winds that produce highly variable loads on them, the deformations they undergo result from two different kinds of response. The first arises from the proportionality between load and deformation, as is commonly assumed in engineering design. The second kind of response involves the vibration of the structure, like a tuning fork, at its critical or resonant frequency as determined by its mass and elasticity. The deformations in this case result from a correspondence between certain frequency characteristics in the impulsive loading produced by the wind gusts and the resonant frequency of the structure vibrating in the plane of the wind direction. Energy transmitted from the wind to the structure keeps it vibrating. The internal energy losses of the structure create a damping effect, which limits the amplitude of vibration to that at which the energy input and the losses will be equal. This dynamic response at resonant frequencies is combined with the response to direct loading to produce deformations that may exceed those from direct loading alone.

Tall, slender buildings and certain other structures that are highly elastic with low damping can be made to vibrate significantly by turbulence. This adds to the deflections and corresponding stresses they must safely withstand, so that it is desirable to check all such structures for possible dynamic effects.

An equation has been developed to express the spectrum of turbulence of typical strong winds with respect to wavelength [**10.16**]. The maximum gust energy is associated with wavelengths of about 600 m which for a wind speed of 30 m/s corresponds to a frequency of 3 cycles per minute [**10.1, p. 98**]. From this spectrum of turbulence it is possible to proceed (with some difficulty) to estimate the influence of wind gusts on the structural oscillations and thus on the maximum deflections and stresses. This subject is discussed in many of the papers that were presented at the International Symposia on Wind Effects on Buildings and Structures [**10.17–10.20**] and elsewhere, and in the texts already referenced. The recognized need to rely on wind-tunnel testing as a basis for design in special cases provides evidence that the prediction of dynamic response from theory alone is not yet adequate. Proper modeling of both wind and structure in a wind-tunnel test confirms the dynamic response to be expected. It is the only way in which directional wind effects and the influence of neighbouring structures can be investigated.

10.15 The Wind Environment around Buildings

The flow of wind around buildings is of interest in many ways. The pressures generated over the building surface have been of prime concern in this chapter. Knowledge of them makes it possible to calculate wind loads and to deal with natural ventilation and air leakage. There are other considerations, however, arising from the ways in which winds are modified by buildings. An important influence is that which individual buildings have on the airflow around neighbouring buildings. There are many others, including the welfare and comfort of pedestrians at street level, the distribution of gaseous effluents discharged from buildings, and the deposition of rain and snow. Winds at street level began to be studied extensively after it was realized that although they were generally moderated by the presence of buildings, they were sometimes markedly increased. In extreme cases, usually involving tall buildings, pedestrians were blown from their feet and vehicle traffic was affected.

The general nature of the flow around a tall building in a boundary layer is shown in Figure 10.14. The highest pressure on the upstream face occurs at the stagnation point at about 80% of the building height. Pressures near ground level may be less than half the stagnation pressure, giving rise to a large pressure difference in support of a downward flow. High wind speeds are produced at the front and at the ends of the building. The higher the building, and in particular the more it extends above the level of upstream obstructions, the greater the winds are likely to be at street level. Only those buildings that are at least twice as high as the upstream obstructions are likely to create serious problems [**10.1, p. 174**]. Penwarden and Wise [**10.5**] have emphasized the interaction of the flow over the top of a low building with the flow down the face of a tall building a short distance downstream to produce augmented wind speeds at the base of the tall building.

The highest wind speeds are to be expected in open arcades and vehicular passages through buildings: negative wake pressures combine with positive pressures on the upstream face to produce large pressure differences leading to high flow rates along such channels.

There is almost no limit to the combinations of wind and buildings that could be presented for consideration. Winds vary in direction, strength, and turbulence, while buildings vary in planform, height, and arrangement. Despite the great difficulties much progress has been made in understanding the effects of wind on and around buildings, largely with the aid of boundary-layer wind tunnels in which various building arrangements have been modelled. Specific cases can always be approached in this way when they are sufficiently important to justify the effort involved. There is also a growing body of knowledge on which useful judgements can be based. Extended discussions can be found in textbooks [**10.1**] and in special reports [**10.5, 10.21**].

10.16 Other Wind-related Problems

There are a number of concerns in addition to the welfare of pedestrians that are now able to benefit from the application of available knowledge of winds around buildings. In some cases, as in the discharge of effluents from chimneys, stacks, and exhaust vents, design recommendations based on data from constant-velocity wind tunnels have been used. Improved information is now being developed from boundary-layer wind tunnel studies and a new chapter on airflow around buildings has been added to the ASHRAE Handbook 1981 [**10.22**].

Snow deposition and drifting create serious problems in areas with appreciable snowfall. In the case of snow loads on roofs, the wind environment determines how much snow is deposited. Exposed roofs may be swept clear of snow, while the snow load on adjacent lower roofs may be greatly increased [**10.13**]. The removal of snowdrifts can be costly as well as inconvenient. Drifting can be controlled in some cases by the special arrangement of buildings and by the use of snow fences [**10.23**]. Snow in suspension in the wind moving over the ground surface will be deposited whenever the wind speed is reduced. This occurs commonly in the wake region of fences, buildings, trees, and other obstructions. Snow fences can be arranged to provide for deposition and storage of the accumulated snow or to increase the wind speed near the ground, thus inhibiting deposition on selected areas. Another problem involving the movement of loose material occurs when the protective gravel covering used on flat roofs is blown from the roofs of high buildings and causes damage to adjacent lower surfaces [**10.24**].

The influence of wind on the rain deposited on flat roofs is not usually of importance. High winds can cause rain penetration of shingled roofs. The situation for walls is somewhat more critical since the angle at which raindrops fall and thus the amount of rain deposited on vertical surfaces is dependent on the wind. Wind can also cause rain water to move diagonally down a wall, producing high flow rates and concentrations at vertical and horizontal discontinuities. Wind pressure is another force that can cause water penetration of walls and joints.

Studies have been made in several countries to determine the coincidence of wind and rain in attempts to obtain indices for various regions of the severity of wind-driven rain [**10.25**]. Extensive research has been done, notably in Norway, on the mechanism of rain penetration, the role of wind, and the design of rain-proof joints [**10.26**]. Limited studies have been made of the flow of wind-driven rain and the wetting of building surfaces [**10.27, 10.28**]. It is evident that, since the frequency and severity of wetting are factors in the deterioration of materials, more use must be made of the relevant information on the wind environment in dealing with this subject.

Wind has a further influence on the environment of materials used in the cladding of buildings since air movement over the building surface affects drying rates and the diffusion of pollutants such as sulphur dioxide. It also affects the convective heat transfer at the building surface, and is a factor in heat transmission through the building enclosure, surface temperatures, and heating and cooling rates, as discussed in Chapter 8.

Natural ventilation by means of wind is of considerable importance for cooling in warm climates when air conditioning is not used. Much depends on the ability to predict the wind pressures at building surfaces. With a knowledge of the pressures and the ventilation openings, the air flows can be calculated, as discussed in Section 10.11. The subject is dealt with at some length by Aynsley et al. [**10.1**], van Straaten [**10.29**], and Givoni [**10.30**].

Air leakage, having characteristics in common with natural ventilation, involves the restricted movement of air through and within buildings. It is of great importance in cold climates when heating must be provided and, since it can occur independent of the wind, it will be discussed in a separate chapter.

10.17 Conclusion

The movement of air over and around buildings as a result of wind is a factor in many building situations, one that is difficult to predict in any detail. The application of theory developed for flow involving undisturbed fluid streams is complicated by the need to deal with a boundary layer. A further complication results because the buildings are part of the roughness that generates the boundary layer. The usefulness of the concept of a boundary-layer profile begins to break down when important features of the wind approaching the building are influenced by individual buildings immediately upstream. This can mean that much of the knowledge based on other known situations may be invalid or inadequate, so that judgements may have to be made without adequate supporting evidence. In such a situation it is possible to use model testing in a boundary-layer wind tunnel. Even this method has its limitations, one of which is that Reynolds-number effects cannot readily be determined. The wind-tunnel approach is always limited by the adequacy of the simulation of the buildings and the wind environment. The design, preparation, and testing of proper models and the evaluation of results are demanding exercises requiring much specialized knowledge, and cannot always be justified for individual projects unless the benefit to be derived is correspondingly great. Nevertheless, there may be no acceptable alternative in the case of important buildings when there is reason to believe that serious wind-related problems may be involved.

Much of the available information on the flow of wind over buildings has been obtained from studies in wind tunnels with constant-velocity profiles. Better information can now be obtained from boundary-layer wind tunnels in which the wind-speed profile and turbulence are properly modeled. The profiles and turbulence conditions most commonly used are for strong winds, which are of interest in connection with wind loads. These may not be the appropriate conditions for some studies in which average or integrated wind effects are wanted, as in the case of ventilation, air leakage, and surface-heat exchange. When wind data are being considered for use in a particular application it is important that the source, the conditions from which they were derived, and the form in which they are presented be carefully reviewed.

References

10.1 Aynsley, R.M., Melbourne, W., and Vickery, B.J., *Architectural aerodynamics*. London: Applied Science Publishers Ltd., 1977.

10.2 Davenport, A.G. *Wind loads on structures*. DBR/NRCC, March 1960. (NRCC 5576)

10.3 Jensen, M., and Franck, N. *Model-scale tests in turbulent wind, Part II*. Copenhagen: Danish Technical Press, 1965.

10.4 Davenport, A.G., and Isyumov, N. The application of the boundary layer wind tunnel to the prediction of wind loading. *Proceedings, International Research Seminar on Wind Effects on Buildings and Structures*, Ottawa, Sept. 1967, Toronto: University of Toronto Press, 1968, pp. 201–30.

10.5 Penwarden, A.D., and Wise, A.F.E. *Wind environment around buildings*. Building Research Establishment Report. U.K. Department of the Environment. London: HMSO, 1975.

10.6 Dalgliesh, W.A. Comparison of model/full-scale wind pressures on a high building. *Journal of Industrial Aerodynamics*. 1975, *1*, pp. 55–66. (NRCC 15130)

10.7 Leutheusser, J.H. *The effects of wall parapets on the roof pressure coefficients of block type and cylindrical structures.* TP 6404, Department of Mechanical Engineering, University of Toronto, April 1964.

10.8 Baines, W.D. Effect of velocity distribution on wind loads and flow patterns on buildings. *Proceedings, International Conference on Wind Effects on Buildings and Structures.* Teddington, U.K., June 1963.

10.9 Houghton, E.L., and Carruthers, N.B. *Wind forces on buildings and structures.* London: Edward Arnold, 1976.

10.10 Macdonald, A.J. *Wind loading on buildings.* London: Applied Science Publishers Ltd., 1975.

10.11 Simiu, E., and Scanlan, R.H. *Wind effects on structures*, New York: John Wiley and Sons, 1978.

10.12 *National Building Code of Canada, 1980.* Ottawa: National Research Council of Canada. (NRCC 17303)

10.13 *Commentaries on Part 4 of the National Building Code of Canada 1980.* Chap. 4, The Supplement to the National Building Code of Canada 1980. Ottawa: National Research Council of Canada. (NRCC 17724)

10.14 Davenport, A.G., and Dalgliesh, W.A. A preliminary appraisal of wind loading concepts of the 1970 National Building Code of Canada. *Proceedings, Third International Conference on Wind Effects on Buildings and Structures*, Tokyo, 1971. Part III, pp. 441–50. (NRCC 13741)

10.15 *Climatic information for building design in Canada.* Chap. 1, The Supplement to the National Building Code of Canada 1980. Ottawa: National Research Council of Canada. (NRCC 17724)

10.16 Davenport, A.G. The spectrum of horizontal gustiness near the ground in high winds. *Quarterly Journal, Royal Meteorological Society*, 1961, *87*, p. 194.

10.17 *Proceedings, International Seminar on Wind Effects on Buildings and Structures*, Teddington, U.K., June 1963.

10.18 *Proceedings, International Seminar on Wind Effects on Buildings and Structures*, Ottawa, Sept. 1967. Toronto: University of Toronto Press, 1968.

10.19 *Proceedings, International Conference on Wind Effects on Buildings and Structures*, Tokyo, Sept. 1971.

10.20 *Proceedings, Fourth International Conference on Wind Effects on Buildings and Structures*, London, Sept. 1975. London: Cambridge University Press, 1976.

10.21 Gandemer, J. et Guyot, A. *Intégration du phénomène vent dans la conception du milieu bâti. Guide méthodologique et conseils pratiques.* Paris: Secrétariat général du group centrale des villes nouvelles, 1976. (Diffusion: La Documentation française, 31 quai Voltaire, Paris.)

10.22 *ASHRAE Handbook 1981 Fundamentals.* Chap. 14.

10.23 Williams, G.P. *Annotated bibliography on snow drifting and its control.* Bibliography 42, DBR/NRCC, 1974.

10.24 Kind, R.J., and Wardlaw, R.L. *Design of rooftops against gravel blow-off.* Ottawa: National Aeronautical Establishment, NRCC, 1976. (NRCC 15544)

10.25 Lacy, R.E. *Driving-rain maps and the onslaught of rain on buildings.* Paper presented at RILEM/CIB Symposium on Moisture Problems in Buildings, Helsinki, 1965.

10.26 Birkeland, Ø. *General report on rain penetration.* Oslo: Norwegian Building Research Institute, 1966.

10.27 Rodgers, G.G., Poots, G., Page, J.K., and Pickering, W.M. Theoretical predictions of raindrop impaction on a slab type building. *Building Science*, 1974, *9*, pp. 181–90.

10.28 Robinson, G., and Baker, M.C. *Wind-driven rain on buildings.* DBR/NRCC, July 1975 (NRCC 14792)

10.29 Van Straaten, J.F. *Thermal performance of buildings.* Barking, U.K.: Applied Science Publishers, 1967.

10.30 Givoni, B. *Man, climate and architecture.* 2d Ed. Barking, U.K.: Applied Science Publishers, 1976.

11

Air Leakage and Ventilation

11.1 Introduction

Air leakage is often of considerable importance to the performance of a building. Leakage can occur through pores in materials, cracks, holes, or other openings. Flow is produced by air-pressure differences that provide the energy to offset friction and other losses. Air from an adjoining room or from outdoors infiltrates into a room or compartment through the enclosing construction. The outward leakage of air to another room or to the outdoors is known as exfiltration. Ventilation (as distinct from air leakage) is the intentional supply or removal of air.

Air leakage is responsible for the transfer of heat, water vapour, smoke, odours, dust, and other pollution products. Infiltration through the building enclosure may provide the outdoor air required to maintain air quality or for fuel-burning appliances, and for the operation of chimneys. Exfiltration may carry water vapour into the building enclosure to condense on contact with cold surfaces. Air is also exchanged between concealed spaces in the construction and the indoors or outdoors, or even between one space and another, with consequent transfer of heat and moisture. It is usually difficult to predict the rate at which air leakage will take place, since both air-leakage paths and pressures producing flow through them are varied and complex. The effect of leakage on building performance justifies careful attention, however.

11.2 Flow through Leakage Openings

In some cases, such as cracks around windows or doors, air-leakage paths can be identified and measured; but in most situations the nature and location of the leakage openings are not known in detail. The openings may be tortuous paths through porous materials, thin laminar passages formed by cracks and joints, or holes of various shapes and sizes.

Each shape and size of opening has its own particular flow characteristic and some can be expected to reflect the influence of Reynolds number, as in the case of flow in pipes. Long paths with small cross-sections may exhibit laminar flow and have a resistance proportional to the velocity. Larger holes may act more like orifices with resistance to flow varying with the square of the velocity. There may be many openings of different kinds contributing to the total leakage across the walls of any compartment, and it is seldom possible to identify, measure, and calculate flow through them individually with due regard for their particular characteristics. It is possible, however, to apply known pressures and measure overall flow rates for the aggregate of openings in any given enclosure element. The results take the form

$$Q = C(\Delta p)^n \tag{11.1}$$

where Q = volume flow rate
C = a coefficient
Δp = pressure difference
n = an exponent varying between 0.5 and 1.0.

For openings assumed to be uniformly distributed over the area of an enclosing element, or for uniform leakage along a joint or crack, Q and C may be defined accordingly to refer to unit area of element or unit length of crack, whichever is appropriate.

Values of the exponent n are found to vary from 0.5 corresponding to orifice flow to as much as 1.0, which is representative of laminar flow, but they may also vary with the pressure difference because many different kinds of openings are involved. A value of $n = 0.65$ represents many cases of wall and window leakage. Results can also be expressed as a volume flow rate at given pressure differences, thus avoiding the concern about the value of n. A pressure difference of 75 Pa has commonly been used in testing windows.

Leakage characteristics can be expressed in terms of the area of an equivalent sharp-edged orifice that will yield the same rate of flow at a particular pressure difference as the element in question. The equation for flow through such an orifice is

$$Q = CA\,[(2/\rho)(\Delta p)]^{1/2} \tag{11.2}$$

where Q = flow rate in m^3/s. C, the orifice coefficient, is commonly taken as 0.60.

11.3 Air Leakage Caused by Wind

The discussion in Chapter 10 provided some understanding of the general nature of the earth's boundary layer and of the flow of wind around buildings. The prediction of air leakage due to wind requires the establishment of a relationship between the character of the wind as determined from available records taken at some site, such as a weather station, and the pressure differences that will be produced across the building enclosure. It is also necessary to have some measure of the openings through which leakage will take place.

It is possible to use information about wind pressures and wind-speed profiles for the prediction of structural wind loads in air-leakage situations. This approach has greatest relevance to buildings that extend well above upstream obstructions under strong wind conditions; but in the absence of suitable alternatives, it is also applied to other situations, as will be discussed later.

It will be recalled from Chapter 10 that pressures over building surfaces can be related to some known reference wind speed by means of dimensionless pressure coefficients. The dynamic pressure of the reference wind is given by

$$q = \tfrac{1}{2}\rho V^2 \qquad \text{(Pa)} \tag{11.3}$$

where ρ = mass density, kg/m^3
V = reference wind speed, m/s.

The surface pressure, p, relative to barometric or freestream pressure is given by

$$p = C_p \cdot q. \qquad \text{(Pa)} \tag{11.4}$$

The reference wind speed is commonly taken as the speed of the approaching wind at the height of the building, found from the assumed or measured wind-speed profile. Values of the dynamic pressure for selected wind speeds are given in Table 11.1.

TABLE 11.1

Wind Pressures for Various Speeds

	Dynamic pressure, q, pascals		
V, m/s	$\rho = 1.20\ kg/m^3$ Dry air 20°C and 101.3 kPa	$\rho = 1.29\ kg/m^3$ Air at 0°C and 101.3 kPa	$\rho = 1.40\ kg/m^3$ air at −20°C and 101.3 kPa
1	0.60	0.65	0.70
2	2.40	2.58	2.80
5	15.0	16.1	17.5
10	60.0	64.5	70.0
V, km/h			
1	0.046	0.050	0.054
5	1.16	1.24	1.35
10	4.63	4.98	5.40
20	18.5	19.9	21.6
30	41.7	44.8	48.6
40	74.1	79.6	86.4

1 m/s = 3.6 km/h

Pressure coefficients obtained from model tests in a wind tunnel in which a boundary layer was simulated by means of spires and blocks are shown in Figure 11.1 [**11.1**]. The wind-speed profile intended to represent urban terrain was given by $V_z = KZ^{0.43}$ where V_z = wind speed at height, Z, above ground and K = a constant. The building model, 76 by 114 mm in plan, was 1/400 of full scale. It was surrounded by a regular array of rectangular blocks of uniform height of 38 mm arranged with a clear space of 76 mm between them to represent surroundings of uniform height separated by streets. Tests were carried out for six model heights varying in steps of 38 mm up to 229 mm corresponding to a maximum building

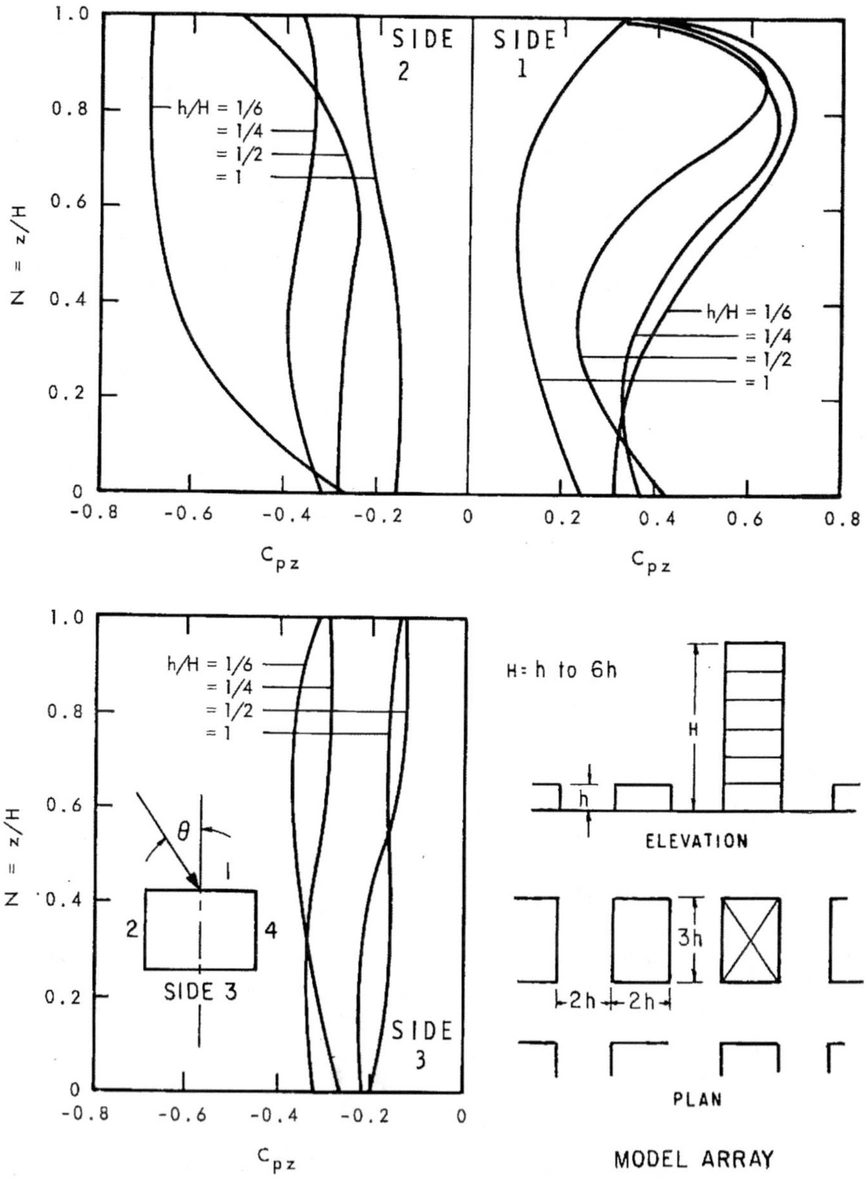

FIGURE 11.1 Vertical distribution of mean wind-pressure coefficients for $\theta = 0°$ and various values of h/H for a model tested in a boundary-layer wind tunnel with reference to wind pressure at building height.
(From Shaw, C.Y., *ASHRAE Trans.*, Vol. 85, Part I, 1979, Fig. 3a, p. 82)[11.1]

height of 92 m. The pressure coefficients are area-weighted averages over the building width and are referenced to the freestream dynamic pressure corresponding to the speed of the approaching wind at building height.

The pressures on the upstream face for $\theta = 0$ as shown in Figure 11.1 are consistent with the discussion in Chapter 10, indicating a maximum pressure at about 80% of height for tall buildings, with a marked reduction toward street level. The pattern persists for decreasing building heights down to $h/H = 1/2$, but changes markedly as might be expected when the building is the same height as its neighbours. Side 4, not shown, gave identical results, as expected, for $h/H = 6$ and $h/H = 4$ but not for $h/H = 2$ and $h/H = 1$. C_{pz} for side 4 had negative values for the case of $h/H = 1$, up to 0.2 greater. This anomalous result was not explained, but may have been due to imperfections in the model or in the air stream. Coefficients on sides 2, 3, and 4 for other building heights appear to be quite stable in a range from -0.2 to -0.4, except for the case of $h/H = 1/6$ for which the suctions on the end walls parallel to the wind rise to -0.7.

It was found that coefficients for $\theta = 45°$, not shown, were generally between 0 and +0.4 on sides 1 and 2, except for $h/H = 1$, which showed a negative value of about -0.2 for side 1. Sides 3 and 4 showed values ranging from -0.2 to -0.5. These and other values were applied to a mathematical model of a building in the calculation of infiltration. Similar calculations for a tall building in a suburban boundary layer with a wind-speed profile represented by $V_z = KZ^{1/3}$ have been described [**11.2**].

11.4 Estimation of Leakage Due to Wind

When the wind is normal to the face of a building, the pressures are positive on the upstream face and negative on the other three. Infiltration takes place on the upstream face, with outward air leakage on the other sides. The interior pressures adjust until the flow in and flow out are equal for each compartment in the building. When there are no interior partitions, each storey becomes a compartment, along with elevator and stair shafts. The relevant pressure differences and flow rates are dependent on the leakage openings in both interior and exterior walls. These must be known or assumed before effective pressure differences and flow rates can be calculated.

It has been found that the flow in tall buildings due to wind is mainly horizontal, with little vertical flow [**11.2**]. This being so, it may be adequate in some cases to consider each floor independently. The interior pressures, already discussed briefly in Chapter 10, can sometimes be estimated with sufficient accuracy from the relationship

$$\frac{p_w - p_i}{p_w - p_\ell} = \frac{1}{1 + \left[\frac{A_w}{A_\ell}\right]^{1/n}} \qquad (11.5)$$

where p = pressure, A = area of openings, n = flow exponent, and subscripts i, w, and ℓ refer to inside and windward and leeward sides of a building.

Equation 11.5 may be applied to the case of a quartering wind on a square building with equal openings on all sides so that $A_w/A_\ell = 1$. In that case $p_w - p_i = 0.5\,(p_w - p_\ell)$, so that the interior pressure is the average of the windward and leeward surface pressures. For the same building with wind normal to one face and equal pressures on the other faces, $A_w/A_\ell = 1/3$ and for $n = 0.65$

$$(p_w - p_i) = 0.84\,(p_w - p_\ell).$$

For more complicated cases with several compartments per storey and with flow from floor to floor it is necessary to set up equations for the flow in and out of each compartment, which must then be solved simultaneously for the interior pressures. When more than two or three compartments are involved, hand methods become laborious and a computer is required [**11.2, 11.3**]. Once interior pressures are known, the pressure differences across compartment-enclosing elements and the corresponding flows can be calculated. Wind-pressure data from building models tested in a boundary-layer wind tunnel have been used in this way in the development of equations describing infiltration in tall, open-plan, block-type buildings. It has been possible, with the aid of simplifying assumptions, to develop calculation procedures suitable for manual use [**11.1, 11.2**].

11.5 Computer Analysis of Tall Buildings

The mathematical representation of air leakage in buildings with analysis by computer has been used in a number of ways. Computers were used first to assist in the analysis and understanding of field measurements [**11.4**]. Later it became possible to determine from field measurements the effective leakage openings in the walls, floors, and shafts of several tall buildings [**11.5–11.7**]. Once such values were established it was possible to extend the mathematical modelling technique to the prediction of air leakage using wind-tunnel data and to develop manual calculation methods as already described. These latter studies also involved consideration of air leakage due to thermal forces and combined wind and thermal forces, which will now be discussed.

11.6 Stack Effect

The action that produces upward flow of heated gas in a chimney or stack can also produce pressure differences and air leakage in buildings. This is commonly referred to as stack effect or chimney effect, and can, under certain conditions, produce pressure differences comparable to those resulting from wind. A simple case of a heated five-storey building with negligible internal resistance to upward flow and with approximately equal leakage openings top and bottom is shown in Figure 11.2. Under winter conditions the outdoor air is more dense than the indoor air; the absolute pressure outdoors changes more rapidly with height, leading to pressure differences across the building enclosure. The location of the neutral pressure level at which indoor and outdoor pressures will be equal is governed by the size and location of leakage openings, since the flows in and out must be equal. Flow can take place when there are openings at different levels, or through a single large opening.

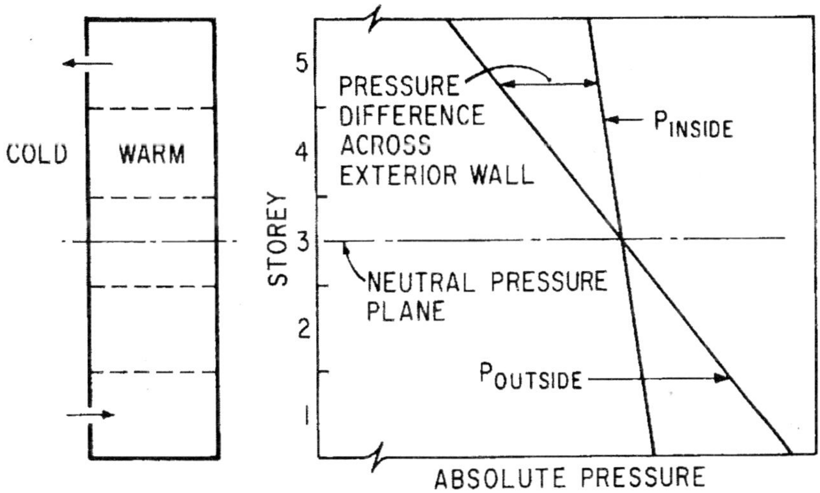

FIGURE 11.2 Stack effect in a building with no internal partitions. (From Wilson, A.G., and Tamura, G.T., Canadian Building Digest 104, DBR/NRCC, 1968, Fig. 1(b))

The pressure at a distance h from the neutral pressure level relative to the barometric pressure at the neutral pressure level is given by

$$p_s = gh(\rho_o - \rho_i) \tag{11.6}$$

where ρ_o and ρ_i are the mass densities outside and inside.

$$\text{But } \rho_o = \frac{p_t}{R_a T_o} \text{ and } \rho_i = \frac{p_t}{R_a T_i}$$

where p_t = barometric pressure, Pascals
R_a = gas constant = 287.1 J/kg • K for air
g = acceleration due to gravity = 9.81 m/s^2
T_o, T_i = outdoor and indoor temperatures, K

so that

$$p_s = g\frac{hp_t}{R_a}\left(\frac{1}{T_o} - \frac{1}{T_i}\right) \tag{11.7}$$

$$= 0.0342\, hp_t\left(\frac{1}{T_o} - \frac{1}{T_i}\right). \quad \text{(Pa)} \tag{11.8}$$

Values of p_s for various distances from the neutral pressure level and for different indoor and outdoor temperatures are shown in Figure 11.3. These may be compared to the dynamic pressures due to wind given in Table 11.1.

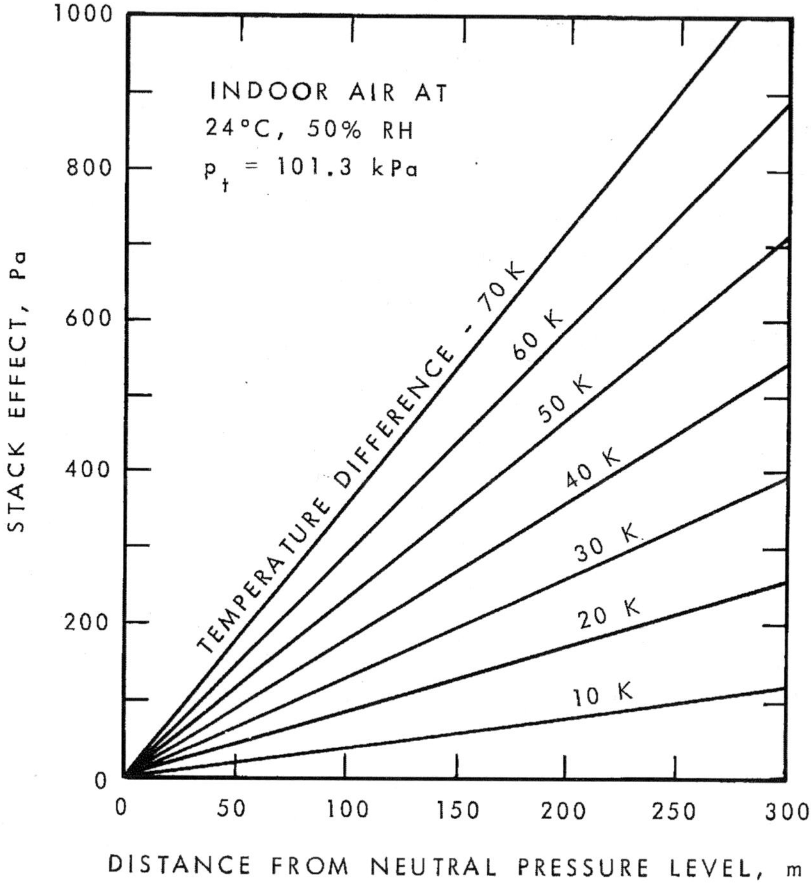

FIGURE 11.3 Stack effect in buildings.

Increasing or decreasing the size of both openings in the simple case shown in Figure 11.2 will change the leakage rate but not the neutral pressure level. When the openings are not equal, the location of the neutral pressure level may be found by equating the orifice flow through the two openings. Let the area of the lower opening be A_1, that of the upper opening be A_2, the vertical distance between the openings be H, and the distance from the centre of A_1 to the neutral pressure level be h. The equation for orifice flow from Chapter 7 can be written for mass flow as

$$Q\rho = \rho CA\,(2/\rho)^{1/2}(\Delta p)^{1/2} = CA\,(2\rho)^{1/2}(\Delta p)^{1/2} \tag{11.9}$$

and, equating flow in = flow out,

$$CA_1\,(2\rho_1)^{1/2}(\Delta p_1)^{1/2} = CA_2\,(2\rho_2)^{1/2}(\Delta p_2)^{1/2}$$

so that

$$\Delta p_1/\Delta p_2 = (A_2/A_1)^2\,\rho_2/\rho_1.$$

Substituting $\Delta p_1/\Delta p_2 = h/(H - h)$, and $\rho_2/\rho_1 = T_o/T_i$

and simplifying

$$h/(H - h) = (A_2/A_1)^2\, T_o/T_i$$

so that

$$h = \frac{H}{1 + (A_1/A_2)^2\, T_i/T_o} \quad \text{(m)} \tag{11.10}$$

where T_o, T_i are the outdoor and indoor temperatures in kelvins and $T_i > T_o$. The neutral pressure level is shown to be dependent mainly on the ratio of opening areas. The ratio T_i/T_o enters into the equation because the air flowing in is at a different density from that flowing out, owing to the difference in temperatures.

In a building with a completely airtight separation at each floor, the pressures will be as shown in Figure 11.4. There will be an exchange of air between each storey and the outdoors, but no vertical flow in the building. In a more realistic situation with a stair shaft or elevator shaft and openings in all floors and walls, the pressures and air flows will be as shown in Figure 11.5. The shaft shows the same pressure gradient as the building in Figure 11.2 since it has no obstructions to vertical flow. The pressure drop across each floor maintains an upward flow through floor openings while the pressure differences between each floor and the shaft, which increase with distance from the neutral pressure level, produce flows into the shaft from lower floors and from the shaft to upper floors.

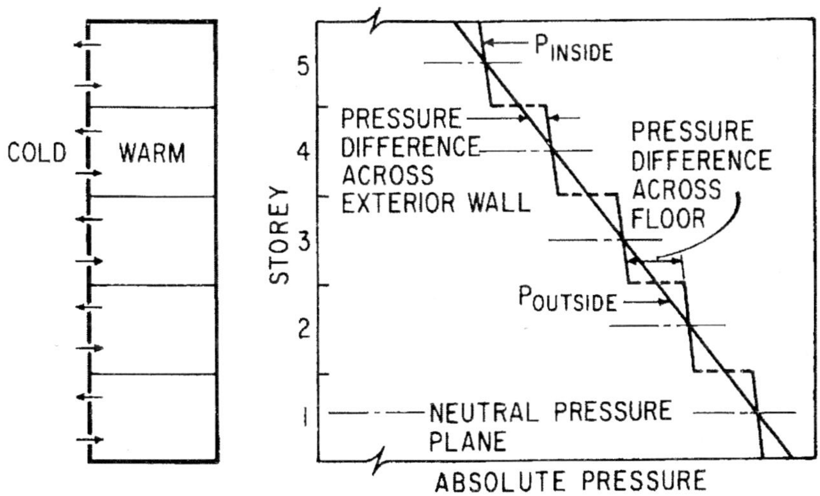

FIGURE 11.4 Stack effect in a building with airtight separation at each floor. (From Wilson, A.G., and Tamura, G.T., Canadian Building Digest 104, DBR/NRCC, 1968, Fig. 2)

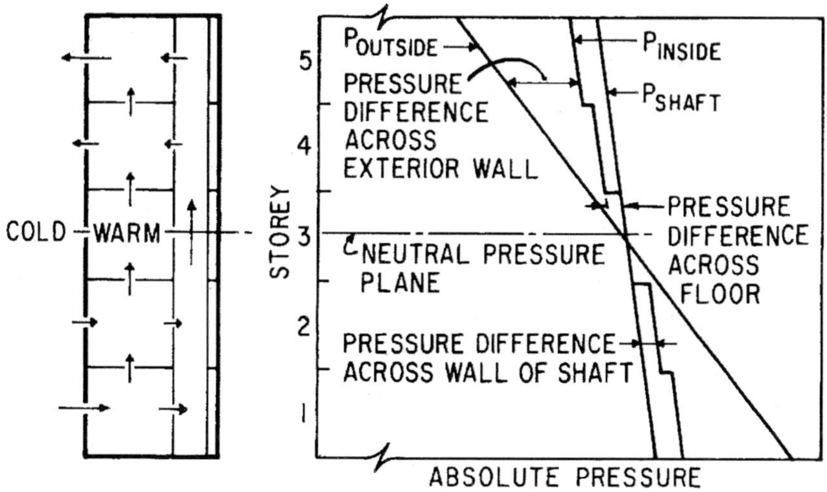

FIGURE 11.5 Stack effect for an idealized building.
(From Wilson, A.G., and Tamura, G.T., Canadian Building Digest 104, DBR/NRCC, 1968, Fig. 2)

11.7 Stack Effect in Real Buildings

Measurements reported for 23 buildings indicate that pressure differences across ground-floor entrances may be from 30 to 70% of the theoretical draft, or stack effect, calculated for the full height of the building [**11.8**]. On the assumption of a linear variation of interior pressure with height, this indicates neutral pressure levels varying from 30 to 70% of building height. Measurements taken on three tall office buildings indicated neutral pressure levels at 35 to 52% of building height [**11.4**]. Measurements in houses have indicated neutral pressure levels as high as 0.87 H to 1.08 H, where H is the height of the ceiling of a single-storey house above grade [**11.9**]. These high values are due, presumably, to the influence of the chimneys (which are, in effect, high-level openings) and to leakage openings in the ceiling. Lower values are to be expected in electrically heated houses without chimneys. Similarly, pressurization by air-handling systems in houses and larger buildings lowers the neutral pressure level, and increasing the rate of exhaust over supply raises it [**11.4, 11.10**].

Tall buildings usually have a large measure of uniformity from floor to floor in the size and distribution of openings. Consequently, pressure changes with height are likely to be linear or nearly so. Marked departures can be produced by unusual openings such as entrances at ground level or by any large variation from floor to floor in the rate of air supply or of exhaust by mechanical systems [**11.4, 11.11**]. When conditions are expected to be reasonably uniform, it is possible to relate the pressure across the exterior walls at any floor to the theoretical stack effect as found from Equation 11.8 by a coefficient, γ, known as the thermal draft coefficient. The pressure across exterior walls at any floor can be found readily when γ and the neutral pressure level are known.

Measurements on three office buildings showed that the actual pressure differences across walls varied from 63 to 82% of the theoretical stack effect. Using a mathematical model of a multistorey building in which leakage areas, pressures, and flow rates were represented, it was shown that the ratio varies with the relative tightness of exterior walls, floors, and shaft walls as given in Figure 11.6 [**11.4**].

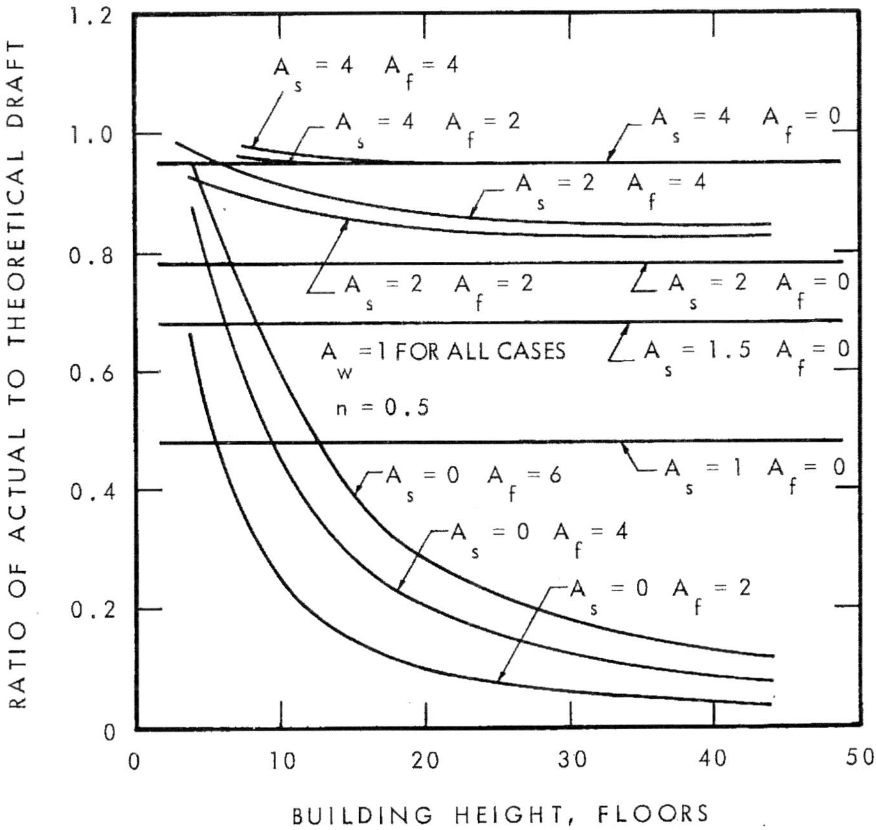

FIGURE 11.6 Effect of openings in walls, floors, and shafts on the ratio of actual to theoretical draft for various building heights.
(From Tamura, G.T., and Wilson A.G., *ASHRAE Trans.*, 1967, *73*, Part II, Fig. 3, p. II.2.3)[**11.4**]

When there are openings in the floors and exterior walls but none into the shaft, the resistance of the floors in series becomes cumulative with height, even with large floor openings, and the pressure difference across the enclosure is decreased. When there are leakage openings into the shaft but none in the floors and no resistance to flow in the shaft itself, the resistance of the flow path is independent of height, so the ratio of actual to theoretical draft is constant for any given value of A_s/A_w. Large values of the ratio, γ, between 0.8 and 1.0, are associated with leakage areas A_s and A_f from two to four times the equivalent leakage area for exterior walls. It follows that, to reduce pressure differences across exterior walls, the resistance to flow via vertical shafts must be increased. The high values of γ measured in the three buildings indicated that they had larger total openings in interior separations than in exterior walls.

11.8 Calculation of Infiltration Due to Stack Effect

A general equation for infiltration due to stack action in multistorey buildings has been derived with the aid of several simplifying assumptions [**11.6**]. The theoretical pressure difference across exterior walls due to stack action is given by Equation 11.8, which may be written as follows:

$$p_s = 0.0342\, h p_t \frac{\Delta T}{T_i\, T_o}. \quad \text{(Pa)} \tag{11.11}$$

For a uniform distribution of openings, the pressure difference across exterior walls is given by

$$\Delta p = \gamma p_s = 0.0342\, \gamma p_t h \frac{\Delta T}{T_i\, T_o} \tag{11.12}$$

where γ = thermal draft coefficient.

The rate of air flow through an incremental area of the exterior wall is given by

$$dQ = C\, dA\, (\Delta p)^n \tag{11.13}$$

where

dQ = leakage rate through an area dA of the exterior wall, m^3/s
C = flow coefficient, $m^3/(s \cdot m^2 \cdot Pa^n)$
n = flow exponent for wall openings.

Combining Equations 11.12 and 11.13 yields

$$dQ = C \left[0.0342\, \gamma p_t h \frac{\Delta T}{T_i\, T_o}\right]^n S\, dh \tag{11.14}$$

where S = perimeter of the building.

When the building has a constant cross-section and uniformly distributed openings, Equation 11.14 can be integrated over that portion of the height below the neutral zone to find the infiltration rate. The neutral pressure level can be expressed as a proportion β of the height H, so that integration must be carried out from $h = -\beta H$ to $h = 0$. The integration yields

$$Q = C\, S \left[0.0342\, \gamma p_t\, \Delta T/(T_i\, T_o)\right]^n \frac{(\beta H)^{n+1}}{n+1} \tag{11.15}$$

where Q is the total infiltration rate in m^3/s caused by stack action.

In a further simplification, by substituting

$$p_t = 101\,300 \text{ Pa},\ T_i = (273 + 21) \text{ K},\ n = 0.65, \text{ and } \beta = 0.5,$$

the relationship becomes

$$Q = 0.96\, C\, S\, (\gamma\, \Delta T/T_o)^{0.65}\, H^{1.65}. \quad (m^3/s) \qquad (11.16)$$

Example 11.1

Calculate the total infiltration due to stack effect for a building 30 × 45 × 60 m high for an indoor temperature of 21°C and an outdoor temperature of −27°C. The leakage coefficient for the exterior walls is $C = 0.93 \times 10^{-4}$, and $\gamma = 0.8$.

From Equation 11.16

$$\begin{aligned} Q &= 0.96\, C\, S\, (\gamma\, \Delta T/T_o)^{0.65}\, H^{1.65} \\ &= 0.96 \times 0.93 \times 10^{-4} \times 150\, (0.8 \times 48/246)^{0.65}\, 60^{1.65} \\ &= 3.44 \text{ m}^3/\text{s}. \end{aligned}$$

Ground floors of multistorey buildings may have many leakage openings because of entrances or for other reasons. Infiltration for them should be calculated separately.

11.9 Seasonal Influence on Stack Effect

The general pattern of air leakage through exterior walls due to stack effect involves air flow in at all openings below the neutral pressure level under winter conditions, and air flow out at all openings above. The pressure differences producing flow vary more or less linearly with distance from the neutral pressure level and with the difference between indoor and outdoor temperatures. The maximum temperature difference may be 50 K or more in winter in northern latitudes, and outdoor temperatures may be consistently lower than those indoors for periods of several months with some heating required over six to eight months of the year. During the summer the outdoor temperatures may exceed those indoors by 8 to 10 K on the hottest days. Under these conditions the flow due to stack effect is reversed, with flow inward at upper levels and flow outward below the neutral pressure level. The daily infiltration from stack effect is much less than in winter, owing to the smaller temperature differences. There may only be a few days each year when the outdoor temperature exceeds that indoors for a full 24 hours. The flow direction will reverse from day to night on those days when outdoor temperature exceeds the indoor temperature during the day but falls below it at night.

11.10 Air Leakage Due to Combined Effects

The pressure differences producing air leakage arise from wind, stack action, and pressurization by air-handling equipment. When two or all three of these effects occur at the same time, the total pressure differences are found from the algebraic sum of the pressure differences produced by each effect acting alone. The corresponding flows cannot be added in the same way but must be calculated from the sum of the pressures. The flow through a given opening can be written

$$Q = CA\,(\Delta p)^n.$$

The flows due to pressure differences Δp_1 and Δp_2 applied separately will be given by

$$Q_1 = CA\,(\Delta p_1)^n \text{ and } Q_2 = CA\,(\Delta p_2)^n.$$

The flow due to the combined pressure differences will be given by

$$Q_3 = CA\,(\Delta p_1 + \Delta p_2)^n.$$

It follows that

$$\frac{Q_3}{Q_2} = \frac{(\Delta p_1 + \Delta p_2)^n}{\Delta p_2{}^n}.$$

Substituting $\Delta p_1 = (Q_1/CA)^{1/n}$ and $\Delta p_2 = (Q_2/CA)^{1/n}$ and rearranging,

$$\frac{Q_3}{Q_2} = \left[1 + \left[\frac{Q_1}{Q_2}\right]^{1/n}\right]^n. \qquad (11.17)$$

From this relationship (shown graphically in Figure 11.7) it can be seen that for an exponent $n = 0.5$ the combined flow Q_3 for two equal separate flows Q_1 and Q_2 is 1.41 times one of them. The corresponding value for $n = 0.65$ is 1.57. For the case of $n = 1$, which applies to laminar flow, the combined flow is equal to the sum of the separate flows. When Q_2 is twice Q_1, the combined flow is 1.12 times Q_2 for $n = 0.5$ and 1.21 for $n = 0.65$.

This characteristic of flow through openings has great significance for the estimation of air leakage. Flows from individual effects cannot be added except in cases when the exponent n approaches unity, but it may often be sufficiently accurate for some purposes to consider only the larger of two separate flows. It follows also that if the pressure difference is reduced, the flow will be reduced less than in proportion. For example, with $n = 0.5$, reducing the pressure by half will result in a flow that is 71% of the original.

Since the pressure patterns for each effect are different, changing their proportions does not produce the same result on all surfaces. Flow may be decreased or reversed in direction for some openings and increased for others for an increase of wind relative to stack effect.

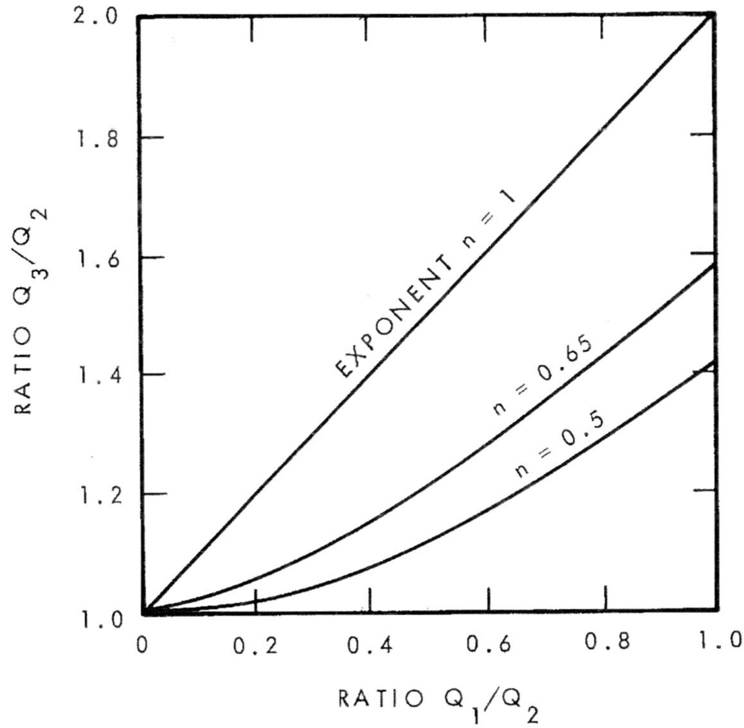

FIGURE 11.7 Combined flow through opening due to sum of pressure differences which produced flow Q_1 and Q_2 separately.

Using computer-model studies of multistorey buildings, it is possible to determine the combined flow for an entire building in terms of the wind and stack action calculated separately. In one study [**11.2**] the relationship was given by the empirical equation

$$\frac{Q_{ws}}{Q_{lrg}} = 1 + 0.24 \left[\frac{Q_{sml}}{Q_{lrg}}\right]^{3.3} \tag{11.18}$$

where Q_{ws} = infiltration rate from combined effects
Q_{lrg} = larger of Q_w and Q_s
Q_{sml} = smaller of Q_w and Q_s

Q_{ws} is shown to be 24% greater than either Q_w or Q_s when $Q_w = Q_s$. This may be compared with the value of 57% for $n = 0.65$ from Figure 11.7. In other words, the total combined flow is roughly equal to the larger of the two flows calculated separately. Similar results were obtained from a second study [**11.1**], but there it was shown that wind direction was also a factor.

11.11 Relative Magnitude of Wind and Stack Action

Stack effect for a building 100 m high operating at 50 K temperature difference with $\gamma = 0.8$ will be about 95 Pa. This pressure difference will be effective in producing infiltration on all four sides at ground level, reducing with height to zero at the neutral pressure level; substantial pressure differences will be maintained over much of the heating season. The corresponding wind pressures for a location such as Ottawa are highly variable with time. The average wind speed is about 15 km/h, producing a pressure of 10 Pa at 100 m at an urban site. The maximum hourly average reached once in ten years is 300 Pa, and gust pressures may be twice as great. The effective pressure difference across the windward wall may be taken to be roughly equal to the dynamic pressure, and on this basis will be up to three times greater than the maximum stack effect for one hour in ten years, but only 1/10 as much if based on the long-time average wind speed. In areas with severe winter weather, stack action is likely to produce much more infiltration than wind for multistorey buildings during the heating season.

The significance of such comparisons depends on whether one is interested in average or in maximum results. Seasonal energy consumption may be related to average infiltration rates, but the heating system may have to be sized to care for the maximum average rate to be expected over a 12- to 24-hour period. For tall buildings, the infiltration rate based on stack action alone may be all that is required for many purposes, and it is less troublesome to estimate than infiltration from wind action. This is likely to be less applicable to low buildings since stack effect is directly dependent on height. Despite this, it has been concluded that even in single-storey dwellings for which the maximum stack effect may be no more than 10 Pa, it may have a greater influence than wind on the average infiltration rate [**11.9**].

11.12 Winds on Low Buildings

The strong-wind profiles that form the basis for estimating wind pressures of structural significance on buildings provide the means of relating wind speeds at various heights at the building site for different terrain conditions to the wind speed recorded at the nearest weather station. These have relevance to tall buildings that project well above their neighbours. It has not been shown that the same procedures are valid for low buildings, which may be surrounded by taller buildings, or for average rather than strong winds. It appears, however, that they may be applicable in part to suburban terrain consisting of many closely spaced low buildings, such as houses.

The data from Table 10.1 can be used with the method in Example 10.2 (Chapter 10) to show that the wind speed over suburban terrain is given by

$$V = 0.37\ V_s\,(Z)^{0.25} \qquad (11.19)$$

where V_s = wind speed at 10 m at the weather station over open terrain and Z is the height in metres at the suburban site. Measurements in Ottawa showed that the mean of many readings at a site in one housing development was 0.65 times the observed wind speed 8 m above ground at the airport 10 km away. The value was consistently about 0.56 for another site [**11.9**]. The value from Equation 11.19 corrected to 8 m at the weather station is 0.64, indicating that adjustments on the basis of strong wind-speed profiles may provide reasonable answers for such situations. The problem of relating house air-leakage rates to wind speed at the site still remains.

As a matter of interest, the corresponding equation for urban terrain with many tall buildings is

$$V = 0.178\ V_s\,(Z)^{0.36}. \tag{11.20}$$

The wind speed at 100 m is 0.94 times the wind speed at 10 m at the weather station. The value for 8 m is 0.4 times the wind speed at 10 m at the weather station, but is of no more than theoretical interest since it refers to a height well below the tops of the surrounding buildings where it is not possible to establish the general wind speed.

11.13 Measurement of Air Leakage in Houses

A method of measuring air leakage using a tracer gas mixed with the air in a space was developed in the U.K. in 1950 [**11.12**]. Helium can be used as the tracer in concentrations of about 0.5% in the house atmosphere. Other gases such as hydrogen and nitrous oxide can also be used. Portable detectors sensitive to the changing thermal conductivity of air with changing concentrations of tracer gas were developed [**11.13**], and have been used in a number of studies [**11.14–11.17**].

The time-decay of concentration of the tracer gas by dilution due to infiltration is given by

$$KC = -V\frac{dc}{d\theta} \tag{11.21}$$

where K = average volume of air infiltration per unit time over the time interval
V = volume of enclosure
C = concentration of tracer gas at time θ.

Integration of Equation 11.14 yields

$$K\theta/V = \log_e(C_o/C) \tag{11.22}$$

where C_o = concentration at $\theta = 0$.

Infiltration may be calculated from the slope of a plot of measured concentration against time on semi-log graph paper.

Extensive measurements made in the course of all four of the studies yielded a wide range of results. All reflected the influence of wind and indoor-outdoor temperature differences, but it was difficult to compare findings. On-site wind measurements, for example, were not all made in the same way. For purposes of comparison, attempts were made to adjust results to a common basis of wind of 16 km/h and a 22 K temperature difference acting together [**11.15, 11.16**]. On this basis, values for three of the studies ranged from 0.37 to 0.99 air changes per hour (ac/h) for electrically heated houses of varied constructions, 0.11 to 0.34 ac/h for two tightly constructed houses, and 0.4 to 0.6 ac/h for two research houses. Wind was generally considered more significant than temperature difference.

The fourth study, conducted in Ottawa, yielded winter results from 0.25 to 0.6 ac/h for temperature differences up to 33 K and winds to 16 km/h. With steady operation of the oil furnaces, the air change based on air going up the chimney was 0.38 when the temperature difference was 22 K and the wind 16 km/h. Stack effect was shown to be more significant than wind in winter, but summer rates with wind at 13 km/h were as high as 0.23 ac/h. A more detailed analysis of the Ottawa houses and an attempt to develop an improved calculation procedure for infiltration have been made [**11.9**].

A subsequent review of the data from these and several other studies has shown a relationship between leakage, stack effect, and wind [**11.18**]. The equation of a form suggested earlier [**11.16**] is

$$Q = A + B(\Delta t) + DV \qquad (11.23)$$

where A, B, and D are coefficients, Q is the air leakage expressed as air changes per hour, Δt is the temperature difference in kelvins, and V is the wind speed in km/h. The values proposed for A, B, and D are 0.1, 0.011, and 0.009 for tightly constructed houses and 0.1, 0.22, and 0.018 respectively for loosely constructed houses. The corresponding values of Q for 22 K and 16 km/h are 0.40 and 0.87 ac/h. (See also **11.19, p. 22.13.**) For V in m/s multiply D by 3.6.

Example 11.2

Calculate the air leakage in m³/s for a single-storey house with basement, 7 m × 14 m in plan, if the infiltration rate is 0.4 ac/h.

Assume the effective height from basement floor to first floor ceiling to be 5 m.

House volume, including basement, is

$$V = 14 \times 7 \times 5 = 490 \text{ m}^3.$$

$$\begin{aligned}\text{Infiltration} &= 0.4 \times 490 = 196 \text{ m}^3/\text{h} \\ &= 190/3600 = 0.054 \text{ m}^3/\text{s or } 54 \text{ L/s}.\end{aligned}$$

11.14 The Crack Method of Estimating Infiltration

A simple approach based on values of air change developed through experience has often been used in the estimation of heating load for individual rooms. Such values are given in Table 11.2. This approach recognizes windows and doors as important factors determining leakage. Another approach, known as the crack method, has been widely used. Calculations are made of the air leakage through cracks around windows and doors under appropriate pressure differences. Leakage rates are given in litres per second per metre of crack length at some selected reference pressure. Crack length is taken as the perimeter for windows and doors, plus the length of meeting rail in the case of double-hung and sliding windows. A pressure difference of 75 Pa is commonly used as a test condition and as a basis for comparison of the air leakage of windows and doors.

TABLE 11.2

Air Changes Occurring under Average Conditions in Residences, Exclusive of Ventilation
(From *ASHRAE Handbook 1981 Fundamentals*, Table 2, p. 22.8) [**11.19**]

Kind of room	Air changes per hour
No windows or exterior doors	0.5
Windows or exterior doors on one side	1.0
Windows or exterior doors on two sides	1.5
Windows or exterior doors on three sides	2.0
Entrance halls	2.0

For windows with weatherstripping or storm sash use two-thirds these values.

Some values for double-hung wood windows are given in Table 11.3. The values for frame leakage are evidence that the crack between sash and frame is not the only leakage path. A review of window-leakage information and additional test data were published in 1965 [**11.20**]. Limits on air leakage are incorporated in many specifications and standards. Values range from 0.58 to 1.16 L/s • m at 75 Pa for most window types. The leakage characteristics of windows may also be presented in terms of an equivalent orifice area, calculated on the basis of Equation 11.2. The flow exponent for windows is commonly taken as 0.65.

When storm windows are used with openable prime windows, the air leakage of the combination may be calculated using the relationship

$$Q_c = Q_p \left[\frac{1}{(C_p/C_s)^{1/n} + 1} \right]^n \qquad (11.24)$$

where

Q_c = flow rate for combined prime plus storm window
Q_p = flow rate for prime window
C_p = flow coefficient for prime window
C_s = flow coefficient for storm window

TABLE 11.3

Infiltration through Locked Double-Hung Wood Windows (Litres per second per metre of crack length for 75 Pa pressure difference)

(From *ASHRAE Handbook 1981 Fundamentals*, Table 3, p. 22.9) [**11.19**]

Condition of window	L/s • m at 75 Pa
Leakage between sash and frame	
—No weatherstrip, loose fit	3.9
—No weatherstrip, average fit	1.5
—Weatherstrip, loose fit	1.5
—Weatherstrip, average fit	0.8
Leakage between frame and wall	
—Frame in masonry wall, not caulked	0.9
—Frame in masonry wall, caulked	0.2
—Frame in wood-frame wall	0.7
Average window has 1.5 mm crack and 1.2 mm clearance.	

Storm windows will not greatly affect the overall air leakage of the combination if they have air-leakage rates when tested alone that are greater than those of the prime window. Condensation on the inside surface of the storm window is reduced if the storm window is relatively leaky [**11.21**], and it is preferable to select a prime window that will be the main contributor to airtightness.

Example 11.3

Calculate the air leakage through a double-hung wood window in a frame with inside dimensions 800 × 1500 mm for a pressure difference of 50 Pa if the leakage rate measured at 75 Pa was 0.8 L/s • m.

The unit leakage rate may be corrected to 50 Pa as follows:

$$C_{50} = C_{75}\left[\frac{50}{75}\right]^{0.65} = 0.8 \times 0.77 = 0.62 \text{ L/s} \cdot \text{m}$$

The crack length is 3 × 800 + 2 × 1500 = 5400 mm
The leakage rate is (5400/1000) 0.62 = 3.35 L/s.

Example 11.4

Find the equivalent orifice area for the window in Example 11.3. From Equation 11.2

$$Q = CA\,[(2/\rho)\,(\Delta p)]^{1/2}$$

where $Q = 3.35$ L/s, $C = 0.6$, $\rho = 1.29$ kg/m³ for dry air at 0°C, and $\Delta p = 75$ Pa, and substituting

$$10^{-3} \times 3.35 = 0.6\,A\,[(2/1.29)\,(75)]^{1/2}$$

and solving

$$A = (10^{-3} \times 3.35)/(0.6 \times 10.8) = 5.2 \times 10^{-4}\,\text{m}^2 = 5.2\,\text{cm}^2.$$

Example 11.5

Calculate the leakage rate if a storm window is added to the window in Example 11.3. The storm window has a leakage coefficient of 1.2 L/s • m at 75 Pa. Substituting in Equation 11.24,

$$Q_C = 3.35\left[\frac{1}{(0.8/1.2)^{1/0.65} + 1}\right]^{0.65}$$

$$= 2.54 \text{ L/s.}$$

It is recognized when using the crack method that air entering on the windward face will leave on the leeward one. In estimating the air leakage for individual rooms with one exposed wall, the total crack length should be considered. If two or more walls are exposed it is recommended to use the wall having the crack that will result in the greatest air leakage but not less than half the total crack [**11.19, p. 25.10**]. This procedure for individual rooms may result in a total infiltration that is too great for the building as a whole, since infiltration due to wind is unlikely to occur over more than half of the total crack length for the building at any one time.

11.15 Air Leakage through Building Enclosures

Laboratory leakage tests on wall panels carried out in much the same way as for windows have produced information such as that given in Table 11.4. Plain brick walls have relatively high air leakage, but plastering reduces this to a low value. Much of the information on leakage through building enclosures has been obtained from pressurization tests.

Pressurization tests of houses have given leakage rates equivalent on the average to an orifice area of 0.14 m². These tests demonstrated that substantial leakage occurs through walls; windows and doors contributed only 20% of the total. Walls with stucco as the exterior finish had the lowest leakage rate at 0.76 to 1.0 L/s • m². Walls of brick veneer or metal siding gave values from 3.4 to 6 L/s • m², and ceilings from 1.4 to 5 L/s • m², all these values being for a pressure difference of 75 Pa [**11.19, p. 22.11; 11.22**].

Leakage of a nine-storey building with masonry walls and operable metal windows was 10 L/s • m² at 75 Pa pressure difference, with windows and doors estimated to account for about one-quarter of the total [**11.23**].

TABLE 11.4

Infiltration through Walls
(Litres per second per square metre
for 75 Pa pressure difference)
(From *ASHRAE Handbook 1981 Fundamentals*, Table 5, p. 22.11) [**11.19**]

Type of wall	L/s • m² for 75 Pa
Brick wall, porous brick, poor workmanship	
—216 mm brick, plain	2.0
—216 mm brick, plaster on brick	0.02
—330 mm brick, plain	1.7
—330 mm brick, plaster on brick	0.008
—330 mm brick, furring, lath, and plaster	0.06
Frame wall	
—With lath and plaster	0.025

Tests on eight multistorey office buildings with sealed glazing and panels of concrete or steel provided the basis for establishing leakage coefficients for tight, average, and loose walls as 0.3×10^{-4}, 0.9×10^{-4} and 1.8×10^{-4} $m^3/s \cdot m^2 \cdot Pa^{0.65}$. When expressed as leakage rates for a pressure difference of 75 Pa for comparison purposes, they become 0.5, 1.5 and 3.0 $L/s \cdot m^2$. Standards of the National Architectural Metal Manufacturers for curtain walls call for the leakage to be not more than 0.3 $L/s \cdot m^2$ for a pressure difference of 75 Pa, yielding a flow coefficient of 0.18×10^{-4} $m^3/s \cdot m^2 \cdot Pa^{0.65}$ [**11.19, p. 22.10; 11.6**].

11.16 Leakage through Floors, Shafts, and Entrances

An average air-leakage coefficient for floors in multistorey buildings has been given as 5.3×10^{-5} $m^3/s \cdot m^2 \cdot Pa^{0.5}$. Values are stated to be 0.075 $m^3/s \cdot Pa^{0.5}$ per car for elevator shafts and 0.03 $m^3/s \cdot Pa^{0.5}$ for stair shafts [**11.1**]. Additional values have been published for leakage rates through elevator-shaft and stair-shaft walls [**11.19, p. 22.11**]. Shafts for stairs and elevators have substantial leakage openings, notably through doors at each floor, and so provide paths for marked upward air flow under winter conditions. They also provide paths for distribution of air carrying smoke from a fire on a floor below the neutral pressure level to all floors above, so that their leakage characteristics become of considerable importance [**11.24, 11.25**].

Exterior doors are critical elements of the building enclosure so far as air leakage is concerned. Stack effect in tall buildings in winter imposes large pressure differences across these doors and, if they are frequently used, can lead to substantial air leakage and correspondingly high heating loads for entrances and adjacent spaces [**11.3; 11.19, p. 22.12**]. A vestibule entrance can reduce air infiltration by a factor of 2 over a single-bank swinging door, while a revolving door may reduce it by a factor of 15 or more.

11.17 Exfiltration and Partial Air Leakage

Attention is usually focused on infiltration as the agent of heat loss in winter or heat gain in summer. The heat is actually carried off by exfiltration in winter, but the difference in heat content of air involved in infiltration and exfiltration determines the heat loss or gain at a given flow rate. Exfiltration can carry moisture as well as heat. Under winter conditions, warm, moist indoor air meets progressively colder surfaces on its way outdoors and, when cooled sufficiently, will deposit moisture as condensation. The need to control this condensation is an important reason to control air leakage in buildings. Exfiltration takes place along many unintentional paths. In houses, for example, openings around electric outlets and lighting fixtures and around plumbing stacks, pipes, and chimneys may allow air leakage through air spaces in interior walls, causing condensation in attic spaces. Such moisture problems will be discussed in Chapter 12.

Air leakage need not take place entirely through a construction in order to be significant. Circulation of air between an air space in a wall and the indoors through leakage openings in the interior cladding at two different levels may be sufficient to influence both heat and condensation in the wall during the winter. Partial air leakage may also take place between an air space and the outdoors, transferring heat and moisture. Air leakage between air spaces on either side of an insulating layer, promoted by stack effect, as discussed in Chapter 8, can produce increased heat transfer. Partial air leakage may be beneficial if it assists in the transfer of moisture to limit condensation or to promote drying.

11.18 Chimney Draft

The draft in domestic chimneys results from temperatures which are generally higher than in buildings in a range up to 200°C or more, depending on furnace operation, extent of mixing with house air at the barometric damper or draft hood, and cooling of gases by heat loss from the chimney. House pressure must be higher than chimney pressure at any opening into the chimney if spillage of gases is to be avoided. Draft is thus dependent on conditions in the house as well as in the chimney. Draft failure and flow reversal can occur under certain conditions [**11.26**].

A house with an inside chimney serving a furnace and a second chimney on an outside wall serving a fireplace or stove is shown in Figure 11.8. The theoretical draft for either chimney is given by the difference between the line of outdoor pressure AB, and the line representing the pressure in the chimney. Line *AC* is for a chimney at room temperature and line AD is for a chimney that is receiving hot furnace gases. Chimney X, much of which is indoors, is unlikely to be cooled below indoor temperature even under no-fire conditions unless flow reversal takes place, bringing cold outdoor air into it. Chimney Y is exposed to the outdoors on three sides and can be cooled under no-fire conditions by heat loss to a temperature approaching that of outdoors, for which the limiting pressures are given by line AB.

If the chimneys were the only openings into the house and were at indoor temperature, the house pressure would be given by *AC*. However, in practice there will be other openings, which will supply the makeup air required to balance the upward chimney flow. The actual house pressure line will be at some position A′C′ with a neutral pressure level that may be near ceiling height because of the chimneys. The effective draft at any level for either chimney is given by the difference between the house pressure line A′C′ and the chimney pressure line for the appropriate chimney temperature. Draft will fail when the pressure in the chimney exceeds that in the house at some critical level, such as level 1 for Chimney X or level 2 for Chimney Y.

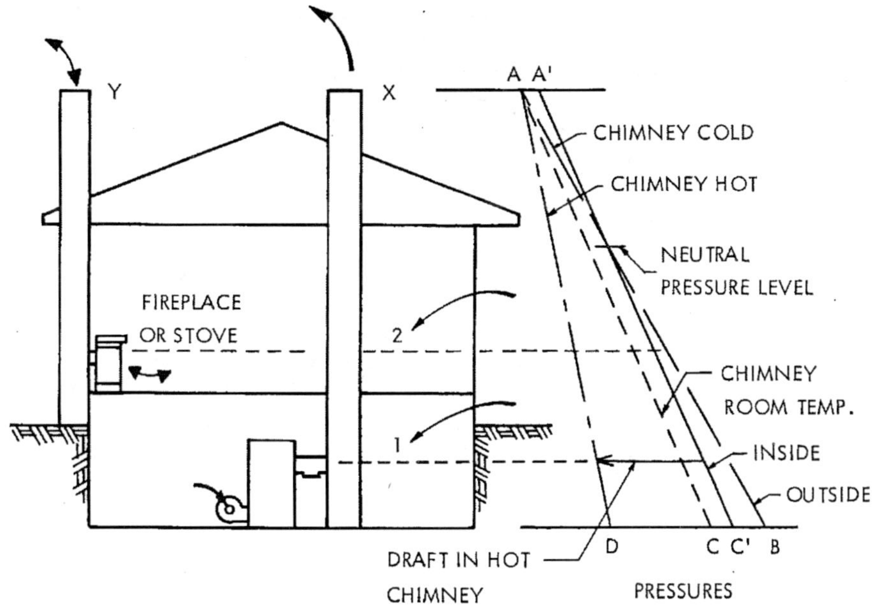

FIGURE 11.8 Draft in chimneys.

Draft failure cannot occur for any level above the neutral pressure level for the house, since no pressure can be developed in the chimney that is higher than that given by line AB corresponding to outdoor temperature. Failure can occur below the neutral pressure level if the chimney gases are cooled sufficiently below indoor temperature, and can occur more readily if the neutral pressure level is high.

Chimney X is unlikely to experience draft failure. Chimney Y, being more exposed, may do so. If Chimney X is operating it will draw air from the house, raising the neutral pressure level and increasing the possibility of failure in Chimney Y. Opening a door or window will have the opposite effect. Adverse wind pressures at the top of a chimney can also contribute to loss of draft. Where there is a fireplace served by an outside chimney and a continuously operated stove or furnace on another chimney, a potential carbon-monoxide hazard arises when the fire in the fireplace is allowed to die down.

A distinction may be made between draft failure and inadequate draft; the latter is often a problem with fireplaces. A mean flow velocity into the open face of not less than 0.24 m/s may be required to prevent smoke from entering the room [**11.27**]. This amounts to a flow rate of 168 L/s for a fireplace opening 900 × 750 mm, which must be offset by a leakage rate equivalent to 1.4 ac/h for a house with a volume of 450 m³.

11.19 Natural Ventilation

Natural ventilation is air leakage resulting from the intentional provision of openings—for example, the use of open windows in houses in summer to carry off heat and moisture and to increase air movement. The principles governing natural ventilation are the same as those already discussed in connection with air leakage [**11.19, p. 21.11**]. The forces to be exploited are wind and stack effect. Natural ventilation is widely used in hot climates when air-conditioning equipment is not installed. It is also an alternative to mechanical ventilation in industrial situations [**11.28**].

11.20 Ventilation Requirements

Ventilation is defined as the process of supplying or removing air by natural or mechanical means to or from any space. Ventilation air may be used to limit by dilution any contaminants added to the air in a space; to replenish oxygen removed by respiration; as a carrier for heat and moisture; or to replace air withdrawn from a space for combustion or other purposes. The rate of ventilation may be determined by the need to maintain certain levels of air velocity or air circulation. Air used for ventilation may consist of outdoor air, or air recirculated from the space, or both, with or without treatment to improve its quality.

The rate at which ventilation air must be supplied to serve as a diluent or as a carrier of heat or moisture can usually be found from relatively simple calculations based on the rate of release of the unwanted product and the difference in content between the air supplied and the air exhausted from the space. Section 6.6 and Figure 6.4 provide an example of situations involving heat and moisture. Properties and tolerable limits of air contaminants have been set out [**11.19, Chaps. 11 and 12**].

11.21 Ventilation for Human Occupancy

People are often the principal source of the heat, moisture, odours, and respiratory wastes that determine the minimum requirements for ventilation. The respiration rate for a seated person at rest will be about 0.15 L/s. The oxygen in respired air will be reduced by about 5% to 16% by volume, and the carbon dioxide will be increased from 0.03% in fresh air to 4%. Respired air becomes mixed with room air, which is replaced in turn at a given rate by ventilation.

When a room is first occupied, the concentration of contaminants rises until a steady state is reached at which the terminal values are determined by the balance between respiration rate and ventilation rate. The time taken to reach these terminal values (sometimes many hours) also depends on the room volume per person [**11.29**]. A ventilation rate of ten times the respiration rate will reduce the changes in carbon-dioxide and oxygen levels in a room to about one-tenth of those for respired air. An oxygen content as low as 16% and a carbon-dioxide content as high as 0.5% are acceptable. Thus, a ventilation rate of 1.5 L/s per person seated at rest is needed to limit the increase in carbon dioxide to 0.5%. The oxygen depletion is much less critical. The volume rate of respiration is related to metabolic rate or physical activity, and the minimum ventilation rate per person must be increased in proportion by four times or more for people engaged in heavy work. The need to control odours may be a limiting factor in determining the minimum ventilation rate.

TABLE 11.5

Ventilation Requirements for Occupants
(Selected items from ASHRAE Ventilation Standard 62-1973)
(From ASHRAE Standard 62-73) [**11.30**]

Type of occupancy	Estimated number of persons per 1000 ft² (93 m²)	Required ventilation per person	
		Minimum L/s	Recommended L/s
Residential			
Single-unit dwellings			
Living areas, bedrooms	5	2.5	3.5–5
Kitchens, bath, toilet rooms*	—	10	15–25
Multiple-unit dwellings			
Living areas, bedrooms	7	2.5	3.5–5
Kitchens, bath, toilet rooms*	—	10	15–25
Garages (L/s • m²)**	—	7.5**	10–15**
Commercial			
Sales floors, lower	30	3.5	5–7.5
Sales floors, upper	20	3.5	5–7.5
Dining rooms	70	5	7.5–10
Cocktail lounges	100	15	17.5–20
Hotel, motel bedrooms	5	3.5	5–7.5
Auditoriums (no smoking)	150	2.5	2.5–5
Auditoriums (smoking)	150	5	5–10
General office space	10	7.5	7.5–12.5
Office conference rooms	60	12.5	15–20
Institutional			
School classrooms	50	5	5–7.5
School auditoriums	150	2.5	2.5–3.8
School gymnasia	70	10	12.5–15
School libraries	20	3.5	5–6

*Installed capacity for intermittent use
**L/s • m² of floor area.

The ASHRAE Ventilation Standard 62-1973 [**11.30**] set out ventilation requirements for spaces intended for human occupancy based on a minimum outdoor air requirement of 2.5 L/s per person. Minimum and recommended values for various occupancies were given. Values for selected occupancies from the standard are shown in Table 11.5. When values higher than 2.5 L/s per person were specified, the amount over 2.5 L/s or over 15%, whichever was greater, could consist of treated, recirculated air. These reductions in the amount of outdoor air became attractive when energy had to be conserved, since the outdoor air brought in by ventilation or by air leakage represented a substantial air-conditioning load.

The sharp rise in oil prices beginning in 1973 led to a growing concern about energy costs and energy conservation in general. The energy used in heating, cooling, and ventilating buildings represented some 30 to 35% of the total national energy consumption. Attention was soon directed to heat transmission losses which could be reduced by insulation, and to the energy required in the heating and humidification of air for ventilation and from infiltration which could be reduced by designing tight buildings to reduce air leakage [**11.31**], and lowering ventilation rates.

Proposals to reduce ventilation rates and reports of difficulties arising from putting them into practice produced countervailing concerns over the quality of indoor air. Concern grew over the hazards to health posed by tobacco smoke and the many chemical contaminants (often present in quantities as low as one part per million) that were being introduced into the air in homes and workplaces. The combined result of these influences has been a rapid increase in the development of measurement techniques, standards, and practices in both the public and the private sectors covering a wide range of subjects related to indoor air quality. This trend is likely to continue for some time to come [**11.32–11.36**].

A revision of the ASHRAE Standard 62-73 did not appear until 1981 [**11.37**]. This standard should be carefully studied in its entirety. Briefly, it confirms a basic minimum ventilation rate of 2.5 L/s per person, adjusted upward for various special situations and increased as much as five times for rooms where smoking is permitted. Increased requirements for persons developing high metabolic rates are recognized. In some situations, such as residential occupancies, rates are specified per room rather than per person. There are provisions to recognize measured concentrations of contaminants in both indoor and outdoor air as a basis for control of air quality by suitable conditioning.

11.22 Air Leakage for Ventilation

The infiltration rates for houses were shown earlier to produce about 0.4 ac/h under winter conditions with outdoor temperatures down to 0°C. Summer rates can be 0.2 ac/h or less. Poorly constructed houses may experience winter rates of 1 ac/h or more. These rates should be compared with the ventilation requirements for human occupancy.

The volume of a single-storey house with basement 120 m^2 in plan is about 580 m^3. An air-change rate of 0.4 means a flow rate of 64 L/s. This will provide sufficient air for ventilation at the minimum rate of 2.5 L/s per person for 25 people, or 13 L/s per person for a family of five, which is well over the recommended value of 5 L/s per person. Infiltration, under heating conditions, could thus be adequate to maintain satisfactory air quality.

Infiltration and ventilation combined can never be less than the rates at which air is lost from the house. Bathroom and kitchen exhaust fans and clothes dryers vented to outdoors may draw off 50 L/s or more when operating. A furnace burning oil at the rate of 3.8 L/h requires about 37 L/s of air for combustion, and the draft-control device may allow additional flow up the chimney for a total of 60 L/s. The same is true for a gas furnace of comparable output. Fireplaces may require air at the rate of 160 L/s when operating. These intermittent requirements have usually been met by air leakage supplemented by vent openings and, on occasion, by opening windows or doors as required. Any excess of supply over requirements is undesirable since it adds substantially to the heating load.

When the air-leakage rate falls below 0.4 ac/h, the moisture given off within the house may raise the relative humidity to levels at which serious window condensation takes place in winter. Increased ventilation is one way of reducing this problem. In summer, when outdoor air is required for cooling, large air flow rates aided by window or attic fans of 2000 L/s or more may be in order.

The operation of warm-air heating systems involves an air circulation within a house at the rate of 2 to 4 ac/h. Further mixing goes on in individual rooms so that both infiltration and contaminants are quickly distributed throughout each room and eventually throughout the house. A similar situation exists in most other buildings equipped with central air-supply and exhaust systems. The rate of air supply, consisting of the required fresh air mixed with a larger quantity of air being recirculated, varies from 2 ac/h for ventilating systems to as much as 10 ac/h for air-conditioning systems. The high rate for air conditioning is required as a carrier of heat and moisture and to promote good air distribution.

The calculation of Example 11.1 yielded a value of 3.44 m^3/s for infiltration due to stack action for a building 30 × 45 × 60 m, having a volume of 81 000 m^3. On the assumption of a combined air-leakage rate for wind and stack effect of 4.4 m^3/s, the infiltration rate is 0.2 ac/h. The minimum ventilation requirement for an office building (from Table 11.5) is 7.5 L/s per person, of which 2.5 L/s must be outdoor air. The assumed density of occupants of 10 per 93 m^2 over a floor area 30 × 45 m yields 145 occupants per floor. The volume per floor at 3.5-m storey height is 4725 m^3, and the fresh-air requirement amounts to 0.28 ac/h. The infiltration rate is thus about 70% of the building fresh-air requirement. Air enters at particular locations in the building and must pass through occupied space before reaching the central plant to be redistributed to other locations not receiving infiltration. It is questionable whether it can be considered "fresh air" for purposes of ventilation. If it can, the fresh-air intake can be reduced by the amount of the infiltration. The infiltration rate is variable, however, with wind and temperature difference, and the fresh air brought into the central plant has to be adjusted accordingly if no more than the required amount is to be brought in. There may be little or no requirement for fresh air in office buildings outside working hours.

11.23 Energy Cost of Infiltration

There is a substantial energy cost associated with infiltration and ventilation. The sensible heat required to bring outdoor winter air to room temperature is

$$H_s = Q\,\rho\,c_p\,(t_i - t_o) \qquad \text{(W)} \qquad (11.25)$$

where H_s = heat required, W
Q = rate of air flow, L/s
ρ = mass density of air, kg/m^3
c_p = specific heat of air, kJ/kg • K
t_o, t_i arc outdoor and indoor temperatures, °C
For ρ = 1.20 kg/m^3 and c_p for air = 1.005 kJ/kg

$$H_s = 1.21\,Q\,(t_i - t_o) \qquad \text{(W)} \qquad (11.26)$$

When the air that comes in by air leakage or by ventilation must be humidified to maintain room conditions, the latent heat that must be provided is given by

$$H_\ell = Q\,\rho\,(W_i - W_o)\,h_{fg} \qquad (11.27)$$

where H_ℓ = heat required, W
Q = air flow rate, L/s
W_i = humidity ratio, kg/kg of dry indoor air
W_o = humidity ratio, kg/kg of dry outdoor air
h_{fg} = latent heat of evaporation of water, kJ/kg
For air density ρ = 1.20 kg/m^3 and h_{fg} = 2465 kJ/kg

$$H_\ell = 3000\,Q\,(W_i - W_o) \qquad (11.28)$$

Example 11.6

Find the sensible and latent heat losses due to infiltration at the rate of 64 L/s in a house that is maintained at 23°C, 30% rh when the outdoor condition is 0°C, 80% rh.

From the psychrometric chart (Figure 6.1), W_i = 0.0052, W_o = 0.003
From Equation 11.28,

$$H_\ell = 3000 \times 64 \times (0.0052 - 0.003) = 422\ W$$

From Equation 11.26,

$$H_s = 1.21 \times 64 \times (23 - 0) = 1781\ W.$$

The heat required in humidification is 24% of the sensible heat required to warm the dry air alone. When ventilation is required to carry off moisture, this would not be counted as a heat loss.

Calculations have been presented showing the proportions of heat loss involved in ventilation, infiltration, and heat transmission through walls in multistorey buildings [**11.6**]. The infiltration loss increases with height. In a 60-m office building, for example, for a wall with average leakage, it was shown that infiltration accounted for 22% of the heat loss due to infiltration and transmission. When ventilation at the rate of 7.5 L/s of outdoor air per person was considered, the ventilation loss was shown to be 67% of the total heating load, and infiltration only 7%. At the reduced fresh-air rate of 2.5 L/s per person the ventilation loss was 40% and the infiltration 12% of the total heat requirement for the conditions assumed.

Various estimates have shown the infiltration heat losses to be from 30 to 40% of total heating requirements for insulated houses of good construction. Increasing insulation thickness in walls and ceilings will decrease transmission losses, so that infiltration will form a still larger part of the total in houses built to conserve energy.

11.24 The Case for Airtightness

Infiltration rates in houses are sometimes inadequate and often more than adequate to meet air-quality ventilation requirements. Increasing demands for energy conservation make it necessary to consider constructions providing greater airtightness. If such measures are implemented the air leakage will be inadequate at times, and it will become necessary to consider ways to increase the ventilation when required. The reduction of air leakage will also reduce condensation difficulties, provided relative humidities indoors are not increased.

Buildings other than houses will already be provided (in most cases) with central systems capable of meeting ventilation requirements. All buildings will experience infiltration rates at times in excess of requirements for air quality. Buildings with large daytime populations may have no need of ventilation at night. Reduction of air leakage will substantially reduce heating and cooling loads. Serious condensation problems caused by exfiltration will be reduced, as will the entry of dust and other contaminants carried by infiltration. Smoke movement in buildings during a fire will be decreased. Finally, many uncertainties in the calculation of heating and cooling loads and the sizing and operation of equipment owing to the variability of infiltration in time and location will be greatly eased if buildings can be made more airtight.

References

11.1 Shaw, C.Y. A method for predicting air infiltration rates for a tall building surrounded by lower structures of uniform height. *ASHRAE Transactions*, 1979, *85*, (Part I), pp. 72–84. (NRCC 18029)

11.2 Shaw, C.Y., and Tamura, G.T. The calculation of air infiltration rates caused by wind and stack action for tall buildings. *ASHRAE Transactions*, 1977, *83*, (Part II), pp. 145–58. (NRCC 16533)

11.3 Sander, D.M. Fortran IV program to calculate air infiltration in buildings. Computer Program No. 37, DBR/NRCC. May 1974.

11.4 Tamura, G.T., and Wilson, A.G. Pressure differences caused by chimney effect in three high buildings, and, Building pressures caused by chimney action and mechanical ventilation. (Companion papers) *ASHRAE Transactions*, 1967, *73* (Part II), pp. II.1.1–II.2.12. (NRCC 9950)

11.5 Shaw, C.Y., Sander, D.M., and Tamura, G.T. Air leakage measurements of the exterior walls of tall buildings. *ASHRAE Transactions*, 1973, *79* (Part II), pp. 40–48. (NRCC 13951)

11.6 Tamura, G.T., and Shaw, C.Y. Studies on exterior wall air tightness and air infiltration of tall buildings. *ASHRAE Transactions*, 1976, *82* (Part I), pp. 122–34. (NRCC 15732)

11.7 Tamura, G.T. Computer analysis of smoke movement in tall buildings. *ASHRAE Transactions*, 1969, *75* (Part II), pp. 81–92. (NRCC 11542)

11.8 Min, T.C. Winter infiltration through swinging-door entrances in multi-story buildings. *ASHAE Transactions*, 1958, *64*, pp. 421–46.

11.9 Tamura, G.T. The calculation of house infiltration rates. *ASHRAE Transactions*, 1979, *85* (Part I), pp. 58–71. (NRCC 18028)

11.10 Wilson, A.G., and Tamura, G.T. *Stack effect and building design*. Canadian Building Digest 107. DBR/NRCC, Nov. 1968.

11.11 Barrett, R.E., and Locklin, D.W. Computer analysis of stack effect in high rise buildings. *ASHRAE Transactions*, 1968, *74* (Part II), pp. 155–69.

11.12 Dick, J.B. Measurement of ventilation using tracer gas technique. ASHVE Journal Section, *Heating, Piping and Air-Conditioning*, May 1950, p. 131.

11.13 Coblenz, C.W., and Achenbach, P.R. Design and performance of a portable infiltration meter. *ASHAE Transactions*, 1957, *63*, pp. 477–82.

11.14 Bahnfleth, D.R., Moseley, T.D., and Harris, W.S. Measurement of infiltration in two residences, Parts I and II. *ASHAE Transactions*, 1957, *63*, pp. 439–76.

11.15 Jordan, R.C., Erickson, G.A., and Leonard, R.R. Infiltration measurements in two research houses. *ASHRAE Transactions*, 1963, *69*, pp. 344–50.

11.16 Coblenz, C.W., and Achenbach, P.R. Field measurements of air infiltration in ten electrically heated houses. *ASHRAE Transactions*, 1963, *69*, pp. 358–65.

11.17 Tamura, G.T., and Wilson, A.G. Air leakage and pressure measurements in two occupied houses. *ASHRAE Transactions*, 1964, *70*, pp. 110–19. (NRCC 7758)

11.18 Peterson, J.E. Estimating air infiltration into houses. *ASHRAE Journal*, 1979, *21* (1), pp. 60–62.

11.19 *ASHRAE Handbook 1981 Fundamentals*. Chap. 22.

11.20 Sasaki, J.R., and Wilson, A.G. Air leakage values for residential windows. *ASHRAE Transactions*, 1965, *71* (Part II), pp. 81–88. (NRCC 9786)

11.21 Sasaki, J.R., and Wilson, A.G. *Window air leakage*. Canadian Building Digest 25. DBR/NRCC, Jan. 1962.

11.22 Tamura, G.T. Measurement of air leakage characteristics of house enclosures. *ASHRAE Transactions*, 1975, *81* (Part I), pp. 202–11. (NRCC 14950)

11.23 Tamura, G.T., and Wilson, A.G. Pressure differences for a nine-storey building as a result of chimney effect and ventilation system operation. *ASHRAE Transactions*, 1966, *72* (Part I), pp. 180–89. (NRCC 9467)

11.24 McGuire, J.H. and Tamura, G.T. *Smoke control in high-rise buildings*. Canadian Building Digest 134. DBR/NRCC, Feb. 1971.

11.25 *ASHRAE Handbook 1980 Systems*, pp. 41.8–41.10.

11.26 Wilson, A.G. Influence of the house on chimney draft. *ASHRAE Journal*, 1960, *2* (12), pp. 63–68. (NRCC 6101)

11.27 *ASHRAE Handbook 1983 Equipment*, p. 27.21.

11.28 *ASHRAE Handbook 1980 Systems*, Chap. 21.

11.29 *ASHRAE Handbook 1982 Applications*, p. 11.7.

11.30 Standard for natural and mechanical ventilation. ASHRAE Standard 62–73. Atlanta: ASHRAE, 1973.

11.31 *Construction details for air-tightness*. Record of DBR Seminar/Workshop, Ottawa, 1977. DBR/NRCC. (NRCC 18291)

11.32 Hollowell, C.D., Berk, J.V., and Traynor, G.W. Impact of reduced infiltration and ventilation on indoor air quality. *ASHRAE Journal*, 1979, *21* (7), pp. 49–53.

11.33 Kusuda, T., Hunt, C.M., and McNall, P.E. Radioactivity as a potential factor in building ventilation. *ASHRAE Journal*, 1979, *21* (7), pp. 30–34.

11.34 Liptak, B.G. Savings through CO_2-based ventilation. *ASHRAE Journal*, 1979, *21* (7), pp. 38–41.

11.35 Woods, J.E. Ventilation, health and energy consumption: A status report. *ASHRAE Journal*, 1979, *21* (7), pp. 23–27.

11.36 Woods, J.E., Maldonado, E.A.B., and Reynolds, G.L. How ventilation influences energy consumption and indoor air quality. *ASHRAE Journal*, 1981, *23* (9), pp. 40–43.

11.37 Ventilation for acceptable indoor air quality. ASHRAE Standard 62-1981. Atlanta: ASHRAE, 1981.

12

Water and Buildings

12.1 Introduction

Water may be present on and in buildings as a vapour, as a liquid, and as a solid. In the vapour phase it forms part of the atmosphere, and thus comes in contact with all building materials and furnishings. Its liquid and solid phases commonly originate outside buildings in the form of rain, hail, or snow or as runoff and ground water. The building must deal with water in these various forms as required in order to maintain the desired indoor environment. Consideration must also be given to the effects of the phase changes occurring through evaporation, condensation, sorption, and freezing.

12.2 Water Vapour

The concentration of water vapour in the atmosphere can be expressed in terms of dewpoint, mass density, vapour pressure, or humidity ratio, as discussed in chapters 5 and 6. Vapour pressure provides a direct indication of the potential for flow; the other forms are more convenient for calculations of quantity. Relative humidity, which is also widely used, provides a direct measure of the degree of saturation, which determines the moisture content of materials in equilibrium with the atmospheric condition. It has limited application as a general measure of moisture conditions when temperatures vary appreciably, since it depends on temperature as well as vapour pressure or vapour concentration.

The vapour pressure outdoors is determined by the invasion and mixing of air masses at different moisture conditions, by evaporation and condensation at the earth's surface, and by vapour exchange accompanying falling precipitation. These processes occur over a wide range of temperatures. The monthly average vapour pressures in Canada vary from 170 Pa in January in the interior plains to as high as 1860 Pa in summer in eastern Canada. The annual range on the west coast is more moderate, from 680 Pa to 1350 Pa [**12.1**]. Vapour pressures outdoors are influenced, as might be expected, by temperature and precipitation.

Relative humidity outdoors approaches 100% during periods of rain or snow. It tends to rise during the night when temperatures are falling, and to fall as temperatures rise during the day. Low outdoor temperatures are usually accompanied by high relative humidity, but the actual water-vapour concentration will be low, owing to the marked decrease in saturation vapour pressure at low temperatures.

Example 12.1

Find the dewpoint temperatures for vapour pressures of 170 Pa and 1860 Pa corresponding to the range of average values given above.

By interpolation from Table 5.1, the temperature for a saturation pressure of 170 Pa over ice is −14.7°C. The temperature for a saturation pressure of 1860 Pa is +16.3°C.

Example 12.2

On an occasion when the temperature outdoors fell to +10°C overnight, the relative humidity increased to 95%. Find the relative humidity at the daytime temperature of +20°C if there was no change in the moisture content of the air in the intervening time.

From Table 5.1, the vapour pressure at 10°C is 1227 Pa and at 20°C it is 2337 Pa.

The night vapour pressure is 0.95 × 1227 = 1166 Pa.
The daytime relative humidity is 100(1166/2337) = 50%.

12.3 Indoor Moisture Levels

The moisture concentrations in the air inside buildings are commonly measured and described in terms of relative humidity, but are more properly given in terms of humidity ratio, vapour pressure, or dewpoint temperature. The humidity ratio for any space will vary until it finds the level at which moisture losses and gains are in steady balance. The mass-balance equation for moisture is

$$M = w\,(W_2 - W_1) \tag{12.1}$$

where w = mass flow rate, dry-air basis, kg/s
W_1 = humidity ratio of entering air, kg/kg dry air
W_2 = humidity ratio of leaving air, kg/kg dry air
M = net moisture gain in the space, kg/s.

The humidity ratio of the leaving air, which is also the humidity ratio of the room air, must change in order to establish a new balance if a change occurs in any of the other variables. A corresponding situation exists for temperature, which is determined by the heat balance.

Practical situations may be more complicated. There may be infiltration, exfiltration, and special exhaust streams as well as supply and return air flows, and many and varied sources of heat and moisture in the space. When control of temperature and humidity is required, suitable means for adjusting the heat and mass balances are possible, ranging from control of both temperature and humidity throughout the year to part-time control of temperature only. All buildings intended for human occupancy in Canada require some form of heating. Some have central air-conditioning systems for both winter and summer use. Most have central heating systems with some means of winter humidification. Some also have unit air conditioners for cooling and dehumidification in summer. When means of control are not adequate, the indoor conditions find natural levels at which heat and moisture losses and gains will be in balance. The variations in temperature and humidity can be, and often are, kept within limits in summer by the use of fans and open windows for ventilation and shades for sun control. Automatic control of temperature is commonly provided in winter. Humidity control is provided in some cases, but in others relative humidity is allowed to vary. The level of indoor humidity in cold weather is an important consideration in design of the building enclosure, and merits further consideration.

12.4 Humidity in Houses

The humidity in houses that are not ventilated intentionally is governed largely by the infiltration rate, the humidity ratio of the entering air, and the moisture gain from the occupancy and any intentional humidification. The moisture gain from the occupancy varies from about 0.3 kg/h for a family of four to as much as 0.9 kg/h on washdays when clothes are dried indoors [**12.2, 12.3**]. The air-leakage rate for a typical single-family dwelling of good construction may be as low as 0.4 ac/h, or 64 L/s, under winter conditions, as discussed in Chapter 11. Using such values it is possible to calculate the humidity ratio and relative humidity in houses for various outdoor conditions.

Example 12.3

Calculate the indoor humidity ratio and relative humidity in Vancouver for the mean January temperature of 4°C and an outdoor relative humidity of 88%. The indoor temperature is 21°C, the average air-leakage rate is 64 L/s, and the moisture gains are 0.4 kg/h. Compare with results for Ottawa having a mean January temperature of -12°C and an outdoor humidity of 88%.

From Table 5.1, p_{ws} at 21° = 2486 Pa, p_{ws} at 4° = 813 Pa.
so that $p_w = 0.88 \times 813 = 715$ Pa outdoors.
Assume air leakage as dry air at 21°C

$$v = R_a T/p_a = 287.1\ (273 + 21)/101\,300$$
$$= 0.833\ \text{m}^3/\text{kg dry air}$$

and air leakage = $64/(1000 \times 0.833) = 0.077$ kg/s.

Humidity ratio outdoors $= 0.622\, p_w/(p_t - p_w)$
$= 0.622 \times 715/(101\,300 - 715) = 0.004\,42$ kg/kg.

Humidity ratio indoors $= 0.004\,42 + 0.4/(3600 \times 0.077)$
$= 0.004\,42 + 0.001\,44$
$= 0.005\,86$ kg/kg dry air.

Substitute in equation for humidity ratio and solve

$$0.005\,86 = 0.622\, p_w/(101\,300 - p_w)$$
$$p_w = 946 \text{ Pa, indoors}$$

rh indoors $= 100 \times 946/2486 = 38\%$

rh indoors with no
moisture gain $= 100 \times 715/2486 = 29\%$.

Repeat calculations for Ottawa and find:
humidity ratio outdoors = 0.001 18 kg/kg dry air
humidity ratio indoors = 0.002 62 kg/kg dry air
rh indoors for W = 0.002 62 is 17%
rh indoors with no moisture gain is 8%.

Example 12.3 illustrates that even well-constructed houses with moderate air-leakage rates may have low humidity ratios and low relative humidities indoors under winter conditions. The relative humidities will be very low if little or no moisture is added to the air during extremely cold weather. Many houses will have higher moisture gains, leading to higher indoor relative humidities, but some may also have substantially higher air-leakage rates than those assumed. Substantial quantities of water may have to be evaporated to maintain moderate levels of relative humidity under conditions of high air leakage and low temperature.

Example 12.4

Find the humidification rate required to maintain 40% rh at 23°C inside and -12°C outside for a house of 580 m³ volume with a leakage rate of one air change per hour. Outdoor humidity is 90%. For simplicity, assume the leakage rate is based on dry air at inside temperature.

air-leakage rate = 580/3600 = 0.161 m³/s dry air
mass flow rate (dry air at 23°C) = 0.161/0.833 = 0.193 kg/s dry air
vapour pressure outdoors at -12°C = 0.90 × 217 = 195 Pa
humidity ratio outdoors = 0.622 × 195/(101 300 − 195)
= 0.0012 kg/kg dry air
vapour pressure indoors at 23°C,
40% = 0.4 × 2809 = 1124 Pa
humidity ratio indoors = 0.622 × 1124/(101 300 − 1124)
= 0.0070 kg/kg dry air
required moisture gain = (0.0070 − 0.0012) 0.193
= 0.001 12 kg/s
= 0.001 12 × 3600 = 4 kg/h

Assuming 0.3 kg/h from occupancy, the required humidification rate is 4.0 − 0.3 = 3.7 kg/h.

Measurements made in a large number of houses at various locations in Canada between 1957 and 1961 produced the results shown in Figure 12.1 [**12.4**]. The humidity ratios dropped as the monthly average outdoor temperatures decreased. They tended to be higher for any given temperature from August to December than for the same temperature in the period from January to July when outdoor temperatures were rising. One explanation is that the house and contents have a substantial capacity for moisture storage: they contribute significantly to the moisture gain while outdoor temperatures and indoor relative humidities are decreasing, and they take up moisture when humidities increase.

Many of the houses were equipped with humidifiers, but this did not prevent indoor relative humidities from dropping as outdoor temperatures decreased. Although the adequacy of the humidifiers in use at the time may be questioned, the principal reason for the decrease is the low humidity ratio outdoors coupled with the removal of moisture by condensation on windows and other cold surfaces. These factors offset the contributions from humidifiers and other sources and limit the level of humidity that can be maintained in cold weather.

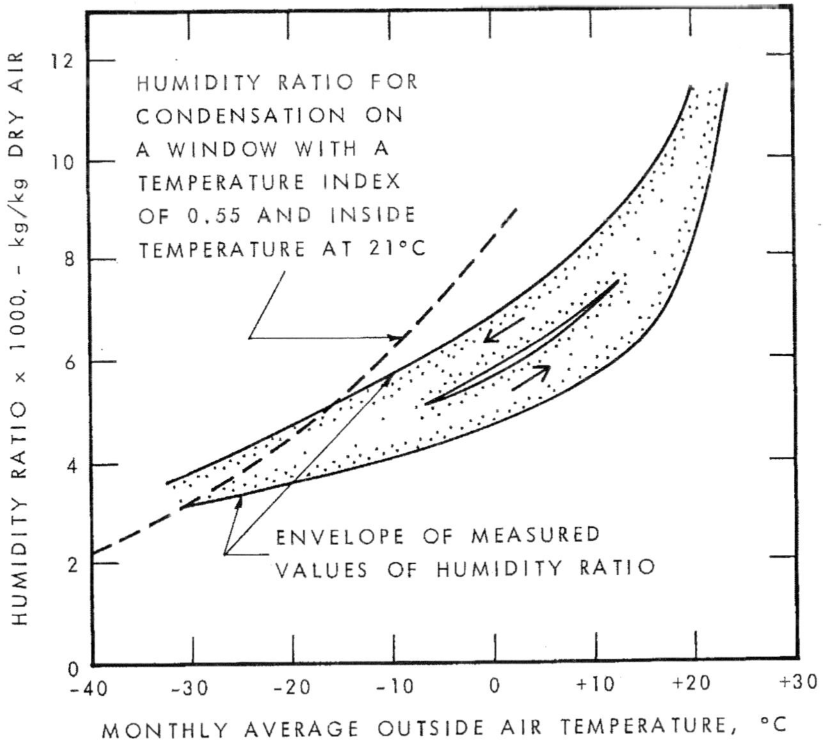

FIGURE 12.1 Monthly averages of humidity ratio in Canadian houses. (From Kent, A.D., Handegord, G.O., and Robson, D.R., *ASHRAE Trans.*, 1966, *72*, Part II, Fig. 8, p. II. 1.6)[**12.4**]

Example 12.5

Find the surface temperature at which condensation will begin when the indoor condition is 40% rh at 23°C.

The saturation pressure at 23°C = 2809 Pa

The partial pressure p_w is 0.40 × 2809 = 1124 Pa

The temperature for a saturation pressure of 1124 Pa is found by interpolation from Table 5.1 to be 8.8°C

(This could also have been read from the psychrometric chart [Figure 6.1] as the dewpoint temperature, 8.7°C.)

Because windows usually have the lowest resistance to heat flow of any part of the building enclosure, they also present the coldest inside surfaces. Such cold surfaces impose limits on the relative humidities that can be carried, and therefore window temperatures require particular attention.

12.5 Window-Surface Temperatures

The thermal properties of windows were explored in Chapter 8, using average values of coefficients which yielded values of average surface temperature. Temperatures are far from uniform over glass, sash, and frame surfaces in most practical situations. There are vertical temperature gradients even when inside and outside air temperatures are uniform over the window height, owing to convection in the air space as shown for an idealized double window in Figure 12.2 [**12.5**]. It is convenient for comparison purposes to express window temperatures as a temperature index, which is given by

$$\text{temperature index} = (t - t_o)/(t_i - t_o) \tag{12.2}$$

where t = temperature of interest, and t_i, t_o are the indoor and outdoor air temperatures on either side of the window. Temperatures are thus reduced to a common basis of an outside temperature of 0° and an inside temperature of 1°, and can later be converted readily to any given set of outdoor and indoor temperatures.

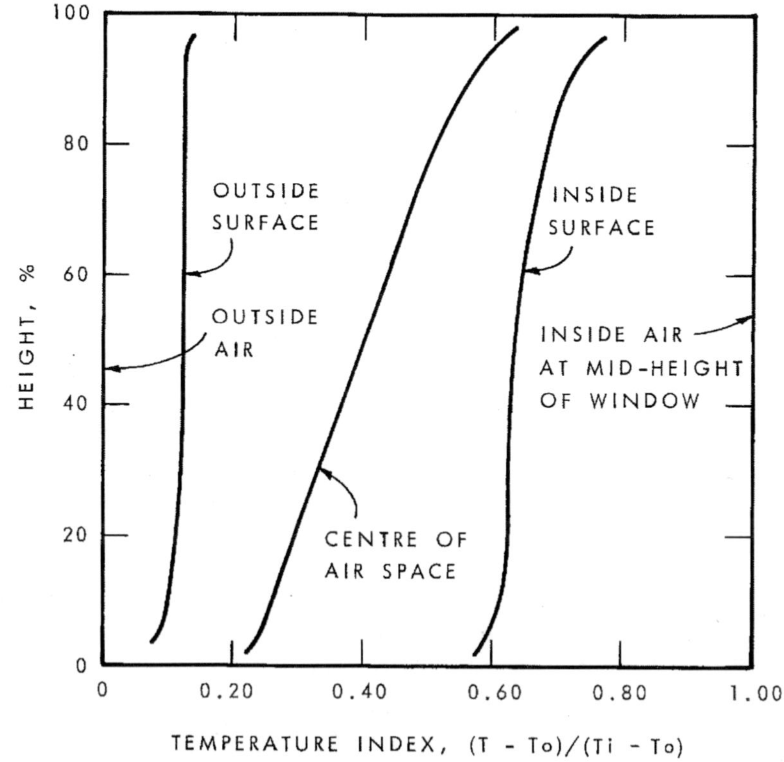

FIGURE 12.2 Temperature index values at vertical centreline of idealized window.
(From Wilson, A.G., and Brown, W.P., Canadian Building Digest 58, DBR/NRCC, 1964, Fig. 1)[**12.5**]

Example 12.6

The temperature index at the centre of a double window was found to be 0.63 when tested with natural convection on the inside and forced convection on the outside. Find the centre temperature for comparable air-flow conditions with an outside temperature of −10°C and an indoor temperature of +23°C.

$$\text{The temperature, } t = -10 + 0.63\,[23 - (-10)] = 11°\text{C}.$$

Double windows and double glazing are in common use in most cold-weather regions, and some triple glazing is also being used. Wood window sashes and frames provide resistance to heat flow comparable to that of the air space between panes. Metal frames and the metal spacers of sealed double and triple glazing offer less resistance, and may cause surfaces exposed to inside air to be colder than the rest of the window. Thermal breaks are provided by the use of two metal frames separated by a spacer of some material having lower thermal conductivity, as shown in Figure 12.3 [**12.5**]. Wood or plastic 10 to 20 mm thick can provide sufficient resistance to raise frame-surface temperatures up to the temperatures of the lower edge of the glass. Stainless-steel frames, with a k-value of 16 W/m • K, are less highly conducting than aluminum, with a k-value of 200 W/m • K, and may meet temperature requirements without a thermal break if otherwise suitably arranged [**12.6**].

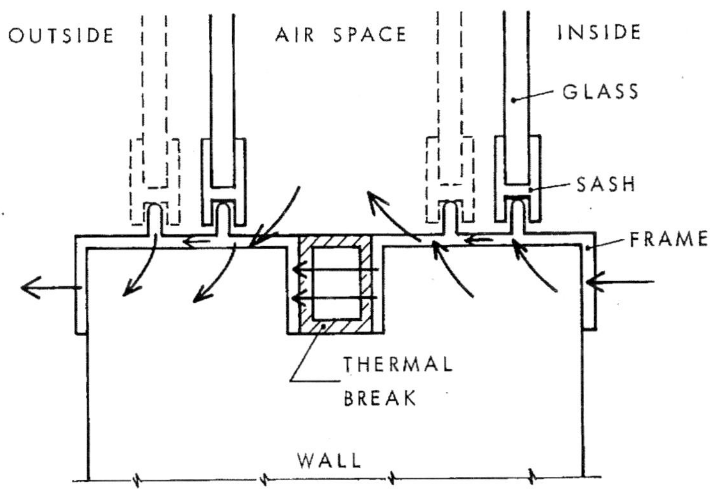

FIGURE 12.3 Heat flow through frame of double window with thermal break. (From Wilson, A.G., and Brown, W.P., Canadian Building Digest 58, DBR/NRCC, 1964, Fig. 3)[**12.5**]

The thermal resistances of double windows and double glazing are made up largely of the air-space resistance plus the resistance to heat flow at the inner and outer surfaces. Window temperatures are therefore readily affected by all the factors that affect the surface-film coefficients, including wind on the outside and the effect of heating outlets, air-supply outlets, and blinds and drapes. Sash and frame temperatures are similarly influenced by these factors as well as by the geometry and thermal conditions created by surrounding construction [**12.7**]. Additional complications are introduced by solar radiation [**12.8**] and by air leakage around sashes and frames.

It is difficult to predict the temperatures of window surfaces accurately because of their sensitivity to environmental conditions. Test results are of necessity based on some selected set of conditions and must be judged accordingly. When warranted, special testing that represents more closely the conditions of a particular application may be carried out.

The temperature index for the centre of the inside surface of a double window with natural convection on the inside and forced convection producing an outside film coefficient of 25 $W/m^2 \cdot K$ is about 0.63. Canadian standards for metal windows call for a minimum temperature index on glass sash or frame which must not be more than 0.14 below the index of the centre of the glass under the prescribed test method [**12.9**], which corresponds to a temperature index of about 0.49 for double windows. The lowest values are usually found at the bottom of a window.

The minimum temperature index for sealed double glazing can be as low as 0.33; with thermally improved frames, it will be in the range of 0.44 to 0.48 with natural convection on the warm side. Residential windows with thermal breaks in metal frames tested under idealized conditions provided minimum values from 0.55 to 0.57. Under-window convectors raised the minimum values to 0.60 to 0.70. Comparable results were obtained with increased air flow on the warm side. Drapes can lower surface temperatures appreciably if they restrict the circulation of room air over the window [**12.5–12.7**].

The relative humidity at 23°C at which condensation will occur on surfaces with various temperature indices is shown in Figure 12.4. Since indoor temperature and relative humidity together define the moisture content of the air, it is possible to add the scale of dewpoint temperature shown on the right-hand side of the illustration. This is also the window-surface temperature at which condensation will begin. Lines have been drawn at a temperature index of 0.22 for a typical single window and at 0.55 for typical residential double windows. A temperature index of 0.48 would be more appropriate for sealed double glazing. The values given in Table 12.1 taken from Figure 12.4 show more clearly the relative humidities at which condensation can be expected on typical single and double windows. These values indicate that window temperatures place a severe limitation on the relative humidity that can be carried indoors in winter without condensation.

The measurements taken in Canadian houses (shown in Figure 12.1) also demonstrate the reduction in indoor humidity ratios at low outdoor temperatures. The curve for a temperature index of 0.55 at 21°C indicates that some of the houses may have been operating under conditions of window condensation, since some measured humidity ratios were above the limiting curve. Interpretation is complicated, however, because the results are based on monthly averages. Selected average values for the coldest month at five locations are shown in Table 12.2. They demonstrate the range of low temperature conditions encountered and the corresponding humidity ratios indoors that would exist if there were no moisture gains other than the moisture present in the infiltrating air. The increase in humidity ratio resulting from the combined effect of moisture added and air leakage plus ventilation is roughly the same for all localities except Vancouver. All the Vancouver houses had single glazing, for which the limiting relative humidity at 21°C indoors and 4°C outdoors is 42%, very close to the 41% calculated from the measured humidity ratios. Condensation was probably occurring on many of the windows. It is probable that markedly higher air-leakage and ventilation rates, not lower moisture gain, were responsible for the smaller increase in indoor humidity ratios in Vancouver houses.

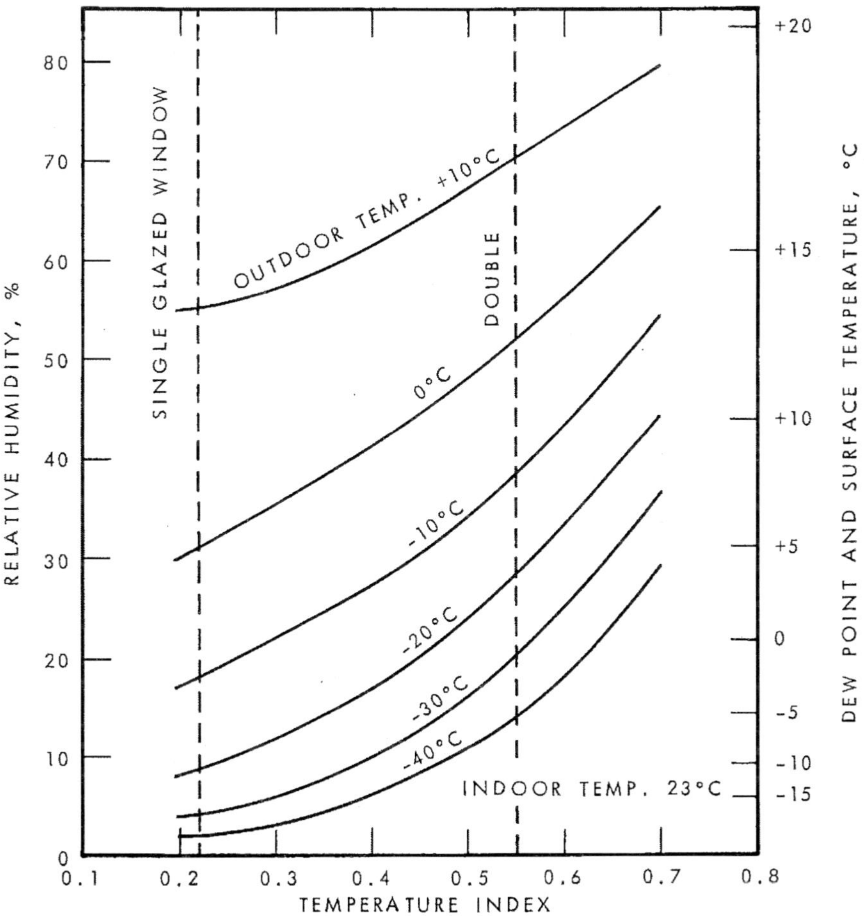

FIGURE 12.4 Limiting conditions for window condensation.
(From Wilson, A.G., and Brown, W.P., Canadian Building Digest 58, DBR/NRCC, 1964, Fig. 2)[**12.5**]

TABLE 12.1

Indoor Relative Humidities at 23°C for Window-Surface Condensation

Outdoor temperature, °C	Limiting relative humidity	
	Single window index = 0.22	Double window index = 0.55
10	55%	70%
0	31	52
−10	18	38
−20	9	28
−30	4	20
−40	2	14

TABLE 12.2

Minimum Indoor Winter Conditions in Canadian Houses

(From Kent, A.D., Handegord, G.O., and Robson, D.R., *ASHRAE Trans.*, 1966, *72*, Part II, Table II, p. II.1.5) [**12.4**]

Location and date	Average conditions for coldest month			
	Temperature, °C	Humidity ratio: Outdoors kg/kg	Humidity ratio: Indoors kg/kg	Relative humidity equivalent indoors at 21°C
Vancouver 1959	+4	0.0043	0.0064	41%
Halifax 1960	−6	0.0020	0.0054	35
Ottawa 1959	−12	0.0011	0.0041	26
Saskatoon 1959	−22	0.0004	0.0043	27
Inuvik 1961	−31	0.0001	0.0034	22

12.6 Condensation on Window Surfaces

Surface temperatures are usually at a minimum over a small area and are markedly higher on adjacent surfaces. When conditions are just sufficient to produce condensation it will occur at the point of lowest temperature. The condensate will be in the form of water if the surface temperature is above freezing, and in the form of frost or ice if it is below freezing. The condensation process releases heat, which increases the surface temperature. Any free water will run down on to the window sill. Frost accumulates on the surface until temperatures rise.

Air circulation in the room produces mixing and distributes the water vapour, bringing some of it into contact with the window surface. When the rate of moisture gain increases, the humidity ratio and the condensation rate also increase until a balance is established at a new level. The condensing area grows to include marginally warmer surfaces, which are now able to produce condensation at the higher moisture content.

The condensation rate varies markedly in practice. Outdoor and indoor air temperatures, wind, and solar radiation can alter the window-surface temperature, and changes in activities such as washing and clothes drying and in the number of occupants influence the rate of moisture gain. Evaporation or melting may occur during part of the day, with condensation forming in the intervening periods. When humidity levels are rising, the rate of condensation will increase until a balance is established at a new level. Higher humidities—above those at which condensation begins—are achieved by adding to the moisture gains by intentional humidification, but only at the expense of increased condensation.

12.7 Condensation on Other Surfaces

Exterior walls and roofs are generally well insulated, so they usually present surfaces to the indoor air that are much warmer than window surfaces. Windows normally provide sufficient dehumidification, so that other surfaces at appreciably higher temperatures do not become involved in the process. But there are exceptions. Thermal bridges formed by metal or masonry members extending through insulated exterior wall constructions can result in low temperatures and interior surface condensation. There may also be cases where the window areas are insufficient to prevent humidity levels from rising until other surfaces become involved. Serious condensation can also occur in unheated rooms that receive moisture-laden warm air from adjacent spaces. Similarly, air leaking past the inner sash of double windows may carry moisture, which condenses on the cold surfaces of the outer window facing the air space, as will be discussed later in this chapter.

12.8 Humidity Levels in Nonresidential Buildings

Many commercial and institutional buildings in Canada experience low humidity levels in winter. These buildings may be occupied only during working hours and may have relatively small moisture gains coupled with high air-leakage and ventilation rates. In such cases, indoor relative humidities are determined largely by the humidity ratio of the outdoor air, and may be as low as 10% or less in cold weather. Buildings may also be equipped with year-round air-conditioning systems or with humidifiers for winter humidification. Indoor relative humidities of 20 to 40% or more may be maintained in midwinter, depending on the severity of the weather, the building construction, and the amount of window condensation that can be tolerated. Buildings exposed to higher humidities indoors under severe winter conditions must be specially designed if severe condensation problems are to be avoided.

Commercial and industrial buildings housing wet processes or other activities that release water vapour at high rates into the indoor air may have high indoor relative humidities, depending on the means used to deal with the high moisture gains. Food-processing plants, laundries, swimming pools, and buildings housing wet industrial processes fall into this category. Buildings such as museums, laboratories, and textile mills may have to add water vapour to hold relative humidities at particular levels to preserve artifacts or to facilitate the activities and processes involved in the occupancy. When relative humidity must be kept constant, the temperature must also be closely controlled.

Indoor moisture conditions can be dealt with in various ways, depending on the technical, economic, and human factors involved. Special exhaust systems and high ventilation rates may be appropriate in some cases to deal with unusual moisture gains. Windows may have to be eliminated to permit high humidities to be maintained without condensation. In some cases, buildings can be specially designed to provide or tolerate necessary moisture conditions. In other cases—for example, in farm buildings—severe condensation may be acceptable, provided the buildings are designed to minimize the associated maintenance costs. The ASHRAE Handbooks, especially *Fundamentals 1981* and *Applications 1982* [**12.10, 12.11**] contain a great deal of information on thermal environmental requirements and practices for a wide range of activities and building occupancies.

12.9 **Vapour Diffusion and Concealed Condensation**

Condensation takes place on cool surfaces within building constructions as well as on cool surfaces exposed directly to inside conditions. The necessary moisture level is obtained through water-vapour diffusion from the warm side of the construction, or, as will be discussed later, by air leakage. The diffusion of water vapour, discussed in Chapter 5, must now be considered in combination with heat transfer.

Example 12.7

Calculate the temperature and vapour-pressure gradients for a simple wall consisting of two layers of fibreboard 12.5 mm thick on either side of wood studs 92 mm thick, with the space between filled with mineral-wool batts. Indoor conditions are 23°C, 40% rh and outdoor conditions −17°C and 90% rh.

Saturation vapour pressures can be found from Table 5.1:

$$p_w \text{ indoors} = 2809 \times 0.40 = 1124 \text{ Pa}$$
$$p_w \text{ outdoors} = 137 \times 0.90 = 123 \text{ Pa.}$$

Heat-transmission coefficients can be found from Tables 8.1, 8.2, and 8.4, and calculations made in accordance with the procedures in Chapter 8 as follows:

Element	k or C	R $m^2 \cdot K/W$	ΔT K	T °C
				−17.0
Outside coefficient	34	0.030	0.3	−16.7
Fibreboard 12.5 mm	k = 0.055	0.23	2.3	−14.4
Mineral wool 92 mm	C = 0.30	3.33	33.9	+19.5
Fibreboard 12.5 mm	k = 0.055	0.23	2.3	+21.8
Inside coefficient	8.3	0.12	1.2	+23.0
		3.94		

The calculations for vapour flow can be made in accordance with the procedures in Chapter 5 as follows:

Element	$\bar{\mu}$ or M	R $s \cdot m^2 \cdot Pa/ng$	Δp_w	p_w
				123
Fibreboard 12.5 mm	M = 2500	0.000 40	299	422
Mineral wool 92 mm	$\bar{\mu}$ = 170	0.000 54	403	825
Fibreboard 12.5 mm	M = 2500	0.000 40	299	1124
		0.001 34		

The vapour-flow rate for continous flow is:

$$(1124 - 123)/0.001\ 34 = 747\ 000 \text{ ng/s} \cdot \text{m}^2.$$

The temperature and vapour-pressure gradients are shown in Figure 12.5.

The low outdoor temperature greatly limits the vapour pressure outdoors so that in cases such as that in Example 12.7 the higher vapour pressure will be found on the warm side of the wall, and the vapour flow will be from the warm side to the cool side in accord with the difference in vapour pressures.

The temperatures and vapour pressures in Example 12.7 and Figure 12.5 are those required to maintain equal flow through each element of a wall. When, as a consequence of reduced temperature, the saturation vapour pressure at any surface is less than that required for continuity of vapour flow, the vapour pressure producing flow to the surface and the corresponding flow rate to the surface will increase, while the pressure difference and the flow rate from the surface at the outside or cold side will decrease. Vapour pressure at the surface will be that for saturation at the surface temperature, and condensation will take place at a rate determined by the difference between the flow to and the flow away from the surface.

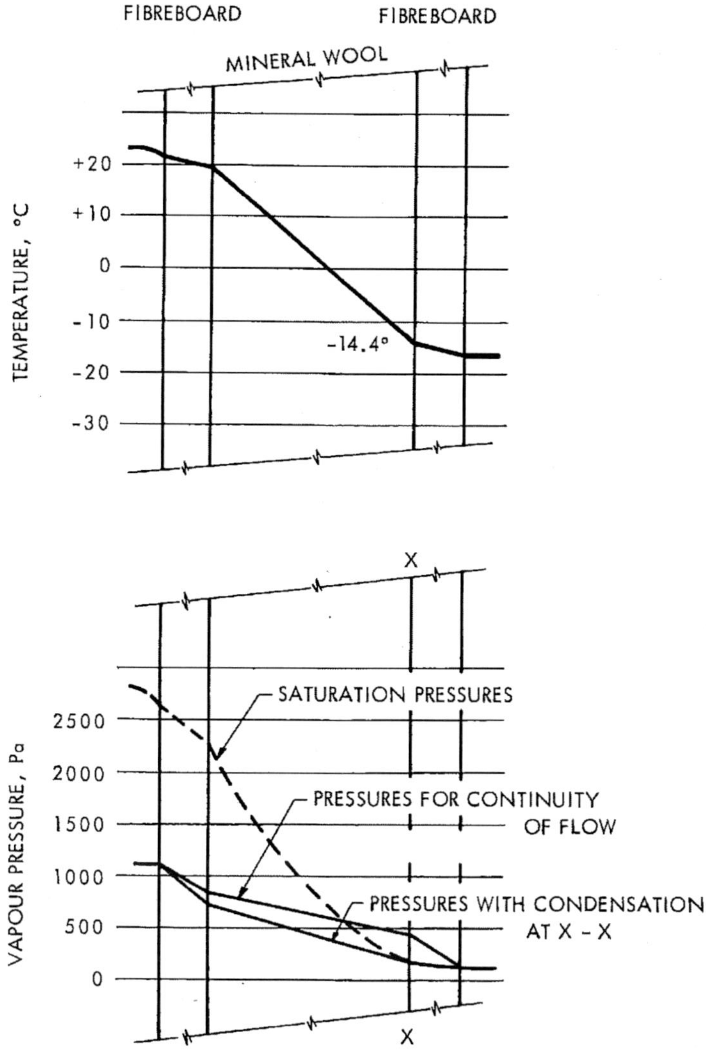

FIGURE 12.5 Temperatures and vapour pressures in simple insulated wall with heat and moisture flow.

The surface *X–X* in Figure 12.5 on the inside of the exterior sheathing is critical. The temperature is –14.4°C, for which the saturation vapour pressure (as found from Table 5.1) is 175 Pa. The pressure required for continuity of flow is 422 Pa. The flow rate to *X–X* will be

$$\text{mass flow} = (1124 - 175)/(0.000\,40 + 0.000\,54)$$
$$= 1\,010\,000 \text{ ng/s} \cdot \text{m}^2.$$

The flow from X–X to the outside will be

$$\begin{aligned} \text{mass flow} &= (175 - 123)/0.000\,40 \\ &= 130\,00 \text{ ng/s} \cdot \text{m}^2 \\ \text{condensation rate} &= 1\,010\,000 - 130\,000 \\ &= 880\,000 \text{ ng/s} \cdot \text{m}^2 \\ &= 8.8 \times 10^{-7} \text{ kg/s} \cdot \text{m}^2 \end{aligned}$$

The inside sheathing of fibreboard is relatively permeable. Since it is a poor barrier to flow of water vapour, the rate of flow into the wall is relatively high while the flow out is limited by the reduced vapour pressure at the critical plane. Condensation may be prevented by limiting the flow into the wall by adding a vapour barrier on the warm side of the insulation. In this case, paint on the interior surface would be quite effective. Two coats of flat paint would reduce the permeance of the interior cladding (and thus the flow into the wall) by a factor of 10. This would permit continuity of flow and eliminate condensation. Another possibility is to select a more permeable exterior cladding allowing greater flow out; here the exterior sheathing of fibreboard is already high permeable.

The predicted condensation rate should always be carefully considered to determine whether it might be tolerated. In the case at hand the rate of 8.8×10^{-7} kg/s • m² may be judged in relation to the moisture-storage capacity of the sheathing. The fibreboard has a mass of 3.6 kg/m². The moisture increase will be:

$$\begin{aligned} \text{increase in 24 h} &= 8.8 \times 10^{-7} \times 24 \times 3600/3.6 \\ &= 0.021 \text{ or } 2.1\% \text{ per day.} \end{aligned}$$

This must be considered to be a very high rate of condensation, capable of producing serious wetting if continued even for a few days.

12.10 Vapour Barriers

Tables 5.5 and 5.6 show a number of coatings and thin sheet materials having permeances of 50 ng/s • m² • Pa or less that might be incorporated in walls and ceilings to provide substantial resistance to the entry of water vapour. Materials used in this way are called vapour barriers. Polyethylene film is one such material that has been widely used in wood-frame construction. A good vapour barrier with a permeance of 50 ng/s • m² • Pa will provide a resistance to vapour flow of 50 times that of the fibreboard in Example 12.7, which may be recalculated for an added vapour barrier of this kind. The heat flow and temperatures will remain unchanged. The resistance to vapour flow is increased by 0.02 to a total of 0.0213. The vapour pressure required at X–X for continuity of flow can be shown to be 142 Pa, which is well below the limit of 175 Pa set by the temperature. The vapour-flow rate through the wall, without condensation, is

$$\begin{aligned} (1124 - 123)/0.0213 &= 47\,000 \text{ ng/s} \cdot \text{m}^2 \\ &= 4.7 \times 10^{-8} \text{ kg/s} \cdot \text{m}^2. \end{aligned}$$

When the indicated condensation rates are low—say 50 000 ng/s • m²—they will result in an increase in moisture content in sheathing, such as fibreboard, of about 0.1% per day. This would lead to an increase of 10% in 100 days of severe weather which might be absorbed by the sheathing and given up later during warmer weather. Such a situation might be tolerated in some cases.

These types of calculations of concealed condensation are approximate. They are based on simple theory and on average coefficients obtained from small samples under isothermal conditions. The existence of temperature gradients along with vapour gradients will complicate the moisture flow at conditions close to saturation, as discussed briefly in Chapter 5. In addition, outdoor temperatures vary and in combination with solar radiation may give rise to temperatures at critical planes for condensation ranging, on a daily or weekly basis, from well below to well above the freezing point. When condensate stored as frost melts, it may either move by gravity or be absorbed directly. Despite these uncertainties, simple theory and available permeability data make it possible to recognize cases in which condensation is likely to occur and are a great help in judging the merits of design solutions.

Recognition of the mechanism of vapour diffusion as a cause of condensation in insulated constructions quickly gave rise to recommendations on the use of vapour barriers. Much has been written about them [**12.10, Chaps. 19 and 20; 12.12–12.19**]. It was eventually realized that air leakage was also an important means of transporting water vapour to come in contact with cold surfaces in walls and roofs [**12.20–12.22**]. Properly installed vapour barriers were effective in many cases primarily because they restricted air leakage, while in other cases condensation occurred despite the use of vapour barriers because air leakage was not controlled. This does not mean that vapour barriers are not needed. They are required where the other construction elements do not provide the required resistance to water-vapour diffusion. They are often essential elements on the warm side of constructions enclosing heated and humidified space and on the outside of the walls and roofs of cold storages where low interior temperatures are to be maintained for long periods. They may also be required where the cold side is highly resistant to vapour flow and it is necessary to limit the entry of water vapour from the warm side. They may serve as air barriers as well as vapour barriers; this aspect of condensation control will now be discussed.

12.11 Air Leakage and Concealed Condensation

Pressure differences across a building enclosure resulting from wind, stack effect, and air-handling equipment produce infiltration over some portions of the building and exfiltration over others. Wind varies greatly in strength and direction. Stack effect persists as long as there is a temperature difference from inside to outside and, when operating alone, produces exfiltration at all upper levels of a heated building in winter. Mechanical equipment can be operated to raise or lower interior pressures by adjusting the difference between the rates of air supply and exhaust for any compartment.

The addition of insulation to an exterior wall or roof construction makes the elements on the warm side warmer and those on the cold side colder. When the wall is well insulated the outer parts of the wall may be within a few degrees of outdoor temperature as shown by figures 8.1 and 12.5. Air leaking from inside along some continuous path provided by openings, cracks, and spaces will always encounter increasingly colder surfaces on the way to the cold side and will be cooled to an extent dependent on the conditions existing for heat transfer. Where there is rapid flow and a large opening to the outside, the air may be cooled very little. Air filtering through a cold exterior masonry wythe of a well-insulated wall may be cooled to within a few degrees of outdoor temperature before reaching the outside. Any excess of moisture carried by the air, above that required for saturation at the temperature to which it is cooled, will be left behind as condensation on the cooling surface.

Condensation may occur as a result of air leakage into air-conditioned buildings in summer, but is much more likely to occur in the case of infiltration into cool and cold storages in summer and of exfiltration from heated buildings in winter. The possibility of condensation will depend on the leakage rate, the initial moisture content, and the extent of cooling of the air stream. Prediction is difficult since the leakage paths, which will seldom be known with any certainty, are most likely to occur at cracks, joints, and similar discontinuities. This presents complicated problems of analysis of the combined fluid flow and heat transfer.

Example 12.8

Find the moisture carried by air leakage into a wall at the top of a building 30 m high when the ratio of actual to theoretical draft is 0.8 with the neutral pressure level at midheight. Indoor conditions are 23°C, 40% rh. Outdoor conditions are -17°C, 90% rh. Assume average leakage.

Stack effect for h = 15 m and 40 K from Figure 11.3 is 27.5 Pa.
Actual stack effect is 0.8 × 27.5 = 22 Pa
Air leakage is given by Equation 11.13,

$$dQ/dA = C(\Delta p)^n \text{ m}^3/\text{s} \cdot \text{m}^2.$$

From Section 11.15 for walls of average leakage

$$C = 0.9 \times 10^{-4} \text{ m}^3/\text{s} \cdot \text{m}^2 \cdot \text{Pa}^{0.65}.$$

$$\begin{aligned}\text{Air leakage rate} &= 0.9 \times 10^{-4} (22)^{0.65} \\ &= 6.7 \times 10^{-4} \text{ m}^3/\text{s} \cdot \text{m}^2.\end{aligned}$$

From the psychrometric chart (Figure 6.1) and calculations,
humidity ratio, W, indoors = 0.007 kg/kg dry air
dewpoint indoors = +4°C
specific volume, v = 0.85 m³/kg dry air at 23°C
vapour pressure indoors = 0.4 × 2809 = 1124 Pa
vapour pressure for saturation at -17°C = 137 Pa
vapour pressure outdoors = 0.9 × 137 = 123 Pa
humidity ratio for saturation at -17°C = 0.000 84 kg/kg.

$$\begin{aligned}\text{mass air flow rate} &= 6.7 \times 10^{-4}/0.85 \\ &= 7.9 \times 10^{-4}\ \text{kg/s} \cdot \text{m}^2 \\ \text{moisture carried into wall} &= 7.9 \times 10^{-4} \times 0.007 \\ &= 5.5 \times 10^{-6}\ \text{kg/s} \cdot \text{m}^2\end{aligned}$$

The air cannot be cooled below outdoor air temperature on its way to the outdoors and so must retain, at a minimum, the moisture required for saturation at outdoor temperature.

$$\begin{aligned}\text{Humidity ratio for saturation at } -17^\circ\text{C} &= 0.000\,84\ \text{kg/kg} \\ \text{moisture leaving the wall} &= 7.9 \times 10^{-4} \times 0.000\,84 \\ &= 0.66 \times 10^{-6}\ \text{kg/s} \cdot \text{m}^2 \\ \text{maximum possible condensation rate} &= (5.5 - 0.66)\ 10^{-6}\ \text{kg/s} \cdot \text{m}^2 \\ &= 4.8 \times 10^{-6}\ \text{kg/s} \cdot \text{m}^2\end{aligned}$$

The result obtained in Example 12.8 may be compared with the flow by diffusion in the highly permeable wall in Example 12.7. The flow rate for continuous flow without condensation was 747 000 ng/s • m^2 or 0.75×10^{-6} kg/s • m^2; the condensation rate was 0.88×10^{-6} kg/s • m^2; the flow rate with a good vapour barrier added was 0.047×10^{-6} kg/s • m^2. The moisture available for condensation due to air leakage is five times greater than the condensation rate by diffusion and 70 times the rate of vapour diffusion with a good vapour barrier. The rate of condensation from air leakage can be substantial even if only a part of the moisture carried is condensed, and may be much higher than that from diffusion even in the case of highly vapour-permeable walls. The building used in the example was 30 m high. The air-leakage rate according to the calculations made will vary as (height)^n, and so will be 4.5 times as great for a building ten times as high, and 0.49 times as great for a building 10 m high, for $n = 0.65$.

The calculations of Example 12.8 were based on an average air-leakage coefficient for multistorey buildings as if the openings for air leakage were uniformly distributed over every square metre. The principal openings may be located at joints, so the condensation may be concentrated there also. In other cases, depending on construction, condensation may not occur until the air leaking outward encounters the cold cladding. Condensation may lead to corrosion of metal elements, to efflorescence, and to spalling owing to wet freezing when absorbed in masonry. Large icicles may form when water from condensation dripping from a wall meets cold outside air. The rapid melting of a large accumulation of frost can produce severe wetting of masonry cladding. These effects are usually undesirable. They can be avoided or at least reduced in severity by reducing air leakage through the building enclosure. As discussed in Chapter 11, this reduction is also desirable for other reasons, and would appear to be an important requirement for all buildings in cold climates that are to be exposed to indoor humidity ratios substantially above those outdoors. The frequency and severity of problems will increase in proportion to the indoor humidity level unless special precautions are taken.

12.12 Condensation between Window Panes

One example of condensation due to air leakage is the condensate that appears on the inner surface of the outer pane of a double window in cold weather. The water vapour is provided by air leaking outward past the inner sash into the air space between panes. The outer glass is much colder than the inner pane, with a temperature index (as shown by Figure 12.2) of 0.1 or less. The humidity ratio for condensation on the inside of the outer pane is consequently much lower than for surface condensation on the inner pane.

When inner and outer sash have equal leakage openings under outward flow conditions, the pressure in the air space will be midway between the indoor and outdoor air pressures. The humidity ratio in the air space will then be the same as that indoors, and will usually be high enough to produce condensation on the cold outer pane. With a perfectly tight inner pane and a loose outer one, the air space will be at outdoor pressure and the overall pressure difference will be resisted by the inner pane. Stack effect will operate between the air space and the outside to produce a flow of air in and out of the air space through openings around the sash perimeter. The humidity ratio in the air space will then be determined by the condition of the outside air, and no condensation will take place on the outer pane.

At some intermediate condition with a relatively tight inner sash there will be a flow of inside air into the air space and an exchange of air between the air space and the outdoors. The humidity ratio in the air space will be determined by the proportion of indoor and outdoor air flowing into the air space. The ratio of outside to inside air flow that will keep the moisture level in the air space below the condensation point can be calculated [**12.23**]. When the indoor moisture level is just high enough to produce condensation on the inner pane, the flow of outdoor air to prevent condensation on the outer sash must be ten times greater than the flow of indoor air when the outdoor temperature is -10°C, and 26 times greater at -30°C.

Windows at low levels on multistorey buildings will normally experience pressures owing to building stack action to produce inward flow, while those above the neutral pressure level will experience outward flow. Building pressurization and wind may alter these conditions, but it is common in cold weather for double windows on lower floors to have no condensation, while those at higher levels have frost covering the inner surface of the outer pane.

Further calculations can determine the leakage characteristics of inner and outer sash, which will produce the relative air flows necessary to control condensation between panes at various pressure differences. The pressures to produce the air exchange with the outdoors around the outer sash are limited to the stack effect over the height of the air space. These pressures are relatively small, since both the height and the temperature difference producing them are limited. The pressures across the inner pane to produce outward flow into the air space are determined by building pressures which in multistorey buildings can be much greater. Since the flow past the outer sash may have to be many times greater than that past the inner one, and since the pressure differences producing flow will be much greater across the inner sash, the inner sash will have to be much tighter than the outer [**12.23**].

When the inner sash is only moderately tight, the outer sash may have to be left loose or provided with additional ventilation openings in order to avoid condensation, even in low buildings. Some increase in window heat transfer and reduction in surface temperatures can be expected with such increased ventilation. For taller buildings having increased stack effect, the inner sash may have to be exceptionally airtight. This discussion of double windows applies also to windows with storm sash and to any double-glazed arrangements in which air leakage past the inner pane enters the space between panes on its way to the outdoors.

Double windows in some multistorey buildings experience no serious condensation between panes because the indoor humidities are low. When condensation does occur it may obscure the view. Wetting of the sill may also occur, and the build-up of ice in severe cold weather may prevent the operation of sliding windows. Repeated wetting by condensation also soils the glass surface, which must be cleaned periodically. Unless it is hermetically sealed, double glazing must be arranged so that all glass surfaces can be cleaned. Sealed double glazing, now widely used, eliminates condensation between panes and the need to clean two of the four glass surfaces. Sealed triple glazing is also available. Reflective coatings that would be damaged by exposure and cleaning can be applied to the inaccessible glass surfaces. Leaving aside breakage, the life of sealed multiple glazing is determined by the ability to resist the entry of water as vapour or as liquid between the panes, leading to condensation, soiling, or scumming. Some units depend on total exclusion of water; others have a desiccant located in the hollow metal spacer to sorb and store the small amounts of water that may penetrate the seal during the life of the units [**12.23–12.25**].

12.13 Moisture in Flat Roofs

Roofs, like walls, must function as part of the building enclosure and are subject to many of the same natural forces. Because they are sloping or flat they are more exposed to precipitation and must have the ability to shed water reliably [**12.26**]. A roof system normally consists of a structural roof deck, an air-vapour control element, insulation for control of heat flow, and a waterproof or water-shedding covering [**12.27**]. A conventional flat-roof arrangement is shown in Figure 12.6.

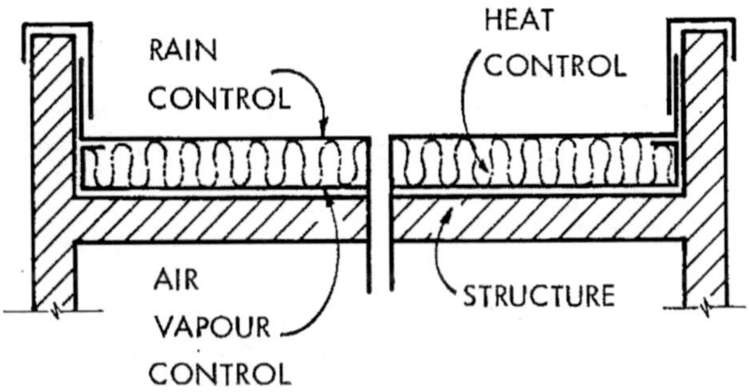

FIGURE 12.6 Conventional flat roof.
(From Baker, M.C., and Hedlin, C.P., Canadian Building Digest 150, DBR/NRCC, 1972, Fig. 1(c))[**12.27**]

Despite their apparent simplicity, the performance of these systems depends on many matters of detail, and careful workmanship and good design are necessary to ensure reasonable freedom from moisture problems.

The waterproof membrane in a flat roof is required to function in a highly exposed location. It may be subjected to severe water, radiation, and temperature conditions as well as to traffic. Leakage can occur through failure of the membrane itself, at joints and junctions with walls and parapets, or at points of penetration by pipes, ducts, machine supports, structural elements, and fastenings for rooftop equipment [**12.27–12.31**]. Once membrane failure occurs, water enters the insulation and migrates by gravity to the lower side where it may be retained by the air–vapour barrier or the deck. In extreme cases the insulation may be almost completely saturated, and will no longer perform as intended. Leakage to the space below is likely to occur, and rapid deterioration of the roofing system may take place.

Wetting of the insulation in flat roofs may also result from condensation from vapour diffusion and air leakage if the air–vapour control is inadequate. Unlike walls, however, the conventional flat roof is covered with a waterproof membrane which prevents the ready movement of water vapour to the outside, while any liquid water is held against the vapour barrier by gravity. The use of more effective water and vapour barriers without other changes means that water that does enter through faults can accumulate and be trapped between them for long periods with little possibility of escape. Temperature and barometric-pressure changes can cause suctions and pressures in the tightly enclosed insulation space which may adversely affect the roofing membrane. Venting of the insulation space is considered desirable, but cannot be relied upon to deal with serious moisture leakage [**12.32**]. Serious rotting can occur when water is held for long periods in contact with natural organic materials such as wood and wood fibre [**12.33**].

These and other difficulties with conventional flat-roof systems have led to designs incorporating some slope for drainage (to avoid ponding) and means for venting and draining the insulation space. In another system the roof membrane is placed on the deck in a protected location under the insulation [**12.27**].

12.14 **Moisture in Sloped Residential Roofs**

Sloped roofs are widely used in residential wood-frame construction. Rain control is provided by shingles on the sloping deck, and insulation and an air–vapour barrier are part of the horizontal ceiling construction, as shown in Figure 12.7. The attic makes it possible to use ventilation for moisture control.

Large sheets of plastic film commonly used as air–vapour barriers are not always installed so as to be completely effective. They are often broken at light fixtures, switches, and convenience outlets, leaving openings for air leakage into air spaces in interior and exterior walls and into roof spaces. Ceiling vapour barriers are usually lapped at the edges with the vapour barrier in exterior walls, but do not prevent air leakage from rooms, basements, and crawl spaces into the attic by way of exterior walls. The vapour barrier may not be made continuous over the plates of interior partitions, and so it is possible for air leakage to take place from indoors into interior partitions and up into the roof space. There are, in addition, many openings into and out of furred spaces around wiring, ducts, piping, plumbing stacks and chimneys that allow leakage of warm, humidified air to the roof space. Some of these paths are indicated in Figure 12.7. The discussions in Chapter 11 of air leakage in houses provide information on the air-pressure differences producing air flow.

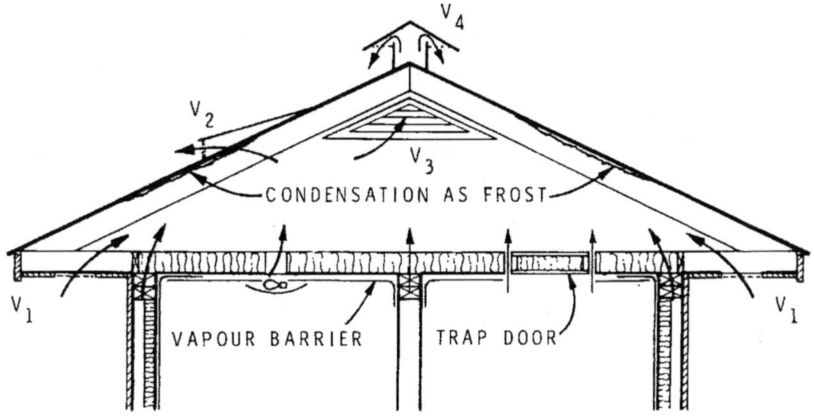

FIGURE 12.7 Ventilated sloping residential roof.

The moisture content of the air in the attic will depend on the moisture content of the air indoors and the relative amounts of air from indoors and outdoors making up the air flow into the attic. Condensation occurs when the dewpoint temperature of the air rises above the temperature of the coldest of the enclosing surfaces.

Attic condensation has always been a common problem in residential construction. It became more serious as houses were better insulated in the interests of economy, making attic surfaces colder. Indoor relative humidities increased as houses were humidified or made tighter. Vapour barriers reduced but did not always prevent serious condensation. Solutions to the problem have been sought by increasing the ventilation of the attic space in order to reduce the moisture content of the attic air. Various locations for openings for this purpose are shown in Figure 12.7. Continuous slots at the eaves coupled with openings in the gable ends provide the possibility of utilizing stack action and wind to create flow. Attic spaces over insulated ceilings are commonly provided with distributed ventilation openings having a total free area of $^1/_{300}$ of the ceiling area [**12.10, p. 20.11; 12.34**].

Attic ventilation reduces attic temperature, making the attic surfaces colder. This in turn makes it more difficult to control condensation and increases the heat transfer through the ceiling. There are problems with the entry of blowing snow; bird and insect screens must be provided.

The ventilation of attic spaces under sloped roofs is beneficial in other ways. It reduces the possibility of prolonged dampness in the attic space and may provide for some cooling under summer conditions. It also assists, in combination with ceiling insulation, in avoiding ice dams at the eaves [**12.35**]. Ice dams develop when water from snow melting on a sloping roof surface runs down and freezes over cold eaves. Once the ice dam forms, water can be backed up until it leaks through the shingles and into the house. Insulation and attic ventilation assist in lowering attic temperatures. This problem is most prevalent in older houses having high heat losses through roofs.

12.15 Ventilated Flat Roofs

Flat roofs are sometimes used in residential wood-frame construction. They can be arranged much as shown in Figure 12.8 with continuous soffit or eave vents for ventilation of the space above the insulation and below the deck. Ventilation can only occur by wind which, to be effective, must produce flow along the individual joist spaces.

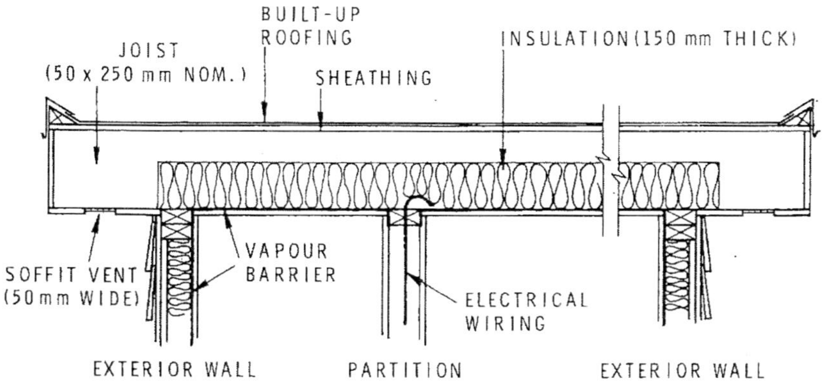

FIGURE 12.8 Ventilated flat roof.

An Ottawa study of a number of electrically heated houses with flat roofs revealed that condensation occurred frequently in individual joist spaces in which there were obstructions to air flow and opportunities for air leakage from below [**12.36**]. It was concluded that the most obvious approach to solving the roof-space condensation problem was to increase the airtightness of ceiling constructions. Canadian residential standards [**12.34**] now require that cross purlins be used across roof joists, that the insulation be at least 25 mm below the top of the joists, and that the unobstructed vent area be 1/150 of the insulated ceiling area, uniformly distributed around the building.

12.16 Roof Condensation under Extreme Winter Conditions

Serious moisture accumulations may occur in roofs under the severe winter conditions of northern Canada [**12.37**]. Frost accumulations resulting from air leakage as high as 5 kg/m^2 over four to five months of sustained low temperatures have been found. Pressurization of spaces in flat roofs by fans was suggested as a possible remedy in some cases.

A study in a northern community of 60 electrically heated houses with attics indicated that the usual recommendations were not effective in preventing condensation [**12.38**]. The absence of chimneys in electrically heated houses leads to lower rates of air leakage with the possibility of higher indoor humidities and a lowering of the neutral pressure level to produce increased stack pressures for flow through the ceiling. It was concluded that exceptional care is needed in locating and sealing all openings for air leakage into the attic.

Contrasting results were obtained from a survey of northern buildings made of insulated plywood sandwich panels [**12.39**]. Condensation did not occur within the panels even though the exterior skins were relatively impermeable. This satisfactory performance was attributed to the absence of openings for air leakage into panels. Presumably, any moisture entering by diffusion in winter was stored in the wood of the panel and escaped by diffusion during the summer.

12.17 Control of Condensation in Walls and Roofs

Control of condensation in walls and roofs is usually accomplished by restricting the entry of water vapour and facilitating the escape of water that has entered the constructions. The principal means of entry and escape are by vapour diffusion and air leakage. Moisture may also be stored in water-sorbing materials and released later under more favourable conditions. Water may also leave the construction by gravity drainage.

Vapour barriers can be used to provide resistance to vapour diffusion on the side of high vapour pressure, usually the warm side, when other materials are inadequate. Materials with high resistance to vapour diffusion can be avoided on the low-pressure side in order to permit the escape of water vapour by diffusion. Air leakage must be reduced as far as possible on the warm side, but can be used as a means of removing water from attics and from air spaces within a construction. Vapour barriers may serve also as air barriers, but they must be applied with tight joints and be properly sealed, and supplemented when necessary by special sealing at all openings formed at penetrations.

The possibility of undesirable air leakage results not only from the existence of openings but also from the air-pressure differences that produce flow through them. Changes in wind, stack effect, and equipment operation can produce marked changes in magnitude and even direction of flow. An operating chimney in a house, for example, may raise the neutral pressure level above the ceiling level, producing flow from the attic through the ceiling into the house. Warm air from the house cannot reach the attic space so long as this condition exists. Similarly, buildings continuously operated under negative pressures during cold weather should not experience condensation due to air leakage.

Pressurized buildings will experience increased air leakage and special precautions should be taken to avoid condensation if they have appreciable moisture gains or if they are humidified. All tall buildings experience strong pressure differences in winter to produce outward air flow as a result of stack action. Pressurization of roof spaces, which has been tried experimentally [**12.36, 12.37**], offers the possibility of avoiding condensation from air leakage in flat roofs, but requires the continuous operation of fans.

Experience shows that ventilation has limited efficacy as a means of removing water from a construction under extreme winter conditions. The temperatures of outer parts of highly insulated constructions are very cold, within a few degrees of outdoor temperature. Air from outdoors entering an air space in the construction can only pick up and carry as much moisture as will saturate it at the temperature of the condensing surface.

Values of humidity ratio for saturation at various temperatures as given in Table 12.3 indicate that as temperatures are reduced the ability of the air to hold water falls off rapidly, and is reduced by a factor of nearly 50 as the temperature drops from 0°C to -40°C. The incremental change per degree is also shown in Table 12.3. The moisture associated with one kilogram of dry air at a room condition of 40% rh at 23°C is 70×10^{-4} kg/kg dry air. Thus, it takes 4 kg of dry air saturated at -10°C to carry as much moisture as 1 kg of dry air at the stated indoor condition. Air drawn from outdoors may be saturated (or nearly so) at outdoor temperature. Its ability to pick up moisture then depends upon its being warmed by a few degrees in the cavity, when it can then take on moisture up to the point of saturation at the condensing surface temperature. Column 3 of Table 12.3 shows that air warmed by one kelvin from saturation at -10°C will be able to pick up an added 1.5×10^{-4} kg/kg, which is only 2% of the moisture carried by 1 kg of indoor air that has reached the cavity. This situation, indicating that the ventilation rate required to forestall or to remove condensation may be many times the leakage rate, was demonstrated by the condensation between the panes of double windows as described in Section 12.12.

TABLE 12.3

Humidity Ratio of Saturated Air at Various Temperatures

Temperature, °C	Humidity ratio at saturation kg/kg dry air	Incremental change per kelvin
20	148×10^{-4}	9×10^{-4}
10	76×10^{-4}	5×10^{-4}
0	38×10^{-4}	3×10^{-4}
−10	16×10^{-4}	1.5×10^{-4}
−20	6.3×10^{-4}	0.6×10^{-4}
−30	2.3×10^{-4}	0.25×10^{-4}
−40	0.8×10^{-4}	0.10×10^{-4}

The ventilation rate necessary to control condensation in an attic or a cavity depends largely on the outdoor temperature and the amount of insulation, and can be very high when temperatures are low and the construction is heavily insulated. These high ventilation rates reduce even further the critical temperatures within the cavity and, as conditions become more extreme, ventilation becomes less effective as a means of preventing condensation. This has serious implications for condensation control in constructions that are to be heavily insulated in order to conserve energy.

The starting point in all these considerations is the moisture content of the air on the warm side of the construction. The possibilities for condensation through diffusion or air leakage are in proportion to the amount by which the humidity ratio indoors in a heated building exceeds that of air saturated at outdoor temperature. When indoor vapour pressures are low, condensation may not be a problem in normal construction. When moisture levels approach those for condensation on the inside of double windows, measures for condensation control will usually be necessary. At still higher levels condensation must be given special consideration in the design of the building enclosure.

12.18 Rain Penetration

The exclusion of rain or melting snow is achieved for flat roofs by impermeable membranes providing, at least in theory, a complete barrier to the movement of water under gravity, wind pressure, and other forces. Thatched or shingled inclined roofs depend on a multiplicity of inclined overlapping units to direct water outward as it moves downward by gravity. They are not tight against either wind or water, and so may leak in severe storms. Walls are wetted only when they receive rainwater or meltwater by drainage from other surfaces or when wind has produced driving rain that is inclined to the vertical as it falls [**12.40, 12.41**]. Wind strength, wind direction, and rainfall intensity determine in a general way the amount of wind-driven rain deposited. Extensive wetting can be expected on windward walls, and little or none on leeward walls. The distribution is determined by the pattern of wind flow around the building, which determines the angle of approach of the raindrops to any surface.

Once rain is deposited on a surface, the direction of its flow will be determined by gravity, the wind flow over the façade, and wall-surface features such as copings, sills, and mullions. Such features can be arranged to direct water along particular paths or even to cause it to drip free of the wall.

Water on a wall surface may be drawn by capillary forces into fine pores and cracks and directed by gravity and wind pressure into larger openings. The wall elements and associated features must be selected and arranged so that water entry is minimized and so that which does enter is dealt with in some acceptable way. The influence of three kinds of forces—capillary, gravity, and wind—must be considered in relation to the many different kinds of openings that can occur. Much insight into these matters has come from the work done by research workers in Norway, where strong wind-driven rains occur frequently along much of the exposed west coast. These researchers have provided information on the design of windows and walls to resist rain penetration [**12.42–12.46**], and many papers on these and related topics have been published [**12.47, 12.48**].

The role of gravity in rain penetration requires little elaboration. An inclined downward path to the inside will allow water to flow inward; a path inclined downward and outward will direct water back to the surface. Shingles, which rely mainly on gravity to prevent the entry of water, are very effective as exterior cladding.

Capillarity was discussed at length in Chapter 5. The suction forces drawing water into fine pores and cracks become large when the pores are small. Water can be absorbed and held in any porous medium that can be wetted by water, but cannot leave as a free liquid unless the capillary suction is opposed by gravity or by wind pressure. It can, however, leave by evaporation.

Under strong wind conditions wind pressures can be well in excess of 500 Pa, which corresponds to a head of 50 mm of water. Wind pressures are an important factor in rain penetration, since they can often be great enough to overcome any restraining influences of gravity and capillarity. Flow depends, however, on the pressure difference produced across the opening, and it is often possible to arrange matters so that large air-pressure differences do not occur across openings, even under high wind pressures. The way in which this can be achieved is shown in principle in Figure 12.9. The unprotected openings in Figure 12.9(a) will leak under combined wind and rain conditions. Figure 12.9(b) shows a closed cavity that has been created on the inside. The pressure in this cavity rises to equal that outdoors so that there is no pressure difference across the leakage opening.

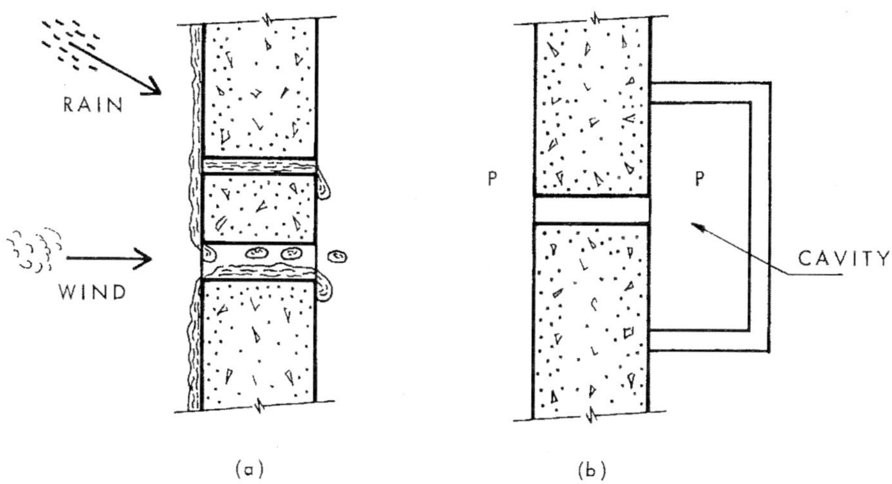

FIGURE 12.9 Air-pressure drop across leakage openings.

In a more realistic situation, there may be a number of openings through the wall with additional openings from the cavity to the indoors. These openings can be represented by equivalent single orifices having areas A_1 and A_2 as shown in Figure 12.10. With an outdoor pressure p_o, an indoor pressure p_i, and a pressure p_c in the cavity, the pressure drop across orifice A_1 will be $\Delta p_1 = p_o - p_c$ and that across A_2 will be $\Delta p_2 = p_c - p_i$. The pressure drop across an orifice varies as the second power of the volume flow rate and it can be shown, by equating flows into and out of the cavity, that

$$\Delta p_1/\Delta p_2 = (A_2/A_1)^2. \tag{12.3}$$

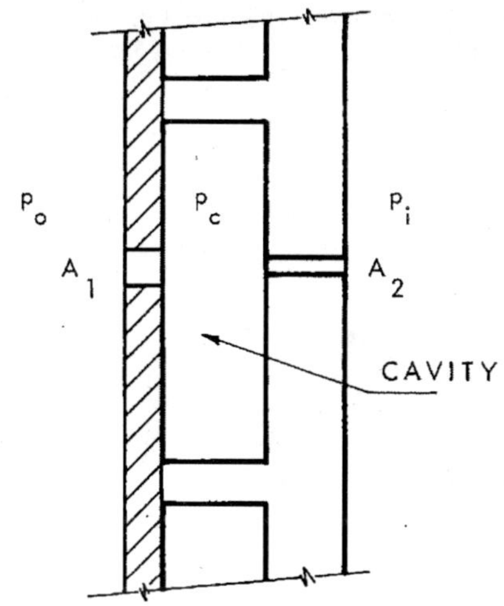

FIGURE 12.10 Air-pressure in a wall cavity with orifices in series.

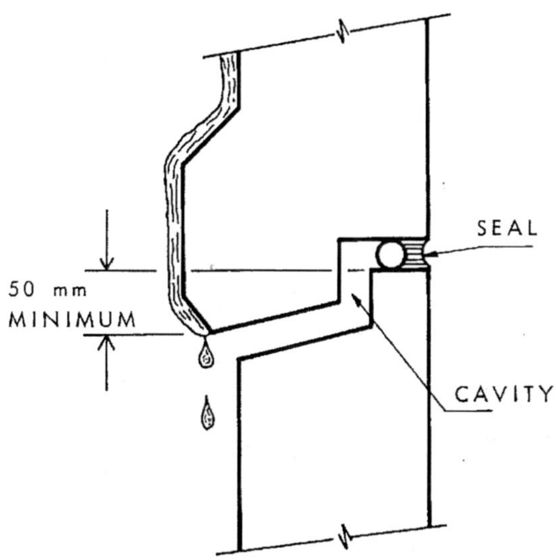

FIGURE 12.11 Drained horizontal joint in concrete panel.

For $A_1/A_2 = 4$ the pressure drop across orifice A_1 is 1/16 of that across A_2 and it is 1/17 of the total pressure difference $p_o - p_i$. For $A_1/A_2 = 10$ the proportion is 100. It is possible to reduce the air-pressure drop across the openings represented by A_1 to a small fraction of the total air-pressure difference across the wall by ensuring that the openings into the cavity from outside are always substantially greater in total area than those leading from the cavity to the inside of the wall.

The cavity that provides the equalization of pressures across critical leakage openings may take various forms: it may consist of an air space behind exterior cladding, the air space between windows, or the protected space in a joint.

The joint shown in Figure 12.11 incorporates several features to make it weatherproof. The projection on the outside diverts water running down the outer surface and causes it to drip free of the joint. The joint clearance is made large to reduce the possibility that it will be bridged by water. The joint is sloped so that any water that enters will drain back to the outside, and the upstand further reduces the possibility that the water will reach the air seal. The pressure in the cavity formed by the joint clearance space will be equal to that outside so that the air-pressure drop is restricted at the air seal. Most important, the opening at which water might enter is relieved as far as possible of any air-pressure difference, while the barrier to air flow, which experiences the air-pressure drop, is protected from wetting. This approach is called two-stage weatherproofing.

These measures can all be used in various ways, along with others, to ensure that exterior walls and windows will resist rain penetration [**12.49–12.51**].

12.19 Basement Walls

Basement walls pose special problems of moisture control. The lower portion of a basement wall may be exposed to hydrostratic pressure, but there may be no great temperature difference. At intermediate levels there will be moderate temperatures and the possibility of contact with damp soil. At the portion above ground, conditions will be those of an exterior wall. It is often difficult to reconcile these different requirements for control of water in arriving at a practical solution. There is an overriding requirement for control of water levels in the soil by means of drains at footing level to ensure that positive water pressures do not develop under basement floors and against walls. Drainage down the wall can be ensured by an external layer of some suitable material, and there are advantages in insulating on the outside of any above-ground portion [**12.52**].

Control of water around basements is closely related to the need for good surface drainage [**12.53**]. Changing water conditions below ground level can lead to swelling and shrinkage of certain soils, and water coupled with freezing conditions can produce damage by frost heaving, as discussed in Chapter 5.

12.20 Buildings Requiring Special Consideration

It is not necessary for the comfort and health of most people that relative humidity be closely controlled. A range from 20 to 70% rh at 23°C is considered acceptable, so humidity levels in occupied buildings may be allowed to vary from winter to summer. Museums, libraries, and art galleries with valuable collections must hold the relative humidity within narrow limits at all times to minimize the dimensional changes due to moisture in natural products such as wood, paper, and textiles. For practical reasons the humidity to be maintained winter and summer is usually set at 50%. A lower relative humidity reduces condensation difficulties in winter but is more difficult to maintain in summer. Temperature must also be controlled in order to keep the relative humidity constant, and artifacts may have to be protected from local heating caused by lamps and solar radiation from windows. Special consideration must be given to the design of a building that has to carry 50% rh indoors in cold winter climates.

Relative humidities as high as 80% may be required in buildings such as textile mills to facilitate the handling of natural and synthetic fibres. Such high levels require that uniformly high surface temperatures be maintained, and air leakage must be strictly controlled. Indoor swimming pools pose a serious humidity problem. They provide large water surfaces for evaporation at temperatures from 23°C to 30°C and act like large humidifiers. The relative humidity must be kept within limits, usually by ventilation, in order to protect the building and to keep spectators comfortable.

Cold storages must be designed to control the entry of water vapour and air from the outside in summer. The possibility that winter outdoor temperatures may drop below storage temperature, reversing the vapour flow, must be considered. Low rates of air leakage no greater than one air change per 24 hours may be required in order to minimize frosting on cooling coils and energy requirements. Cool storages held at temperatures down to freezing will experience a reversal in temperature and vapour-pressure gradients from winter to summer; they pose special problems.

Buildings that are cooled in summer for human occupancy may also experience some reversal in conditions. If the dewpoint temperature outdoors rises appreciably above the indoor temperature, condensation may occur on the outside of windows and, in extreme cases, within walls. This will be a problem mainly in hot humid climates. When the corresponding winter conditions are not severe, control of inward air leakage and vapour diffusion may be provided and the possibility of the reverse condition ignored.

An interesting short-time reversal in vapour flow can occur under summer conditions when a wall of porous brick is wetted by rain and then heated by solar radiation. The moisture is driven inwards to condense within the wall, and can produce wetting. Serious deterioration of natural organic materials may occur, since the combination of high temperature and high humidity are conducive to rotting [**12.54**].

References

12 1 Thomas, M.K., *Climatological atlas of Canada*. Meteorological Division, Department of Transport and DBR/NRCC, 1953. (NRCC 3151)

12.2 Hite, S.C., and Dray, J.L. Research in home humidity control. Research Series No. 106, Engineering Experiment Station, Purdue University, Nov. 1948.

12.3 *ASHRAE Handbook 1983 Equipment*, Chap. 5.

12.4 Kent, A.D., Handegord, G.O., and Robson, D.R. A study of humidity variations in canadian houses. *ASHRAE Transactions*, 1966, *72*, (Part II), pp. II.1.1–II.1.8. (NRCC 9648)

12.5 Wilson, A.G., and Brown, W.P. *Thermal characteristics of double windows*. Canadian Building Digest 58. DBR/NRCC, Oct. 1964.

12.6 Sasaki, J.R. Condensation performance of metal-framed double windows with and without thermal breaks. *Specification Associate*, 1971, *13* (1), pp. 25–31. (NRCC 11913)

12.7 Sasaki, J.R. Effect of indoor shading and heater configurations on the surface temperature performance of a sealed, double-glazed window. *Specification Associate*, 1973, *15* (3), pp. 12–17. (NRCC 13565)

12.8 Sasaki, J.R. Measurement of thermal break potential of solar-control sealed glazing units. *Specification Associate*, 1974, *16* (3), pp. 11–13, 16–20, 22. (NRCC 14259)

12.9 *Windows, extruded aluminum, vertical and horizontal sliding, standard duty.* CGSB Standard 63-GP-3M. Ottawa: Canadian Government Specifications Board, 1976.

12.10 *ASHRAE Handbook 1981 Fundamentals.* Chaps. 20 and 21.

12.11 *ASHRAE Handbook 1982 Applications.*

12.12 Rowley, F.B., Algren, A.B., and Lund, C.E. Condensation within walls. *ASHVE Transactions*, 1938, *44*, pp. 95–130.

12.13 Babbitt, J.D. The diffusion of water vapour through various building materials. *Canadian Journal of Research*, 1939, *A*, *17*, pp. 15–32.

12.14 Teesdale, L.V. Condensation in walls and attics. *Journal Royal Architectural Institute of Canada*. 1938, *15* (8), pp. 185–88.

12.15 *Moisture condensation in building walls.* Building Materials and Structures Report 63. Washington: National Bureau of Standards, 1940.

12.16 Britton, R.R. *Condensation in walls and roofs.* Tech. Paper No. 8. Washington: Housing and Home Finance Agency, 1948.

12.17 Joy, F.A., Queer, E.R., and Schreiner, R.E. *Water vapour transfer through building materials.* Bulletin 61, Engineering Experiment Station, Pennsylvania State College, 1948.

12.18 Handegord, G.O. *Vapour barriers in home construction.* Canadian Building Digest 9. DBR/NRCC, Sept. 1960.

12.19 *Vapour barrier, sheet, for use in above-grade building construction.* National Standard of Canada, CAN2-51.33-M80. Ottawa: Canadian General Standards Board, 1980.

12.20 Wilson, A.G. *Air leakage in buildings.* Canadian Building Digest 23. DBR/NRCC, Nov. 1961.

12.21 Hutcheon, N.B. *Humidified buildings.* Canadian Building Digest 42. DBR/NRCC, June 1963.

12.22 Latta, J.K. *Vapour barriers: What are they? Are they effective?* Canadian Building Digest 175. DBR/NRCC, March 1976.

12.23 Wilson, A.G. *Condensation between panes of double windows.* Canadian Building Digest 5. DBR/NRCC, May 1960.

12.24 *Insulating glass units*. CGSB Standard 12-GP-8M. Ottawa: Canadian Government Specifications Board, 1975.

12.25 Solvason, K.R., and Wilson, A.G. *Factory-sealed double-glazing units*. Canadian Building Digest 46. DBR/NRCC, Oct. 1963.

12.26 Baker, M.C. *Drainage from roofs*. Canadian Building Digest 151. DBR/NRCC, (1972).

12.27 Baker, M.C., and Hedlin, C.P. *Protected-membrane roofs*. Canadian Building Digest 150. DBR/NRCC, June 1972.

12.28 Handegord, G.O., and Baker, M.C. *Application of roof design principles*. Canadian Building Digest 99. DBR/NRCC, March 1968.

12.29 Turenne, R.G. *Shrinkage of bituminous roofing membranes*. Canadian Building Digest 81. DBR/NRCC, July 1976.

12.30 Baker, M.C. *Flashings for membrane roofing*. Canadian Building Digest 69. DBR/NRCC, Sept. 1965.

12.31 Turenne, R.G. *Joints in conventional bituminous roofing systems*. Canadian Building Digest 202. DBR/NRCC, Jan. 1979.

12.32 Baker, M.C., and Hedlin, C.P. *Venting of flat roofs*. Candian Building Digest 176. DBR/NRCC, May 1976.

12.33 Baker, M.C. *Designing wood roofs to prevent decay*. Canadian Building Digest 112. DBR/NRCC, April 1969.

12.34 *Residential Standards Canada 1977*. Ottawa: National Research Council of Canada, p. 54 (NRCC 15563).

12.35 Baker, M.C. *Ice on roofs*. Canadian Building Digest 89. DBR/NRCC, May 1967.

12.36 Tamura, G.T., Kuester, G.H., and Handegord, G.O. *Condensation problems in flat wood-frame roofs*. Paper presented at second International CIB/RILEM Symposium on Moisture Problems in Buildings, Rotterdam, 1974. (NRCC 14589)

12.37 Dickens, H.B., and Hutcheon, N.B. *Moisture accumulation in roof spaces under extreme winter conditions*. Paper presented at RILEM/CIB Symposium on Moisture Problems in Buildings, Helsinki, 1965. (NRCC 9132)

12.38 Orr, H.W. *Condensation in electrically heated houses*. Paper presented at Second International Symposium on Moisture Problems in Buildings, Rotterdam, 1974. (NRCC 14588)

12.39 Platts, R.E. Condensation control in stressed-skin and sandwich panels. *Forest Products Journal*, 1962, *12* (9), pp. 429-30. (NRCC 7057)

12.40 Robinson, G., and Baker, M.C. *Wind-driven rain and buildings*. DBR/NRCC, July 1975. (NRCC 14792)

12.41 Rodgers, G.G., Poots, G., Page, J.K., and Pickering, W.M. Theoretical predictions of raindrop impaction on a slab-type building. *Building Science*. 1974, *9*, pp. 181–90.

12.42 *Rain penetration investigations*. CIB Report No. 1, 1963. (Also published as Report No. 36, Norwegian Building Research Institute, Oslo, 1963).

12.43 Isaksen, T. *Rain penetration in joints: influence of dimensions and shape of joints on rain penetration*. Paper presented at RILEM/CIB Symposium on Moisture Problems in Buildings, Helsinki, 1965.

12.44 Isaksen, T. *Rain leakage tests on open joints in ventilated claddings*. Paper presented at CIB Symposium on Weathertight Joints for Walls, Oslo, 1967.

12.45 Isaksen, T. *Rain leakage tests on joints*. Paper presented at CIB Symposium on Weathertight Joints for Walls, Oslo, 1967.

12.46 Isaksen, T. *Driving rain and joints*. Report 61. Oslo: Norwegian Building Research Institute, 1972.

12.47 Bishop, D., Webster, C.D.J., and Herbert, M.R.M. *The performance of drained joints*. Paper presented at CIB Symposium on Weathertight Joints for Walls, Oslo, 1967.

12.48 *Proceedings, CIB Symposium on Weathertight Joints for Walls*, Oslo, 1967. Report 51. Oslo: Norwegian Building Research Institute, 1968.

12.49 Latta, J.K. *Walls, windows and roofs for the Canadian climate*. DBR/NRCC, Oct. 1973. (NRCC 13487)

12.50 Garden, G.K. *Rain penetration and its control*. Canadian Building Digest 40. DBR/NRCC, April 1963.

12.51 Garden, G.K. *Glazing design*. Canadian Building Digest 55. DBR/NRCC, July 1964.

12.52 Crocker, C.R. *Moisture and thermal considerations in basement walls*. Canadian Building Digest 161. DBR/NRCC, 1974.

12.53 Williams, G.P. *Drainage around buildings*. Canadian Building Digest 156. DBR/NRCC, 1972.

12.54 Baker, M.C. *Decay of wood*. Canadian Building Digest 111. DBR/NRCC, March 1969.

13

The Indoor Thermal Environment

13.1 Introduction

A building is usually required to provide an indoor environment that can be maintained within certain limits as required by the occupancy. In Canada, this means that most buildings must be heated in winter and, sometimes, cooled in summer. To facilitate this, walls, windows, floors, and ceilings must provide resistance to the passage of radiation, heat, and air as required. In previous chapters it has been shown that water vapour must also be controlled to avoid difficulties with the materials of the building and in the interests of the occupancy, and must be considered as a factor in the thermal environment in many situations.

Thermal environment can be defined as "those characteristics of the environment that affect a person's heat loss". The presence of people often determines the choice of the environmental conditions to be maintained. In many commercial and industrial buildings, however, economic and technical considerations may require the maintenance of conditions suitable for animals, plants, products, or processes, leading to a broader definition of thermal environment to include the exchange of energy between the surroundings and animate or inanimate objects.

13.2 Elements of Thermal Exchange

Thermal exchange between any object and its surroundings takes place through the mechanisms of radiation, convection, and conduction (Figure 13.1). The individual components of energy exchange at the surface of a body were shown in Figure 4.9 for the general case where both short-wave and long-wave radiation were involved. When there is a net exchange, the surface temperature changes, and conduction takes place internally to make up the balance, with a corresponding change in the heat stored. Evaporation and condensation at a surface contribute to the energy exchange and can lead to a change in the moisture content of the body. "Body" in this discussion means any object, not just human bodies. Clearly, different conditions may exist at different parts of a body's surface. The energy exchange becomes complex when environmental conditions are nonuniform and change with time. It is expedient to deal initially with the simpler cases involving uniform, steady-state conditions.

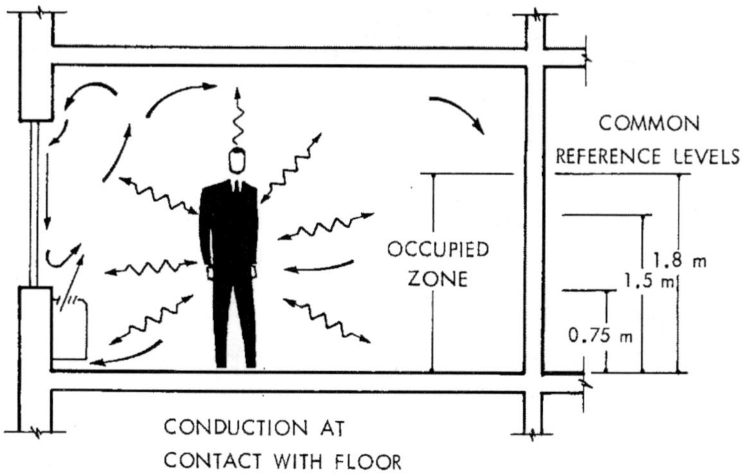

FIGURE 13.1 Body exchanging energy with its surroundings by radiation and conduction and with room air by convection and evaporation.

A simple case of energy exchange occurs when a body is in thermal and moisture equilibrium with the steady conditions in an enclosure having all black surfaces at the same uniform temperature, equal to air temperature. The body is neither gaining nor losing heat or moisture, so there is no change in storage of heat or moisture within the body, and the net exchanges of energy by radiation, convection, and evaporation are zero.

It is also possible for the body to be in thermal equilibrium with its environment when the air temperature differs from that of the surroundings: there will be net exchanges by radiation and convection, and the steady body temperature will have some value between the air temperature and the temperature of the surroundings. This brings to mind the limitations of thermometers for measurement of air temperature. They will read air temperature correctly only if they are suitably shielded from radiation from surfaces that are at other than air temperature [**13.1, p. 13.6**]. The significance of depth of immersion and stem correction for mercury thermometers is related to conduction effects, which must be standardized. The wet-bulb thermometer provides a useful measure of the wet-bulb depression only when radiation and evaporation are suitably controlled by the conditions of the test.

The thermal environment cannot be measured by air temperature alone, except in some special cases. A better measure, taking into account both radiation and convection, is provided by the globe thermometer in which the temperature is measured at the centre of a blackened hollow sphere. The globe thermometer can also be used to provide a measure of the mean radiant field.

13.3 The Globe Thermometer

The globe thermometer as commonly used consists of a thin-walled copper sphere of 150 mm dia. with a mercury thermometer inserted so that it registers the temperature within the sphere. A lighter version with a smaller response time can be made using a plastic sphere or a balloon. The exterior surface will have an emissivity of about 0.94 when coated with a flat black paint.

Consider the globe thermometer located in a black enclosure having a uniform surface temperature t_s, with an air temperature t_a, and air approaching the globe at a speed u. The rate of energy exchange by convection for a sphere 150 mm dia. at the air speeds encountered in rooms is given by

$$q_c/A = 13.5\sqrt{u}\,(t_g - t_a) \qquad (\mathrm{W/m^2}) \qquad (13.1)$$

where u = air speed, m/s

t_g, t_a = globe and air temperatures, °C.

The rate of energy exchange by radiation is given by:

$$q_r/A = \sigma\epsilon\,(T_s^4 - T_g^4) \qquad (\mathrm{W/m^2}) \qquad (13.2)$$

where σ = radiation constant = 5.670×10^{-8} $\mathrm{W/m^2 \cdot K^4}$

ϵ = emissivity of globe surface.

T_s, T_g are temperatures of surroundings and globe respectively, K.

Under steady-state conditions the net energy-exchange rate is zero, so that

$$\sigma\epsilon\,(T_s^4 - T_g^4) = 13.5\sqrt{u}\,(t_g - t_a) \qquad (13.3)$$

This equation may be solved for T_s, the temperature of the surroundings, when the globe and air temperatures and the air speed past the globe are known. The air speed can be measured with a hot-wire anemometer or other suitable instrument [**13.1**]. An extended discussion of the rate of convective heat transfer for a sphere can be found in [**13.2**].

Use of the globe thermometer to measure the surface temperature of a uniform black environment would not appear to be particularly valuable, since the temperature could be measured directly. However, the globe thermometer can also be used to provide a measure of the radiant field when the surroundings are not black and are not at a uniform temperature, and this can be of considerable value. The departure of the surrounding surfaces from blackbody conditions may be ignored with little error, since most common surfaces have emissivities of about 0.9; in any event, in the case of a large enclosure and a relatively small body, little reflected radiation is returned.

The fact that various surfaces may be at different temperatures is of considerably more significance, since surfaces exchange energy at a rate in proportion to the fourth power of temperature and to the respective angle factors. This means that the thermal response of the globe is peculiar to its geometrical relationship to the various surfaces at different temperatures; despite this, a useful measure can be obtained of a "mean radiant temperature" having application to the globe but also providing an approximate value for other shapes and sizes, including the human body, when no better indicator is available. The globe temperature may also be used more directly as an approximate value of the temperature that would be attained by other inert bodies not generating heat or moisture.

Mean radiant temperature (MRT) as given by the globe can be found from a rearrangement of Equation 13.3 as

$$T_{MRT}^4 = T_g^4 + 13.5\sqrt{u}\,(t_g - t_a)/\sigma\epsilon \tag{13.4}$$

substituting for $\sigma = 5.670 \times 10^{-8}$ and $\epsilon = 0.94$

$$T_{MRT}^4 = T_g^4 + C\sqrt{u}\,(t_g - t_a) \tag{13.5}$$

where C has the value 2.53×10^8.

Other sources provide different values of C, depending on the values assigned to the convective heat transfer for spheres. One source [**13.3**] gives a linearized form of the equation:

$$\bar{t}_r = t_g + k\sqrt{v}\,(t_g - t_a)\ (°C) \tag{13.6}$$

where $\bar{t}_r$ = mean radiant temperature, °C
k = 2.2
v = air speed, m/s.

Example 13.1

A globe thermometer gives a reading of 20°C when the air temperature is 23°C and the air speed is 0.15 m/s. Find the mean radiant temperature.

Substituting in Equation 13.5

$$(T_{MRT}/100)^4 = [(273 + 20)/100]^4 + 2.53\sqrt{u}\,(20 - 23)$$

$$T_{MRT} = 290\ \text{K}$$

$$t_{MRT} = 290 - 273 = 17°\text{C}$$

Another instrument, called a two-sphere radiometer, measures MRT by means of two small spheres, one gold-plated and the other blackened, which are heated electrically to the same temperature. The difference in electrical input provides a measure of the radiation from the blackened sphere [**13.1, p. 13.10**]. The radiant field may also be measured by traversing with directional radiometers. Values of MRT can also be obtained through calculation methods, which vary widely in the amount of computation required depending on the simplifying assumptions made.

13.4 Mean Radiant Temperature

The concept of MRT has been developed up to this point in relation to the globe as the body receiving radiation. It can be extended to any body, however, leading to the definition of MRT as "the temperature of a uniform black enclosure in which a solid body or occupant would exchange the same amount of radiant heat as in the existing non-uniform environment" [**13.1**].

The radiant environment can now be described in terms of a single temperature, MRT. The net radiant exchange rate can be written as follows:

$$q_r = \epsilon_b \sigma A_b (T_{MRT}^4 - T_b^4) \qquad \text{(W)} \qquad (13.7)$$

where q_r = net radiant exchange rate, W
ϵ_b = emissivity of the body
T_{MRT} = mean radiant temperature, K
T_b = surface temperature of body or of clothed human body, K.

The complexity of radiant exchange in a real enclosure such as a room has not been avoided but is reflected in the MRT and its determination. Equation 13.7 can be rearranged and solved for MRT when the total net radiation exchange rate is known, either from calculation or from direct measurement with radiometers.

The general equation for radiation exchange rate between two surfaces as developed in Chapter 4 is

$$q_{1-2} = \sigma A_1 F_A F_E (T_1^4 - T_2^4). \qquad (13.8)$$

This equation can be used for calculating MRT. Although formerly widely used and still useful as a starting point for certain kinds of problems, it has now been replaced by equations more appropriate for machine calculation based on concepts of irradiation and radiosity, which are described elsewhere [**13.1, p. 2.10; 13.4**]. When the temperatures of the enclosing surfaces are not known, an analysis of the room thermal conditions is indicated, taking into account the convective exchanges with the room air at each surface, in addition to radiation, so that the surface temperatures can be calculated.

A complete analysis should take into account all hot and cold bodies and surfaces entering into the thermal exchange, including surfaces warmed by solar radiation, furniture, heating devices, people, and lights as well as infiltration and ventilation. The requirements for angle factors and other data describing the various elements entering into the thermal exchange and the computation involved are so formidable that complete analyses are seldom attempted. Various simplifying assumptions are usually made in dealing with some of the simpler room situations. Fanger [**13.4**] describes in some detail the analysis of a room situation when the enclosing surfaces are the only elements in the radiation exchange.

For many purposes a substantial reduction in computation can be made, without serious loss in accuracy, by ignoring the multiple reflections and in effect considering the room surfaces to be black, but not necessarily all at the same temperature. This means that F_E in Equation 13.8 can be greatly simplified and reduced to ϵ_b, taking account only of the grey surface characteristic of the subject body.

It follows from Equation 13.8 that the expression describing the radiant exchange rate between the subject body and surface 1 is given by

$$q_{b-1} = \epsilon_b \sigma A_b F_{b-1} (T_b^4 - T_1^4). \quad (13.9)$$

Similar equations can be written for all N room surfaces and summed to give the total net radiant exchange rate. This sum must also equal the expression given by Equation 13.7, namely,

$$\text{total net radiant exchange rate} = \epsilon_b \sigma A_b (T_{MRT}^4 - T_b^4).$$

Equating, rearranging, and simplifying yields

$$T_{MRT}^4 = T_1^4 F_{b-1} + T_2^4 F_{b-2} + \ldots + T_N^4 F_{b-N}. \quad (13.10)$$

This equation requires only that the surface temperatures and the angle factors between the body and each surface in turn be known. Even this can pose difficulties, however, since angle factors may be difficult to establish for many body shapes. Fanger has developed methods and data that make it possible to take into account a variety of body sizes, clothing, and postures and to establish appropriate angle factors [**13.4**].

A further simplification can be introduced by linearizing Equation 13.10 so that

$$T_{MRT} = T_1 F_{b-1} + T_2 F_{b-2} + \ldots + T_N F_{b-N}. \quad (13.11)$$

A rough approximation may be made when angle factors are not known by calculating the area-weighted average of the surface temperatures, so that

$$T_{MRT} \simeq \frac{T_1 A_1 + T_2 A_2 + \ldots + T_N A_N}{A_1 + A_2 + \ldots + A_N}. \quad (13.12)$$

The weakness in this last approximation is that it does not reflect the effect of position and posture of the subject body in relation to various cold and hot surfaces, as does the angle factor. The same MRT is shown for a heated ceiling panel as for a heated floor panel, and for a position close to a cold wall and for one farther away. Nevertheless, it has some value in giving a quick, though rough, indication of the effect of hot and cold surfaces on MRT and thus on the thermal exchange. It may be misleading and should be used with caution.

13.5 Angle Factors for the Human Body

Angle-factor data developed and published by Fanger [**13.4**] are presented in 13 figures for persons seated or standing. Six of them are for walls; three refer to ceiling and floor surfaces; four, shown here as figures 13.2 to 13.5, are for a person rotated about a vertical axis to be used when his facing direction relative to the surfaces is not known. The figures in all cases are for rectangles oriented so that the subject is located on a perpendicular drawn from one corner. It is thus necessary to divide each rectangular room surface appropriately into four parts and to sum the values of the angle factor found for each part. The reference height for the seated person is 0.6 m and for the standing person is 1 m.

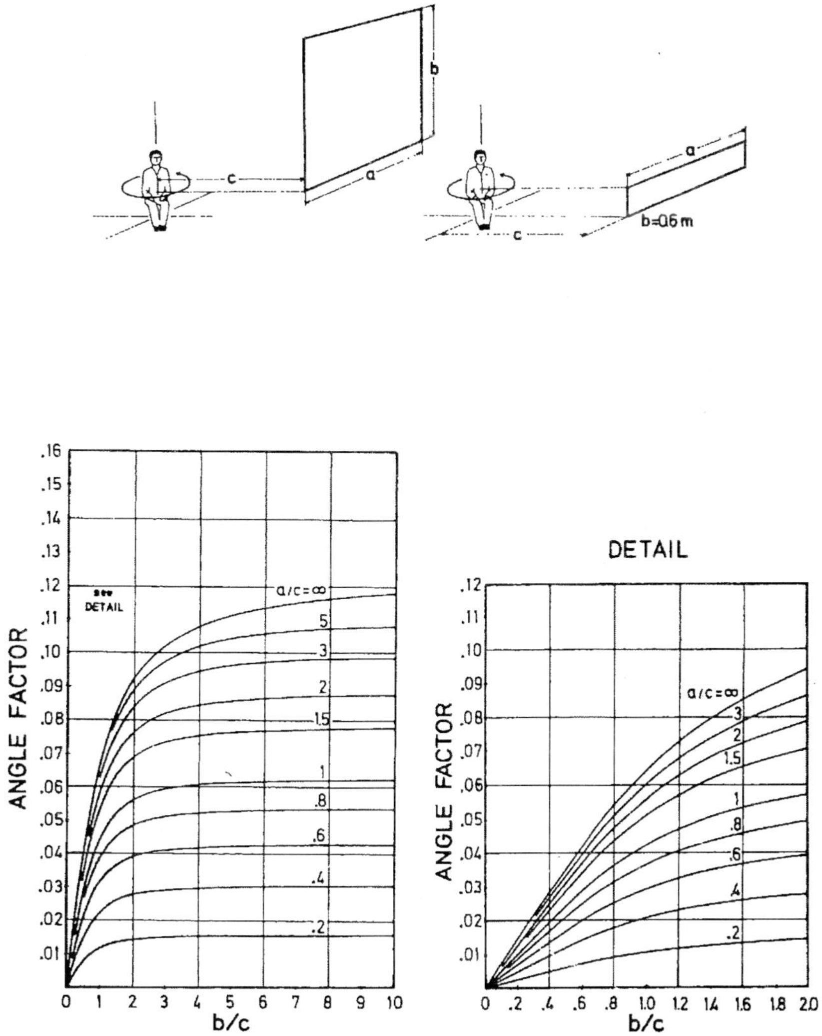

FIGURE 13.2 Mean value of angle factor between a seated person and a vertical rectangle (above or below his centre) when the person is rotated around a vertical axis. To be used when the location but not the orientation of the person is known. Example: a = 4 m, b = 3 m, c = 5 m, b/c = 0.6, a/c = 0.8; F_{P-A} = 0.029. (From Fanger, P.O., *Thermal comfort*, New York: McGraw-Hill, 1972, Fig. 49)[**13.4**]

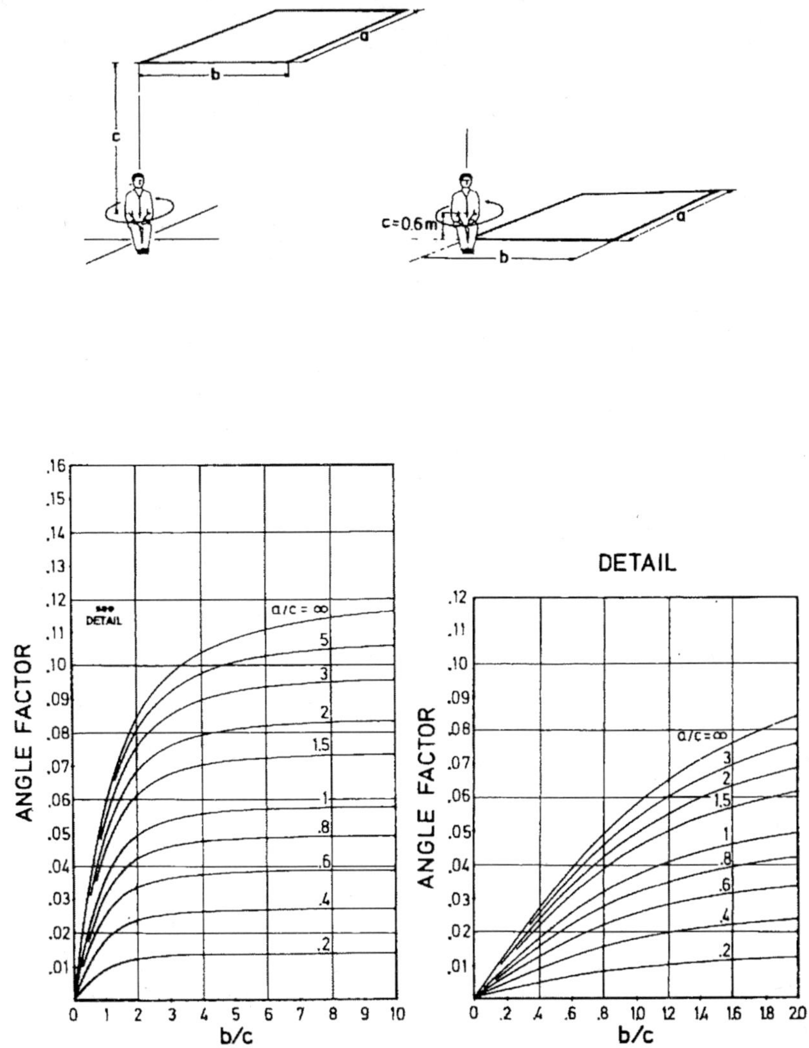

FIGURE 13.3 Mean value of angle factor between a seated person and a horizontal rectangle (on the ceiling or on the floor) when the person is rotated around a vertical axis. To be used when the location but not the orientation of the person is known. Example: a = 3 m, b = 6 m, c = 2 m, b/c = 3.0, a/c = 1.5; $F_{P-A} = 0.067$.
(From Fanger, P.O., *Thermal comfort*, New York: McGraw-Hill, 1972, Fig. 50)[**13.4**]

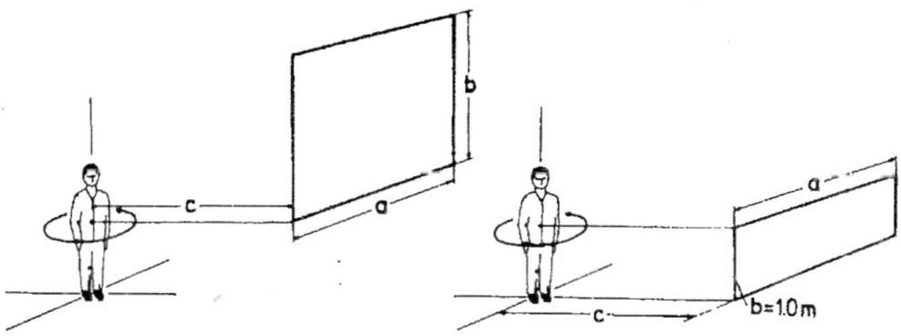

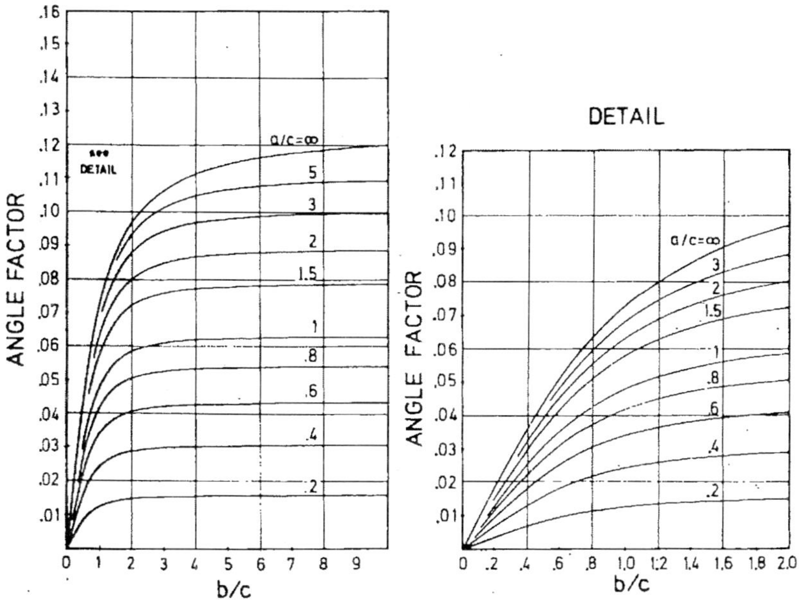

FIGURE 13.4 Mean value of angle factor between a standing person and a vertical rectangle (above or below his centre) when the person is rotated around a vertical axis. To be used when the location but not the orientation of the person is known. Example: a = 4.5 m, b = 2.0 m, c = 3.0 m, b/c = 0.67, a/c = 1.5; $F_{P\text{-}A}$ = 0.047.
(From Fanger, P.O., *Thermal comfort*, New York: McGraw-Hill, 1972, Fig. 55)[**13.4**]

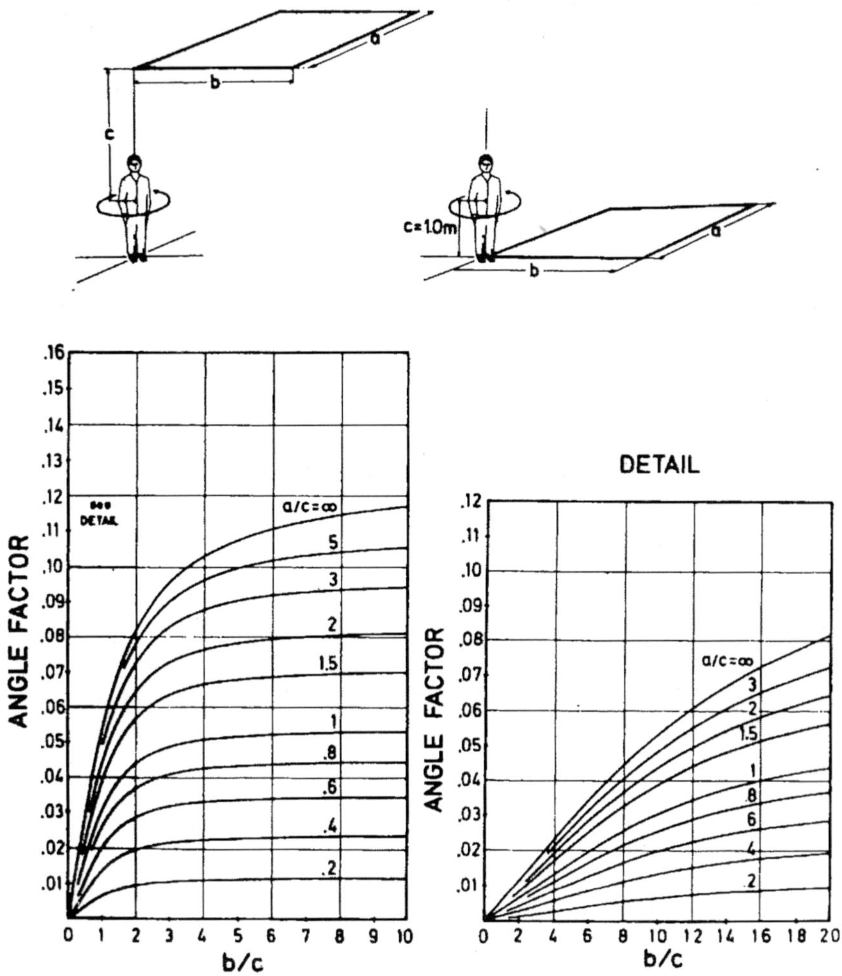

FIGURE 13.5 Mean value of angle factor between a standing person and horizontal rectangle (on the ceiling or on the floor) when the person is rotated around a vertical axis. To be used when the location but not the orientation is known. Example: a = 1.0 m, b = 15 m, c = 1.5 m, b/c = 10, a/c = 0.67; F_{P-A} = 0.039.
(From Fanger, P.O., *Thermal comfort*, New York: McGraw-Hill, 1972, Fig. 56)[**13.4**]

Example 13.2

Find the angle factor for a person relative to an exterior wall with a window if he is standing 2 m away and opposite the vertical centreline of the window, which is in turn centred on the wall. The window is 1.5 m × 1.0 m with its sill 1 m above the floor and the wall is 5 m × 3 m.

Note that, since angle factors can be added, the angle factor for the wall can be found by subtracting the angle factor for the window from the angle factor for the wall without a window. Because of symmetry and for sill elevation at the body reference level, the number of different cases is reduced to three.

For the wall above 1 m, left or right

$$a = 2.5 \text{ m},\ b = 2 \text{ m},\ c = 2 \text{ m},\ b/c = 1,\ a/c = 1.25 \text{ and } F = 0.052.$$

For the wall below 1 m, left or right

$$a = 2.5 \text{ m},\ b = 1 \text{ m},\ c = 2 \text{ m},\ b/c = 0.5,\ a/c = 1.25 \text{ and } F = 0.035.$$

For the wall with no window

$$F_{p-0} = 2 \times 0.052 + 2 \times 0.035 = 0.174.$$

For the window above 1 m, left or right

$$a = 0.75 \text{ m},\ b = 1.0 \text{ m},\ c = 2 \text{ m},\ b/c = 0.5\ a/c = 0.375 \text{ and } F = 0.015.$$

For the window alone

$$F_{p-1} = 2 \times 0.015 = 0.030.$$

For the net wall area

$$F_{p-2} = F_{p-0} - F_{p-1} = 0.174 - 0.030 = 0.144.$$

It is now possible to calculate mean radiant temperatures for the human body.

Example 13.3

Calculate the MRT for the room and subject position of Example 13.2, if the room is 4 m deep, the inside surface of the exterior wall is at 15°C, the window surfaces are at 6°C and all other surfaces are at 22°C. It will not be necessary to calculate angle factors for the warm room surfaces separately, since they are all at the same temperature. The sum of all angle factors must be unity, so the angle factor for all warm surfaces together can be found by difference, as the angle factors for the cold wall and window are already known from Example 13.2.

The angle factor for all warm surfaces is given by

$$F_{p-3} = 1 - (F_{p-1} + F_{p-2}) = 1 - (0.030 + 0.144) = 1 - 0.174 = 0.83.$$

From Equation 13.11, and noting that temperatures may be in degrees Celsius,

$$t_{MRT} = 22 \times 0.83 + 15 \times 0.144 + 6 \times 0.030 = 20.5°C.$$

As a matter of interest, had the cold wall been all glass at 6°C, the MRT would have been

$$t_{MRT} = 22 \times 0.83 + 6 \times 0.174 = 19.3°C.$$

A rough check using Equation 13.12 yields for the wall with window

$$t_{MRT} \simeq (22 \times 79 + 15 \times 13.5 + 6 \times 1.5)/94 = 1949/94 = 20.7°C.$$

Similarly, for the case of the all-glass wall,

$$t_{MRT} \simeq (22 \times 79 + 6 \times 15)/94 = 19.4°C.$$

The approximate method applied in Example 13.3 yielded results within a fraction of a degree of the more exact method of Equation 13.11. This could not be expected in a case where a person is close to a large warm or cold area.

The value of F for one-quarter of a large cold surface cannot exceed 0.125, so in the extreme case of a person very close to a large panel the angle factor will not exceed 0.5 and the MRT will not be lower than the average of the warm and cold surface temperatures.

It is not known what the air temperature was in Example 13.3, but it could be expected to be a little above the surface temperature of the warm wall owing to the radiant cooling effect of the cold wall and window on the warm surfaces. This would mean that the MRT was more than 1 K below air temperature in the case of the window and more than 3 K in the case of the cold window wall. Values of MRT that differ from air temperature by 4 K or more need only be expected in situations involving relatively large hot or cold surfaces, as in the case of panel heating, cold window walls, high-temperature radiant heating, or solar radiation.

13.6 Thermal Exchange by Conduction

Thermal exchange by conduction is always present to some degree, since an object must be supported by connections or contacts with other bodies or with walls, ceilings, or floors. It is significant when there is good thermal contact and of little importance if the thermal connection is poor. It must be considered carefully in temperature-sensing devices such as mercury thermometers, thermocouples, and thermostats. Goods in storage in contact with cold walls or cold warehouse floors will often be at a different temperature from those in other positions, owing in part to conduction. Finally, the human body experiences an energy exchange by conduction through the feet; this can under certain conditions be a significant factor in thermal comfort.

13.7 Air Movement and Temperature Differences

Air movement and air temperature are key factors in convection, and are therefore important features of the indoor thermal environment. Air also serves as a diluent and as a carrier of gaseous and particulate contaminants such as smoke, dust, and odours as well as respiratory wastes, and for this reason the rate and pattern of air movement can also be of considerable importance.

The particular temperature to be provided in the occupied zone must be decided in relation to the needs of the occupancy. When active means for control of the thermal environment are used, the conditions at some selected control point can be monitored and adjusted as required. The conditions at other locations may differ, depending on many factors.

Air speeds up to 0.15 m/s can develop across floors owing to the downward convective flow immediately adjacent to cold walls and windows in winter. Speeds up to 0.5 m/s or higher can develop above hot objects such as radiators, stoves, and heaters; the speed depends on the extent to which the temperature of the air is increased above room temperature [**13.5**]. Rising and falling air streams, redirected to flow horizontally at ceilings and floors, contribute to a general air movement in the space, since any shift of air from one point to another must be accompanied by a compensating flow to provide a replacement. Rising and falling air streams also result in vertical temperature gradients.

Forced air streams produced by fans in or external to the space to deliver air and to promote air distribution and air movement may have speeds up to 2 m/s or even higher. Air noise associated with high air speeds is often a limiting factor, and air impinging on occupants at high speeds is usually objectionable, except under hot conditions. All streams of moving air will promote some general circulation through shearing at stream boundaries, entrainment, and displacement.

Vertical air-temperature gradients are a natural consequence of the vertical movement of heated and cooled air. An extreme example is provided by the air-space temperatures in walls and double windows as shown in figures 8.6 and 12.2. Large temperature gradients are produced, with the vertical movements taking place largely within 10 mm of the glass or wall surfaces. Vertical temperature differences are seldom so extreme in complete rooms since there are other factors influencing the room conditions, but the same general tendency exists. Temperature gradients from floor to ceiling vary from less than 1 K/m with some types of heating devices to as much as 10 K/m in extreme cases when air is heated to a high temperature in passing over very hot surfaces. Vertical gradients in rooms of 3 K/m are not uncommon.

It is not to be expected that vertical gradients in heated rooms can be reduced to zero without the use of fans, because of the natural tendency of warm air to rise; but they can be as little as 1 K/m. They are a complication in maintaining uniform conditions, and lead to high temperatures above the 2-m level where added heat is not needed, often resulting in increased heat losses. Additional difficulties in maintaining acceptable conditions arise when there are occupancies at two or more levels with open connections to high spaces, as in mezzanines and atriums.

The situation is different for summer cooling. The cooled air being introduced tends to fall to the occupied zone, which is where it is wanted. Gradients leading to high ceiling temperatures can often be tolerated in such cases, at least in rooms of normal ceiling height.

The air-temperature differences from point to point in a room are dependent on the condition, movement, and extent of mixing of air streams at various temperatures that result from convection, infiltration, and air supply. These and other matters related to the indoor thermal environment can best be examined further in the context of the various ways in which the heating and cooling of rooms are accomplished.

13.8 Adjustment of the Thermal Environment

It is commonly assumed that heating and cooling devices are intended to adjust the air temperature in a space. It should now be clear that although air temperature is often the most important single index of the indoor thermal conditions, the thermal exchange of the object (and thus its temperature) can be significantly influenced by MRT and air movement as well. This raises some interesting considerations when temperature must be closely controlled and known [**13.6**]. The thermal exchange and thermal comfort for human subjects must take into account the heat and moisture they generate as well as their ability to control body temperature within a wide range of conditions. Evaporation enters into the energy exchange, and so relative humidity must also be taken into account. These matters will be dealt with more fully in the next chapter.

The means used to adjust the thermal environment to suit a particular occupancy may be an open fire; space heaters; radiators or convectors; the use of heated or cooled panels; infrared heaters; or the conditioning of an air stream as to heat and perhaps moisture, which is then introduced into the room, displacing an equal amount of room air. Partial measures may involve humidifiers or dehumidifiers in the space to control the relative humidity, fans to modify air-temperature gradients and air speed, and the introduction of air for ventilation. Combinations of these measures may also be used.

The equipment and systems used may be adjusted manually, as in the case of open fires and stoves, to satisfy conditions judged subjectively or measured by a thermometer at some fixed location. In most other cases the adjustment may be automatic to satisfy the conditions called for by a thermostat or hygrostat or both. The thermostat may be influenced by radiation and conduction as well as temperature. If so, it may have a body temperature somewhat different from air temperature and may be sensitive to air speed as well. The hygrostat intended to read relative humidity will also be affected by the temperature and radiation. Both kinds of instruments can only respond to the conditions where they are located, which may be on a wall at some fixed height, usually about 1.6 m, in a duct outside the space, or in the air-conditioning devices. Conditions elsewhere may vary widely, depending on differences in the contributory factors.

Engineering practice in the estimation of loads and the design and operation of equipment is dealt with at length in many books and in the current editions of the ASHRAE Handbooks and Supplements [**13.7–13.9**]. Some consideration of how the indoor thermal environment is influenced by the combination of the building and its equipment is useful here.

13.9 Open Fires, Fireplaces, and Small Heaters

Open fires fed by some form of solid fuel are the oldest manmade method of heating. The heat contained in the stream of heated air and gases of combustion rising over the fire provides no warmth in the immediate environment. Only that heat which is emitted radially outward as radiation, mainly from the glowing carbon and flames, is useful; it will be greatest from glowing coals in a freestanding basket. The intensity of radiation varies with distance from the source in accord with the inverse-square law. Unheated air moving in toward the fire can produce unwanted cooling, and positions downstream from the fire may be polluted by smoke and fine ash.

When open fires are used indoors, some of the radiant energy that would otherwise be lost is intercepted by the room surfaces and redirected, leading to some improvement in the thermal environment. The recirculation of smoke and combustion gases is usually a problem, and can be combated by large vent openings above the fire which inevitably cause a substantial inflow of outdoor air.

Open fireplaces with chimneys provide some assurance that smoke will be vented to the outdoors. They radiate directly through the front opening. The proportion of the energy in the fuel that is delivered directly as radiation, known as the radiation efficiency, will seldom be higher than 25% for well-designed grates burning coal [**13.10**] and may be no more than 12 to 20% for fireplaces burning wood fuel. A fire burning 1 kg of coal per hour exhibits a radiant intensity of 650 W/m^2 at a distance of 1 m; the intensity of direct normal radiation from the sun is about 850 W/m^2.

Some of the energy radiated and some of the heat in the rising stream of air and combustion gases is intercepted by the body of the fireplace. A limited portion of this may be returned to the room; the amount depends greatly on many features of construction and operation. Arrangements may be made to allow room air to circulate through channels provided at the back and sides of a metal fireplace, but the proportion of the heat in the fuel that can be recovered in this way is limited by the large uncontrolled flow of room air over the fire and up the chimney, which dilutes and cools the gases and cools the fireplace.

It is reported that the average air speed through the open face must be at least 0.24 m/s to ensure complete venting of the gases [**13.11**], requiring a flow of 180 L/s for a fireplace opening 1000 × 750 mm. This rate will produce 6.5 ac/h in a room 8 × 5 × 2.5 m. As many as six to ten air changes per hour were estimated in rooms heated by the open domestic grates commonly used at that time in Great Britain [**13.12**]. The air flow that leaves the room by way of the chimney must be made up by air from outdoors. It will travel across the floor and is unlikely to be warmed appreciably on its way to the fireplace. It was acknowledged in the 1925 British studies that fireplaces can only produce acceptable conditions for thermal comfort in a limited area in front of the fire, and then only if the outdoor temperature is above about 7°C [**13.10**].

When houses are tightly constructed, the supply of air to fireplaces may be a problem. One possible solution other than opening doors or windows is to provide an opening from outdoors directly into the fireplace. The problem of chimney-draft failure was discussed in Chapter 11.

Small unvented oil- and gas-fired heaters deliver to the room all the heat released from the fuel burned. They can be designed for maximum output by radiation, which can be made highly directional. The gases of combustion released into the space along with any vaporized unburned fuel are well diluted and are usually tolerable, but may be hazardous if the device for any reason produces relatively high amounts of carbon monoxide, as is possible when flames impinge on cold surfaces. Adequate ventilation should always be provided in spaces so heated. Combustion gases may be objectionable to some people who are allergic to them. Electric heaters produce no combustion gases, and can be made to deliver as much as 80% of their output by radiation.

13.10 Stoves and Heaters

Stoves provide a marked improvement over open fireplaces, since they make it possible to control the amount of air and to direct it to the burning fuel. This results in better control over combustion and also means that a smaller volume of much hotter gases is produced. The walls of the stove are heated by the direct radiation from the fire and by the hot combustion gases, and transmit energy to the room by radiation and convection. By providing an extended path for the hot gases and increasing the heat-transfer surface, it is possible to extract a large part of the heat in the gases before they reach the smoke pipe. Practical considerations limit the extent of the heat-transfer surface provided, and the total heat recovered seldom exceeds 50% of the energy of solid fuel burned in small stoves. Higher efficiencies are possible with furnaces and with liquid and gaseous fuels.

Since stoves are usually simple in shape and limited in size to suit individual rooms, they often present relatively hot surfaces to the space. This leads to high radiation intensities, varying inversely as the square of the distance from the stove, and to highly heated convection streams which rise rapidly, producing high vertical temperature gradients. High temperatures at ceiling level have often led to warm floors and to a kind of panel heating in the room above. Air flowing down cool exterior wall surfaces or air entering as infiltration will move toward the stove, leading to reduced temperatures at floor level and the possibility of discomfort from drafts. Conditions can be modified by enclosing the stove so that it gives up most of its heat by convection. High vertical and horizontal temperature gradients are likely to persist unless the rate of air circulation and mixing with room air are increased and the temperature of the stream of heated air correspondingly decreased.

Space heaters burning gaseous or liquid fuels are usually enclosed and designed to provide a maximum of heat by convection with air delivered through suitable openings in the enclosure at moderate temperatures acceptable to the touch. Vertical temperature gradients in rooms so heated may be little more than 3 K/m, and may be reduced even further when blowers are incorporated to promote still higher circulation rates and greater directional control of heated air streams.

13.11 Radiators, Convectors, and Baseboard Units

Central heating involving the burning of fuel, extraction of heat, and venting of combustion products in a single unit serving several rooms eliminates the need to service stoves in individual living spaces. A fluid is circulated to carry heat to individual locations throughout a building as required. When air is used as the heat carrier it is delivered through ducts into each room to be exchanged for room air. Hot water, used in wet systems, is circulated through supply and return piping to serve various heat-distributing units. When steam is used, it is condensed in the heat-distributing units and returned as condensate to the central boiler for reuse. Steam is commonly supplied at pressures a little above atmospheric to facilitate the purging of air from the system, at a temperature of about 103°C. Water-supply temperatures may be as high as 100°C but are more commonly 80°C or less. Lower temperatures lead to smaller vertical air temperature gradients, but require larger heating units for the same output. Radiators, convectors, and baseboard and finned-tube units are used as the heat-distributing devices.

Radiators commonly consist of an assembly of vertical, cast-iron, tubular sections, which in use are filled with the heating fluid. Air in contact on the outside is warmed and rises in the vertical channels formed by the heated sections. A substantial portion of the output is in the form of radiation, since hot surfaces see the room. Temperatures of the rising air streams are less and air circulation rates greatest when radiators are long and low. The influence of radiator shape on vertical temperature gradients is shown in Figure 13.6. Room-air circulation with three radiator applications is shown in Figure 13.7 [**13.13**]. Locations at the sources of cold air such as windows and exterior walls are preferable, since the falling convection and window-infiltration streams can be interrupted and mixed with warmer air before they reach the floor. The radiant output also compensates for the radiant cooling effect of the window and wall surfaces.

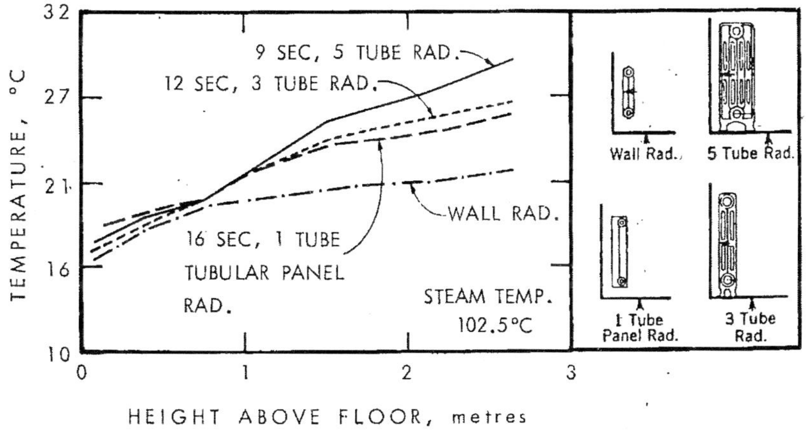

FIGURE 13.6 Room thermal conditions produced by radiators.
(From Severns, W.H., and Fellows, J.R., *Heating, ventilating and air conditioning fundamentals*, New York: John Wiley and Sons, 1949, Fig. 93, p. 217)[**13.25**]

FIGURE 13.7 Room-air circulation with radiator (a) under window (b) at opposite wall (c) partially enclosed.
(From Willard, A.C., Kratz, A.P., Fahnestock, M.K., and Konzo, S., *Investigation of Heating Rooms with Direct Steam Radiators Equipped with Enclosures and Shields*, University of Illinois Engineering Experiment Station Bulletin 192, 1929, pp. 66, 67.)

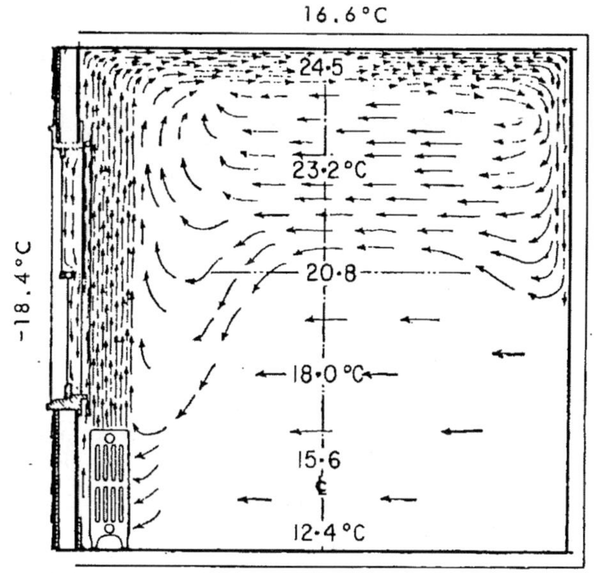

(a)

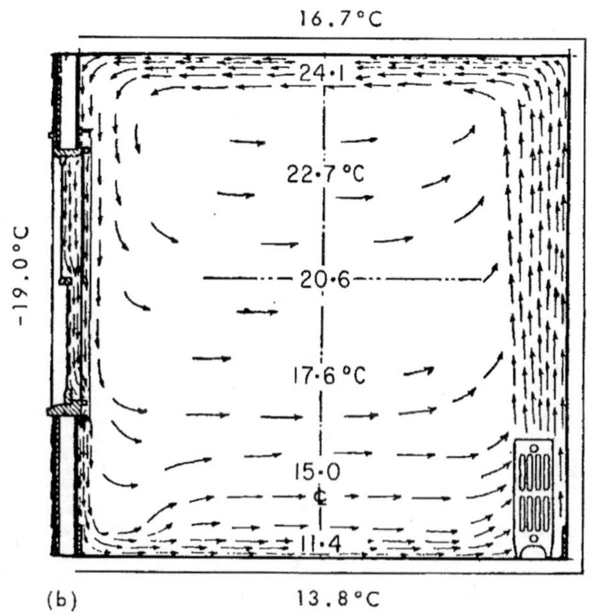

(b)

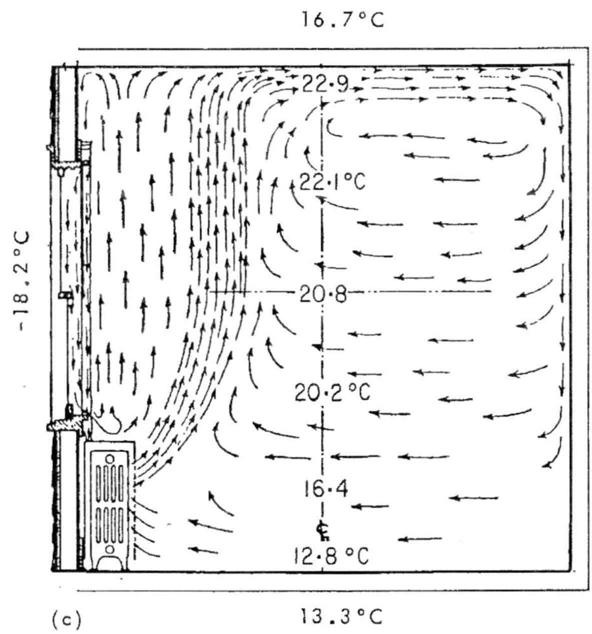

Convectors have heating elements of two or more horizontal tubes fitted with transverse fins or plates to provide a greatly extended surface for heat transfer. These elements are enclosed in metal cabinets. Air enters at the bottom, contacts the heating surface, and rises in the cabinet to leave by openings in the top or front. The output is largely by convection. Convectors can be freestanding, wall-hung, or recessed (Figure 13.8).

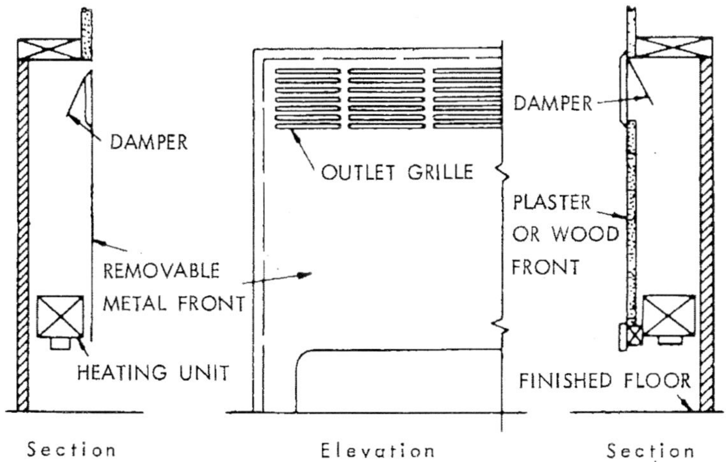

FIGURE 13.8 Typical recessed convector.
(From *ASHRAE Handbook 1983 Equipment*, Fig. 1, p. 29.2)[**13.11**]

Baseboard heating units are small in vertical section and very long for a given output. They are made to be located along exterior walls, taking the place of the traditional baseboard. The radiant-convector type is of steel or cast iron and is arranged for air circulation up the back side next to the wall; the exposed heated face provides some exchange by radiation. The finned-tube type consists of an extended surface finned tube in a metal enclosure. Baseboard units deliver heat near the floor and can be arranged to cover the full length of cold walls. They lead to low vertical temperature gradients which may be less than 1 K/m [**13.14**].

Finned-tube elements can also be used in a variety of ways with enclosures designed for special purposes. They may, for example, be incorporated in window sills to offset the cold downward air flow.

13.12 Unit Heaters and Ventilators

Heating elements can be combined with fans or blowers in an enclosure for use in space heating; such devices are called unit heaters. They are commonly designed for steam or hot water but may also be indirectly gas-fired or even oil-fired. When provided with air connections to the outdoors they become unit ventilators and can be arranged for combined heating and ventilating. The various configurations and uses are well described in the ASHRAE Handbooks; significance of forced-air streams is of decided interest here.

The earliest unit heaters consisted of an extended surface coil heated by steam or hot water and arranged in a simple casing for horizontal blow by propeller fans (Figure 13.9). They are usually located above head level with the directional

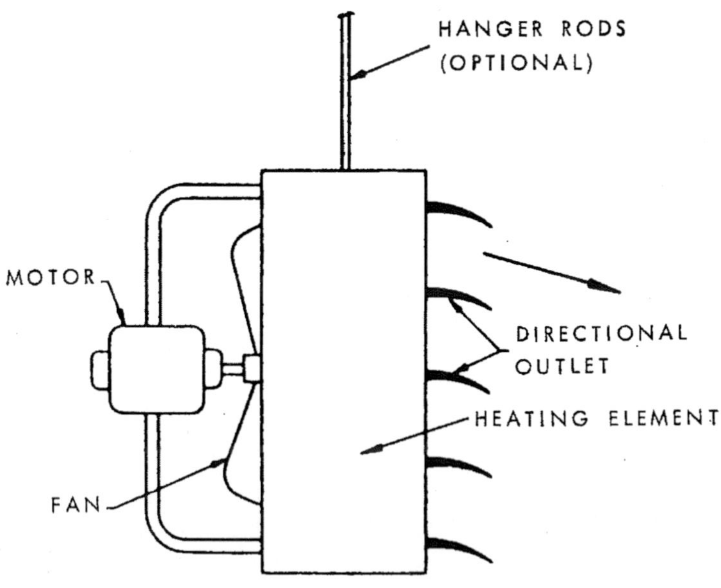

FIGURE 13.9 Propeller fan unit heater — horizontal-blow type. (From *ASHRAE Handbook 1983 Equipment*, Fig. 6, p. 28.4)[**13.11**]

outlet vanes adjusted to project the heated air downward at an angle into the occupied zone. Unless the cool air collecting at floor level was stirred vigorously, temperatures could remain low there. Down-blow types of unit heaters with more powerful propeller fans produced strong jets of air to scour the floor and break up the cool layers, but they also produced hot spots, with high air speeds in the occupied zone at the heater locations. It was discovered that the accumulation of cool air at floor level could be avoided by a combination of withdrawal, displacement, and mixing, and the cabinet types of unit heaters equipped with centrifugal fans drawing air from floor level and discharging it at high speed above head level began to be used (Figure 13.10). It is mainly these cabinet types that are adapted to provide ventilation along with heating.

Unit heaters can provide high output and large area coverage in a single unit. They are of great use in commercial and industrial applications. Cabinet units are used extensively in school classrooms, which require high ventilation rates when occupied (Figure 13.11).

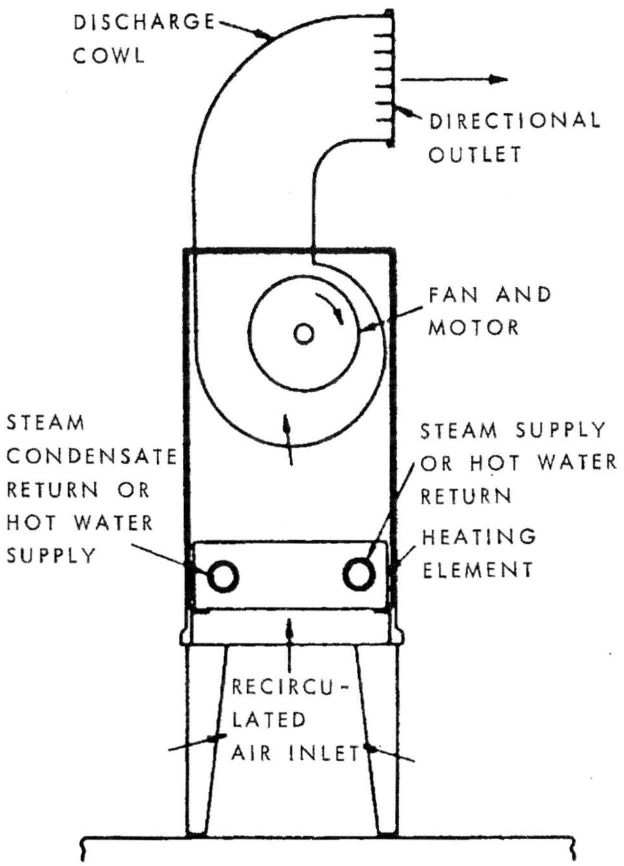

FIGURE 13.10 Floor-mounted centrifugal fan unit heater — industrial type. (From *ASHRAE Handbook 1983 Equipment*, Fig. 11, p. 28.5)[**13.11**]

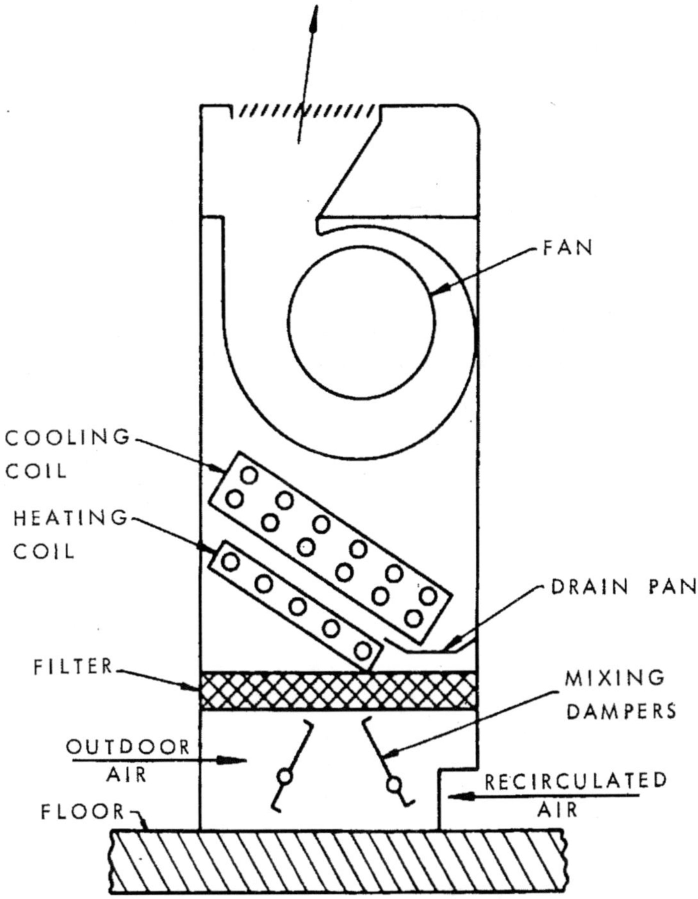

FIGURE 13.11 Typical air-conditioning unit ventilator with separate coils. (From *ASHRAE Handbook Educational Supplement 1978*, Fig. II-1, p. 162)[**13.8**]

13.13 Buoyancy Forces and Fans

The energy required to overcome buoyancy forces in moving heated or cooled air as much as 10 m vertically through an air layer at some different temperature is only a small fraction of the differences in enthalpy at the two temperature conditions. It requires far less energy to move heated air in to replace cold air than it does to heat the cold air that is already there. This helps to explain why vertical gradients develop so readily, and it also suggests that the use of fans to move large quantities of air from one level to another in order to reduce gradients can be accomplished using relatively little energy. For example, a small disk fan located at floor level discharging upward in a room heated by a stove can do much to overcome the high vertical air-temperature gradient and thus to conserve fuel and improve comfort.

Fans can also be used in conjunction with unit heaters or other heat sources to ensure good vertical distribution in many industrial applications. Since they can be arranged to draw air from floor level, the difficulties with drafts associated with blowing air into the occupied zone are avoided. The kinetic energy in the fan or duct discharge can also be effective in producing mixing above the occupied zone.

13.14 High-Intensity Infrared Heating

The sun qualifies as a high-intensity radiant heater. It is well known that a person feels warmer in the sunshine than out of it, but it is not always appreciated how much heating the sun provides. The equivalent sol-air temperature described in Chapter 9 gives some indication of this; for plane surfaces normal to the sun it can be as much as 50 K above air temperature. The direct normal radiation at the earth's surface is about 850 W/m^2. High-intensity infrared heaters are available in a range of types and outputs (Figure 13.12). Typical electric heaters provide from 600 to 1500 watts per steradian (W/sr), and typical gas-fired heaters provide from 2300 to 5500 W/sr. The lower values for each type are for broad-beam heaters, and the higher ones for narrow-beam devices. Output data for a typical broad-beam gas-fired heater are shown in Figure 13.13, and for an electric heater in Figure 13.14. These devices can be used indoors or outdoors. Single units can provide an augmented MRT for fixed work stations, and an array of heaters can provide acceptable thermal conditions over large areas.

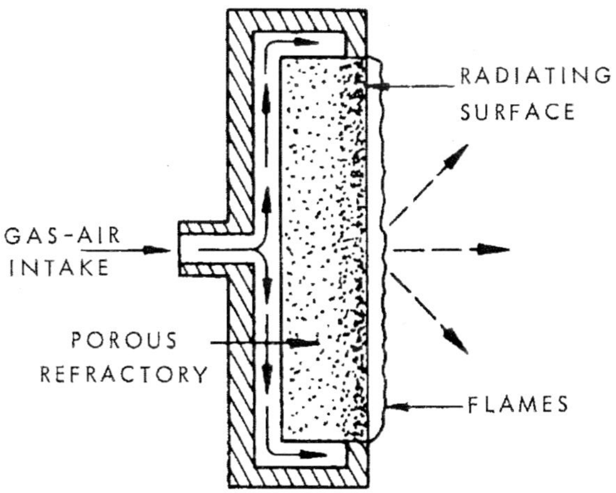

FIGURE 13.12 One type of gas-fired infrared burner.
(From *ASHRAE Handbook 1983 Equipment*, Fig. 1(c), p. 30.2)[**13.11**]

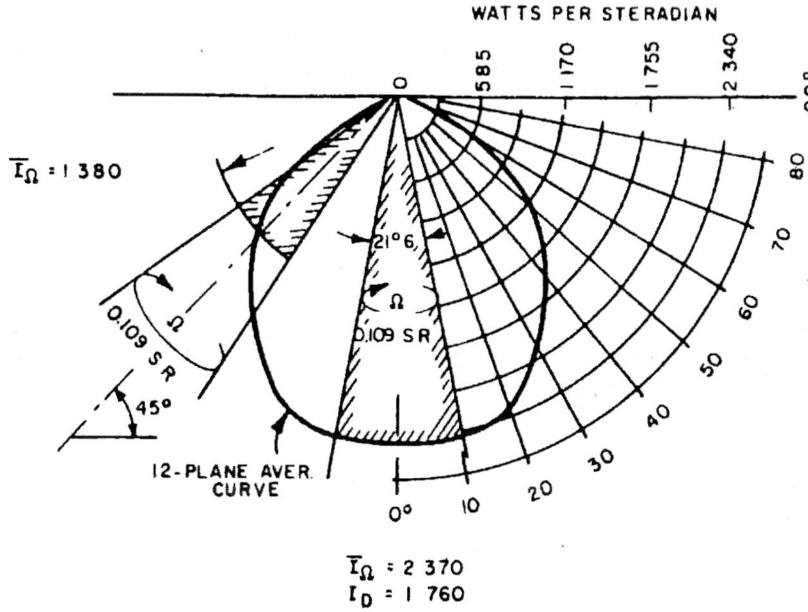

FIGURE 13.13 Radiant flux distribution curve of typical broad-beam high-intensity atmospheric gas-fired infrared heaters.
(From *ASHRAE Handbook 1980 Systems*, Fig. 6, p. 19.3)[**13.23**]

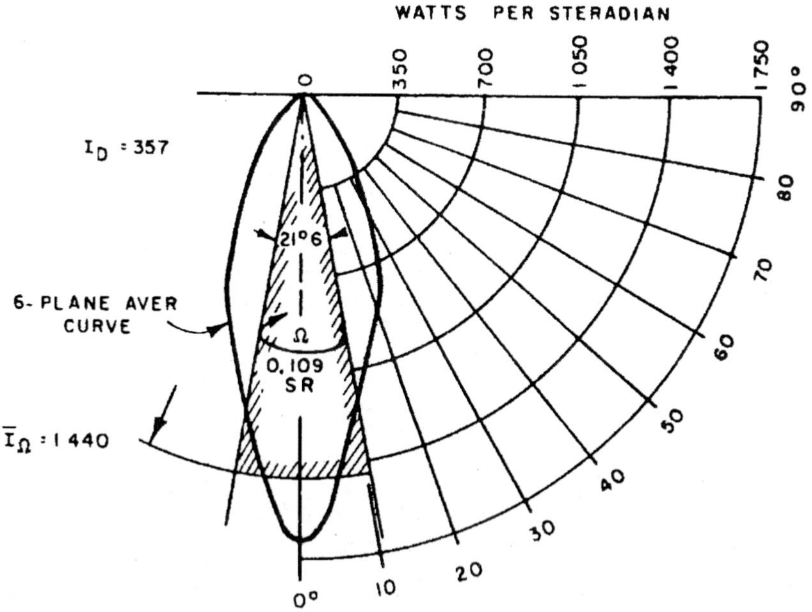

FIGURE 13.14 Radiant flux distribution curve of typical narrow-beam high-intensity electric infrared heaters.
(From *ASHRAE Handbook 1980 Systems*, Fig. 3, p. 19.2)[**13.23**]

The sun is so far away that a few metres more or less make little difference in its radiation intensity, which can be expressed as watts per square metre. Infrared heaters, however, will be located within a few metres of the receiving objects and surfaces, and distance from the heater becomes very important in determining the intensity of radiation since it varies inversely as the second power of the distance from the source. It is convenient to express the intensity of radiation, I_{sr}, as watts per steradian. The steradian is the unit of measure of solid angles. For present purposes the solid angle subtended by a given area may be taken with sufficient accuracy as the area normal to the radiation divided by the square of the distance to the radiating source. The intensity at any distance r is given by

$$I = I_{sr}/r^2 \qquad (\mathrm{W/m^2}) \qquad (13.13)$$

The use of the steradian in dealing with visible light has already been described in Chapter 4.

Example 13.4

The intensity of radiation from a heater is 1500 watts per steradian. Find the radiation that will be received on an area of 0.5 m² normal to the radiation at a distance of 3 m from the source.

Given $I_{sr} = 1500$ W/sr, the intensity I at 3 m will be

$$I = I_{sr}/r^2 = \frac{1500}{3^2} = 167 \mathrm{\ W/m^2}.$$

The radiation received on 0.5 m² will be $0.5 \times 167 = 83$ W.

Infrared heaters can, in principle, be used to provide a greatly augmented MRT for a particular work station or room position to compensate for high convection losses and a high net radiation loss to surroundings at low temperature. Small gas and electric radiant heaters and open fires have long been used in this way. When comfort conditions must be provided indoors over a wide area, an overhead array of heaters is required. The radiation received by the low-level surroundings (and particularly the floor) tends to raise both air and surface temperatures. It becomes difficult when designing for such situations to estimate the radiant exchange with the surrounding surfaces which, in combination with the direct contribution from the source, determines the MRT for the subject.

13.15 Adjustment of Mean Radiant Temperature

The MRT required to provide comfort for a given set of conditions can be determined from comfort considerations (to be discussed in the next chapter). It is necessary also to know the projected area, A_p, for the human body, which is presented to the radiating source. Values of A_p have been measured by Fanger for the full range of positions of the radiating source expressed in azimuth and altitude for the standing and seated human body [**13.4, p. 169**]. It was found that these could be expressed as projected area factors to be applied to the surface area of the human body as found by the Dubois equation, and that differences due to clothing and sex could be disregarded. The Dubois equation is

$$A_{Du} = 0.203\, w^{0.425} \times h^{0.725} \tag{13.14}$$

where

A_{Du} = body surface area, m²
w = body mass, kg
h = height of person, m.

A further simplification of the Fanger data is shown in Table 13.1, which gives values of the ratio A_p/A_{Du}. Fanger has also proposed that the effective area of the human body, for radiation from all directions, be expressed as

$$A_{eff} = f_{eff} f_{cl} A_{Du} \quad m^2 \tag{13.15}$$

TABLE 13.1

Area Ratio A_p/A_{Du} for Different Radiant-Source Positions
(From *ASHRAE Handbook 1980 Systems*, Table 3, p. 19.4) [**13.15**]

Seated posture

Azimuth angle, ϕ, degrees	Altitude angle, α, degrees			
	0 or 75	15 or 60	30 to 45	90
0 to 15	0.25	0.27	0.29	0.21
30 to 45	0.24	0.26	0.27	0.21
60 to 75	0.23	0.24	0.24	0.21
90	0.21	0.21	0.21	0.21
105 to 120	0.20	0.19	0.19	0.21
135 to 150	0.19	0.17	0.16	0.21
165 to 180	0.18	0.16	0.14	0.21

Standing posture

ϕ, degrees	Altitude angle, α, degrees						
	0	15	30	45	60	75	90
0 to 180	0.30	0.29	0.27	0.24	0.20	0.15	0.10

The clothing area factor f_{cl} was found to be 1.1 for seated persons and 1.2 for standing persons for a "standard" clothing ensemble. The effective radiation area factor, f_{eff}, was found to be about 0.7 for sedentary body posture and 0.73 for standing posture.

The absorptance of the skin and clothing, α_m, varies with the temperature of the radiating source, the proportion of exposed skin to clothed area, and skin colour: it is 0.95 for the nude body and 0.90 for clothing at source temperatures up to 1200 K, and decreases to as low as 0.6 for white nude subjects and radiation at the sun temperature of about 5800 K [**13.15, p. 19.1**].

The rate of heat loss by radiant exchange between the clothed human body and the augmented radiant environment can be set equal to the rate of heat loss from the body to the surroundings, decreased by the contribution from the radiant heating source in an equation [**13.15, p. 19.4**] that follows closely that proposed by Fanger [**13.4, p. 149**].

$$\epsilon_p \sigma A_{eff} (T_{cl}^4 - T_{MRT}^4) = \epsilon_p \sigma A_{eff} (T_{cl}^4 - T_{SMRT}^4) - \alpha_m I_{sr} (A_p / r^2) \quad (13.16)$$

where

ϵ_p = emissivity of the person (about 0.97)
σ = Stefan Boltzmann constant, 5.67×10^{-8}
T_{cl} = average surface temperature of the clothed body, K
T_{MRT} = mean radiant temperature, K
T_{SMRT} = mean radiant temperature of surroundings, less the radiant source, K
α_m = absorptance to the source radiation
I_{sr} = intensity of heater radiation, W/sr, in the body direction
A_p = projected body area, m^2
r = distance from radiant source to the person, m

Rearranging and simplifying Equation 13.16 yields

$$T_{MRT}^4 = T_{SMRT}^4 + (\alpha_m I_{sr} A_p)/(\epsilon_p A_{eff} \sigma r^2) \quad (13.17)$$

Examples of the use of this equation are given in the ASHRAE Handbook [**13.15, p. 19.4**].

A field experiment [**13.16**] provides some interesting performance results on a comparison of convective and radiant heating systems in two buildings with ceiling heights of 4.9 and 9.2 m. Convective systems, both gas and electric, used about 15% more energy than the radiant systems when operated to provide the same air temperatures at the 1.5 m level. This could be attributed in one case to the increased combustion efficiency in the radiant devices compared with the convective gas-fired equipment. In the other case (with electric heating) the increase in energy with convective heat could be accounted for by the floor-to-ceiling temperature differences up to 10°C leading to higher ceiling and wall losses. Ceiling temperatures were almost the same as those at the 1.5 m level with both radiant systems.

13.16 **Panel Heating**

Entire walls, floors, or ceilings can be heated to serve as heat-disseminating units in panel-heating systems. Panels can be warmed by electric-resistance cables or by hot-water pipe coils embedded in them [**13.17**]. Ceiling panels can be operated at temperatures up to 55°C in rooms 2.5 m high; they contribute heat mainly by radiation exchange, with a small convective output. Temperatures above 55°C lead to deterioration of gypsum plaster and high mean radiant temperature at head level, but can be used successfully with appropriate design in rooms with high ceilings.

Floor panels have a high convective and a high radiant output, and operate at surface temperatures up to 29°C, the limit based on foot comfort [**13.17**]. When carpets are used, panel temperatures can be increased to produce carpet-surface temperatures up to the tolerable level.

Panel areas may often have to be extended to cover the entire floor or ceiling in order to generate sufficient output for severe heating requirements. Ceiling panels promote little general air circulation and do not prevent the accumulation of cool air at floor level from cool wall and window surfaces. Both air temperature and MRT increase appreciably with height above the floor. Floor panels promote more active general circulation and can deal with cold air at floor level, but not necessarily more effectively than some other systems.

The design of panel systems can be greatly simplified by the assumption that all the unheated surfaces enclosing the heated panel are at a surface temperature equal to their area-weighted average surface temperature (AUST). The radiant output of the heated panel can then be found from the general equation 4.14 developed in Chapter 4.

$$q_r/A = \sigma F_A F_E [(t_p + 273)^4 - (t_{AUST} + 273)^4]. \qquad (13.18)$$

The value of $F_A F_E$ combined is commonly found from an expression developed by Hottel, and for average conditions has a value of about 0.87 [**13.17**]. Panel output by convection can also be calculated using the values given in Chapter 8, and the combination of radiant and convective output can be equated to the room-heat requirements. Account must be taken of heat loss from the back of the heated panels, a loss that does not contribute to room heating but has to be made up by the embedded heating cables or pipes.

Experiments with covered hot-water-heated floor panels were carried out in a special test facility having insulated floor slabs on grade [**13.18**]. Two of the four rooms had only one exposed wall, facing north, 4.6 m long, with a single-glazed window amounting to 10% of the floor area. Rooms were 3.7 m deep. The other two rooms, similar in size, had twice as much glass area, and one of them had a second exposed wall. Walls and ceilings were insulated.

Average room-air temperatures differed by 2 K from 76 mm above the floor to 76 mm below the ceiling for the rooms with 20% glass and by 1.3 K for those with 10% glass, for a 44 K indoor–outdoor temperature difference. Cool air dropped at the exposed wall, producing speeds across the floor as great as 0.2 m/s, decreasing to 0.15 m/s toward the back of the room at an outdoor temperature of -1°C. These levels were considered to be capable of causing drafts under extreme conditions. The MRT measured at the centre of one room 76 mm above the floor closely approximated the area-weighted average surface temperature. Panel surfaces were operated at temperatures up to 8.5°C above air temperature.

An extensive study of a ceiling-panel-heated room was carried out in the ASHRAE Environment Room with controlled panel-surface temperatures. The room was 7.5 m × 3.7 m × 2.9 m high [**13.19**]. When one long wall was held at -2.6°C and the ceiling area was heated over an area extending 2.44 m back from the cold wall to 70°C, the room-air temperature 1.5 m above the floor was 21°C, and floor and wall surfaces (which were effectively thermally neutral since the panels were unheated but insulated) were at 25°C. Air temperatures 50 mm above the floor and 300 mm from the cold wall dropped by as much as 5 K below those at the 1.5 m level, and were generally from 2 K to 3.5 K lower across the floor at the 305 mm level. Air falling at the cold wall formed a layer 150 to 300 mm thick across the floor with speeds as high as 0.25 m/s near the cold wall, decreasing to about half that value 2.7 m away, as shown in Figure 13.15. Globe and MRT values measured at various locations were relatively high at the 1.5 m level and above. Many tests were conducted under other conditions.

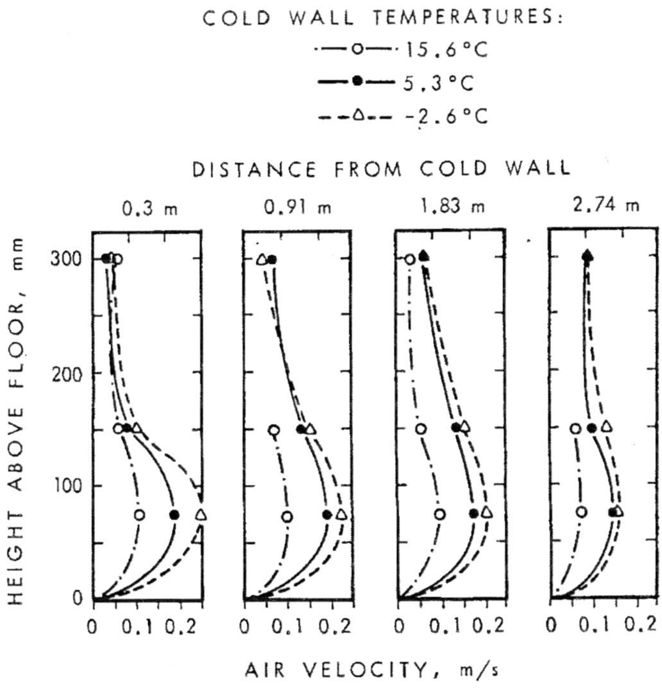

FIGURE 13.15 Air velocity near floor for three cold-wall temperatures, 8 ft (2.4 m) panel width.
(From Schutrum, L.F., and Min, T.C., *ASHRAE Trans.*, 1957, *63*, Fig. 6, p. 194)[**13.19**]

The window-wall surface temperature of -2.6°C in the ceiling-panel test corresponds to an outdoor temperature of about -13.5°C for single glazing. The flow of cold air leaving over the length of the cold wall (shown in Figure 13.15 at an average speed of 0.2 m/s for a depth of 150 mm) can be shown by a simple calculation to amount to 30 L/s for each metre of wall length, which is enough to produce a circulation rate of ten air changes per hour in the room. The ventilation rate through a ceiling diffuser with air at 21°C was only 1.5 ac/h. Further simple calculations show that the convective heat loss at the cold wall would cool a stream of air flowing at the rate calculated above by about 9.5 K.

The wall-surface temperature of -2.6°C represents an extreme condition attainable only with single glazing. Another test (Figure 13.15) at a cold-surface temperature of +5.3°C provided a more reasonable representation of double glazing at an extreme outdoor temperature of about -36°C or of single glazing at -1°C. The average air speed in the 150 mm layer was then found to be 0.15 m/s, which simple calculations show to be able to produce 7.5 ac/h and a temperature drop of 7.5 K in the falling air stream. A less severe case was illustrated by the test at +15.6°C surface temperature, which might be produced by an outside air temperature of 1°C with double glazing or 13°C with single glazing. The test showed an average air speed of 0.10 m/s over the 150 mm depth capable of producing 5 ac/h and an air-stream temperature reduction of 2.5 K.

These results cannot readily be related to the actual conditions over the cold wall owing to variations in air temperature, flow pattern, and mixing, but they should make clear that falling convection streams at cold walls and windows can have a marked influence on air speeds and air temperatures at floor level, and on room air circulation.

The tests with the heated floor panels [**13.18**] already described also showed air flows across the floor at about the same air speeds, despite the possibility that the heated floor might have dispersed these quickly. The measurements were made opposite the centre of the exposed wall that had single-glazed windows amounting to 20% of the floor area. In this case the conditions were not uniform over the whole length of the wall, as in the ceiling-panel tests discussed above.

13.17 Residential Warm-Air Heating

All heating systems warm the air. In systems used for residential warm-air heating the air is drawn from the room, heated by passing it over the hot surface in a furnace in some central location, and returned to the room. The furnace may be fired with gaseous, liquid, or solid fuel or equipped with electric heating coils.

The pipeless furnace was an early development; it was essentially a stove removed to a central basement location and equipped with a double-walled enclosure and a large supply-and-return register in the floor above. More uniform distribution of heated air throughout a house became possible with the gravity warm-air furnace equipped with individual supply ducts feeding registers in each room. This was followed by the forced warm-air system, which had a fan that permitted a wider choice of furnace location, smaller ducts, and the use of filters to clean the air. An air connection to the outdoors to provide fresh air could also be used. Humidification was either added or provided from separate units in the living space. There were advantages to a basement location for the storage of solid fuels, but the use of gas, oil, and electricity made it possible to erect basementless houses on concrete slabs placed on or near grade level. This led to the development of perimeter duct systems, where ducts were buried in or under the slab; they delivered air through small openings in the floor at perimeter locations fed from downflow furnaces located above grade level. The perimeter ducts also provided heating to the floor slabs, which offset a tendency to cold floors; thus, the system provided a combination of radiant and warm-air heating.

Various locations and arrangements of supply-and-return ducts and the associated diffusers, registers, and grilles were used with forced-air systems. The arrangement most widely used in Canada, where single-family houses with full basements are common, is the small-pipe forced-air system which delivers air through small floor registers located beneath windows and at exterior walls. Air can be raised as much as 50 K in temperature to accommodate maximum heating loads and circulated at the rate of about three air changes per hour, more or less, to carry heat and ensure good room-air distribution. Control is usually by on–off operation of the furnace, but the blower can be run continuously.

Cooling units can be added or included for summer use, using the same air-distribution system (which must be suitable for both heating and cooling). Basic central heating and cooling equipment is being changed with the addition of new equipment such as solar devices, heat pumps, thermal storage, and heat exchangers, but the problem of room air distribution is basically much the same for all conditioned air systems, whether for houses or for larger buildings.

13.18 Air Conditioning

Heating can be accomplished through radiation or by the provision of heated air. Cooling is generally achieved with suitably conditioned cooled air. Thus, air conditioning can be used to provide both heating and cooling as required. Humidification is usually required during the heating season, while summer cooling is commonly accompanied by the need for dehumidification, which is a natural consequence of cooling air below its dewpoint. Ventilation and air cleaning may be required in all seasons.

Air conditioning is defined as the process of treating air to control simultaneously its temperature, humidity, cleanliness, and distribution in order to meet the requirements of the occupancy of the space being conditioned. Its purposes have now become so varied that it can be further qualified as comfort, industrial, winter, or summer air conditioning. It is the responsibility of the heating, ventilating, and air-conditioning (HVAC) engineer to ensure that so far as possible the condition and quantity of the air delivered will meet the requirements. Although space conditioning is dependent in the first instance only on the condition, quantity, and distribution of the air, the overall result—including costs, energy requirements, control and degree of refinement of space conditions, ease of operation, space for equipment, and noise—can be greatly influenced by the combination of the systems and equipment used and the characteristics of the space and its occupancy. Thus, many of the decisions of the building designer also influence the final result.

Air may be suitably conditioned in a central plant and circulated to all rooms through supply-and-return duct systems as the principal means of controlling room conditions. In other systems both air and water may be used as carriers, with chilled and hot water supplied to provide additional heating or cooling of the air, as required, in individual zone units or terminal units. There are also many types of packaged and unit mechanical-refrigeration conditioners, which are more or less self-contained apart from electric power, water, and outdoor-air connections. Extensive information on most aspects of HVAC can be found in the ASHRAE Handbooks. Factors that affect the quality of the thermal environment will be discussed further here.

13.19 Heat and Moisture Balances for a Room

Air conditioning involves adjusting the condition of the supply air appropriately. The problem is illustrated in Figure 13.16 [**13.20**]. The space to be maintained at a temperature t and relative humidity r has heat gains and moisture gains from all sources except from the incoming air, amounting to H kilowatts and M kilograms of moisture per second. The incoming air must be deficient in its heat and moisture content, so that heat and moisture balances will be established as follows:

$$w(h_2 - h_1) = H \tag{13.19}$$

$$w(W_2 - W_1) = M. \tag{13.20}$$

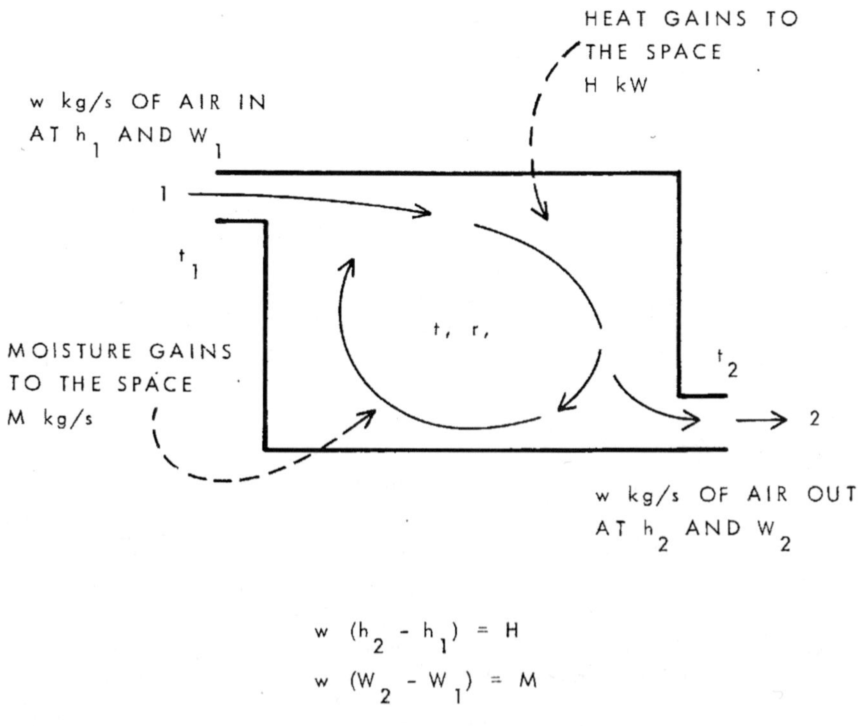

FIGURE 13.16 Heat and moisture balance for a room.

The condition of the air exhausted at condition 2 will therefore be close to that maintained in the occupied zone, that is, at temperature t and relative humidity r. The next consideration is how to achieve the proper heat and moisture relationship with a single-supply air condition. The ratio of heat to moisture to be removed is given by

$$\frac{H}{M} = \frac{h_2 - h_1}{W_2 - W_1}. \tag{13.21}$$

Since the psychrometric chart (Figure 6.1) is based on uniform linear scales for enthalpy h and humidity ratio W, the ratio of incremental differences of these two quantities is constant along any straight line and is given by the slope of the line as discussed in Chapter 6. It is possible, therefore, to find a line through the state point for the room condition at a slope that represents the ratio of heat to moisture loads for the room. Every kilogram of air supplied to the room at any state point on this condition line for the space will be deficient in heat and moisture in exactly the right proportion. It remains only to decide on the state point for the supply air and the mass flow rate of air to be provided. Several factors will have to be taken into account in this decision but they are not part of the present discussion.

The sensible-heat ratio scale can be used to establish the slope of the room condition line for a given sensible-heat ratio as discussed in Chapter 6. The ratio of heat to moisture load can be converted to a sensible-heat ratio, using the enthalpy for water vapour at the room air temperature as the latent heat, or can be used directly with charts that provide this facility [**13.1, pp. 5.5 and 5.9**].

Example 13.5

A room maintained at 24°C and 60% rh has a sensible-heat gain of 30 kW and a moisture gain of 3 g/s. The conditioned air to be introduced should not be colder than 15°C in order to avoid problems with drafts. Find the air condition at delivery and the mass flow of dry air required.

The total heat gain is 30 kJ/s
The moisture gain is 0.003 kg/s

Find the latent heat as the enthalpy for water vapour at 24°C. From Equation 6.10, Chapter 6,

$$h = 2500 + 1.86\,\mathrm{t} = 2500 + 1.86 \times 24 = 2545 \text{ kJ/s}$$

Latent heat gain is

$$0.003 \times 2545 = 7.63 \text{ kJ/s}.$$

The sensible-heat factor is $\dfrac{30 - 7.63}{30} = 0.75.$

From the room condition state point at 24°C and 60% rh, draw a line parallel to the sensible-heat factor line for a sensible-heat factor of 0.75 to intersect the air temperature line for 15°C. Find the rh to be 93%. Find h at 15°C and 93% to be 40 and at 24°C and 60% rh to be 53 − 0.2 = 52.8 (Figure 13.17).

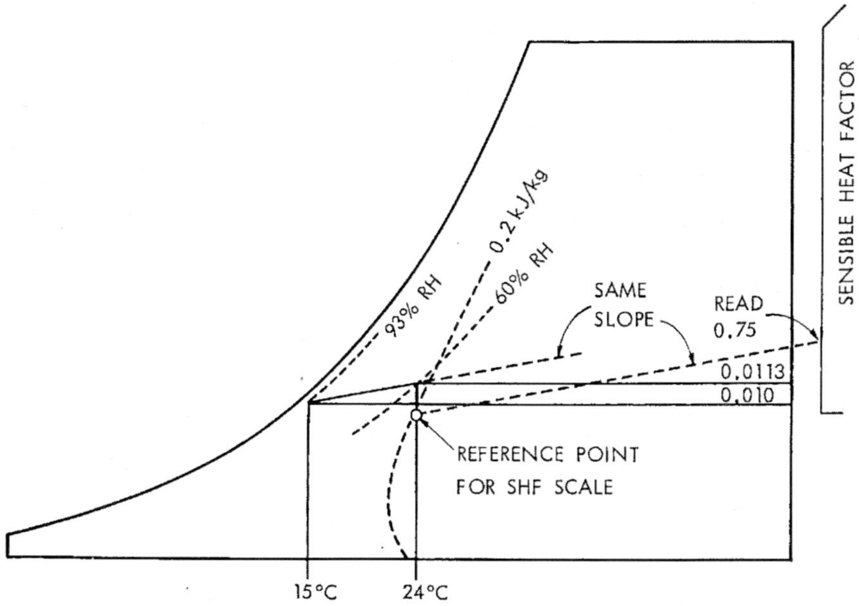

FIGURE 13.17 Outline of psychrometric chart illustrating Example 13.5.

The heat absorbed by 1 kg of dry air at the supply condition taking up heat until it reaches the room condition will be 52.8 − 40 = 12.8 kJ/kg, and the mass flow of air to absorb 30 kW (30 kJ/s) = 30/12.8 = 2.34 kg/s.

The volume read from the chart for the room condition of 24°C and 60% rh is 0.86 m³/kg, so the volume flow rate will be 2.34 × 0.86 = 2.01 m³/s.

A check for moisture removed can be made as follows:

$$W \text{ at } 24°\text{C and } 60\% = 0.0113$$
$$W \text{ at } 15°\text{C and } 93\% = 0.0100$$

Moisture removed per kilogram of dry air supplied = 0.0013 kg, and for 2.34 kg/s = 2.34 × 0.0013 = 0.0030 kg/s = 3.0 g/s.

The preceding discussion has not been concerned with the source of the air that is to be conditioned or the overall system cycle of operation. The state point and the air-condition line were set by the room requirements and were not dependent on the source or original condition of the supply air. The energy balance for the overall system is, however, very much influenced by the original condition of the ventilating air.

13.20 Typical Heating and Cooling Cycles

A schematic diagram of a typical system is shown in Figure 13.18, together with the cycles for a typical cooling situation and for heating and humidifying operation [**13.21**].

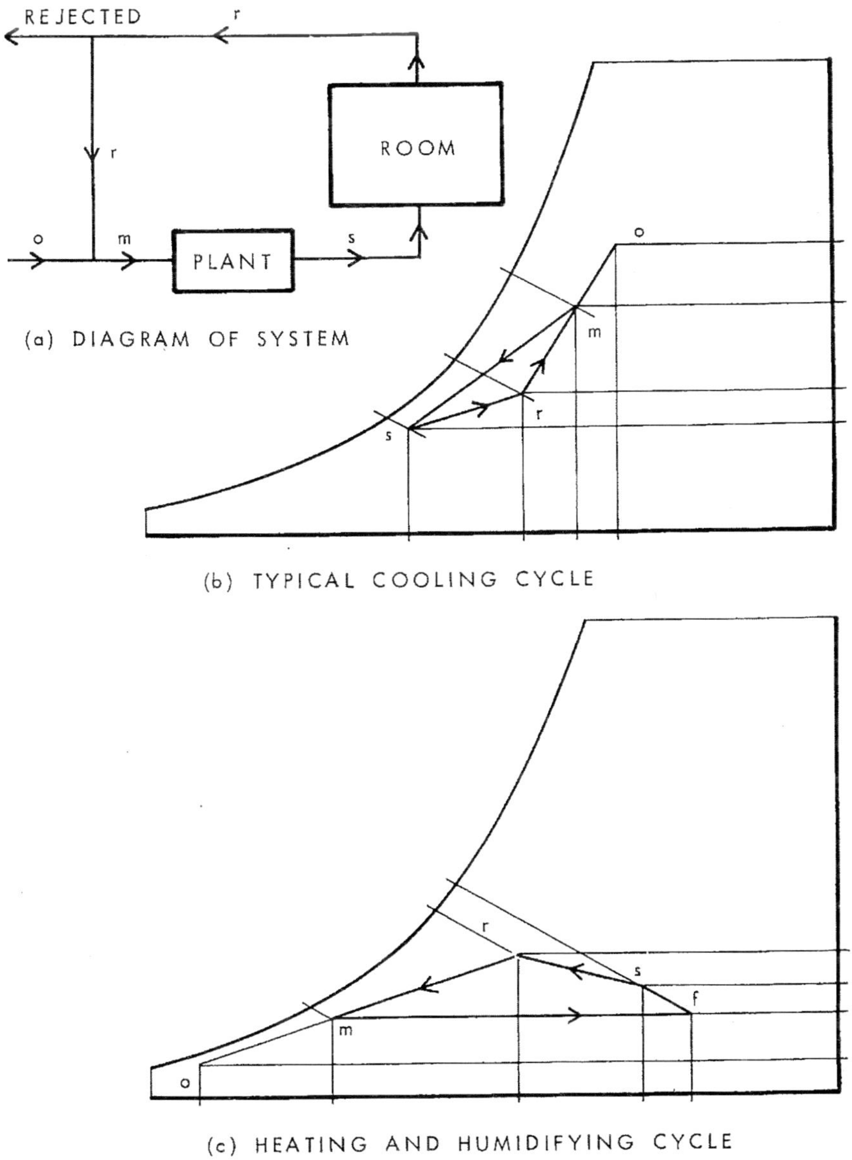

FIGURE 13.18 Air-conditioning systems and typical cycles.
(From *ASHRAE Handbook Educational Supplement* 1978, Fig. I-18, p. 61)[**13.8**]

Air is brought to condition *s* for delivery to the room during the cooling cycle, as determined by the room heat and moisture loads. It takes on heat and moisture in the room until it reaches room condition *r*, and is then exhausted from the room. Part of it is rejected to the outdoors to make way for fresh outdoor air for ventilation. The remaining part, which will be recycled, is mixed with the outdoor air at condition *o* to produce a mixture condition *m* in a typical mixing process. This is the condition of the air entering the conditioning plant, which must bring it to the proper entering condition at *s*. It is assumed in this example that the air gains both heat and moisture in changing from *s* to *r*; in other words, the room requires both cooling and dehumidifying.

The typical winter air-conditioning cycle is also shown in Figure 13.18. It is assumed here that the room requires heat, but does contain some moisture to be removed. The ventilating air requires humidification, however, so that the plant is required to heat and humidify the supply air. Following the cycle as before, the supply air at condition *s* gives up heat and takes on moisture in the room, bringing it to room condition *r*. The mixing with outdoor air at condition *O* creates condition *m*, and the mixture must then be brought to condition *s* by the air-conditioning plant. This requires heating and humidifying and is done in two stages. The first involves heating to condition *f* so that when it is sprayed with water in an evaporative process the supply air is brought to condition *s* as required.

These discussions of processes and cycles have not been concerned with the fact that room heat and moisture loads usually vary continuously with time. The air-conditioning system must be adaptable under control to these varying conditions if temperature and humidity are to be controlled. Systems for simultaneous close control of both temperature and humidity in each one of a number of rooms can become quite elaborate. The precise control of conditions in test cabinets also involves the air-conditioning problem and, as has been shown, the characteristics of the space as well as the system [**13.6**].

It is common, in air conditioning for general occupancy, to design to a condition line representing average room load and then to operate to control temperature, allowing relative humidity to vary over a range. The effect of a change in the sensible-heat factor on the room relative humidity, other conditions remaining constant, is shown in Figure 13.19. There are many other possibilities [**13.22, 13.23**]. Some systems may consist of combinations of room heating and cooling units with conditioned air from central plants. Still others may have dual air supplies, which are mixed as required to match the loads for each room.

13.21 Room-Air Distribution

The basic situation in room-air distribution is shown in Figure 13.20. A stream of cooled air projected horizontally will fall, as shown, and a heated one will tend to rise. Heated air may be introduced at temperatures as much as 50 K above room air. Cooled air will produce drafts, so it is necessary to ensure that it does not strike the occupants before adequate mixing occurs and to limit the temperature to which it is cooled. For this reason, a lower limit of 5°C is usually adopted, because it is not practical to produce lower temperatures without causing frost accumulation on the cooling coil. The temperature difference between room air and

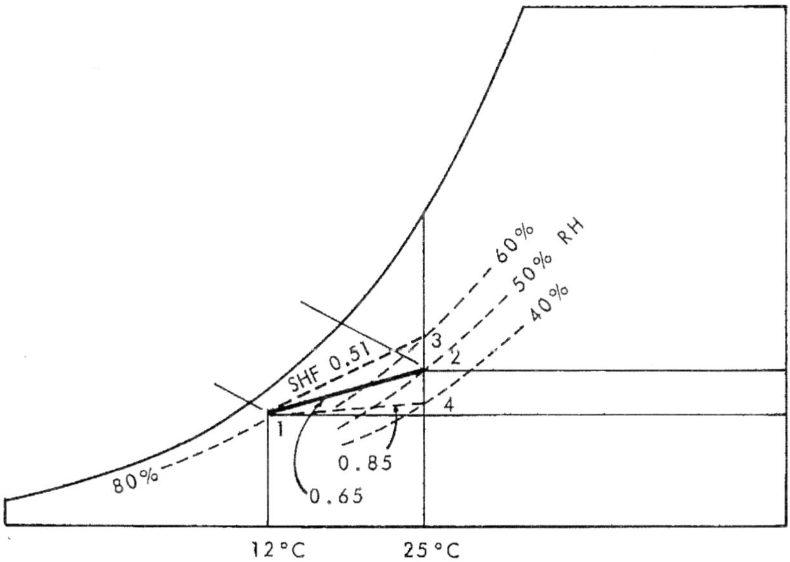

FIGURE 13.19 Outline of psychrometric chart showing effect of change in sensible-heat factor with fixed supply condition and flow rate adjusted to produce the same room temperature. Room relative humidity changes from 40 to 60% as sensible-heat ratio changes from 0.85 to 0.51.

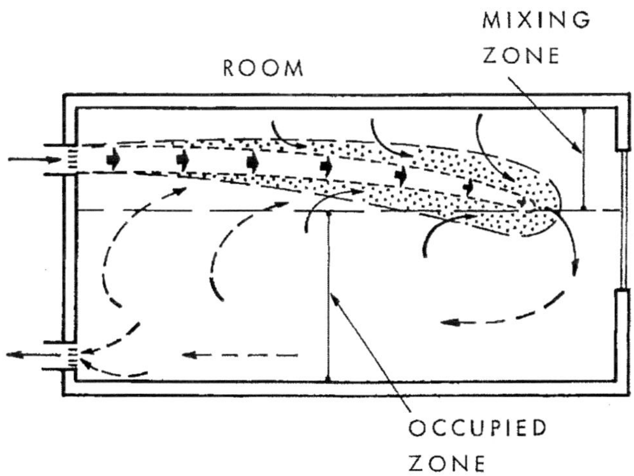

FIGURE 13.20 The basic air distribution problem.
(From Hutcheon, N.B., Canadian Building Digest 106, DBR/NRCC, 1968, Fig. 1)[**13.20**]

cooled air will thus be limited to a range from 10 K to 20 K depending on conditions. The air-flow rate may have to be enough to produce ten air changes per hour or more in order to provide the necessary cooling capacity, whereas for heating, with much higher temperature differences, the rate may be no more than 2 or 3 ac/h. The increased rate of flow does assist, however, in promoting good room-air distribution. Rates substantially higher than 10 ac/h are usually justified only in special cases.

In normal air-conditioning practice, great advantage is taken of the fact that of an entire space only the part that is to be occupied must be held within acceptable limits for occupancy. That part of the room volume above the 1.8 m level can be used to advantage as a distribution and mixing zone (Figure 13.20). This requires that the primary air be projected with sufficient velocity to entrain room air by induced secondary circulation in amounts up to three to five times its own volume before it enters the occupied zone, and at the same time provide the best possible distribution and movement of air throughout the occupied zone. This is not easy to accomplish, and presents a great challenge to the designer of the air-conditioning system. Clearly, the extent to which uniform air temperature, air movement, and relative humidity can be achieved throughout the occupied zone is greatly dependent on the distribution and mixing achieved.

Distribution of the air is accomplished by three factors: velocity energy of the primary stream, gravity effects due to differences in temperature, and displacement resulting from the general movement caused by the continuous overall introduction and withdrawal of air. It is necessary to take account of all three in designing for adequate room-air distribution. The diffusers and registers used at the supply-air outlets must also be designed to assist in air distribution. Their size, location, and spacing, the amount of air handled, and the direction given to the air are all important factors.

If the "throw" of a register is too great, the primary air stream may be projected directly into the occupied zone or may "splash" on a wall and be deflected downward. Heated air streams rise when they are projected horizontally; cooled streams fall and must be given an appropriate initial deflection.

It is hardly necessary to point out that a cooled primary stream striking anyone will give rise to an objectionable draft. The secondary circulation (Figure 13.20) induced by the primary stream, however, can also produce drafts if room conditions are generally on the cool side. Note that the induced flow making up the secondary air circulation is in the same direction as the primary stream when it is near it, but in the opposite direction when it is farther away. Examples of such "reverse flow" accompanied by marked drafts can be produced by the hot primary streams of car heaters or warm-air heating registers in low side-wall locations where the air temperature is generally below the comfort level, even though the primary stream itself is quite warm. Discomfort due to drafts has been extensively examined [**13.1–13.4**]. Much research has also been done on various aspects of room-air distribution, including ways of achieving the desired results [**13.1, p. 32.1**].

13.22 Room Supply and Return Locations

The supply-air stream projected into a room provides kinetic energy that can be used to promote room-air circulation. The speed and mass flow rate of the primary air stream determine the rate of supply of energy which, in combination with the direction and character imparted by the diffusers to the issuing stream, has an important influence on the throw, entrainment, general distribution, and circulation of the air in the room. The location of the supply-air outlets in relation to the principal sources of heating and cooling loads therefore becomes of great importance in determining room conditions. The return-air inlet locations are less important since they can affect room-air distribution in a general way only, to the extent that their location determines the point of withdrawal of return air.

Practical considerations related to such matters as room geometry, requirements of the occupancy, and economy will usually limit the choices of supply and return inlet locations. Four configurations in common use are shown in figures 13.21 to 13.24, designated here as schemes W, X, Y, and Z; their essential features are indicated in the captions.

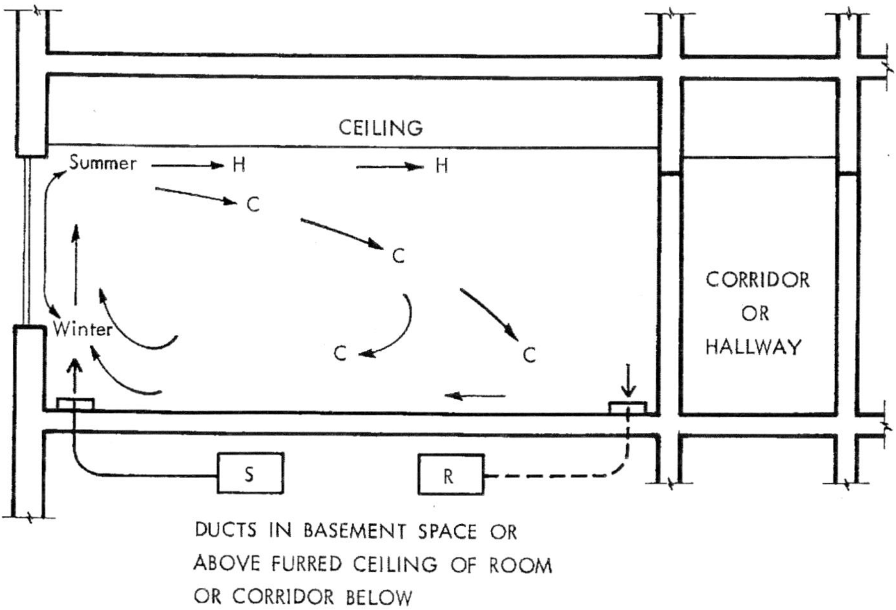

FIGURE 13.21 Scheme W. Floor level supply and exhaust. Cooling: good; heating: moderate vertical air-temperature gradients depending on rate of air change.

Schemes W and X and the high side-wall arrangement of Figure 13.20 have been commonly used in warm-air heating systems for single-family houses. It has been found in practice that one large return inlet centrally located in the common hallway can serve several rooms. The long return ducts required for Scheme X were often ineffective, and the air found its way back by other routes.

Scheme Z shows a fan-coil unit conditioner as the source of the heated or cooled air. Ventilation air is shown coming from a separate duct system. A unit ventilator with a direct outdoor-air connection might have been used with the attendant risk of freezing of the coils in cold weather, or ventilating air could have been fed to the unit from a supply duct to be mixed with other air before distribution.

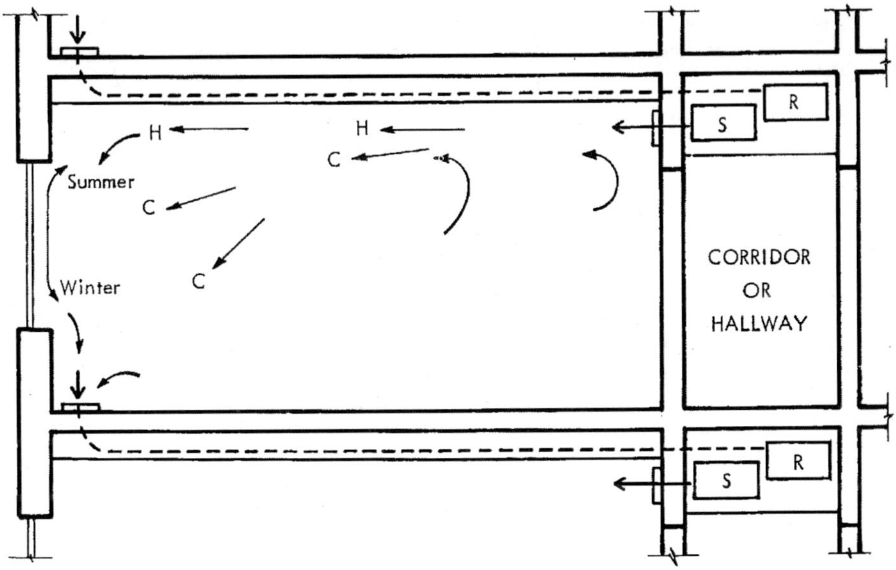

FIGURE 13.22 Scheme X. High-end wall supply. Return at outside wall. General circulation weak. Fair for heating. Fair to good for cooling if lighting and exterior wall and windows principal sources of loads. Return at inner end wall better for cooling.

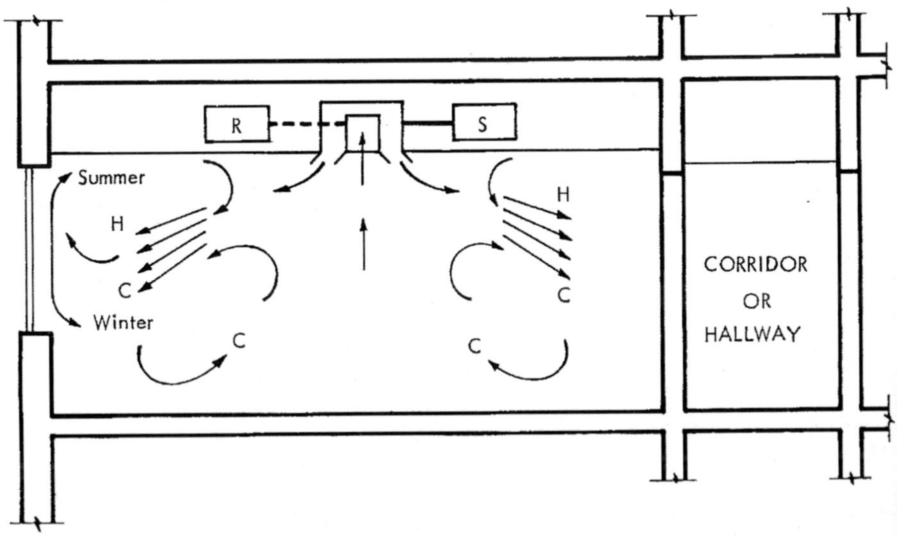

FIGURE 13.23 Scheme Y. Ceiling supply and return. Good for cooling. High air-temperature gradients with heating. Good for ventilation with convector heating at cold wall in winter.

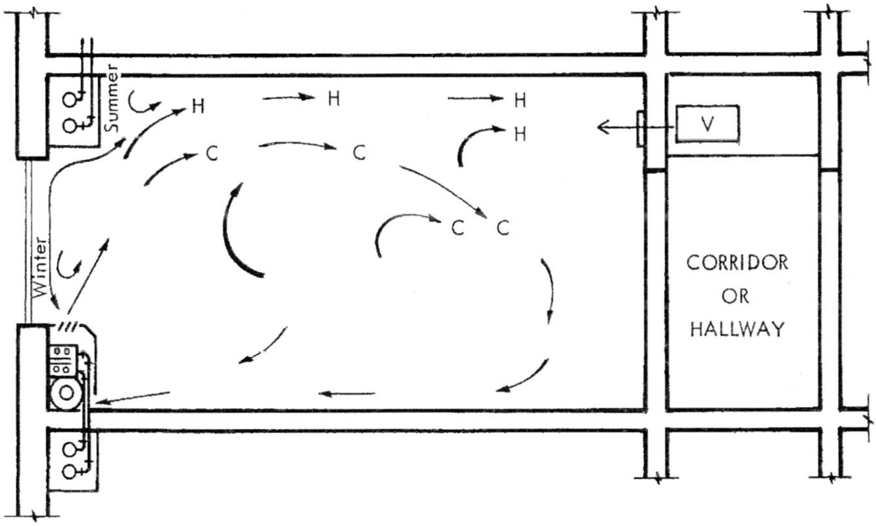

FIGURE 13.24 Scheme Z. Unit conditioner at outside wall. Ventilation air supply shown but no return. Good for both heating and cooling. Moderate vertical air-temperature gradients with heating depending on air change rate. Unit provided with hot water or chilled water as required.

These schemes have tended to emphasize exterior walls and windows as the principal sources of heating and cooling loads. When high-density occupancies are involved, the heat and moisture from people may be the principal concern, while in many other commercial and industrial situations the lighting, requiring electrical energy inputs up to 50 W/m^2 or more, may be the largest single element of cooling load. Rooms in interior locations may require cooling for the entire year. When substantial lighting loads are involved, much of the heat from the lights can be intercepted before it enters the room by drawing off the return air through a pierced ceiling.

13.23 Vertical Temperature Differences with Heating

Vertical air-temperature differences are a sensitive indicator of the adequacy of the room-air distribution and circulation and are likely to be of concern more frequently in heating than in cooling. The evidence obtained from the two studies on panel heating (already discussed) was most significant. It seems clear that even a heated floor panel cannot prevent the development of a strong flow of cool air at floor level as a result of the convection streams at cool exterior wall and window surfaces. This being so, there is an explanation for the common experience of return-air inlets at outside walls often failing to prevent cool air flows across the floor: they did not have enough capacity to handle the flow or were not suitably located to intercept it. With these possibilities in mind, it is interesting to examine some of the evidence available on air-temperature differences with various heating arrangements as summarized and presented in Table 13.2. Little additional comment is required. The substantial improvement in design and heating made possible by research and the use of insulation and double glazing are reflected in the results. The outstanding performance of baseboard radiation in respect of uniformity in vertical air temperatures can be attributed to its ability to intercept cool air streams over the full length of exterior walls and to neutralize them by dispersion and mixing.

Table 13.2

Vertical Temperature Differences in Heated Rooms

Source and description	Δt-floor* to 1.5 m	Δt-floor* to ceiling
1. National Bureau of Standards: tests in small uninsulated test bungalow with basement, $t_i - t_o = 21$ K.		
(a) Various types of heaters and floor furnaces centrally located	8 K to 10.5 K	13 K to 17 K
(b) Gravity hot-water system	3 K	5 K
(c) Gas-fired floor furnace with return ducts from perimeter locations	5 K	5 K
(From Dill, R.S., and Achenbach, P.R., BMS Report 108, U.S. Dept. Commerce 1947) [**13.24**]		
2. University of Illinois: radiator test room with exterior of cold wall held at −18°C. $t_i - t_o = 39$ K. No insulation, single glazing.		
(a) Steam radiator tests. Room 2.75 m high.	4.5 K to 9 K	7 K to 13 K
(From University of Illinois Eng. Exp. Sta. Bulletin 192) [**13.13**]		
(From Severns, W.H., and Fellows, J.R., *Heating, ventilating and air conditioning fundamentals*, New York: John Wiley & Sons, 1949) [**13.25**]		
3. Equation for vertical temperatures for general use: values for $t = 21$°C at 1.5 m. Room 2.5 m high.	2.75 K	5.25 K
(From Severns, W.H., and Fellows, J.R., *Heating, ventilating and air conditioning fundamentals*, New York: John Wiley & Sons, 1949) [**13.25**]		
4. University of Illinois: Warm-air heated test houses. 92 mm insulation. Storm windows. $t_i - t_o = 39$ K		
(a) High sidewall registers [**13.26**]	2.5 K to 4 K	7 K to 9.5 K
(b) Perimeter system [**13.26**]	1 K to 2.5 K	1.5 K to 2.5 K
(c) Perimeter loop and radial systems [**13.27**]. Slabs on grade.	1 K to 4 K	2 K to 6.5 K
(From Bahnfleth, D.R., Gilkey, H.T., and Chen, C.F., *ASHAE Trans.* 1956, *62*) [**13.26**]		
(From Gilkey, H.T., Roose, R.W., Childs, M.E., *ASHVE Trans.*, 1953, *59*) [**13.27**]		

TABLE 13.2 Continued (2)

Source and description	Δt-floor* to 1.5 m	Δt-floor* to ceiling
5. Baseboard radiation in 5 occupied houses: full and partial insulation. $t_i - t_o$ = from 19 to 29 K	1 K to 1.5 K	1.5 K to 2 K
(From MacLeod, G.S., and Eves, C.E., *ASHVE Trans.*, 1950, *56*) [**13.14**]		
6. ASHAE: floor-slab laboratory. Walls and ceilings insulated, single-glazing 10% to 20% of floor area. Tests of covered hot-water floor panels, $t_i - t_o$ = 44 K. Floor panel surface temperature 6.5 to 8.5 K above air temperature.		1.3 K to 2 K
(From Sartain, E.L., and Harris, W.S., ASHAE Trans., 1957, *63*) [**13.18**]		

*Readings at 50 to 75 mm above floors and below ceilings

The equation of item 3 in Table 13.2, which has been widely used in the past in the estimation of design air temperatures under ceilings and at midheight of walls and windows, no longer reflects the results that can be expected under favourable conditions. Complications in its converted SI form have been avoided here by providing the values for 21°C as the temperature at 1.5 m. In its original form it showed temperatures increasing by 1°F/ft up to 5 ft and at the rate of 2% of the Fahrenheit temperature at 5 ft for each additional foot up to a height of 20 ft, above which no change was expected. The values in SI for 21°C at 1.5 m are 2.5 K/m above 1.5 m up to 32.5°C at 6 m.

13.24 Thermal Comfort with Heating

A comparison of the room thermal conditions produced by nine different heating systems in a 2.4 - by - 4.8-m insulated test room has been reported [**13.28**]. Particular attention was given to the conditions for thermal comfort of people. It was concluded that all the systems examined produced acceptable comfort conditions within the occupied zone in accord with currently accepted criteria. The vertical air-temperature differences in the occupied zone were always less than 2.9 K. The highest differences floor-to-ceiling were produced by a heated ceiling and by a warm-air system with a high-level outlet in the back wall. Test conditions were not particularly severe; they corresponded to an outdoor temperature of −5°C. Heat losses through the substantial area of double-glazed window were increased as much as 10% by a convector and by warm-air outlets at the cold wall, which produced increased air flow and therefore increased heat losses at window surfaces. Two floor-heating systems and a baseboard system had low heat requirements.

13.25 Room-Air Distribution for Heating and Cooling

An extensive systematic study involving more than 100 test runs was carried out at the University of Illinois on air distribution for systems to be used for both heating and cooling [**13.29, 13.30**]. The first 43 tests were concerned with supply outlets at one high side-wall location. They showed generally that the return location had only a localized effect in determining the air distribution. Free supply openings were generally unsatisfactory for summer cooling, but the use of horizontal and vertical vanes in the supply diffuser greatly improved air distribution. Vertical temperature variations were much greater in the occupied zone with heating than they were with cooling. An increase in flow rate during heating with an accompanying decrease in supply-air temperature led to decreased temperature variations, indicating that the higher flow rate required for summer cooling might be used satisfactorily for winter heating also.

During the second series of 60 test runs it was found that for year-round operation floor outlets produced more satisfactory air distribution than high side-wall outlets. Floor diffusers produced more uniform air motion and a lower temperature variation during heating than floor registers, but required a higher supply-air velocity for optimum air distribution during cooling. It was concluded finally that "a floor outlet located at the centre of an exposed wall has excellent possibilities of providing good air distribution during year 'round air-conditioning, particularly if the flow rate used during summer cooling is also used during winter heating" [**13.30**].

13.26 Conclusion

This chapter has been concerned with an examination of the nature of the indoor thermal environment, how it is influenced by the characteristics of the building, and how it can be adjusted by the building equipment, with emphasis on the conditions produced in the occupied space. The occupancy of a building may consist of animals, plants, products, or processes, and the thermal environment may have to be adjusted to suit them as well as the human occupants. The thermal exchange of a body was shown to be influenced by radiation (and in some cases by evaporation) as well as by convection and conduction, so that the temperature of surrounding surfaces and the relative humidity of the air become involved in the thermal environment along with air temperature and air motion, commonly regarded as the principal agents. The rates of energy exchange by these various modes have been examined. With this background it is now possible to consider the thermal environment in relation to the comfort and other requirements of the occupants.

References

13.1 *ASHRAE Handbook 1981 Fundamentals*.

13.2 McIntyre, D.A. *Indoor climate*. London: Applied Science Publishers, 1980. Chap. 8.

13.3 *Thermal environmental conditions for human occupancy*, ASHRAE Standard, ANSI/ASHRAE 55-1981.

13.4 Fanger, P.O. *Thermal comfort*. New York: McGraw-Hill, 1972.

13.5 Hatch, T.F., and Barron-Oronzco, D. Air flow in free convection over heated bodies. *ASHRAE Transactions*, 1957, *63*, pp. 275–90.

13.6 Solvason, K.R., and Hutcheon, N.B. Principles in the design of cabinets for controlled environments. Vol. 2, *Proceedings, International Symposium on Humidity and Moisture*. Washington, 1963. Chap. 30, pp. 241–48. (NRCC 8603)

13.7 *ASHRAE Handbooks: 1982 Applications, 1983 Equipment, 1980 Systems*, and *1981 Fundamentals*. (One revised each year, in rotation.)

13.8 *Environmental control principles*. Educational Supplement to ASHRAE Handbook 1977 Fundamentals.

13.9 *Cooling and heating load calculation manual*. GRP 158. ASHRAE, 1979.

13.10 Fishenden, M.W. *The heating of rooms*. Technical Paper No. 12, Fuel Research Board. London: HMSO. 1925.

13.11 *ASHRAE Handbook 1983 Equipment*.

13.12 Fishenden, M.W. *The domestic grate*. Technical Paper No. 13, Fuel Research Board. London: HMSO. 1925.

13.13 Willard, A.C., Kratz, A.P., Fahnestock, M.K., and Konzo, S. *Investigation of heating rooms with direct steam radiators equipped with enclosures and shields*. Bulletin 192, Engineering Station, University of Illinois, 1929.

13.14 MacLeod, G.S., and Eves, C.E. Baseboard radiation performance in occupied dwellings. *ASHVE Transactions* 1950, *56*, pp. 157–74.

13.15 *ASHRAE Handbook 1980 Systems*, pp. 19.1 and 19.4.

13.16 Bailey, H.R. An experimental comparison of energy requirements for space heating with radiant and convective systems. *ASHRAE Transactions*, 1980, *86* (Part 1), pp. 73-81.

13.17 *ASHRAE Handbook 1980 Systems*. Chap. 8.

13.18 Sartain, E.L., and Harris, W.S. Performance of covered hot water floor panels. Part II—Room conditions. *ASHAE Transactions, 1957, 63*, pp. 239–54.

13.19 Schutrum, L.F., and Min, T.C. Cold wall effects in a ceiling-panel heated room. *ASHAE Transactions*, 1957, *63*, pp. 187–208.

13.20 Hutcheon, N.B. *The basic air-conditioning problem*. Canadian Building Digest 106. DBR/NRCC, Oct. 1968.

13.21 *Environmental control principles*. Educational Supplement to ASHRAE Handbook 1977 Fundamentals, p. 61.

13.22 Solvason, K.R. *Air conditioning systems*. Canadian Building Digest 109. DBR/NRCC, Jan. 1969.

13.23 *ASHRAE Handbook 1980 Systems*.

13.24 Dill, R.S., and Achenbach, P.R. *Temperature distribution in a test bungalow with various heating devices*. Building Materials and Structures Report BMS 108. Washington: U.S. Dept. of Commerce, 1947.

13.25 Severns, W.H., and Fellows, J.R. *Heating, ventilating, and air-conditioning fundamentals*. 2d Ed. New York: John Wiley and Sons, 1949, p. 112.

13.26 Bahnfleth, D.R., Gilkey, H.T., and Chen, C.F. Small-pipe perimeter heating in a residence. *ASHAE Transactions*, 1956, *62*, pp. 373–90.

13.27 Gilkey, H.T., Roose, R.W., and Childs, M.E. Performance of warm-air perimeter-loop and perimeter-radial systems in a residence. *ASHVE Transactions*, 1953, *59*, pp. 473–94.

13.28 Olesen, B.W., Mortensen, E., Thorshauge, J., and Berg-Munch, B. Thermal comfort in a room heated by different methods. *ASHRAE Transactions*, 1980, *86* (Part 1), pp. 34–48.

13.29 Gilman, S.F., Straub, H.E., Hershey, A.E., and Engdahl, R.B. Room air distribution research for year 'round air conditioning. Part I–Supply outlets at one high sidewall location. *ASHVE Transactions*, 1953, *59*, pp. 151–72.

13.30 Straub, H.E., and Gilman, S.F. Room air distribution research for year 'round air conditioning. Part II–Supply outlets at three floor locations. *ASHVE Transactions*, 1954, *60*, pp. 249–70.

14

Requirements and Characteristics of Occupancies

14.1 Introduction

When establishing thermal requirements for occupancies it is necessary to be clear about the objectives; these may include the maintenance of conditions promoting comfort, efficiency, learning, health, survival, growth, or preservation. Comfort considerations are usually paramount in most building situations involving people, and require the bringing together of knowledge of physics, physiology, and psychology.

Thus far, the human body has been considered as an inert object exchanging energy with its environment through radiation, convection, and conduction. It also experiences heat loss by evaporation and is capable of adapting to conditions in order to regulate body temperature. The human body is in some respects like a power plant. Food is metabolized to produce energy. Some external work may be done. The balance of the energy in the form of heat must either be stored (leading to a rise in body temperature) or lost from the body. A steady state with no change in storage is possible when the body is able to make suitable adjustments. This may take place over a wide range of conditions which at the hot and cold extremes may be unpleasant or even painful. Thermal comfort is associated with conditions to which the body can adjust readily. It is defined as "that condition of mind which expresses satisfaction with the thermal environment" [**14.1**].

ASHRAE has long had a substantial interest in conditions for human health, efficiency, and thermal comfort, and has provided a focal point for studies on these matters as described in the ASHRAE Handbook [**14.2**]. Such studies in turn have provided the basis for the influential ANSI/ASHRAE Standard 55-1981 [**14.1**]. Books by Fanger [**14.3**] and McIntyre [**14.4**] provide extensive, clear treatments of the thermal environments for people, and merit detailed study. Fanger's comfort equation, which is now widely adopted and referenced, was presented first in an ASHRAE paper [**14.5**]. The topics chosen for presentation here have been selected for their potential contribution to an enhanced appreciation of the thermal environment. Much material on occupancies involving animals, plants, products, and processes can be found in the *ASHRAE Handbook 1981*, chapters 9, 10, 30, and 31, and in the 1982 *Applications* volume [**14.6**].

14.2 Physiology

The human body is able to maintain a basic temperature in its core within close limits in a range from 36 to 38°C. The principal mechanism by which this control is accomplished is the adjustment of the blood circulation in the outer shell and extremities of the body by constriction or dilation of the blood vessels. When body temperature tends to rise, blood flow to the surface is increased and skin temperatures increase to values closer to deep-body temperature. If the condition persists, perspiration is secreted to wet the skin surface; this process will increase evaporative heat loss if conditions are such that evaporation can occur. Sweat rates up to one litre per hour or even higher can develop under extreme heat and work conditions. When the temperature of the air and surroundings is higher than body temperature, there is a sensible-heat gain rather than a loss, and this must be dissipated along with the metabolic heat by evaporation alone.

When the body is unable to attain a sufficiently high heat loss, some heat is stored and body temperature increases. The limit of storage is about 335 kJ for a rise in body temperature of 1.4 K for a man of average size. At double these values, a person may collapse from heat exhaustion.

When the body must adjust to conserve heat the blood circulation to the shell and to the extremities is reduced, and skin and extremity temperatures fall. A mean skin temperature of 29°C, or 4 K below a comfortable condition, corresponds to sensations rated as extremely cold. Hand-surface temperatures may drop to 15°C, which is sensed as being extremely cold. A drop in deep-body temperature to 28°C is considered critical. As the body cools below comfort levels, shivering leads to increased heat production. When violent shivering takes place, heat production may be up to six times greater than at resting levels.

The human body, unlike inert objects, must always lose heat to avoid a rise in body temperature. The rate at which heat must be lost depends on the metabolic rate, which varies with the level of physical activity. The ease with which the body can adjust to maintain a thermal balance depends on the amount of clothing worn and on the thermal environment. A general thermal balance equation may be written as follows:

$$S = M - (\pm W) \pm E \pm R \pm C \tag{14.1}$$

where

S = rate of heat storage
M = metabolic rate
E = rate of evaporative heat loss
W = rate of mechanical work done
R = radiant heat-exchange rate with surroundings
C = convective heat-exchange rate with surroundings.

All items except metabolic rate may be positive or negative, depending on whether they represent losses or gains of energy. It is now possible to begin to identify the ways in which various factors will affect these processes, making it possible to predict the environmental conditions that will produce thermal comfort.

14.3 Internal Heat Production

The metabolic rate, M, can be related directly to the rate of oxidation of food within the body. Part of the energy may be converted to external work, W, but most of it goes to make up the body heat, H, so that

$$M = H + W \tag{14.2}$$

The mechanical or work efficiency, η, may be written

$$\eta = W/M \tag{14.3}$$

and

$$H = M(1 - \eta). \tag{14.4}$$

It is often convenient to deal with the heat released per unit body-surface area, and

$$H/A_{Du} = (M/A_{Du})\,(1 - \eta) \tag{14.5}$$

A_{Du} is the body-surface area given by the Dubois formula presented in Chapter 13.

$$A_{Du} = 0.203\ w^{0.425}\ h^{0.725} \tag{14.6}$$

where

A_{Du} = body surface area, m^2
w = body mass, kg
h = height, m.

Metabolic rates for several levels of activity are given in Table 14.1 in met units. The quiet, seated, resting condition is the basis, being rated as 1 met, which equals 58.2 W/m^2. Reference to a more complete listing [**14.2**] shows that the rate can vary from 0.7 met in sleeping to as much as 8.7 met for intensive physical activity. Fanger [**14.3**] gives a more extensive table which includes estimates of mechanical efficiency and the air speeds over the body that are associated with particular activities. The values of the heat production H can be taken as equal to the metabolic rate, except for cases in which hard physical work is involved, when the mechanical efficiency, W/M, may be as high as 0.25.

TABLE 14.1

Metabolic Rates of Typical Tasks
(From ASHRAE Standard 55-1981, Table 3, p. 8) [**14.1**]

Activity	Metabolic Rate, met units
Reclining	0.8
Seated, quietly	1.0
Sedentary activity (office, dwelling, lab, school)	1.2
Standing, relaxed	1.2
Light activity, standing (shopping, lab, light industry)	1.6
Medium activity, standing (shop assistant, domestic work, machine work)	2.0
High activity (heavy machine work, garage work)	3.0

14.4 Clothing

Transfer of heat through clothing may be influenced by body movement and posture as well as external air movement, which can induce ventilation within the clothing. Measurements are usually made with the actual clothing on heated manikins, and the results expressed as a thermal resistance. The resistance of the clothing for a man in a business suit somewhat heavier than those now commonly worn was found to be 0.155 $m^2 \cdot K/W$ and was designated as a basic unit to be called a clo. Clothing is thus described in terms of a dimensionless ratio given by

$$I_{cl} = R_{cl}/0.155 \tag{14.7}$$

where R_{cl} = total heat-transfer resistance from skin to outer clothing surface, but not including the surface-film resistance, $m^2 \cdot °C/W$. The sensible-heat transfer from skin to outer surface of the clothing per unit area of body surface is given by

$$K = (t_{sk} - t_{cl})/0.155\, I_{cl} \quad (W/m^2) \tag{14.8}$$

Clo values for a number of individual items of clothing are given in Table 14.2. The clo value for a particular clothing ensemble can be found from:

$$\text{clo for an ensemble} = 0.82\ (\text{sum of clo units for individual items}) \tag{14.9}$$

Typical light clothing ensembles worn indoors in summer have insulation values ranging from 0.35 to 0.60 clo; during winter a range from 0.8 to 1.2 clo is more representative [**14.1**].

TABLE 14.2

Clo Units for Individual Items of Clothing and Formulae for Estimating Total Intrinsic Insulation
(From ASHRAE Standard 55-81, Table 2, p. 7) [**14.1**]

Clo = 0.82 (Σ individual items)

MEN		WOMEN	
Underwear			
Sleeveless	0.06	Bra and Panties	0.05
T-shirt	0.09	Half Slip	0.13
Briefs	0.05	Full Slip	0.19
Long underwear, upper	0.10	Long underwear, upper	0.10
Long underwear, lower	0.10	Long underwear, lower	0.10
Torso			
Shirt		Blouse	
Light, short sleeve	0.14	Light	0.20
long sleeve	0.22	Heavy	0.29
Heavy, short sleeve	0.25		
long sleeve	0.29	Dress	
(Plus 5% for tie or turtleneck)		Light	0.22
		Heavy	0.70
Vest		Skirt	
Light	0.15	Light	0.10
Heavy	0.29	Heavy	0.22
Trousers		Slacks	
Light	0.26	Light	0.26
Heavy	0.32	Heavy	0.44
Sweater		Sweater	
Light	0.20	Light	0.17
Heavy	0.37	Heavy	0.37
Jacket		Jacket	
Light	0.22	Light	0.17
Heavy	0.49	Heavy	0.37
Footwear			
Socks		Stockings	
Ankle length	0.04	Any length	0.01
Knee high	0.10	Panty hose	0.01
Shoes		Shoes	
Sandals	0.02	Sandals	0.02
Oxfords	0.04	Pumps	0.04
Boots	0.08	Boots	0.08

14.5 Evaporative Heat Losses

Evaporative heat loss is dependent upon the water-vapour pressure in the ambient air or, more properly, the difference between the saturation water-vapour pressure at the evaporating temperature and the actual water-vapour pressure in the air. Evaporation from the skin surface will be influenced by air movement and by clothing. The perspiration must evaporate, taking up latent heat, in order to contribute to body-heat loss. Under extreme conditions with high sweat rates perspiration may run off and be lost without evaporating. Under more moderate conditions it is reasonable to assume that all the perspiration will be evaporated at the skin surface.

There is a small contribution to evaporation loss from diffusion through the skin. A more substantial evaporative loss is involved in respiration, since the air drawn into the lungs is usually increased in temperature and moisture content before being exhaled. The discussion by Fanger [**14.3, 14.4**] of the condition of exhaled air and the amounts of sensible and latent heat involved is of particular interest where low outdoor air temperatures and low indoor relative humidities are common in winter.

The body-heat loss by evaporation and its relation to the thermal environment are discussed at greater length in the *ASHRAE Handbook* [**14.2**] and in other texts, e.g., Fanger [**14.3**] and McIntyre [**14.4**]. The concepts of skin wettedness and the related index of effective temperature (ET*) as measures of the potential for heat loss by evaporation will be introduced later.

14.6 Heat Loss by Radiation

The energy exchange by radiation at the outer surface of the clothed human body was discussed in Chapter 13 in connection with infrared heating. Complications arising from posture and surface temperatures of surroundings were dealt with by the use of an effective radiation area for the human body and the concept of a mean radiant temperature. The rate of exchange can be expressed as

$$R = \epsilon\sigma\,(T_{cl}^{\,4} - T_{MRT}^{\,4}) \qquad (\text{W/m}^2) \qquad (14.10)$$

where
- R = radiant-energy exchange rate per square metre of effective body area, W/m²
- ϵ = emittance of the outer clothing surface
- σ = Stefan Boltzmann constant (5.67×10^{-8}), W/m² · K⁴
- T_{cl} = temperature of outer clothing surface, K
- T_{MRT} = mean radiant temperature, K.

The mean radiant temperature can be found by measurement or by calculation. The value of A_{eff}, the effective body area, can be found when required from Equation 13.15 in Chapter 13 or from the more extensive data presented by Fanger [**14.3**].

The sensible heat-flow rate making up the radiant-energy exchange at the surface of the clothed body can also be expressed in terms of the linear equations of Chapter 8:

$$R = f_{cl}\,h_r\,(t_{cl} - t_r) \qquad (\text{W/m}^2) \qquad (14.11)$$

where h_r = mean radiant heat-transfer coefficient, W/m² • K
t_{cl} = temperature of outer clothing surface, °C
t_r = surface temperature of surroundings, °C
f_{cl} = ratio of surface area of clothing to area of body.

14.7 Heat Loss by Convection

The rate of heat loss by convection from the surface of the clothed body can be expressed as

$$C = f_{cl} h_c (t_{cl} - t_a) \qquad (\mathrm{W/m^2}) \qquad (14.12)$$

where f_{cl} = ratio of surface area of clothing to area of body (1.1 seated to 1.2 standing)
h_c = convective heat-transfer coefficient, W/m² • K
t_{cl} = temperature of outer clothing surface, °C
t_a = air temperature, °C.

The convection coefficient is highly sensitive to air motion. At air speeds up to 0.1 m/s, regarded as the "still air" condition, the transfer will be by natural or free convection, and h_c will be given by

$$h_c = 2.38\,(t_{cl} - t_a)^{0.25} \qquad (\mathrm{W/m^2 \cdot K}) \qquad (14.13)$$

When the air speed is greater than 0.1 m/s, but not more than 2.6 m/s, a formula for forced convection is more appropriate, such as

$$h_c = 12.1\sqrt{v} \qquad (\mathrm{W/m^2 \cdot K}) \qquad (14.14)$$

where v = air speed over the body, m/s.

Other equations are listed elsewhere [**14.2, 14.4**]. Fanger [**14.3**] gives values of the air speed appropriate to various activities and situations.

14.8 Indices of Thermal-Environmental Conditions

The thermal environment can be described as to its ability to promote an exchange of energy with the human body in terms of air temperature, mean radiant temperature, air speed, and water-vapour pressure. Singly or in combination these factors affect the components of body-heat exchange, namely, convection, radiation, and evaporation. Various attempts have been made to develop and promote the use of single indices to describe the combined effect on the body-heat loss of all four variables in combination, so that all situations described by the same value of the index might be expected to produce the same thermal condition.

The most widely used index is air temperature alone. The possibilities in describing the combined influence of air temperature, air speed, and radiation in terms of the globe temperature are evident from the discussions in the previous chapter. A wet-bulb globe temperature, used mainly for high-heat stress situations, includes the effect of humidity. The convenience of being able to describe a complex situation by a single number has been well demonstrated in the case of mean radiant temperature. An extended discussion of various indices is given in the *ASHRAE Handbook* [**14.2**]. Two indices, operative temperature and effective temperature, are in current use in the definition of comfort conditions.

14.9 Operative Temperature

Operative temperature has been defined as "the uniform temperature of a radiantly black enclosure in which an occupant would exchange the same amount of heat by radiation plus convection as in the actual non-uniform environment" [**14.1**]. The sensible-heat loss from the surface of the clothed body can be expressed in terms of Equations 14.11 and 14.12, arranged to apply to unit area of clothed surface, as

$$R + C = h_r(t_{cl} - t_r) + h_c(t_{cl} - t_a). \qquad (14.15)$$

Rearranging

$$R + C = (t_{cl} - t_o)\,h \qquad (\mathrm{W/m^2}) \qquad (14.16)$$

where $t_o = (h_r t_r + h_c t_a)/(h_r + h_c)$
and $h = h_r + h_c$.

Operative temperature may also be defined as "the average of the mean radiant and ambient air temperatures weighted by their respective heat-transfer coefficients. At air speeds 0.4 m/s or less and MRT less than 50°C, operative temperature is approximately the simple average of the air and mean radiant temperatures" [**14.1**]. This average is also called the adjusted dry-bulb temperature.

14.10 Effective Temperature

Effective temperature, ET*, is defined as "the uniform temperature of a radiantly black enclosure at 50 percent relative humidity in which the occupant would experience the same comfort, physiological strain and heat exchange as in the actual environment with the same air motion" [**14.1**].

The asterisk used by ASHRAE as a superscript in ET* differentiates it from the earlier version of effective temperature developed many years ago and still in use. The present form, which is based on physiological and physical considerations, has been derived from the energy balance of Equation 14.1 and provides values that can be related directly to the proportion of body surface wetted by perspiration, w, the net heat exchange, and the skin temperature, t_{sk}. Figure 14.1, drawn on an outline of the psychrometric chart, shows geometrically the various relationships. The air temperature at the operating condition *OP* is t_a, the vapour pressure is p_a, and the saturation vapour pressure is p_{sa}. The saturation vapour pressure at the skin temperature t_{sk} is p_{ssk}, and t_{act} is the hypothetical air temperature at which all the body-heat loss would be as sensible heat without the aid of sweating. A line of constant ET* as shown defines the relationship for various combinations of the external conditions of temperature and humidity that will produce the same heat loss as that experienced at 50% humidity at an air temperature equal to effective temperature ET*, and, when based on operative temperature rather than air temperature, also accounts for departures of MRT from air temperature.

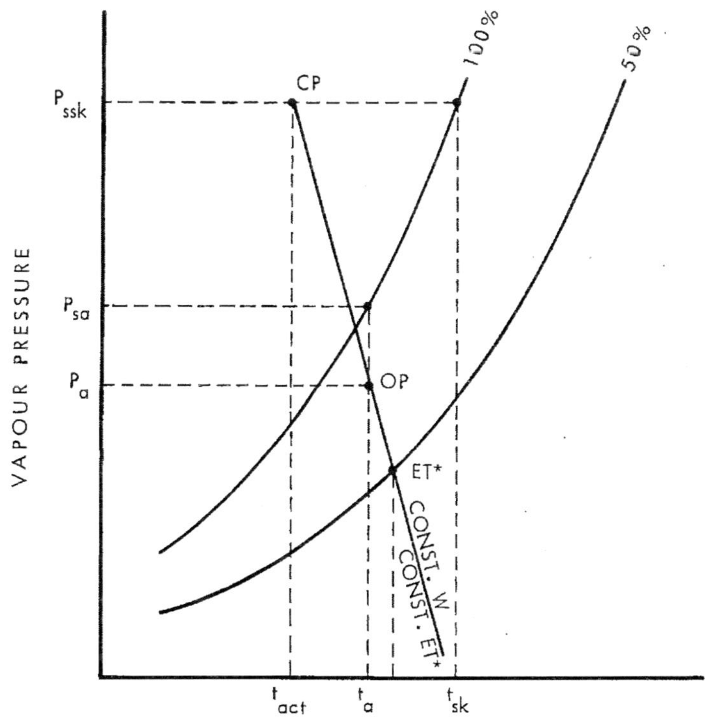

FIGURE 14.1 Graphical presentation of relation of heat-balance equation to effective temperature for operative temperature equal to air temperature and heat to storage S = 0. Line of constant ET* also a line of equal wettedness w, equal skin temperature and equal physiological strain or thermal sensation for a given level of activity, air speed, and clothing.
(From *ASHRAE Handbook 1981 Fundamentals*, Fig. 2, p. 8.7)[**14.2**]

A set of lines for a range of values of ET* is shown in Figure 14.2 for sedentary activity, a clothing level of 0.6 clo, and "normal ventilation". Changes in these parameters require a different set of lines having different slopes to define new values of net heat exchange, skin temperature, and wettedness consistent with the changed conditions.

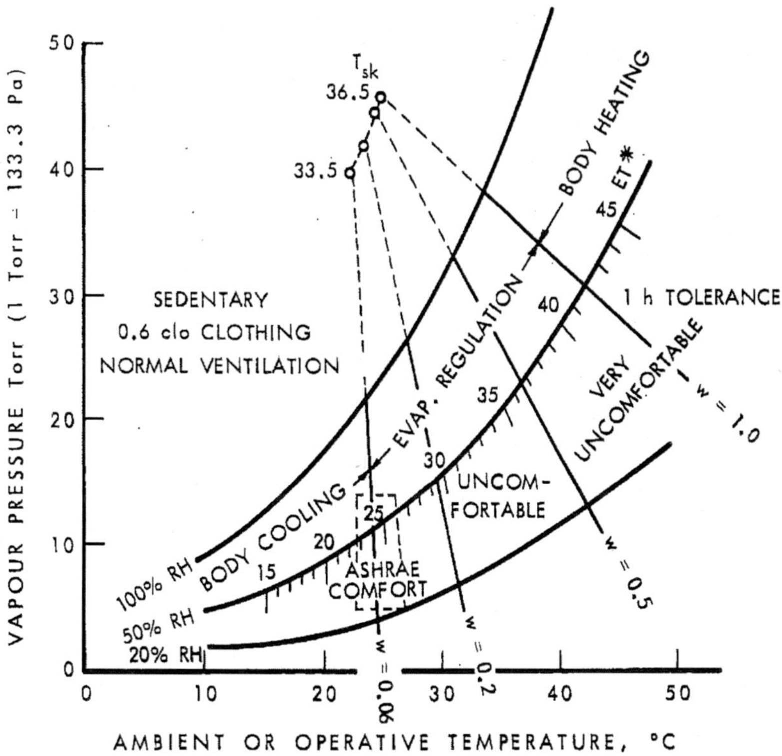

FIGURE 14.2 Comfort and heat tolerance of a sedentary subject. (From *ASHRAE Handbook 1981 Fundamentals*, Fig. 3A, p. 8.8)[**14.2**]

The ET* lines of Figure 14.2 corresponding to skin wettedness from 0.06 to 1.0 define the complete range over which the body is able to influence its thermal condition through evaporation from the skin. Thermal neutrality occurs for sedentary subjects when regulatory sweating is zero. The value assigned to skin-wettedness for this condition is 0.06, to account for the evaporative loss by skin diffusion which is independent of sweating. For conditions falling to the left of this line, the body is no longer able to maintain its temperature and will cool unless a change is made in clothing or activity. Changes to progressively higher ET* conditions produce increasing discomfort until the value for $w = 1.0$ is reached. Above this line increased evaporation is possible only if there is a rise in skin temperature, which in turn requires a rise in body temperature. Heat stroke and collapse may develop at or before this point.

These effects are predicted from heat-balance considerations and do not always occur exactly as indicated, owing to various factors (including individual differences in people). Deviations also occur in the correspondence between conditions of thermal neutrality as defined by $w = 0.06$ and the sensations of thermal comfort. It is necessary to establish the relationship more fully by testing the subjective reactions of many subjects to measured thermal conditions. A good correspondence is shown when the subjectively based ASHRAE comfort chart is superimposed on the effective temperature scales of Figure 14.2. Extended discussions of ET* and thermal comfort are given in many texts, including the *ASHRAE Handbook* [**14.2, p. 8.1**].

14.11 Prediction of Thermal Comfort

The original comfort index based on effective temperature as developed by ASHRAE many years ago is widely known and still in use. In its basic form it took account of air temperature, relative humidity, and air motion. Surrounding surfaces were assumed to be at air temperature. The index was developed from the subjective reactions of a number of test subjects who moved back and forth between test rooms maintained at slightly different conditions. Conditions producing the same thermal sensations were assigned the same value, this being taken as the air temperature at 100% rh—the effective temperature. The widely known comfort chart based on these results was then produced for air speeds 0.075 to 0.125 m/s, and so effectively combined only air temperature and relative humidity.

Later work showed that the effect of relative humidity on sedentary subjects engaged only in light work had been exaggerated, and that the comfort reactions of people were changing. Over the years, extensive studies were undertaken at Kansas State University by Nevins, McNall, Rohles, and others, and at the John B. Pierce Foundation by Gagge and colleagues. These provided a background for subsequent revisions to the comfort chart and for the ASHRAE Standard 55-74. The new effective-temperature scale (ET*) for lightly clothed sedentary subjects and low air motion [**14.2, p. 8.21**] and the 1981 revision of the ASHRAE comfort chart [**14.1**] are shown in Figure 14.3.

14.12 Fanger Comfort Charts

The studies on thermal comfort begun by Fanger at Kansas State University in 1966 were continued in Denmark. A comprehensive heat-balance equation based on the various elements of the energy exchange was developed. It was recognized that subjects who felt comfortable at a given activity level had much the same sweat rates and skin temperatures. Two empirical equations relating evaporative heat loss and skin temperature to metabolic rate were developed for extensive test data from measurements on subjects reporting comfortable conditions. The heat-balance equation, with these relationships introduced, became a comfort equation.

The Fanger comfort equation takes into account all four environmental factors of air temperature, MRT, air speed, and water-vapour pressure, as well as the variables of clothing and metabolic rate. The equation has been solved by computer to produce a series of comfort charts which are restricted to three variables at a time, the other three being held constant. Charts covering the four environmental factors for three levels of activity and three levels of clothing insulation have been published by Fanger [**14.3**] and by ASHRAE [**14.2, 14.5**]. Five are reproduced here as Figure 14.4. Fanger has included charts for the nude condition.

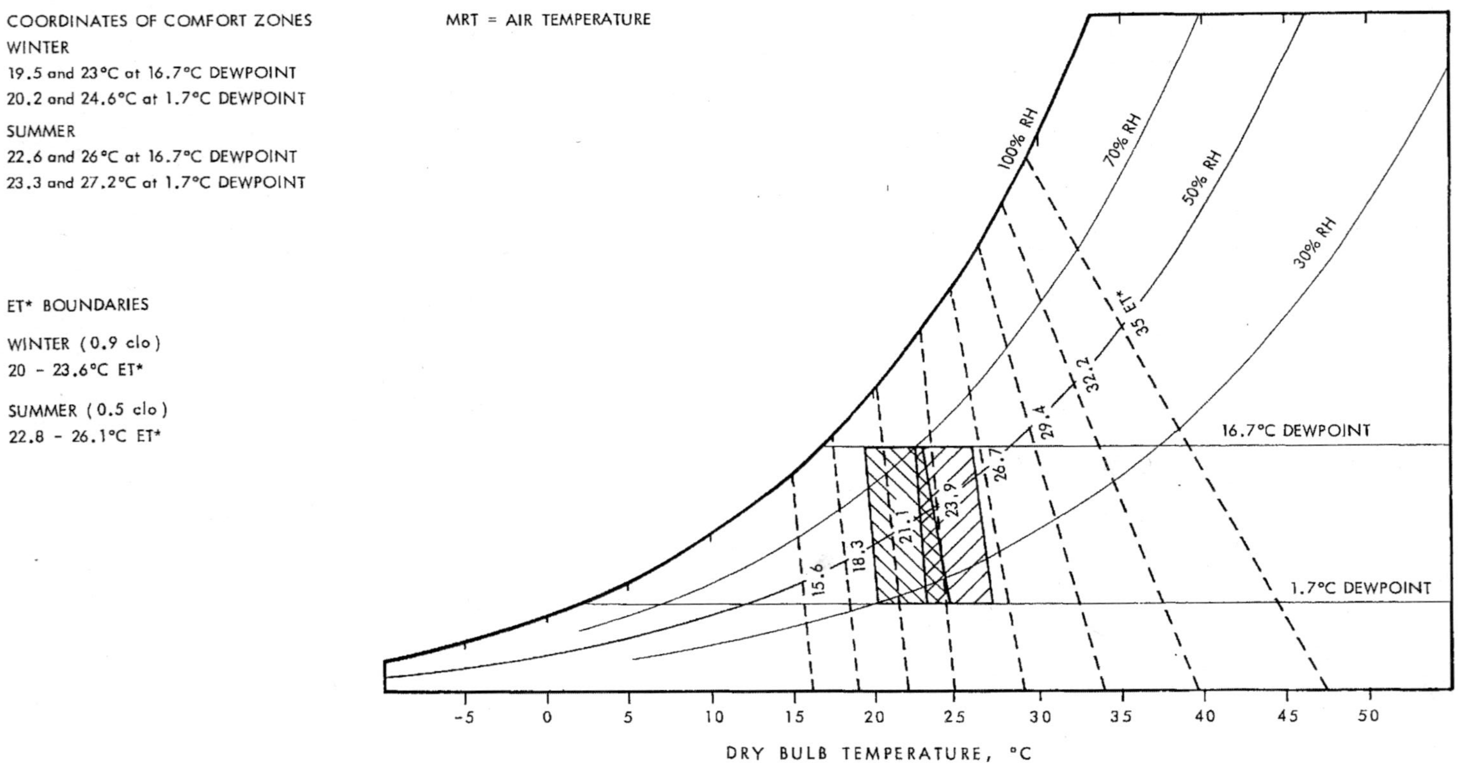

FIGURE 14.3 ASHRAE effective temperature scale and comfort chart replotted on outline of Carrier psychrometric chart. (From ASHRAE Standard 55-1981 [**14.1**] and *ASHRAE Handbook 1981 Fundamentals*, Fig. 16, p. 8.21)[**14.2**]

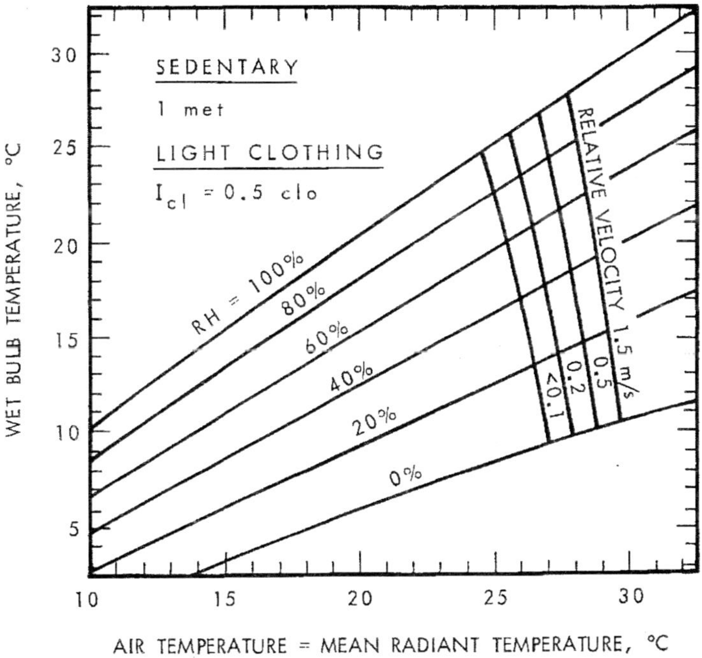

FIGURE 14.4 (a)

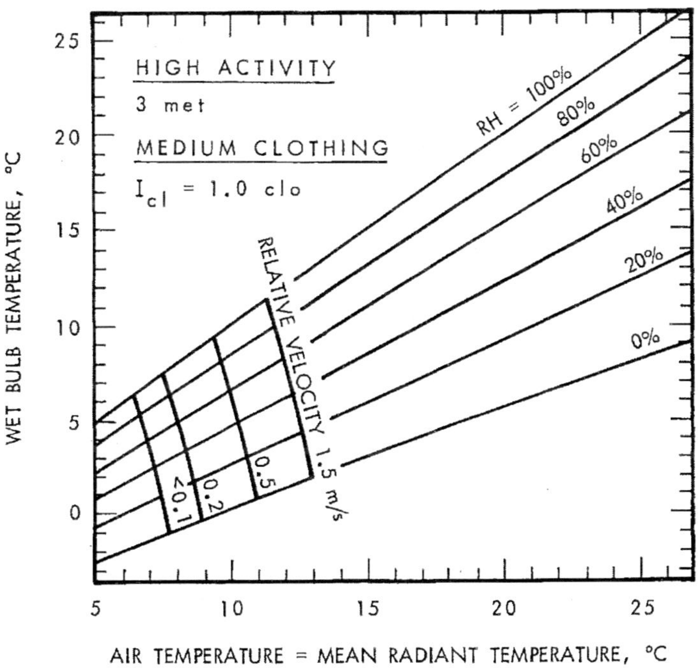

FIGURE 14.4 (b)

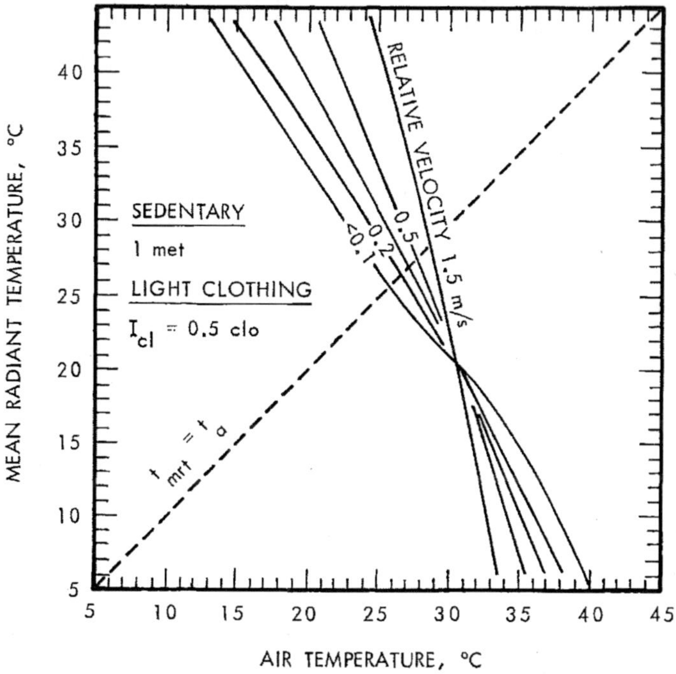

FIGURE 14.4 (c)

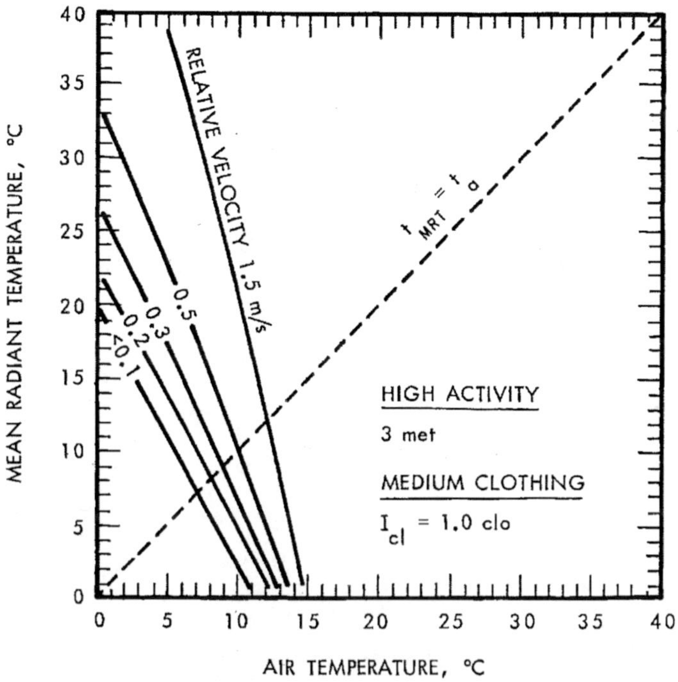

FIGURE 14.4 (d)

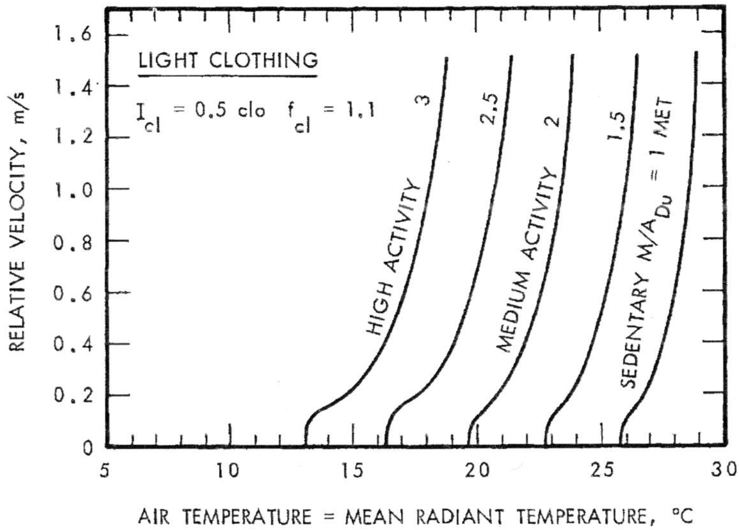

FIGURE 14.4 (e)

FIGURE 14.4 Comfort lines from the Fanger equation.
(a) Sedentary: wet-bulb vs. air temperature
(b) High activity: wet-bulb vs. air temperature
(c) Sedentary: MRT vs. air temperature
(d) High activity: MRT vs. air temperature
(e) Light clothing: relative velocity vs. air temperature
(From *ASHRAE Handbook 1981 Fundamentals*, 14.4(a) and (b), Fig. 18, p. 8.23; 14.4(c) and (d), Fig. 19, p. 8.24; 14.4(e), Fig. 20, p. 8.25)[**14.2**]

The Fanger comfort charts provide for the evaluation of a range of levels of clothing, MRT, and activity, along with air temperature, air motion, and relative humidity. Fanger has also devised a means of estimating a predicted mean vote (PMV) of the subjects in a space in which there are deviations from optimal in the thermal conditions. From this it is possible to predict the percentage of people dissatisfied (PPD) by which the acceptability of practical situations can be assessed.

14.13 ASHRAE Standard 55-81

The 1981 revision of the ASHRAE Standard, like the 1974 version, is based on a range of conditions defined in terms of the effective temperature, ET*. It applies mainly to sedentary or slightly active (1.2 met) building occupants wearing typical indoor clothing, and sets out conditions as given in Figure 14.3, which should be thermally acceptable to 80% or more of the occupants within the limits specified (Figure 14.5). Effective temperature in this case is related to operative temperature rather than to dry-bulb temperature as would normally be the case. In this way, the influence of MRT is taken into account. Operative temperature equals dry-bulb temperature when MRT does not differ from dry-bulb temperature.

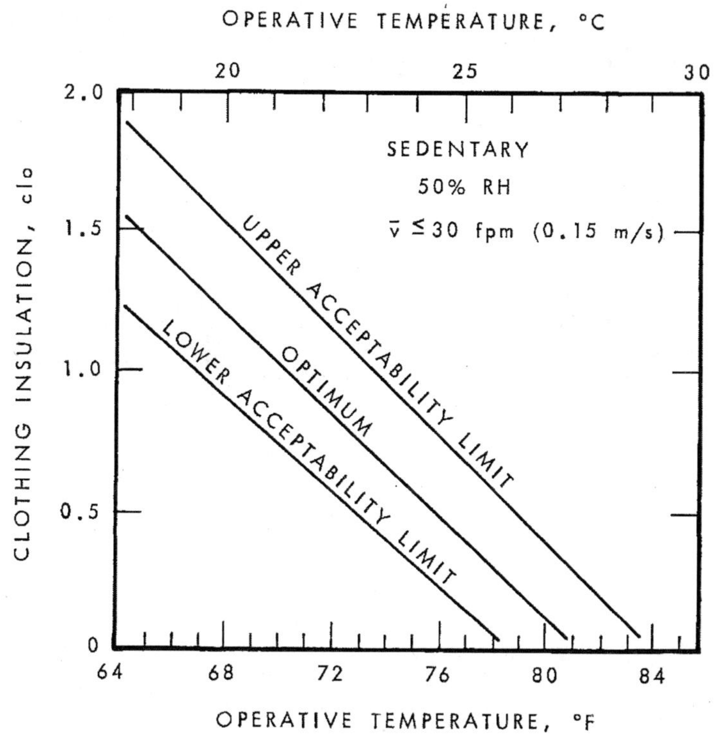

FIGURE 14.5 Clothing insulation necessary for various levels of comfort at a given temperature during light, mainly sedentary activities (1.2 met).
(From ASHRAE Standard 55-1981, Fig. 1, p. 3)[**14.1**]

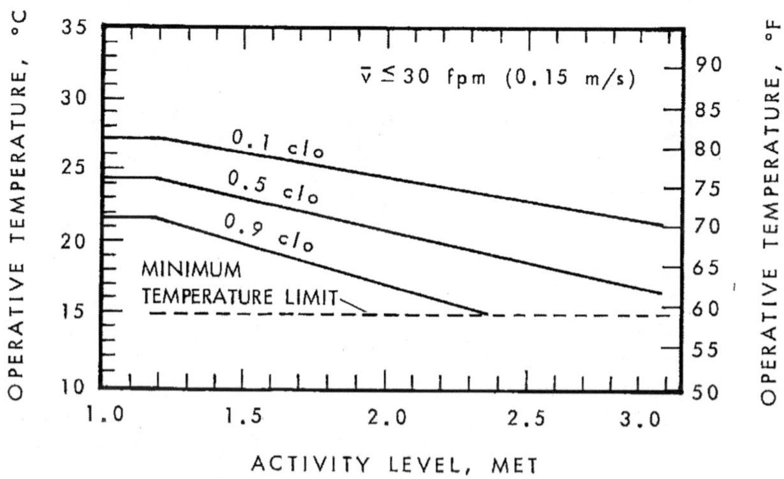

FIGURE 14.6 Optimum operative temperatures for active people in low air movement.
(From ASHRAE Standard 55-1981, Fig. 4, p. 9)[**14.1**]

Different ranges are now given for winter and summer, requiring clothing to provide 0.9 clo in winter and 0.5 clo in summer. Adjustments to operative temperatures for other levels of activity and clothing as shown in Figure 14.6 and a special range for minimal clothing (0.05 clo) are provided. An adjustment permitting a reduction in operative temperature by 0.6 K for each 0.1 clo of increased clothing is permitted for sedentary subjects.

No minimum for air movement is specified for comfort, but winter values shall not exceed 0.15 m/s. Summer rates shall not exceed 0.25 m/s, except that operative temperatures in excess of the normal maximum may be used if higher values of air motion are provided, as shown in Figure 14.7.

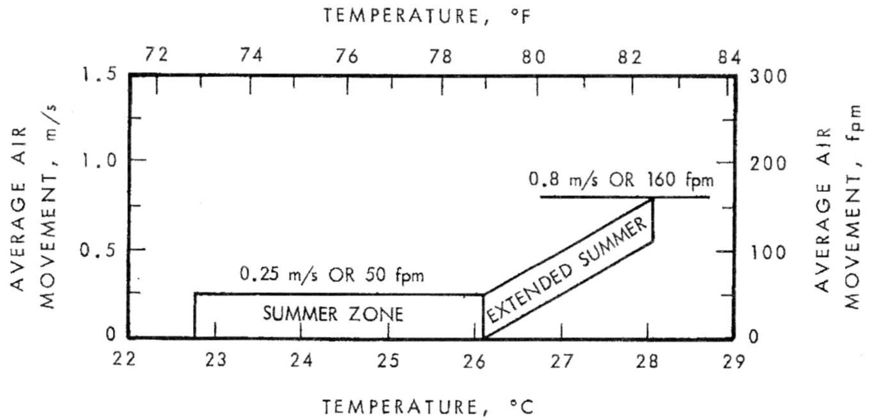

FIGURE 14.7 Range of average air movements permitted in the summer and extended summer zones.
(From ASHRAE Standard 55-1981, Fig. 3, p. 6)[**14.1**]

14.14 Nonuniformity of Conditions

ASHRAE Standard 55-81 specifies limits on nonuniformity of thermal conditions throughout the occupied zone, which is defined as "the region within a space, normally occupied by people, generally considered to be between the floor and 1.8 m above the floor and more than 0.6 m from walls or fixed air-conditioning equipment". Vertical air-temperature difference is to be limited to 3 K between the 0.1-m and 1.7-m levels to avoid local discomfort. The surface temperature of the floor is required to be between 18°C and 29°C for people wearing appropriate indoor footwear.

Local discomfort may be caused by marked differences in the radiation exchange being experienced in two opposite directions, as produced by combinations of hot and cold room surfaces. Radiant temperature asymmetry in the vertical direction such as might be produced by a heated ceiling is to be less than 5 K and in the horizontal direction less than 10 K.

The radiant-temperature asymmetry in the vertical direction is defined as the difference in plane radiant temperature of the upper and lower parts of the space with respect to a small horizontal plane 0.6 m above the floor. For the horizontal direction, the radiant-temperature asymmetry is the difference in plane radiant temperature in opposite directions from a small vertical plane area 0.6 m above the floor.

The plane radiant temperature can be found by calculation from the temperatures of the relevant room surfaces and the angle factors between the surfaces and a plane element. The procedure is essentially the same as that required for calculation of MRT, but is based on the radiation exchange with surfaces on one side of a plane only.

The subject element is a small plane area in a reference plane, and the room surfaces to be included for any one direction are those facing or enclosing the reference plane, as shown in Figure 14.8. In calculating vertical asymmetry, the horizontal reference plane is specified to be 0.6 m above the floor. All surfaces above the reference plane, including the ceiling and portions of walls, contribute to the upper-side radiation exchange, while those below determine the lower-side exchange with the plane element. For horizontal asymmetry the distance of the reference plane from the wall and the horizontal location of the plane element in the plane are determined by the position being examined, while the related room surfaces are those enclosing the reference plane on the side in question.

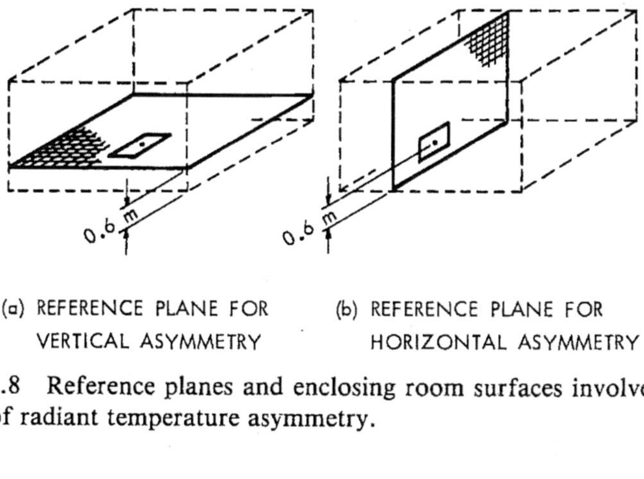

FIGURE 14.8 Reference planes and enclosing room surfaces involved in determination of radiant temperature asymmetry.

Plane radiant temperature can be found from the linear equation given in the standard, using suitable angle factors as given in figures 14.9 and 14.10. It is not necessary to know the area of element e.

$$t_{pr} = F_{e-1}t_1 + F_{e-2}t_2 + \ldots + F_{e-N}t_N \qquad (14.17)$$

where t_{pr} = plane radiant temperature, °C

$F_{e-1}, F_{e-2} \ldots F_{e-N}$ are angle factors from the element to the various surfaces involved

$t_1, t_2 \ldots t_N$ are the respective surface temperatures, °C.

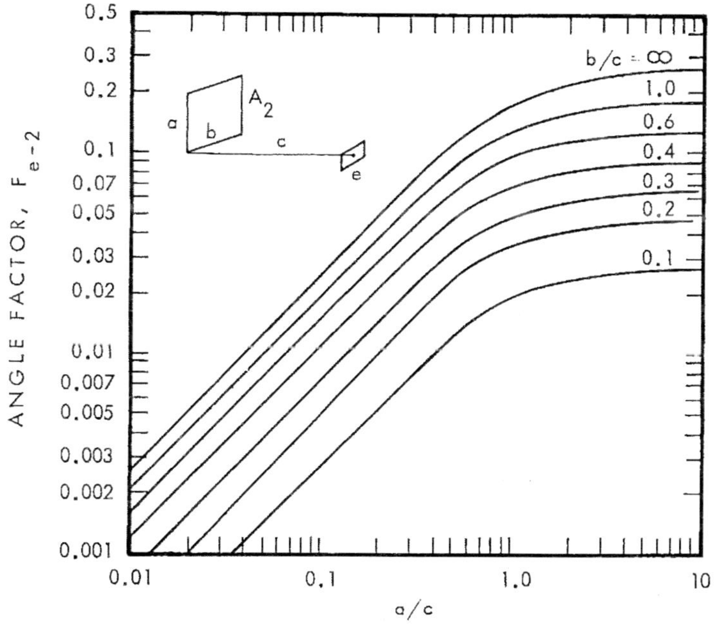

FIGURE 14.9 Angle factor from element to a parallel surface. (From ASHRAE Standard 55-1981, Fig. 9, p. 13)[**14.1**]

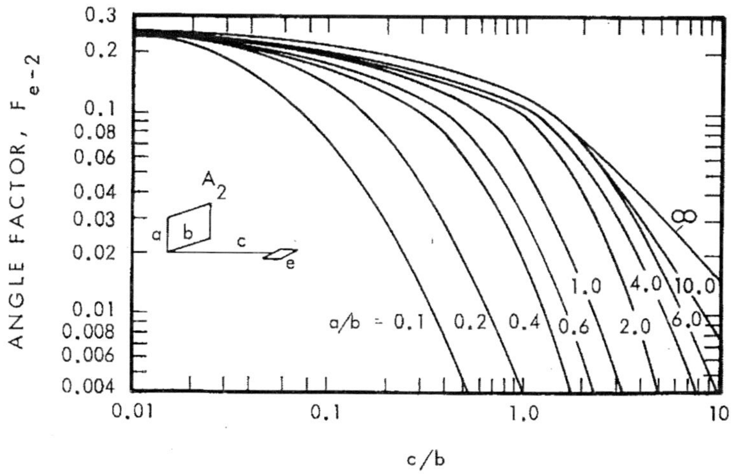

FIGURE 14.10 Angle factor from element to perpendicular surface. (From ASHRAE Standard 55-1981, Fig. 10, p. 14)[**14.1**]

The plane radiant-temperature asymmetry, Δt_{pr}, is the difference between plane radiant temperatures in opposite directions:

$$\Delta t_{pr} = t_{pr1} - t_{pr2}. \tag{14.18}$$

Example 14.1

A person is to be seated 1 m in front of an all-glass outside wall with a minimum surface temperature of 0°C. The room is 5 × 4 × 3 m. Other surfaces are all at 21°C. Find the least favourable radiant-temperature asymmetry.

The largest angle factor for the cold wall will be for a position opposite the midpoint of the window.

The person seated is assumed to be 0.6 m above the floor.

All surfaces on the room side of the reference plane 1 m in front of the window will be at 21°C, so $t_{pr1} = 21°C$.

The angle factor for the window wall will be found from Figure 14.9 as the sum of four areas, two at 2.5 × 0.6 m and two at 2.5 × 2.4 m. $c = 1$, $b = 2.5$, and $a = 0.6$ and 2.4.

$$F_{e-w} = 2(0.11) + 2(0.19) = 0.60.$$

The angle factor for the wall, ceiling, and floor surfaces, all at 21°C, on the window side will be $1 - 0.60 = 0.40$.

$$t_{pr2} = 0.60 \times 0 + (1 - 0.60)21 = 8.4°C$$
$$\Delta t_{pr} = 21 - 8.4 = 12.6°C.$$

14.15 Determination of Mean Radiant Temperature

Mean radiant temperature must be known in applying the Standard, and is taken into account through the operative temperature. Its determination for human subjects was discussed in Chapter 13. The MRT is found by a procedure identical to that just discussed for plane radiant temperature, except that it applies not just to one side of a plane, but to a human body which is exchanging radiation with surfaces forming a complete enclosure. The appropriate angle factors are therefore different and more complicated because of variable body size, shape, posture, and orientation. Angle factors developed by Fanger for use in calculating MRT for the human body are given in the Standard and are reproduced in Chapter 13.

From Equation 13.10 of Chapter 13,

$$T_{MRT}^4 = T_1^4 F_{b-1} + T_2^4 F_{b-2} + \ldots + T_N^4 F_{b-N}.$$

This can be simplified to the linear form given as Equation 13.11 in Chapter 13 and in the Standard

$$t_{MRT} = t_1 F_{b-1} + t_2 F_{b-2} + \ldots + t_N F_{b-N}.$$

14.16 Air Motion

Air motion is almost always variable throughout any space, and is frequently a cause of drafts, which have been defined elsewhere as "any localized feeling of coolness or warmth of any portion of the body due to both air movement and air temperature, with humidity and radiation considered constant". (McIntyre refers to sensations arising from radiant losses to a cool surface as "radiation draughts" [**14.4**].)

It was believed at one time that some minimum level of air motion was desirable to avoid subjective reactions of "stagnation". The ASHRAE Standard 1981 recognizes no such lower limit so far as thermal comfort is concerned, and sets an upper limit of 0.15 m/s in winter and 0.25 m/s in summer, with certain exceptions, allowing higher values in combination with higher summer temperatures.

In discussing his comfort charts, Fanger points out that when the air speed is close to zero, free convection controls the convective heat-transfer process. Further, he says, "the necessary temperature for comfort is independent of the air velocity, when it is low. It will be seen that there is inflection for the comfort lines at 0.1–0.2 m/s and that changes in velocity, especially in the interval 0.1–0.3 m/s are important. An increase from 0.1 to 0.3 m/s must thus be compensated for by an increase in temperature of 1.5–3 K.

"The considerable temperature changes necessary as compensation for changes in air velocity within the interval 0.1–0.3 m/s possibly give a partial explanation for the phenomenon of draught. One can compensate for an increase in the air velocity by increasing the temperature, but since it is very difficult to create a uniform velocity field in a room, considerable differences in thermal sensation from point to point may easily arise. In order to obtain a uniform thermal climate in the occupied zone of a room, one normally attempts to keep the air velocities <0.1 m/s where the heat transfer to the air is by free convection and therefore independent of the velocity."

The upper limits on air speed needed to avoid marked variations in comfort place increased responsibility on the HVAC engineer, who relies on air motion to achieve air distribution and dilution of contaminants in the occupied zone. The engineer is not limited in the same way in the space above 1.8 m from the floor and within 0.6 m of walls and fixed air-conditioning equipment, which is not, by definition, part of the occupied zone.

14.17 Radiation from Electric Lighting

The radiant-energy output from electric lighting includes infrared and visible radiation in varying proportions, depending on the type of lamp. It will be recalled from Chapter 4 that all wavelengths in the visible range do not provide the same sensation of light to the eye. The total radiant output required to produce a given illuminance may in some cases be high enough to be a significant factor in comfort conditions as well as in air-conditioning loads.

The proportion of the total energy input delivered as useful visible light is expressed as luminous efficacy, and varies from 10 to 18 lumens per watt for incandescent lights, 30 to 80 lumens per watt for fluorescent, and over 100 lumens per watt for some high-pressure sodium lamps. Since the visible output is measured in terms of the response of the human eye, it cannot be expressed readily in energy terms as a simple proportion of the energy input.

McIntyre [**14.4**] has provided the information shown in Table 14.3. An illuminance of 1000 lux, which is a relatively high level of illumination, can be produced by an array of fluorescent ceiling fixtures (total electrical input of about 50 W/m^2), leading, according to McIntyre, to a thermal irradiance of about 8 W/m^2 and an increase in MRT of 0.7 K. The illuminance can be raised to 7000 lux before the plane radiant-temperature asymmetry exceeds 10 K. The relatively small amount of direct radiation from fluorescent fixtures is unlikely to be a complicating factor in maintaining thermal comfort. Incandescent lighting, as can be seen from Table 14.3, is in a different category because of its relatively low efficacy, and may often have to be considered as an important radiation source.

Part of the energy from lights is absorbed by the lighting fixtures and by adjacent construction, to be given up by reradiation and convection to the occupied space, and, in the case of recessed fixtures in suspended ceilings, to the space above the ceiling. Ceiling systems may therefore become secondary radiators to the occupied space, a function influenced greatly by their integration with the air-conditioning system [**14.2, 14.7**].

TABLE 14.3

Thermal Irradiance from Different Lamp Types
(From McIntyre, D.A., *Indoor climate*, London: Applied Science Publishers, 1980, Table 8.2, p. 227) [**14.4**]

Lamp type	Irradiance W/m^2 per 1000 lux		Increase in MRT K per 1000 lux	Compensating decrease in T_a K per 1000 lux	Illuminance for vrt = 10 K, lux
	Total	Short-wave			
Fluorescent white	8	2.7	0.7	0.3	7000
Tungsten	70	55	6	2.6	800
Tungsten halogen	50	40	5	2.2	1000
Sodium LP	6	2.7	0.5	0.2	10,000
Sodium HP	9	5	0.8	0.4	7000
Mercury fluorescent	15	6	1.4	0.6	4000

The figures are appropriate for diffuse radiation from an extended source.

14.18 Comparison of ASHRAE and Fanger Comfort Charts

There is a remarkable similarity in the comfort evaluations shown for the particular environmental conditions in the ASHRAE comfort chart and the Fanger charts. The present ASHRAE chart applies only to sedentary subjects with specified limits of clothing; the Fanger comfort equation and the Fanger charts also take account of differing levels of physical activity and clothing. Both are reasonably consistent in the weight given to relative humidity as a factor in the comfort of sedentary subjects, which is much less than that found from the earlier work. An increase from 20% to 80% rh is equivalent to a temperature change of about 1.5 K for the same comfort, according to the Fanger equation; the ASHRAE chart shows a corresponding temperature decrease of about 1 K. This is in marked contrast to the 4.5 K found from the original Houghton effective temperature chart.

Fanger does not attempt to set limits on relative humidity. The ASHRAE comfort chart and the ASHRAE standard limit rh indirectly by requiring that the dewpoint temperature be between 1.7°C and 16.7°C, based on "considerations of comfort, respiratory health, mold growth and other moisture-related phenomena". This sets the range of rh in winter between extremes of about 23% and 84% and in summer between 20% and 70% (Figure 14.3).

14.19 Heat Stress and Effective Temperature

The upper limits of environmental conditions under which people can perform required tasks (or, in the extreme, survive) have been the subject of much study and have been of particular concern in physically demanding industrial occupations such as deep mining and in military operations [**14.4**]. Such needs have promoted the development and use of single indices such as wet-bulb-globe temperature (WBGT) which has been found to be a useful indicator for many kinds of situations [**14.2, 14.4**]. The WBGT combines air temperature, air motion, radiation, and vapour pressure with the weighting given to each, depending on the instrumental characteristics.

The ASHRAE ET* scale is by definition a measure of physiological strain and heat exchange, and it ought therefore to provide a rational basis for establishing limiting conditions. The basic scale is fixed on 50% rh, but the slopes of the lines relating conditions to the basic scale which define conditions of constant stress are specific to each combination of activity level, clothing, and air motion. One analysis of military experience with heat deaths under light activity and clothing conditions suggests that 35 ET* or wettedness $w = 0.45$ is a practical working limit. Other data suggest that for sedentary persons the limit could be as high as 41 ET*, for a wettedness more nearly equal to $w = 1.0$. Acclimatization and duration of activity can have a marked influence and are complicating factors [**14.2, 14.4**].

14.20 Heat and Moisture from People

It is possible to establish from physiological data the heat and moisture given off by building occupants; these are often important factors in air-conditioning heating and cooling loads. The amount and proportion of heat and moisture given up per person are determined mainly by activity and by the proportion of men, women, and children in the occupancy. Rates recommended for design use are given in Table 14.4 [**14.2, p. 26.25**].

Table 14.4

Heat and Moisture Contribution from People
(From *ASHRAE Handbook 1981 Fundamentals*, Table 18, p. 16.26) [**14.2**]

Activity and application	Heat gain per person, W Total, male adults	Total heat adjusted*	Sensible heat, watts	Latent heat, watts
Seated at rest. Theatre, movie	115	100	60	40
Seated, very light work, writing. Offices, hotels, apartments	140	120	65	55
Seated, eating. Restaurant	150	170	75	95
Seated, light work, typing	185	150	75	75
Standing, light work, walking slowly. Retail store, bank	235	185	90	95
Light bench work. Factory	255	230	100	130
Bowling. Bowling alley	350	280	100	180
Moderate dancing. Dance hall	400	375	120	255
Walking 1.3 m/s, light machine work. Factory	305	305	100	205
Heavy work, heavy machine work, lifting. Factory	470	470	165	300
Heavy work, athletics. Gymnasium	585	525	185	340

See Ref. **14.2**, p. 26.25 for further explanatory notes.
*Adjusted for proportion of men, women, and children.

14.21 Health and Efficiency

There is a widely held belief that conditions for thermal comfort are also best for health and efficiency. It appears, however, that health, at least, may not be endangered for normal subjects by substantial departures from the conditions of comfort. Relative humidity has been perhaps the most widely discussed factor in relation to health. Proof that high or low relative humidity is either beneficial or hazardous to health has been difficult to obtain. Recent experiments appear to confirm that humidification to avoid the low indoor relative humidities that accompany cold weather can reduce the incidence of upper respiratory infections. McIntyre [**14.4**] provides a discussion of recent studies.

The aged and those in poor health benefit by the maintenance of conditions that will reduce the physiological stress of extreme temperatures and humidities.

The beneficial influence of controlled comfort conditions upon work efficiency is still being debated, despite many efforts to prove conclusively that there is a close relationship [**14.2**]. It appears that there are often complicating psychological factors involved [**14.4**].

14.22 Plant and Animal Growth and Production

The effects of environmental conditions on plant and animal growth and production are reasonably well known [**14.2, Chap. 9; 14.6, Chap. 22**]. Economic return is usually the main criterion when adjustment of the environment is under consideration. The effects of temperature on the production and growth of animals are shown in figures 14.11 and 14.12; heat and moisture dissipation rates for dairy cattle are shown in Figure 14.13 [**14.2**].

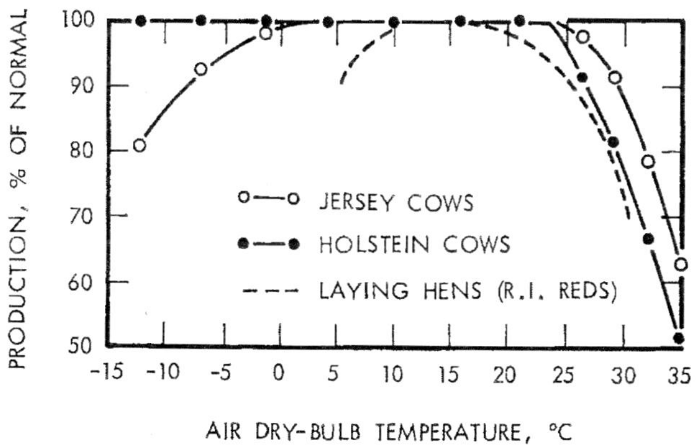

FIGURE 14.11 Comparative effect of air temperature in mature animal production.
(From *ASHRAE Handbook 1981 Fundamentals*, Fig. 1, p. 9.2)[**14.2**]

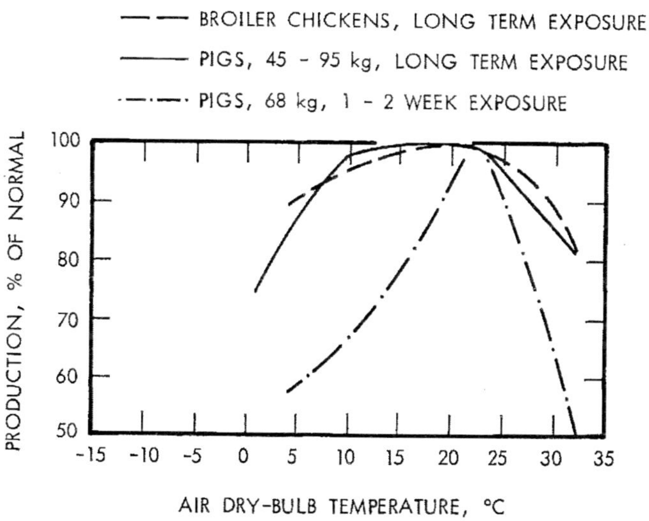

FIGURE 14.12 Comparative effect of air temperature on growing animals.
(From *ASHRAE Handbook 1981 Fundamentals*, Fig. 2, p. 9.2)[**14.2**]

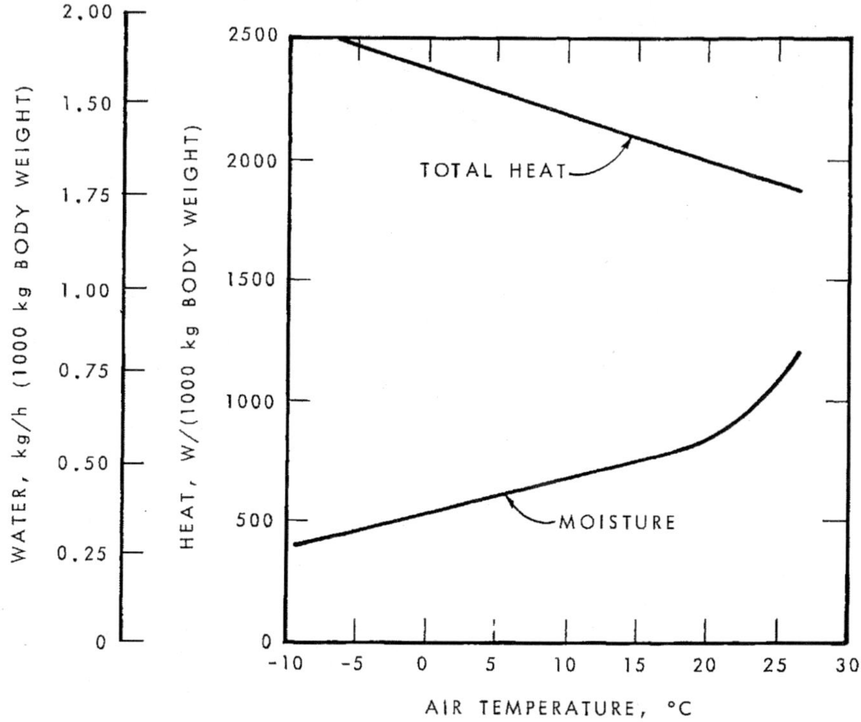

FIGURE 14.13 Stable heat and moisture-dissipation rates with stanchioned dairy cattle.
(From *ASHRAE Handbook 1981 Fundamentals*, Fig. 10, p. 9.5)[**14.2**]

Animal barns pose special problems. They almost always have high-moisture occupancies, which can create difficulties in cold climates. This does not necessarily mean high relative humidity indoors, but it does usually mean excessive condensation on windows and other cold surfaces. Moisture can be controlled if a substantial proportion of animal heat can be conserved or if additional energy can be provided for use in warming outdoor air for ventilation to carry off the moisture.

Plants also constitute a high-moisture occupancy, because of the water evaporated from them and from their soil. They also need light for growth, which must come from sunlight or from electric lights, and control of the heat associated with this radiation becomes a major consideration. It will be recalled from Chapter 9 that the solar heat gain through single glazing may reach a peak of almost 800 W/m^2 for south-facing walls in January and horizontal surfaces in July at 45°N latitude. As much as 24 W/m^2 of radiation in the range 400 to 850 nm may be used as supplementary lighting in greenhouses, while a minimum of 50 W/m^2 (400 to 850 nm) is usually required in growth chambers and growth rooms [**14.2, p. 9.13**]. The outputs of photosynthetically active radiation (PAR) in the range 400 to 700 nm and in the range 400 to 850 nm (which stimulates plant growth) are shown for various kinds of light sources in Table 14.5. About 59% of solar radiation is in the range 400 to 850 nm [**14.2, p. 9.12**].

There are other cases in which the operation involves conditions that range far beyond the limits for comfort. Restaurant cooking and many operations in heavy industry pose such severe heating and cooling loads that it is impractical to consider means for the provision of comfort conditions throughout the space. Partial measures are widely used. Under hot conditions, when worker mobility can be sacrificed, cooled helmets and cooled suits can be used. Cranes are often equipped with air-conditioned cabs. Jets of cool air can be directed at fixed work stations, and high rates of general air circulation can be used to promote body-heat loss. In extremely cold work situations, protective clothing is widely used. Spot heating of fixed work stations can be accomplished with directional radiant heating.

Discussions of the equipment, systems, and applications used for a wide range of occupancies are to be found in the ASHRAE Handbooks.

TABLE 14.5

Radiation Power Distribution of Light Sources Per 100 Watt Total Radiation

(From *ASHRAE Handbook 1981 Fundamentals*, Table 8, p. 9.12) [**14.2**]

Light sources	UV 400 nm W	PAR + FR 400-850 nm W	IR 850-2700 nm W	Thermal 2700 + nm W	Total radiation W
FCW	2	36	1	61	100
HG/DX	3	19	18	60	100
MH	4	41	8	47	100
HPS	0.4	50	12	38	100
LPS	0.1	56	3	41	100
INC	0.2	17	74	9	100
SUN	6	59	33	2	100

FCW = Cool white fluorescent
HG = Mercury
MH = Metal halide
HPS = High-pressure sodium
LPS = Low-pressure sodium
INC = Incandescent
SUN = Solar radiation
PAR = Photosynthetically active radiation

14.23 Buildings for Products and Processes

Buildings used for storing products usually require conditions that will prevent deterioration and preserve (or even enhance) quality. Chemical, physical, and biological changes during the storage period may have to be suitably controlled by the maintenance of appropriate levels of temperature, humidity, or atmospheric composition within the storage space. The most common example is the use of refrigeration to produce cool stores (down to freezing) and cold stores (below freezing) for food preservation. Another example is the controlled-atmosphere storage (CAS) used for apples and other fruits in which carbon dioxide, oxygen, and sometimes other gases are maintained at levels that will produce the desired degree of ripening or preservation.

Occupancies involving processes requiring special considerations can be divided into two broad classes, namely, those that require closely controlled environmental conditions in the interests of the process and those that discharge relatively large amounts of heat or moisture or both into the air in the space. In a smaller number of cases the troublesome output may be other gaseous or particulate matter. As will be shown later, the quantity of water released and the level of relative humidity required may have serious implications in buildings constructed for cold-weather conditions.

Textile mills and tobacco factories are good examples of buildings with occupancies that require special conditions. Textile mills need a high relative humidity, often up to 85%, to facilitate the handling of fibres. Tobacco products must be maintained in a high-humidity atmosphere during manufacture and while in storage to prevent loss of moisture.

Laundries and dairies, where large quantities of steam and hot water are used freely, are good examples of high-moisture occupancies, with no special requirement for close control of the thermal environment. It has been common to regard copious condensation and fogging in such buildings as inevitable, and to accept high rates of deterioration and frequent and costly maintenance. Such conditions need not always be tolerated. In the case of laundries, for example, the provision of adequate ventilation to carry off moisture at suitable locations may produce a low-humidity condition throughout much of the building.

Indoor swimming pools pose a similar problem, since they present a large exposed water surface. In addition, the temperature of the water may be as high as 29°C, so that the pool surface serves as a large-capacity humidifier [**14.8**]. Some occupancies presenting large water surfaces create no difficulty, because the temperature is at or below the tolerable limit of dewpoint.

In hospital operating rooms, the control of bacterial infection is of paramount importance. There is also a requirement for controlled temperature and humidity. In clean rooms required for certain high-technology operations, people may be the principal source of contamination. When the requirements are extreme it becomes necessary to consider isolating people from the clean environment.

A number of occupancies require close temperature and humidity control. When precise measurements are being made, it is necessary to minimize variations due to ambient temperature and relative humidity changes, or to facilitate fine control of other conditions. This requirement arises frequently in research laboratories.

Museums, art galleries, and libraries housing valuable collections should be carefully climate-controlled to minimize the rate of deterioration. Controlled relative humidity is necessary in order to avoid conditions favourable to biological growth and chemical attack and to minimize dimensional change due to changes in moisture content. Close temperature control is also necessary, not only to minimize dimensional changes due to temperature but also because relative humidity is temperature-dependent. This must include consideration of solar and other radiation sources that will cause the temperature of an object to differ from air temperature, as discussed in Chapter 13. It is also desirable to maintain constant conditions throughout the year; to accomplish this, it is necessary to select a relative humidity level that is feasible to maintain during summer and winter. Low humidities may be difficult to maintain in summer; high relative humidities may cause damage to the building in winter.

There are other cases in which the operation involves conditions that range far beyond the limits for comfort. Restaurant cooking and many operations in heavy industry pose such severe heating and cooling loads that it is impractical to consider means for the provision of comfort conditions throughout the space. Partial measures are widely used. Under hot conditions, when worker mobility can be sacrificed, cooled helmets and cooled suits can be used. Cranes are often equipped with air-conditioned cabs. Jets of cool air can be directed at fixed work stations, and high rates of general air circulation can be used to promote body-heat loss. In extremely cold work situations, protective clothing is widely used. Spot heating of fixed work stations can be accomplished with directional radiant heating.

Discussions of the equipment, systems, and applications used for a wide range of occupancies are to be found in the ASHRAE Handbooks.

14.24 Multiple Occupancies

In many cases, several activities, each requiring different conditions, will take place in the same building. The compartmentation separating two or more conditions so that they do not interact unduly will depend on the activities involved. The household refrigerator is one familiar example of a small compartment suitably enclosed to restrict the exchange of heat, moisture, and air. Special rooms for storage, curing, growth, conditioning, and photoprocessing are other examples.

Multiple occupancies in buildings pose a major problem in preparing rules for the maintenance of safe environments in the routine operation and use of buildings [**14.9**]. Such regulations are set out in Canada in the model National Fire Code which, when adopted, becomes the basis for the maintenance of fire safety. It is a companion document to the National Building Code, which states the requirements to be met in new construction to provide safety in respect of structure, fire, and health. One of the potential difficulties with both documents is that a building constructed with one kind of occupancy in mind may during its lifetime be used to house occupancies that were never originally envisaged. Even if a building is safe for a new occupancy, it may not be efficient or economical. This is a problem for the owner and the tenant to resolve.

References

14.1 *Thermal environmental conditions for human occupancy*. ASHRAE Standard ANSI/ASHRAE 55-1981.

14.2 *ASHRAE Handbook 1981 Fundamentals*. Chaps. 8, 9, and 10.

14.3 Fanger, P.O. *Thermal comfort.* New York: McGraw Hill, 1970.

14.4 McIntyre, D.A. *Indoor climate.* London: Applied Science Publishers, 1980.

14.5 Fanger, P.O. Calculation of thermal comfort: Introduction of a basic comfort equation. *ASHRAE Transactions*, 1967, *73*, (Part II), pp. III.4.1–4.18.

14.6 *ASHRAE Handbook 1982 Applications.*

14.7 Mitalas, G.P. Cooling load caused by lights. *Transactions, Canadian Society for Mechanical Engineering*, 1973–74, *2* (3), pp. 169–74. (NRCC 14660).

14.8 Garden, G.K. *Indoor swimming pools.* Canadian Building Digest 83. DBR/NRCC, Nov. 1966.

14.9 Hutcheon, N.B. *Safety in buildings.* Canadian Building Digest 114. DBR/NRCC, June 1969.

15

Buildings for a Cold Climate

15.1 Introduction

Previous chapters dealt with various atmospheric and weather factors and the importance of radiation, heat, air, and water as they affect buildings. These were followed by discussions of the interactions between the basic elements and the building. Discussions of the indoor thermal environment served as preparation for the treatment in Chapter 14 of the thermal- and moisture-related characteristics and requirements of various kinds of occupancies. In this final chapter it is possible to consider these topics together and to attempt to identify what they tell us about the requirements and the performance of buildings for a cold climate. The discussion will be confined to the passive aspects of a building in its major role as a separator of environments. The emphasis will be on the separation of indoor and outdoor thermal and moisture conditions, and will not be concerned directly with the equipment and accessories by which the indoor environment can be adjusted in an active way.

15.2 Characteristics of a Cold Climate

The question of what constitutes a cold climate for building purposes must now be considered. The answer will be somewhat arbitrary, since the individual elements that make up cold climates are themselves complex, and vary widely with time and geographic location. They have different effects on different parts of a given building, depending on the nature of the building and its occupancy. The best that can be expected is that some index of cold-weather severity will be proposed, beyond which serious consideration of cold-weather conditions will be required in the design, construction, operation, maintenance, repair, and use of buildings.

Cold climates in different countries have some basic features in common. The climate of Canada, which has given rise to much of the experience on which this book is based, will be used for reference in these discussions. The variations in climate that can occur over any large continental land mass are well illustrated by the climatic data for Halifax, Ottawa, Saskatoon, Vancouver, and Inuvik. These are the centres for the five regions selected for the administration and organization of building research in Canada. Their climates typify the five sets of climatic conditions that must be considered separately for many building purposes. A brief summary description is given in Table 15.1, and annual values of pertinent meteorological data for the five centres are shown in Table 15.2. The monthly mean temperatures, precipitation, and hours of bright sunshine are listed for the five centres in Table 15.3 and shown graphically in Figures 15.1 to 15.3.

The climatic regions in Table 15.1 are as given by Boughner and Thomas [**15.1**]. The characteristic vegetation is in accordance with Miller's "quograph" [**15.2**], which delineates on a world basis the relationship between climate and vegetation (Figure 15.4).

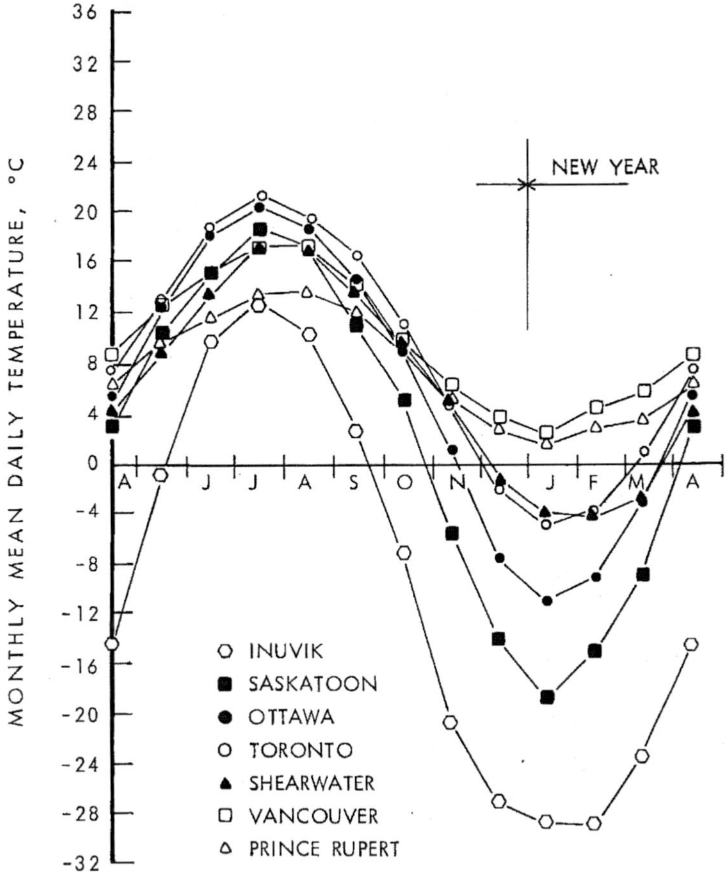

FIGURE 15.1 Monthly mean daily temperature.

The vegetation of a region reveals much valuable information about the region's climate, and may be studied with profit. Vegetation reflects not only the temperature and precipitation, as the simple plot in Figure 15.4 based on mean annual values shows, but also reveals something of the coincidence of temperature and precipitation. Grassland, for example, results from low annual precipitation coupled with low evaporation during cold winters. Savannah reflects high temperatures, high evaporation rates, and high precipitation. These conditions, which influence the type of vegetation, will also determine the drying, wetting, and moisture storage of exposed building materials. The Canadian cities of Halifax, Ottawa, and Vancouver are rated as having climates that will support forests. Halifax has an Atlantic-coast climate, cold and very wet. Ottawa has much colder winters and does not have as much precipitation, but merits the designation "cold and wet". Saskatoon, which is classed as having grassland vegetation, and Inuvik, which has tundra, are dry and very cold.

Inuvik is on the border between the Arctic region, which is very dry, and the northern region, which extends southward in the east into Ontario and Quebec with enough precipitation to support forest growth. Vancouver, in contrast, is wet and cool, with a west-coast marine climate. It has a mean annual temperature of 10°C and a mean daily temperature in January of 2.4°C, yet it has a heating index of 3031 degree days, a heating-design temperature that occurs 2½% of the hours in January of -7°C, 52 cm of snow, and 56 days when the temperature falls to the freezing point (Tables 15.2 and 15.3). Vancouver is clearly an important case to consider in setting criteria for cold climates. All of the other four centres are substantially colder.

The heating-design temperature, also called the 2½% January temperature, is now proposed here as the most useful simple criterion. Winter climates having a design-temperature value more than 7 K below freezing should be classed as cold. In Canada the design temperature is based on 2½% of the hours in January (2½% × 31 × 24 = 18.6). The practice in the United States, as reflected in the climatic data in the *ASHRAE Handbook 1981*, is to base the percentage on the

TABLE 15.1

Summary Description of Climate of Five Locations in Canada

(From Boughner, C.C., and Thomas, M.K., *The climate of Canada*, Canada Year Book, 1959 and 1960 [**15.1**] and Miller, A.A., *The skin of the earth*, London: Methuen, University Paperbacks, 1964, p. 173 [**15.2**]

Regional centre	Climatic region [**15.1**]	Significance for building	Vegetation [**15.2**]	Regions with similar climate
Halifax	Southeastern	wet and cold	forest	Atlantic provinces
Ottawa	Southeastern	moderately wet, cold	forest	Southern parts of Ontario and Quebec
Saskatoon	Prairie	dry, very cold	grassland	Southern parts of Manitoba, Alberta and Saskatchewan
Vancouver	Pacific	wet, cool	forest	West coast
Inuvik	Arctic	very dry, very cold	tundra	Arctic coast, Arctic Islands

three months December, January, and February, totalling 2160 hours, and to express the percentage as the percentage of hours in which the temperature was higher than the stated value. The 99% value means that 1% of the hours will be colder (1% × 2160 = 21.6). Thus, the 99% value based on three months will be roughly the same as the 2½% value for January, as used in Canada on the assumption that January is the coldest month.

The criterion of more than 7 K below freezing for the 2½% January temperature means that all Canadian locations, except Vancouver and some locations on the west coast and on the adjacent islands, will rate as having a cold climate for building purposes [**15.3**].

The Pacific coast is wet and cool. Few problems attributable to low temperatures can be clearly distinguished from those arising from relatively high levels of precipitation. At the other extreme, very low temperatures produce severe cold-weather problems in buildings even when precipitation is low, so that it is necessary to consider carefully the influence of temperature and moisture both separately and in combination.

TABLE 15.2

Meteorological Data for Five Selected Stations in Canada

	Mean annual temp., °C	Mean daily temp. range, K	Degree days, re 18°C	2½% Jan. temp., °C	2½% July DB temp., °C	Mean annual precipitation, mm	Mean annual snowfall, cm	Annual total hours bright sunshine	Mean wind, km/h and m/s	Annual days freezing	Altitude, m
Shearwater*	6.8	—	4104	—	—	1381	210	—	—	—	—
Halifax 44°43′N 63°34′W	—	9	—	−16	26	—	—	—	16.7/4.6 24% NW	134	25
Ottawa 45°24′N 75°43′W	5.8	11	4635	−25	30	851	216	2010	16.6/4.6 19% NW	165	125
Saskatoon* 52°08′N 106°38′W	1.6	12	6053	−35	30	353	113	2367	18.2/5.1 16% S	205	501
Inuvik*	−9.7	—	10183	−46	25	260	174	260	—	261	—
Aklavik 68°14′N 135°00′W	—	9	—	—	—	—	—	—	10.5/2.9 24% N	—	61
Vancouver* 49°17′N 123°05′W	9.8	8	3031	−7	26	1068	52	1784	13.0/3.6 33% E	56	5

*Station at airport

TABLE 15.3

Temperature, Precipitation, and Sunshine for Five Selected Stations in Canada

	Jan.	Feb.	March	April	May	June	July	Aug.	Sept.	Oct.	Nov.	Dec.
Shearwater												
Mean daily temp., °C	−3.8	−4.2	−0.7	4.0	9.0	13.7	17.6	17.8	14.8	9.9	4.9	−1.1
Precipitation, mm	147	129	112	105	110	85	92	94	94	113	152	148
Bright sunshine, h	110	129	148	176	217	217	220	214	172	155	98	90
Ottawa												
Mean daily temp., °C	−10.9	−9.5	−3.1	5.6	12.4	18.2	20.7	19.3	14.6	8.7	1.4	−7.7
Precipation, mm	60	57	61	68	70	73	81	82	79	66	79	77
Bright sunshine, h	96	115	150	175	231	245	277	243	171	138	76	78
Saskatoon												
Mean daily temp., °C	−18.7	−15.1	−8.7	3.3	10.6	15.4	18.8	17.4	11.3	5.0	−5.8	−14.0
Precipitation, mm	18	18	17	21	34	57	53	45	33	19	19	18
Bright sunshine, h	99	129	192	225	279	280	341	294	207	175	98	84
Vancouver												
Mean daily temp., °C	2.4	4.4	5.8	8.9	12.4	15.3	17.4	17.1	14.2	10.1	6.1	3.8
Precipitation, mm	147	117	94	61	48	45	30	37	61	122	141	165
Bright sunshine, h	55	93	129	180	253	243	305	255	188	116	70	44
Inuvik												
Mean daily temp., °C	−29.3	−29.4	−23.8	−14.6	−0.8	9.8	13.3	10.3	2.7	−7.2	−20.6	−27.1
Precipitation, mm	20	10	17	14	18	13	34	46	21	34	15	19
Bright sunshine, h	9	68	173	254	289	366	314	208	110	53	19	0

From Dept. of Environment, Atmospheric Environment Services, Leaflet 0064-0009.

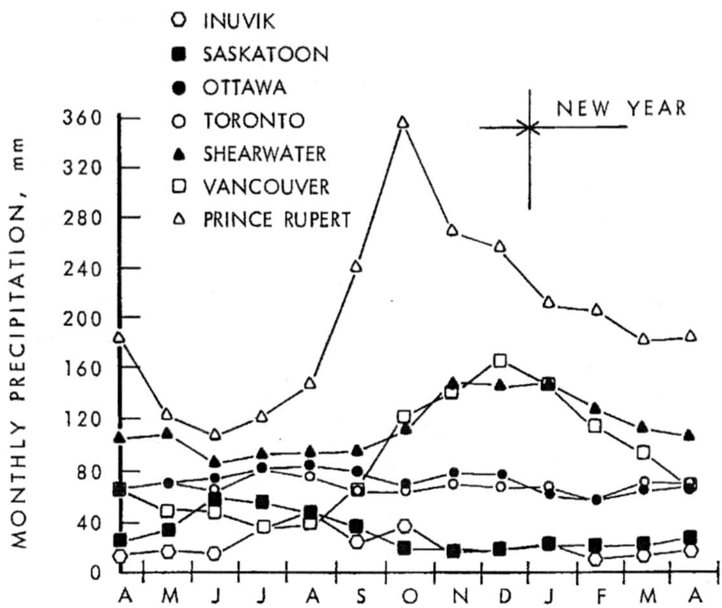

FIGURE 15.2 Average monthly precipitation, mm.

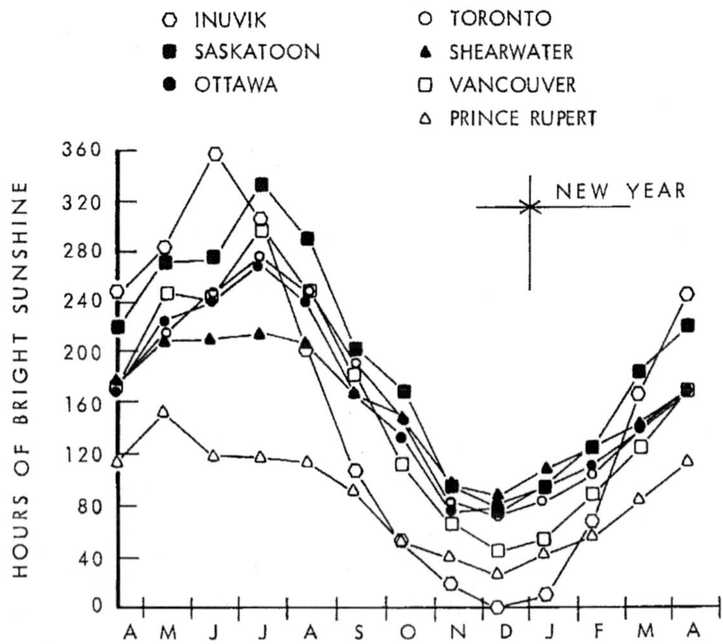

FIGURE 15.3 Average monthly hours of bright sunshine.

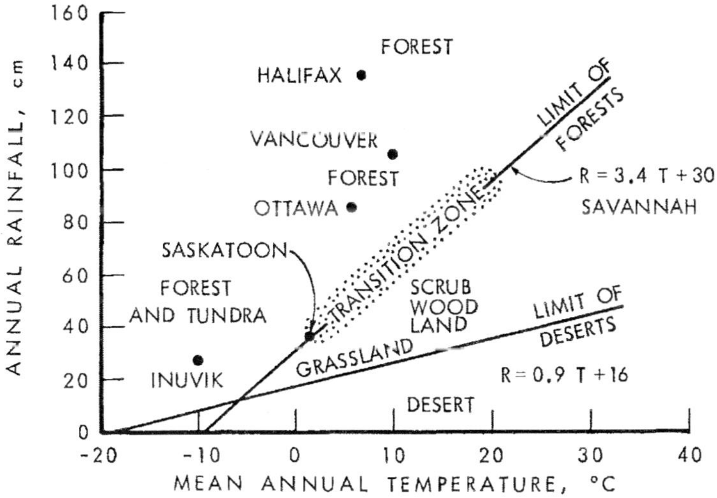

FIGURE 15.4 Canadian locations shown on graph of vegetation types versus climate based on 160 world stations.
(From Miller, A.A., *The skin of the earth*, London: Methuen, University Paperbacks, 1964, p. 173)[**15.2**]

TABLE 15.4

Annual Water Surplus (AWS) by Provinces and Territories in Canada
(From Sanderson, M.E. and Phillips, D.W., Climatological Studies No. 9, Toronto: 1967, Department of Transport, Meteorological Branch, 1967) [**15.4**]

Province or territory	No. of stations	No. of stations with surplus	Maximum surplus, mm	Minimum surplus, mm
Newfoundland and Labrador	35	35	1046	183
New Brunswick	30	30	734	315
Nova Scotia	36	36	950	467
(Halifax AWS = 810 mm)				
Prince Edward Island	3	3	472	305
Quebec	139	139	886	94
Ontario	161	160	546	0
(Ottawa AWS = 318 mm)				
Manitoba	43	15	91	0
Saskatchewan	88	3	56	0
(Saskatoon AWS = 0 mm)				
Alberta	87	19	226	0
British Columbia	150	108	3901	0
(Vancouver AWS = 470/856 mm, 2 stations)				
Northwest Territories	18	0	0	0
(Inuvik AWS = 0 mm)				
Yukon	10	2	38	0

A further measure of the differences between Canada's climatic regions is provided by the annual water surplus, or annual runoff [**15.4**]. It represents the difference between precipitation and evapotranspiration and may be taken as a rough measure of the ability of the climatic conditions to maintain exposed materials in a dry state. The data of Table 15.4 confirm that the Prairie provinces, the Yukon, and the Northwest Territories are "dry", as concluded earlier. Coastal British Columbia has a great excess of precipitation over evaporation, while some valleys in the interior are very dry. The remaining provinces from Ontario eastward are all moderately "wet".

15.3 Cold Climates Are Also Hot

All of the five Canadian centres selected for consideration have dry-bulb temperatures of 25°C or over, which are exceeded for 2½% of the hours in July. Two of them, Ottawa and Saskatoon, have 2½% July temperatures of 30°C, while January 2½% temperatures are −25°C and −35°C respectively (Table 15.2). The sun shines brightly on clear days in northern and southern latitudes as well as at the equator. The intensity of the incident solar beam is only moderately diminished by the absorption of the longer atmospheric paths, while that incident on a surface is influenced by the angle of incidence, which depends on the sun's altitude and the orientation of the surface.

The solar energy falling on the outside of buildings in summer in countries with cold climates is almost as great as for areas much closer to the equator. The use of added insulation and multiple glazing and shading will determine how much the solar energy contributes to the cooling load.

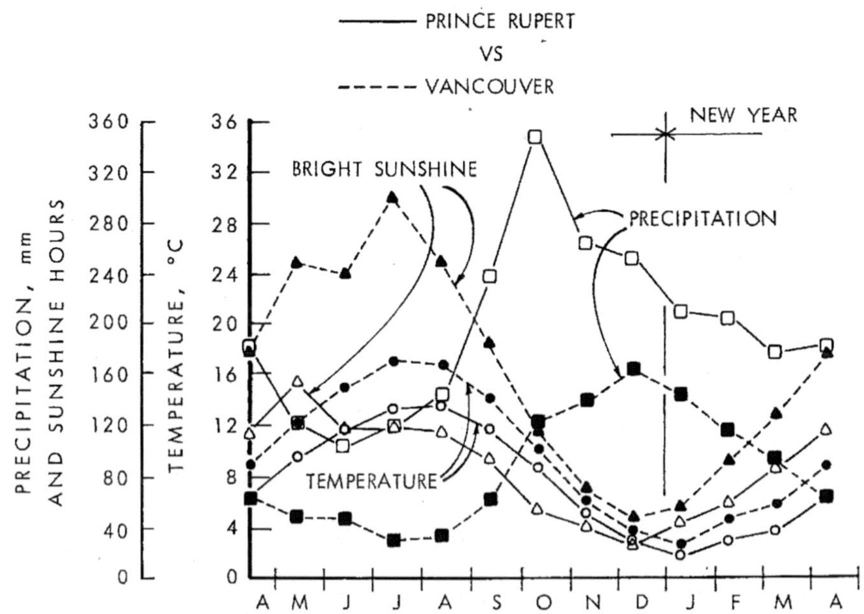

FIGURE 15.5 Monthly hours of bright sunshine, precipitation, and mean daily temperature for Prince Rupert and Vancouver.

Cloud cover is a major factor in determining the amount of solar radiation reaching the earth. The hours of bright sunshine experienced by the five reference locations in Canada are shown in Figure 15.3. Inuvik is north of the Arctic Circle, and so has a decrease in hours of sunshine to zero in December. The clear skies and long hours of sunshine in summer combine to produce a high value of 360 hours in June. Shearwater (Halifax) has the lowest summer values of 220 in June and July, owing to relatively high precipitation and consequent high incidence of cloud cover in summer. Ottawa, Shearwater, and Saskatoon at latitudes of 45°, 44°, and 52° N respectively show similar winter patterns of 80 to 100 hours in December; Vancouver, with 160 mm of precipitation in December, has a little over 40 hours of bright sunshine.

Some interesting differences between cities are evident from the comparisons of figures 15.5 to 15.8. Figure 15.5, for example, compares Prince Rupert on the west coast at 54°N latitude with Vancouver at 49°N. The precipitation for Prince Rupert is high and the sunshine hours in summer are correspondingly lower than for Vancouver. The graph for Saskatoon versus Inuvik (Figure 15.6) shows the influence of latitude and the correspondence between decreasing sunshine hours

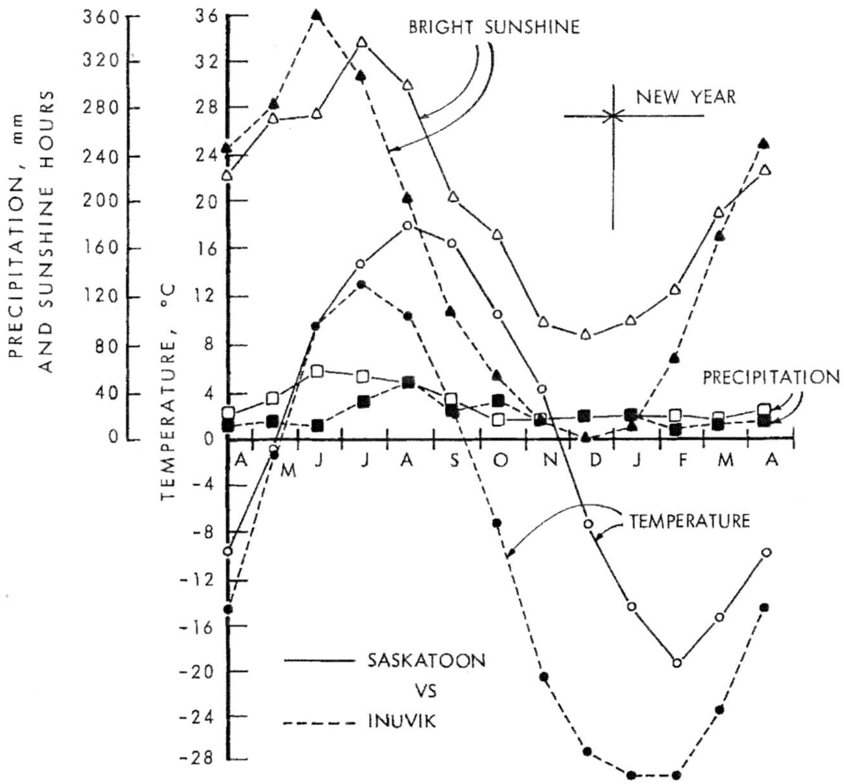

FIGURE 15.6 Monthly hours of bright sunshine, precipitation, and mean daily temperature for Saskatoon and Inuvik.

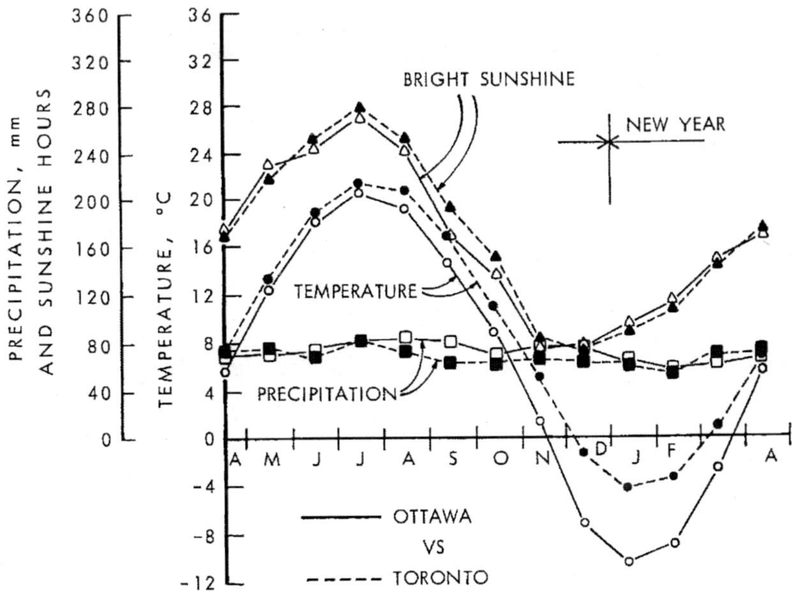

FIGURE 15.7 Monthly hours of bright sunshine, precipitation, and mean daily temperature for Ottawa and Toronto.

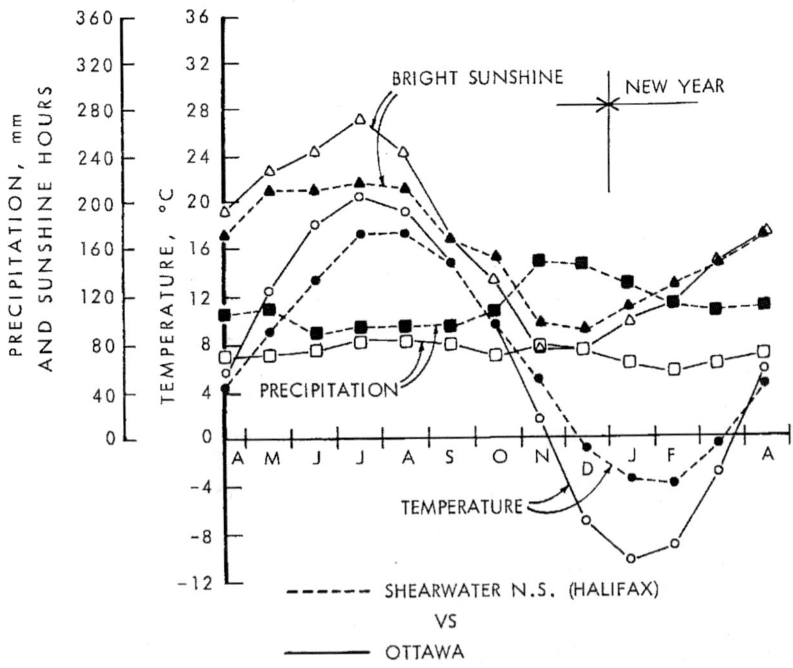

FIGURE 15.8 Monthly hours of bright sunshine, precipitation, and monthly mean daily temperature for Shearwater and Ottawa.

and decreasing temperatures. Ottawa and Toronto, shown in Figure 15.7, are remarkably similar, except for a substantial moderation of winter temperature in Toronto giving a mean daily value in January of −4.4°C compared with Ottawa at −10.9°C. Shearwater (Figure 15.8) has less extreme temperatures, more precipitation, and less sunshine than Ottawa.

15.4 Effects of Cold Weather

The principal effects of cold weather have already been identified and discussed in previous chapters:

1. Temperatures are lowered progressively relative to indoor temperature from inside to outside through the enclosing components of a building.
2. Temperature gradients are established which promote heat and moisture flow through a building enclosure in the direction of decreasing temperature.
3. Temperature changes occurring through an enclosure create differences in temperature and differential contraction, leading to distortion and loading of building elements and components.
4. Wetting by condensation is promoted on cool indoor surfaces and on cooled surfaces within the construction.
5. Increased moisture content of materials owing to increased relative humidity from reduced temperatures, rain, or condensation can promote expansion and accelerated deterioration resulting from chemical, physical or biological action.
6. Differences in air density due to differences in temperature between indoors and outdoors give rise to stack effect, which promotes air leakage through a building enclosure and a generally upward movement of air within a building. Exchange of air may also develop with vertical air spaces in walls and windows, and even with spaces filled with very low-density fibrous insulation.
7. Cold outdoor air entering through a building enclosure because of wind, exhaust fans, or stack effect will usually be at a high relative humidity, but at a low humidity ratio.
8. Exfiltration occurring at rates determined by the available leakage paths and the pressures resulting from the combined influence of wind, stack effect, and supply fans can carry moisture to cold parts of the construction where condensation may occur.

These effects have important implications for the design and operation of buildings (and particularly of building enclosures) that can now be identified.

15.5 Influence of Temperature Gradients

A plain concrete wall is shown in Figure 15.9. The winter gradient is drawn for the steady-state condition, which would only be approximated after two or more days at the steady outdoor temperature. The summer condition with solar radiation is always dynamic in nature, as shown in Figure 15.10, varying over a 24-hour period [**15.5**].

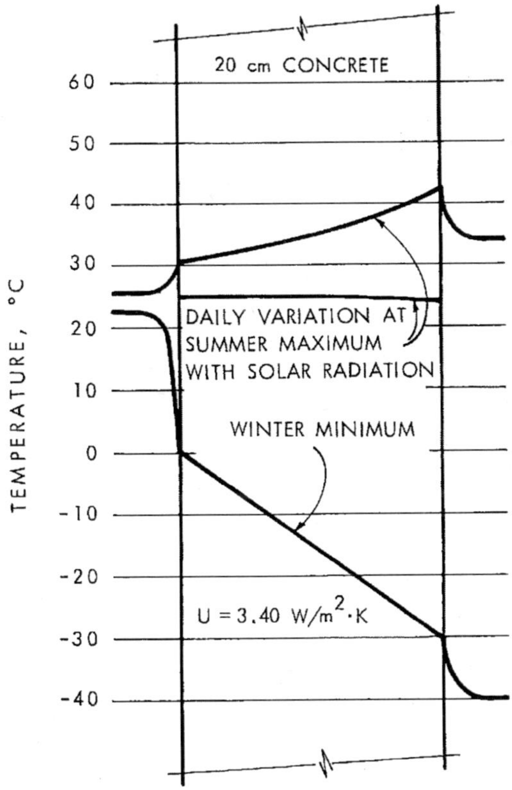

FIGURE 15.9 Yearly temperature range throughout a plain concrete wall for air temperature range -40° to +35°C.
(From Hutcheon, N.B., Fundamental considerations in the design of exterior walls for buildings, *Engineering Journal*, 1953, *36*(6), Fig. 1, p.7 NRCC 3057) [**15.5**]

The temperature gradient for the winter condition is 1.5 K/cm, which will produce a camber in a 15-m high wall, if unrestrained, of 56 mm. If fully restrained so that no shortening occurs, the resulting tension on the cold face will be nearly 7000 kN/m^2. Such length changes often lead to unforeseen distortions, cracking, and loads [**15.6**] within and between components.

The temperature gradients produced in walls, windows, floors, and roofs by lowered outdoor temperatures also promote the flow of heat to the cold side, resulting in a loss of heat to the outdoors and the possibility of condensation of water vapour on surfaces that are reduced in temperature.

15.6 Use of Insulation

The rate of heat loss can be reduced by the use of insulation in walls, floors, ceilings, and roofs and by double and triple glazing. The addition of 50 mm of an insulating material will more than double the resistance to heat flow of basic wall

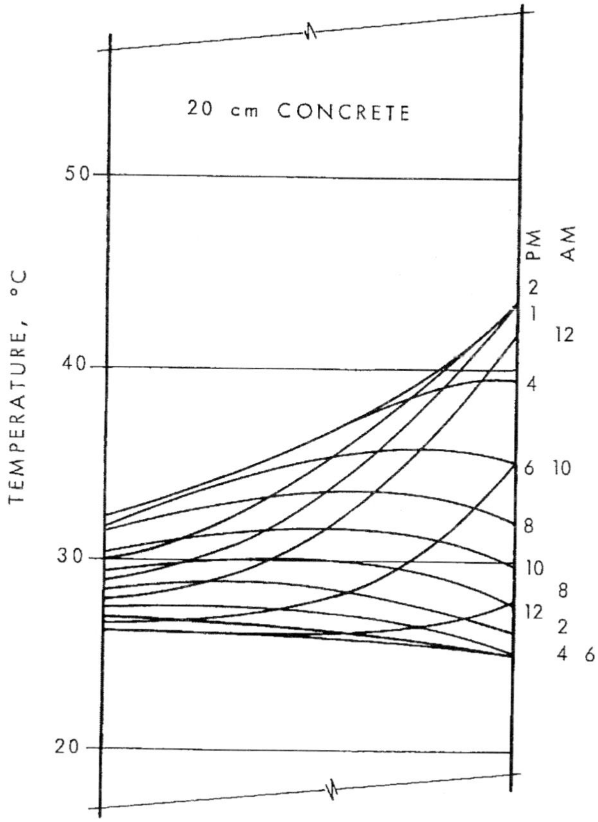

FIGURE 15.10 Variation of temperature with time in an 8-in. plain concrete wall due to solar radiation in summer.
(From Hutcheon, N.B., Fundamental considerations in the design of exterior walls for buildings, *Engineering Journal*, 1953, *36*(6), Fig. 5, p. 11 NRCC 3057) [**15.5**]

constructions. The use of 100 mm or more of insulation with a resistance in SI units of RSI-4 is now common, and amounts providing RSI-12 or more are now being used where a high degree of energy conservation is desired.

The addition of insulation causes a readjustment of the temperatures throughout a construction under a given overall temperature difference. This effect is shown by comparing the insulated concrete wall in Figure 15.11 with the uninsulated wall in Figure 15.9. The surface temperature on the warm side is increased, an effect that can be exploited in the use of insulation to reduce or eliminate surface condensation.

A different situation is created at the inner surface of the basic concrete wall in Figure 15.11, which is now very much colder than it was before the insulation was added. This illustrates the basic problem of condensation control within walls, floors, and roofs, one of the principal difficulties arising under cold-weather conditions. Condensation occurs on interior surfaces whenever water vapour at a concentration with a sufficiently high dewpoint comes in contact with them. Moisture may be transported to the condensing surfaces by capillary flow and water-vapour diffusion, but is more commonly carried as a component of air flowing to and over them. (Vapour diffusion has already been discussed at length in Chapter 5; air leakage was discussed in Chapter 11 and condensation in buildings in Chapter 12.)

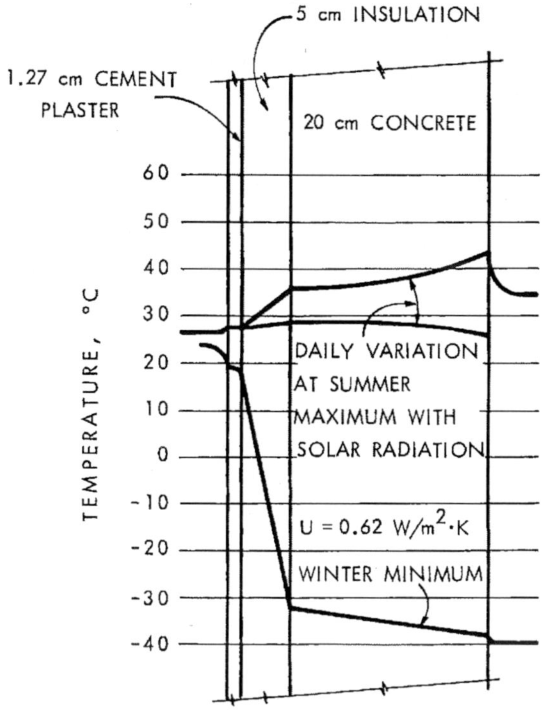

FIGURE 15.11 Yearly temperature range throughout a concrete wall insulated on the inside, for air temperature range −40° to +35°C.
(From Hutcheon, N.B., Fundamental considerations in the design of exterior walls for buildings, *Engineering Journal*, 1953, *36*(6), Fig. 2, p. 8 NRCC 3057) [**15.5**]

15.7 Control of Water Vapour

The practice of insulating residential buildings grew steadily from about 1920 and was accompanied by an increase in wetting and rotting of wood-frame walls. Laboratory investigations showed that such wetting could occur as a result of vapour diffusion from inside to the surface of sheathing, which was made colder in winter because insulation was being used. Over the next 20 years, this led to the widespread practice of incorporating vapour barriers on the warm side of insulation. There were many apparent successes—and baffling failures—in attempts to control condensation in this way. The advent of tall buildings drew attention to the role played by stack effect along with wind and fan action in producing air leakage. It is now well established that the water vapour that condenses on the colder outer layers of building enclosures can be carried by air leaking from the building. This effect, rather than vapour diffusion, is now judged to be by far the most common means by which condensation occurs within the building enclosure and in attics and other unheated spaces.

Vapour barriers, it is now realized, were effective in many cases because they controlled air leakage, not because they provided resistance to the flow of water vapour by diffusion through materials [**15.7**]. The resistance of many common interior finishes is often sufficient protection against vapour diffusion. Added protection by a vapour barrier may be desirable where there is high indoor relative humidity and with certain interior finishes that provide little resistance to vapour diffusion. This does not rule out the current practice of using continuous plastic films in particular types of construction as an appropriate way of controlling air leakage. They resist vapour flow and air flow and, although they are often called vapour barriers, they serve as air barriers that provide added protection against water-vapour diffusion. The sealing of all joints and holes is important for air barriers but is not required for protection against water-vapour diffusion.

Surface condensation on cool interior room surfaces can be reduced or eliminated by raising the surface temperature through the use of insulation or by heating. Failing this, the other choices are to accept the occurrence of condensation or to reduce the relative humidity of the indoor air. The difficult matter of windows, which are often the critical elements for surface condensation, was discussed in Chapter 12.

15.8 Control of Air Flow

Uncontrolled air flow in and out through the building enclosure, usually referred to as air leakage, is a cause for concern in the design and operation of modern buildings. Air flow occurs through pores, cracks, joints, and other openings that are frequently difficult to identify and are of unknown size. The forces producing flow through these openings are now becoming much better known, and efforts are being made to develop means of predicting the air-leakage rates to be expected, as discussed in Chapter 11. Control of air leakage is now seen as an essential part of the control of condensation within constructions. Air leakage can also be a complicating factor in the distribution and control of fresh air for ventilation and in the spread of smoke and other contaminants.

Leakage of indoor air into the enclosure is generally undesirable, because there will be a high risk of condensation in cold weather unless the leakage is confined to the warmer parts of the enclosure by a suitable air barrier. Leakage of outdoor air may be tolerated or even desirable as a means of carrying water vapour out of the wall. The flow of air through or around an insulating layer adds to the heat flow since the insulation is being bypassed. The direction and extent of air penetration and its influence on heat and moisture flow is illustrated in Figure 15.12. Air flow within insulated walls may have to be controlled as well as air flow into and out of them.

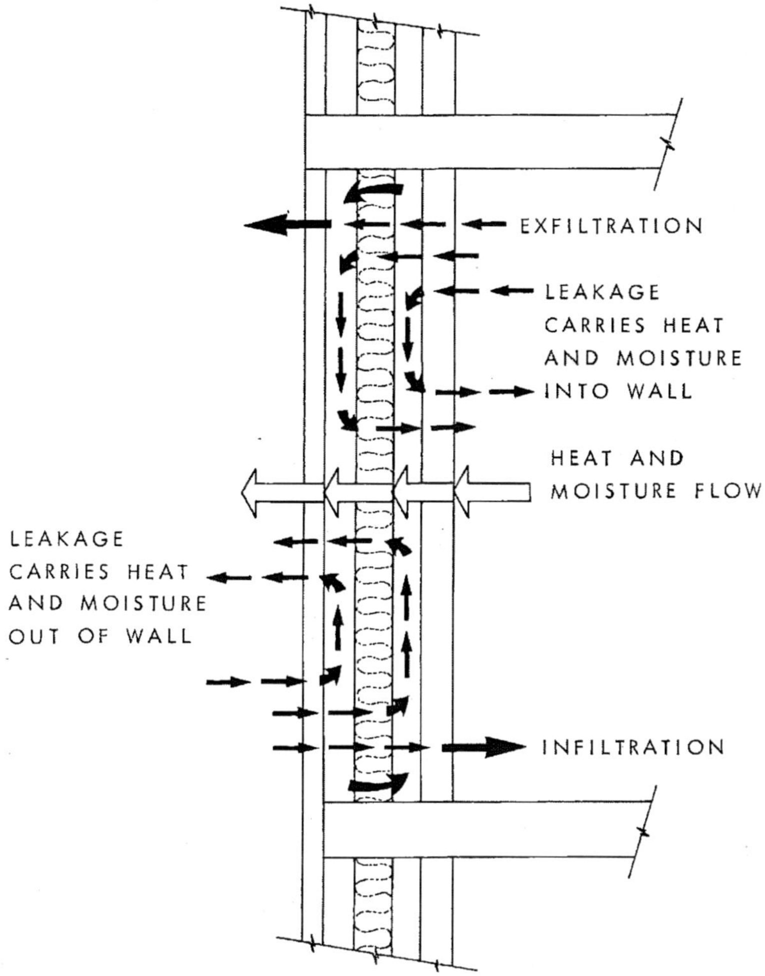

FIGURE 15.12 Air leakage in relation to heat and moisture flow in walls.

There is no easy way to control air leakage. The potential leakage paths must be identified, and steps must be taken to provide features through design and close control of construction which will block the air leakage [**15.8**]. There is a greater requirement for air-leakage control in buildings with high indoor relative humidities, since the condensation will probably be more severe and its effects more serious. Tall buildings will have increased air-pressure differences across the enclosure owing to increased stack effect with height, which will promote higher leakage rates. Increasing severity and persistence of low outdoor temperature may result in large accumulations of condensation as frost over long periods of time; this condensation is then released suddenly when the temperature rises.

Many of the difficulties arising from increased relative humidity and inadequate air-leakage control are well illustrated in the performance of a ten-storey building in Ottawa that has been under study for many years. It was designed as an office building but was converted to temporary service as an art gallery. The relative humidity was maintained at 50% all year [**15.9, 15.10**]. Deterioration of masonry and corrosion of metal ties resulted from excessive condensation. A major reconstruction of the cladding failed to eliminate the problem, since serious new air-leakage paths were not identified.

15.9 Influence of Cold Weather on Materials

Changes in temperature that take place slowly usually have little effect on dry homogeneous building materials. Rapid changes resulting in internal temperature gradients can produce stresses and strains because of thermal expansion and contraction and, in extreme cases, can lead to fracture. Similar results may be obtained if, under more or less steady conditions, one part of a building element is held at one temperature while different temperatures are produced elsewhere in such a way that the material is restrained. The breakage of window glass owing to cooling at the edges is a good example of this [**15.11**]. Masonry materials, being brittle, will often crack upon cooling if they are restrained, since their tolerable strain in tension is relatively low. Plastic materials generally have large coefficients of thermal expansion and so shorten substantially under reductions in temperature; but, because they have low moduli of elasticity, they do not fail when restrained [**15.6**].

The freezing of most porous building materials when they are wet can be destructive, as discussed in Chapter 5. The degree of saturation at the time of freezing is critical. In the case of exposed brick masonry the incidence of wetting and freezing may be dependent not only on the weather but also on the directional orientation for a particular site and on the location on the building [**15.12**]. Results obtained for one brick type exposed in a simple four-way rack at two locations [**15.13**] are given in tables 15.5 and 15.6. They show that materials may not be highly saturated at the time of freezing, and that marked directional effects in respect of degree of saturation can be expected. Local wetting on a building followed by freezing due to melting by the sun of snow accumulations on sills, ledges and roofs may occur much more frequently than in the case of rainfall followed by freezing. The wetting of cold outer layers of masonry by condensation due to air leakage is likely to be even more severe and to occur more regularly, causing damage over a period of years to metal ties by corrosion as well as to the masonry cladding [**15.9**].

Corrosion will take place only when moisture is present, so that the rate of atmospheric corrosion of architectural metals is related to the length of time they are wet [**15.14**]. A surface may be considered to be wet for these purposes if the relative humidity is as high as 80 to 90%. Corrosion, a chemical reaction, is also temperature-dependent, but the influence of cool weather in producing wetting conditions means that atmospheric corrosion can be a problem in cold climates, particularly when pollutants such as sulphur dioxide are present.

Metals can hold water only as a film on their exposed surfaces; many porous materials can hold large quantities of water. They may be wetted as a result of condensation or rain penetration in cool weather and may hold enough water to support rotting in warmer weather. This can be a problem in some wood roofs [**15.15**] and insulated wood-frame walls, and may result indirectly from cold weather when condensation is the main cause of wetting, and drainage and drying are inhibited.

Table 15.5

Number of Cycles of Freezing and Thawing at Ottawa and Halifax

(From Ritchie, T., and Davison, J.I., *Journal of Materials ASTM*, 1968, *3* (3), NRCC 10297, Table 2, p. 667) [**15.13**].

Copyright, ASTM, 1916 Race Street, Philadelphia, PA. 19103.
Reprinted with permission.

Brick facing	Number of freeze-thaw cycles: Ottawa 1963–64	Ottawa 1964–65	Halifax 1963–64	Halifax 1964–65
N	47	65	65	81
E	51	70	66	83
S	81	98	86	108
W	63	79	87	88

Table 15.6

Freeze-Thaw Cycles of a Brick Corresponding to Its Moisture Content when Frozen

(From Ritchie, T., and Davison, J.I., *Journal of Materials ASTM*, 1968, *3* (3), NRCC 10297, Table 4, p. 669) [**15.13**].

Copyright, ASTM, 1916 Race Street, Philadelphia, PA. 19103.
Reprinted with permission.

Moisture content*	Number of freeze-thaw cycles corresponding to moisture content indicated: Ottawa East	Ottawa South	Halifax East	Halifax South
0 to 9	0	36	0	0
10 to 19	1	43	0	0
20 to 29	1	2	1	6
30 to 39	0	0	0	34
40 to 49	6	0	7	11
50 to 59	11	0	11	6
60 to 69	9	0	17	3
70 to 79	8	0	12	2
80 to 89	13	0	10	2
90 to 99	2	0	7	1
Over 100	0	0	1	0
Total cycles	51	81	66	65

*Per cent of moisture absorbed by 24 h immersion.

15.10 Considerations in Design

One of the intriguing possibilities recognized in the building-science approach to enclosure design is that the specific environments in which various materials with different properties and functions must perform can be varied by adjusting the order in which the materials are arranged from warm side to cold side. This is shown by Figure 15.13, where the temperature conditions to which the main concrete wall is exposed are greatly changed if the insulating layer is placed on the outside rather than the inside. The main wythe, which is required to provide strength and stability and to be durable, is protected against low temperatures and temperature changes and the possibility of wetting by rain. The other possible benefits from such an arrangement will be discussed later.

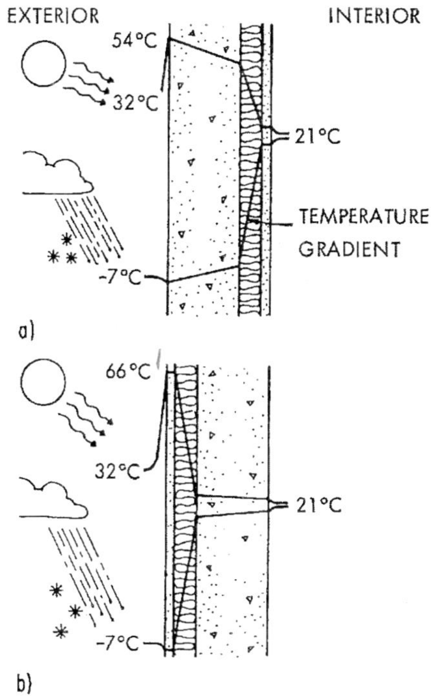

FIGURE 15.13 Wall section.
(From Garden, G.K., Canadian Building Digest 120, DBR/NRCC, 1969, Fig. 1)

The conclusions relevant to the design of building enclosures that can be drawn from building-science considerations following the discussions in this chapter are as follows:

1. Environments for particular materials can be adjusted by the order in which the materials are arranged to form building enclosures. In particular, the structural members and inner wythes can be protected from serious temperature and moisture changes if the insulating layer is located outward from them, on the cold side.

2. Wetting can be a serious problem under cold-weather conditions.
3. Control of air leakage is essential to control condensation.
4. Barriers to air flow should be located where the temperature is above the dewpoint of the indoor air.
5. Barriers to vapour flow by diffusion should be located where the temperature is above the dewpoint of the indoor air.
6. Air exchange between the two sides of insulating layers must be prevented.
7. Protection against the effects of any condensation that does take place can be provided by features that allow drainage or drying or both.

A simple wall designed to take account of these considerations might have the idealized form shown in Figure 15.14.

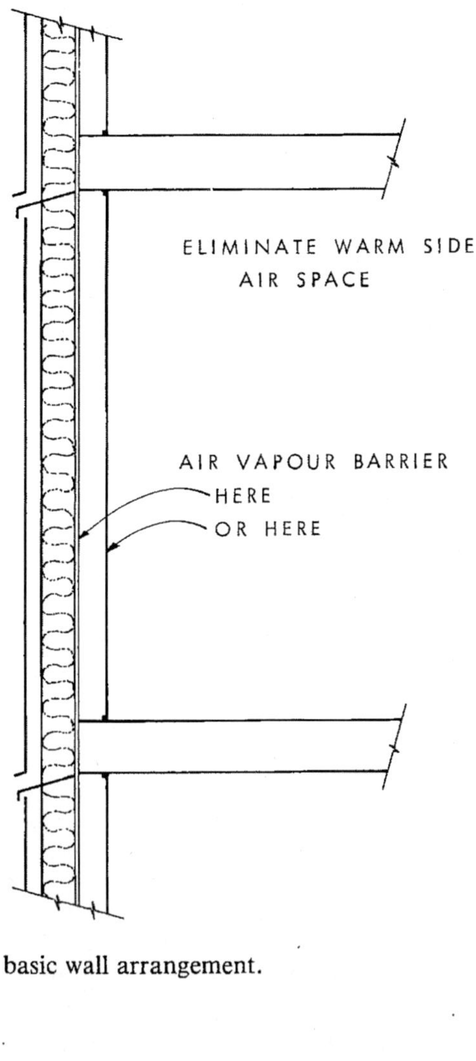

FIGURE 15.14 A basic wall arrangement.

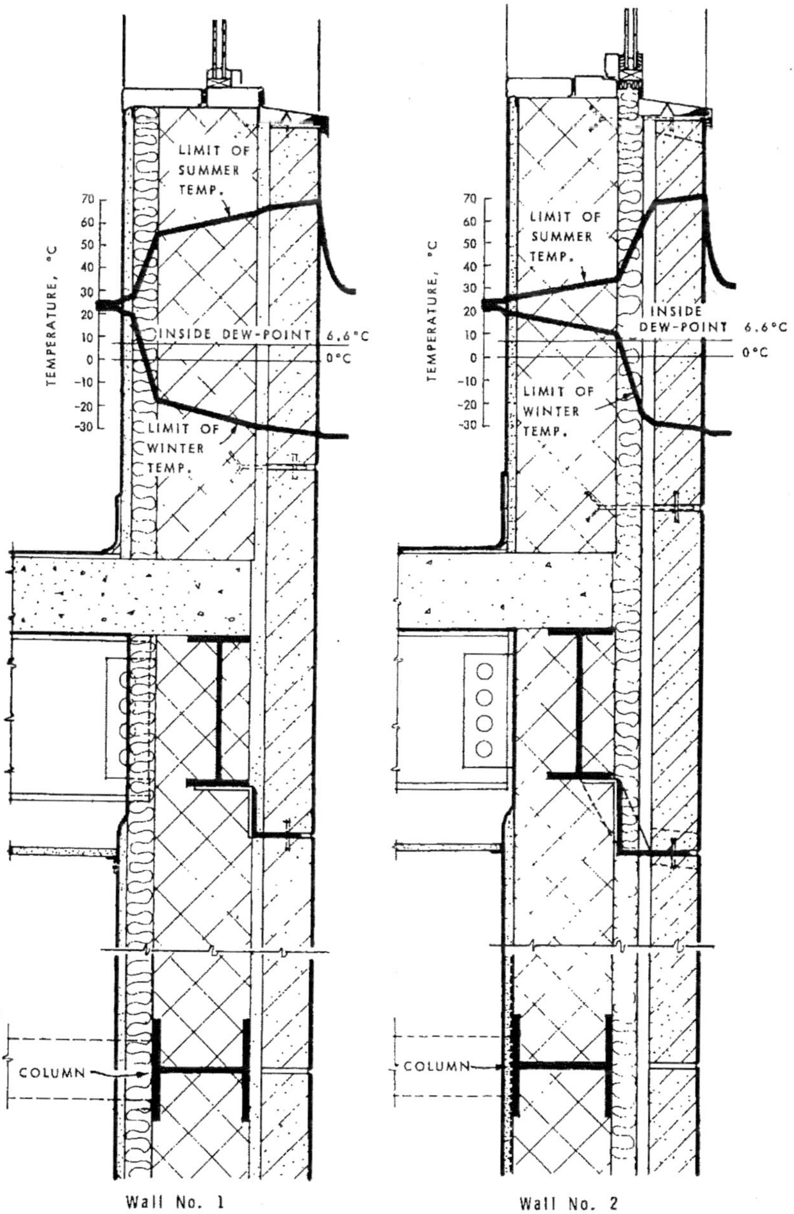

FIGURE 15.15 A masonry wall as commonly constructed compared with a revised form taking account of building science considerations.
(From Hutcheon, N.B., Canadian Building Digest 50, DBR/NRCC, 1964)[**15.16**]

15.11 Principles Applied to a Masonry Wall

Many of the practical requirements and features of real walls cause them to depart from the idealized form shown in Figure 15.14. A masonry wall as commonly constructed on a steel frame, and a revised version, are shown in Figure 15.15 [**15.16**].

Wall 1 has many weaknesses, some of which quickly become evident if the indoor relative humidity is increased. The steel frame and face and back-up wythes are all subjected to large temperature excursions and to corresponding dimensional changes both daily and seasonally, with temperatures in winter falling much below the dewpoint of 7°C for an indoor condition of 23°C and 35% rh and also much below freezing at times. Other constructions connected to these cold parts, including floor beams and windows, are also cooled by contact with them and may exhibit surface condensation. Barrier protection afforded by the interior finish is frequently omitted above the suspended ceiling.

In the revised form, wall 2, all the main structural members and the backup masonry are kept at temperatures well above freezing and above the indoor dewpoint temperature. The face wythe is now subjected to the large annual temperature excursion, but there is no reason why it cannot be articulated to adjust readily to this with open or ventilated and drained joints, as discussed in Chapter 12. Drainage can also take place readily by way of the vented air space between the insulation and the face wythe. The insulating value of the air space, face wythe, and surface film is reduced, but this may be acceptable in the case of heavily insulated walls where the insulation provides most of the resistance to heat flow. There is also the possibility in wall 2 of applying the air and vapour barriers on the outside of the frame and backup wythe, where there are fewer obstructions. The window frames are now in contact with warm construction.

15.12 Reversal of Temperature Gradients

Wall 2 in Figure 15.15 is arranged for winter conditions, with air and vapour barriers on the warm side. In summer the temperature gradients may be reversed, with higher temperatures on the outside and with heat and water vapour tending to flow from outside to inside. The air–vapour barrier should then in theory be on the outside of the insulation, and condensation might be expected on the outside of windows and within walls in air-conditioned buildings under extreme hot and humid conditions. In very hot climates solutions must be developed to deal with these reversed flow conditions.

Reference to the weather data for Canada will show that the dewpoint developed outdoors seldom exceeds 27°C in summer [**15.17, p. 21.19**], while the indoor dry-bulb temperature may be about 26°C. The reversed vapour-flow gradient, if one develops, will be small and usually will persist for only a few hours.

A marked reversal of conditions is encountered in cold storages, which may be cooled to -20°C or even to -40°C. The air and vapour barriers become very important to reduce refrigeration loads, and must be located on the outside of the enclosures. Questions may arise as to the practice to be followed in very cold regions where the outdoor temperature may fall below the storage temperatures.

Cool storages operating at or near 0°C pose a much more difficult problem, since the reversed flow conditions may exist for substantial periods in cold regions [**15.18**].

A special case of reversed gradients was encountered in experiments with test huts in Ottawa. The construction involved a masonry wall made of a rather porous brick, strapped and insulated on the inside, with a vapour barrier under the gypsum board finish. Following a rain in which the brick absorbed a substantial amount of water, hot sun on the brick wythe caused water vapour to migrate inward and to condense on the back of the vapour barrier. The wood strapping rotted away in a few years. This experience revealed the value of installing saturated sheathing paper between the insulation and the brick wythe to provide some resistance to the reversed flow of water vapour, but it may also be taken as evidence that cladding materials capable of absorbing large amounts of water should be avoided.

15.13 Alternative Design Approaches

The joint arrangement shown in the outer wythe of the revised wall design in Figure 15.15 provides a continous connection between the joints and the air space. This arrangement is called the "open rain screen". This is not the only form that a drained, ventilated joint may take, since it may have its own restricted air chamber for air pressurization as shown in Figure 12.11. When a continuous air space is used, as in Figure 15.15, there is a possibility that the variation in air pressure over the façade due to wind will create pressure differences and strong air flows in the air space. Compartmentation of the air space by seals or gaskets at all edges and corners and at suitable intervals over any large areas would appear to be a sensible precaution to minimize the pressure differences that can develop to force rain through joints. This is one of many practical considerations that the designer should keep in mind to arrive at a successful design [**15.6**, **15.19**].

Many designers instinctively want to make a wall that is exposed to driving rain as tight as possible on the outside. This requires an absence of rain-leakage paths, which is very difficult to achieve with the materials and methods normally used in wall construction [**15.20**]. Complete watertightness can be achieved in ships, but when the hulls must be insulated on the inside to conserve energy there is a problem of control of vapour flow. Since no vapour can escape outwards, it is necessary to consider ways in which a build-up of moisture in the insulation can be limited.

Built-up flat roofs pose a similar problem, since it has been common practice to install a highly impermeable roofing membrane on the outside. This membrane effectively prevents the escape of water vapour by diffusion from the insulating layer below while resisting the entry of water from above. Long-time experience with this form of construction has demonstrated the extreme difficulty in maintaining perfect seals against water entry. Porous insulation in sealed packages may not be the solution, since changes in moisture, temperature, and barometric pressure can produce pressure differences leading to bulging and even rupture.

The problem of moisture control in roofs has been discussed in Chapter 12, in several Canadian Building Digests [**15.21–15.23**], and in a detailed study by Baker [**15.24**]. Various design approaches are possible for both walls and roofs. Those for walls may differ from the solutions adopted for roofs: walls can be made to drain freely while roofs may have to withstand ponding of water. When the seasonal changes in conditions are such that escape of moisture during one season can more than offset its entry in the opposite season, it may be possible to use an insulated construction with high moisture-storage capacity and a high tolerance for wetting, capable of storing the moisture that accumulates in the insulation from one season to the other. Platts found that plywood stressed-skin panels in northern service successfully resisted the build-up of moisture to objectionable levels [**15.25**].

One of the more interesting design approaches now being widely and successfully used with flat roofs is the use of the protected membrane. The main waterproof membrane, which also serves as the vapour barrier, is placed below the insulation, which must then withstand some periodic wetting when it rains until drainage and drying by evaporation can take place [**15.23**]. A more conservative roof design involves the use of a continuous vapour and water barrier below and a waterproof membrane above the insulation. The insulation space is suitably graded and arranged for drainage and for venting to relieve pressure differences [**15.22**].

When the insulation is located between two impermeable layers it becomes possible, in principle at least, to control air and vapour flow in walls and roofs by pressurizing this intermediate space mechanically using dry air. The impermeable layers may have to have a high degree of airtightness to avoid loss of energy by leakage. This approach is attractive in extreme situations such as high-humidity textile mills and swimming-pool buildings where heated ventilating air under pressure is needed and can be supplied by way of the intermediate space.

None of these approaches is infallible or capable of providing solutions without effort on the part of the designer. They are only suggestive of ways in which the principal requirements can be met, with all the usual details of joints, fastenings, penetrations, edges, and corners and the overriding requirements of construction feasibility remaining to be dealt with. Joints in particular are always difficult, since they usually involve some capabilities in addition to those already required of the basic wall, roof, or floor sections [**15.6, 15.19, 15.20, 15.26**].

15.14 Energy Conservation

The costs of oil, electricity, and natural gas in heating and power have resulted in great attention being paid to the energy requirements of buildings. There is hardly any energy-related aspect of building for a cold climate that has not been reviewed and revised in attempts to reduce energy requirements. Design solutions are being adopted in many cases on the basis of theoretical considerations without adequate proof that they will perform as expected. The final test is whether the predicted energy consumption is achieved. The uncertainty of future oil prices, interest rates, and the weather add to the difficulties of sound economic analysis.

Building energy requirements range widely in practice from as little as 100 kW • h/m² to as much as 3000 kW • h/m² depending on design, type, climatic region, occupancy, and method of operation [**15.27**]. One recently designed building in Calgary anticipates an annual energy requirement of only 120 kW • h/m² with a heating-season severity of 5400 degree days (18°C basis). The average annual energy consumption for many types of buildings is about 600 kW • h/m² in moderate to cold climates.

It has been common practice to provide uniformly high lighting levels throughout many commercial office buildings, often using up to 50 W/m² of electrical energy for the purpose. For 2000 working hours per year, this amounts to 100 kW • h/m² per annum. It is widely believed that low-energy buildings can be designed to operate in Canada with total annual consumptions between 100 and 200 kW • h/m² of floor area. These estimates, which are usually based on design calculations, are not always achieved in practice. Attainment of such a design target requires a substantial reduction in energy requirements through careful attention to more efficient use of lighting, increased thermal insulation, reduced air leakage and minimum ventilation rates, recovery and storage of heat for reuse from interior spaces and from exhaust air and waste water, energy-saving operational features, and, perhaps, passive and active use of solar energy.

15.15 Solar Energy

The solar energy available through windows can be compared to the amounts required for heating with the aid of some simple calculations.

Example 15.1

Find the solar heat gain factor (SHGF) for a south-facing double window at 45°N latitude on January 21.

From Table 9.5, the SHGF daily total is 4798 W • h/m²; from Table 9.4, the shading coefficient for double glazing is 0.83.

$$\begin{aligned} \text{Solar heat gain per day} &= 4798 \times 0.83 \\ &= 3982 \text{ W} \cdot \text{h/m}^2 \\ \text{Average over 24 hours} &= 3982 \div 24 \\ &= 166 \text{ W/m}^2. \end{aligned}$$

If window area is 10% of the room floor area, the average energy to the room is

$$166 \times 0.10 = 16.6 \text{ watts per square metre of floor area.}$$

The solar energy available on a clear day on January 21 through a south-facing double window is shown in Example 15.1 to amount to an average hourly rate over 24 hours of 166 W/m² of window area or 16.6 W/m² of floor area, which may be compared with the electrical energy rate for lighting of 50 W/m². When the sun is not shining the double-glazed windows in this case have a heat loss much greater (by a factor of 10) than a well-insulated wall.

Mitalas has calculated the net annual heat loss factors (NAHLF) for test-year weather conditions at several locations across Canada. Using these, the net annual heat loss of small, well-insulated, single-family houses can be estimated [**15.28**]. The factors for double-glazed windows are shown in Figure 15.16 for various severities of heating season in degree days 18°C basis for different directions. The south-facing window has a modest net gain over the heating season, while windows facing other directions show net losses that are large for all directions except southeast and southwest. This means that even south-facing double windows acting as solar collectors can have losses during the hours when the sun is not shining that are comparable on a 24-hour basis to the energy collected. They may have to be suitably insulated during non-sunny periods if a substantial net gain is to be obtained.

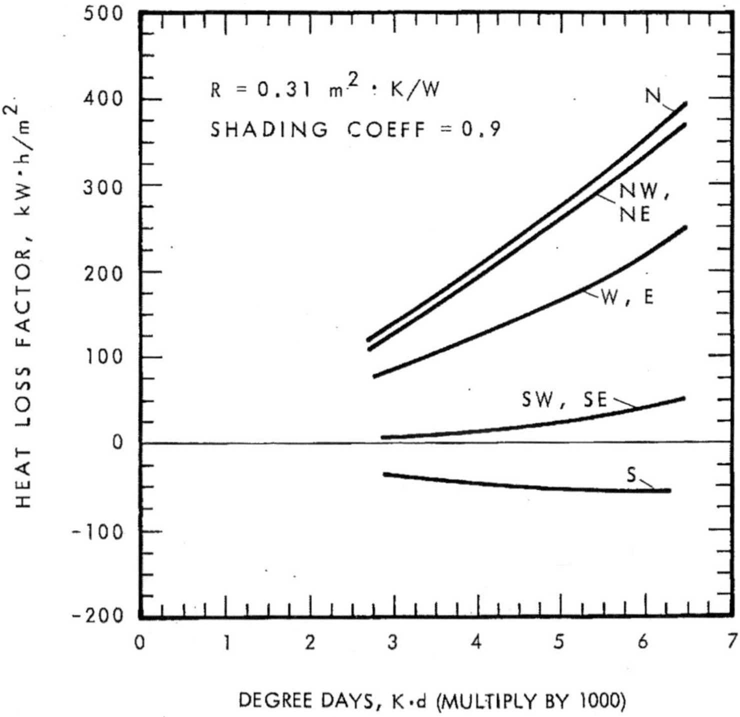

FIGURE 15.16 Window heat-loss factors for double glazing. (From Mitalas, G.P., Building Research Note No. 117, DBR/NRCC, 1976, Fig. 2)[**15.28**]

When the mass of the house is the principal means of storage, the use of large window areas to increase the energy collected may result in marked overheating in some rooms when the sun is shining. Windows can be shaded in summer but they may still add substantially to the heat gains during the cooling season. Windows are not the only type of collector, and there are kinds of storage other than the mass of the building, but all have similar difficulties differing only in degree:

1. Solar energy must be gathered when available, and that which is not immediately required must be stored for use when it is needed.
2. The losses from the collector and storage when the sun is not shining can be very large and must be kept within limits.
3. Storage may be desired for half a day, half a week, or half a year. The capacity required increases greatly with increase in the storage period.
4. The thermal environment in occupied spaces must not be allowed to vary beyond tolerable limits.
5. The use that can be made of energy from storage depends on the temperature at which it is available and on the means of storage.

The requirements following from these difficulties are not readily met in buildings without some considerable sacrifice by way of added features and equipment, loss of convenience, and design constraints. It is possible in dry moderate climates with high sunshine hours and reduced heating to devise solar houses that require no added energy for heating. When special collectors and added mechanical equipment requiring electric power are used, the system becomes an "active" one, and must be integrated with the building and the building's other energy systems. Domestic hot-water heating is the simplest and most widely used application of solar systems to date.

15.16 Low-Energy Buildings

The more successful developments in low-energy buildings for single-family residential use involve the provision of substantial amounts of insulation, measures to reduce air leakage, a judicious use of windows, and the reduction of ventilation rates to minimum acceptable levels. Difficulties with odours and other contaminants may be encountered if ventilation rates are reduced much below 0.5 air changes per hour in residential buildings. At this rate the energy needed to heat and humidify the air for ventilation accounts for 50% or more of the energy consumption under heating conditions. Heat exchangers can be used to warm the incoming fresh air for ventilation and to recover the heat from the air being exhausted.

Heat exchangers have long been used in buildings such as hospitals, which require high ventilation rates and 100% exhaust. They are now being used more extensively in other types of buildings. The quality of indoor air is becoming an important issue as rates of ventilation are reduced and as more toxic contaminants from the outgassing of various products are identified.

Large block-plan buildings conditioned for human occupancy require heating in winter and cooling in summer for perimeter spaces, and cooling all year in interior spaces. Heat pumps lend themselves to this kind of service, and are thus essential components in systems for simultaneous heating and cooling in different parts of the building, with any imbalance of energy delivered to or withdrawn from thermal storages in the form of water in large tanks.

15.17 Airtightness in Buildings

There is clearly an increasing requirement for improved performance of buildings in respect of radiation, heat, air, and water. The increased cost of energy and the prospect of future energy shortages have made energy conservation an urgent matter. Indoor conditions must be held within close limits to avoid the waste of energy associated with poor control of environmental factors. As ventilation rates are reduced, delivery of supply air at the proper rates is required to ensure the maintenance of air quality and the control of contaminants. This means that the unpredictable, disturbing factor of air leakage must be eliminated as far as possible by providing tighter construction so that the air-handling systems will perform as intended. There is an added benefit: the rate of spread of smoke throughout a building under fire conditions is reduced.

Building enclosures must be made as airtight as possible. This has been shown to be essential in the attainment of moisture control, which is the most important single requirement for buildings for a cold climate. Serious condensation in building enclosures and its costly consequences cannot readily be avoided unless outward air leakage is prevented under cold-weather conditions [**15.8**].

Condensation difficulties will increase in frequency and severity unless suitable changes in current practices are made, since it is certain that indoor relative humidities will increase in future as a result of control over ventilation rates in the interests of energy conservation and the demand by occupants for reasons of health, comfort, and other benefits. There is already ample evidence that raising indoor relative humidities can lead to serious condensation and deterioration in buildings as normally constructed.

There are also likely to be other factors at work. The use of more insulation will increase the likelihood of condensation and will promote changes in accepted practices in wall and roof construction, which may lead to further problems.

15.18 A Total Approach

The proper design of buildings, and particularly of building enclosures, to perform satisfactorily under cold-weather conditions must take into account all the principal requirements. Satisfactory solutions are unlikely to be found if the factors are considered in isolation, since there are strong interactions between and among them. Dealing with one but not the others is likely to create problems.

The computer has become an essential tool in making the extensive computations now needed to deal with thermal storage, transient and cyclical heat flow, equipment characteristics, and solar and other weather factors as they interact in various designs. The computer assists in determining the environmental conditions that will be created for people and products, and the influence on annual energy requirements. Many of the programs based on the present state of the art are inadequate to meet these new needs and must be improved or replaced.

New computer programs are not the only requirement, and may not even be the most urgent one. More and better knowledge about buildings and how they perform has given us the means to identify what is needed and has made it possible to identify some promising solutions. It remains for the knowledge and skill of the designer to be applied in devising satisfactory solutions in particular cases. It must not be assumed that perfect solutions can be created easily. There is much room for invention and intuitive contributions, but difficult new situations, each with many interacting factors to be considered, are unlikely to be properly dealt with by intuition alone. There is a great need to combine and exploit the experience and the ingenuity of the craftsmen and technicians who manufacture and assemble building components on the job and in the factory, the contractors who undertake the responsibility for construction, and the designers in devising the best possible solutions. Finally, it is essential that experience in the field, which is the real and final test, be recorded and communicated to others as a contribution to the body of knowledge essential for continued improvement in building.

References

15.1 Boughner, C.C., and Thomas, M.K. *The climate of Canada.* Reprinted from *Canada Year Book*, 1959 and 1960.

15.2 Miller, A.A. *The skin of the earth.* London: Methuen, University Paperbacks, 1964, p. 173.

15.3 *The supplement to the National Building Code of Canada 1980.* Ottawa: National Research Council of Canada. (NRCC 17724)

15.4 Sanderson, M.E., and Phillips, D.W. *Average annual water surplus in Canada.* Climatological Studies No. 9. Toronto: Department of Transport, Meteorological Branch (now Atmospheric Environment Services, Environment Canada), 1967.

15.5 Hutcheon, N.B. Fundamental considerations in the design of exterior walls for buildings. *Engineering Journal*, 1953, *36* (6), pp. 687-98. (NRCC 3057)

15.6 *Cracks, movements and joints in buildings.* Record of DBR Building Science Seminar, 1972. (NRCC 15477)

15.7 Hutcheon, N.B. Researchers break through the vapour barrier. In *Moisture Control.* Don Mills: Southam Business Publications, April 1979, pp. 24–25.

15.8 *Construction details for air tightness.* Record of DBR Seminar/Workshop, 1977. (NRCC 18291)

15.9 Brand, R.G. High-humidity buildings in cold climates–a case history. *Proceedings, First International Conference on Durability of Building Materials and Components*, Ottawa, 1980. Special Technical Publication 691, ASTM, pp. 231-38. (NRCC 18636)

15.10 Burn, K.N., and Schuyler, G.D. Applications of infrared thermography in locating and identifying building faults. *Journal of the International Institute for Conservation–Canadian Group*, 1979, *4* (2), pp. 3–14. (NRCC 19211)

15.11 Sasaki, J.R. *Potential for thermal breakage of sealed double-glazing units.* Canadian Building Digest 129. DBR/NRCC, Sept. 1970.

15.12 Crocker, C.R. *Influence of orientation on exterior cladding.* Canadian Building Digest 126. DBR/NRCC, June 1970.

15.13 Ritchie, T., and Davison, J.I. Moisture content and freeze-thaw cycles of masonry materials. *Journal of Materials, ASTM*, 1968, *3* (3), p. 658–71. (NRCC 10297)

15.14 Sereda, P.J. *Atmospheric corrosion of metals.* Canadian Building Digest 170. DBR/NRCC, 1975.

15.15 Baker, M.C. *Designing wood roofs to prevent decay.* Canadian Building Digest 112. DBR/NRCC, April 1969.

15.16 Hutcheon, N.B. *Principles applied to an insulated masonry wall.* Canadian Building Digest 50. DBR/NRCC, Feb. 1964.

15.17 *ASHRAE Handbook 1981 Fundamentals.*

15.18 *ASHRAE Handbook 1982 Applications.*

15.19 Garden, G.K. *Look at joint performance.* Canadian Building Digest 97. DBR/NRCC, Jan. 1968.

15.20 Garden, G.K. *The problem of achieving weathertight joints.* Paper presented at CIB Symposium on Weathertight Joints for Walls, Oslo, 1967. (NRCC 9874)

15.21 Handegord, G.O. *Moisture considerations in roof design.* Canadian Building Digest 73. DBR/NRCC, Jan. 1966.

15.22 Handegord, G.O., and Baker, M.C. *Application of roof design principles.* Canadian Building Digest 99. DBR/NRCC, March 1968.

15.23 Baker, M.C., and Hedlin, C.P. *Protected membrane roofs.* Canadian Building Digest 150. DBR/NRCC, June 1972.

15.24 Baker, M.C. *Roofs: Design, application and maintenance.* Montreal: Multi-Science Publications Ltd., 1980.

15.25 Platts, R.E. Condensation control in stressed skin and sandwich panels. *Forest Products Journal*, 1962, *12* (9), pp. 429–30. (NRCC 7057)

15.26 Sasaki, J.R. Evaluating the rain-tightness of joints between exterior wall components. *Proceedings, Fifth CIB Congress, Research into Practice: the Challenge of Application*, Versailles, 1971, pp. 485–87. (NRCC 12579)

15.27 Spielvogel, L.B. Energy performance data. *ASHRAE Journal*, 1980, *22* (1), pp. 46–50.

15.28 Mitalas, G.P. *Net annual heat loss factor method for estimating heating requirements of buildings.* Building Research Note 117. DBR/NRCC, 1976.

INDEX